HOME CARPENTRY

Improvements and Repairs

HOME CARPENTRY

Improvements and Repairs

Written and Illustrated by
Thomas H. Jones

MEREDITH® PRESS
New York, New York

Because of differing conditions, tools, and individual woodworking skills, Meredith® Press and the author assume no responsibility for any injuries suffered, damages or losses incurred during, or as a result of, the construction of these projects.

Before starting on any project, study the plans carefully, read and observe all the safety precautions provided by any tool or equipment manufacturer and follow all accepted safety procedures during construction.

Published by Meredith® Press
150 East 52nd Street
New York, NY 10022

Meredith® Press is an imprint of Meredith® Books:
President, Book Group: Joseph J. Ward
Vice-President, Editorial Director: Elizabeth P. Rice

For Meredith® Press:
Editorial Project Manager: Barbara Machtiger
Associate Editor: Guido Anderau
Production Manager: Bill Rose

Cover Photograph: Bill Hopkins Associates
Cover Design: Ernie Shelton, Lynda Haupert

Produced by Roundtable Press, Inc.
Designed by Jeff Fitschen

Distributed by Meredith Corporation,
Des Moines, Iowa

ISBN: 0-696-11125-X (hardcover)
ISBN: 0-696-11139-X (softcover)
Library of Congress Card Number: 91-060640

Manufactured in the United States of America

10 9 8 7 6 5 4 3 2 1

IN MEMORY
of my daughter
Barbara Elizabeth Jones, RN

Contents

Preface

This is a book for the first-time homeowner who is faced with the new responsibility of taking care of all the carpentry that is part of moving in and owning a house or townhouse—from shelves and built-in storage to the repairs that will have to be made from time to time. By necessity you will become a do-it-yourselfer. The cost of hiring a professional is just too high. Learning how to take care of these jobs yourself will not only save you a lot of money, but also give you a real sense of satisfaction and independence.

This is also a book for the buyer of an older house that is in need of repair, renovation, or remodeling. With an older house, you may be faced with more involved projects—patching walls, replacing doors and windows, even building or tearing out partitions.

There are times when hiring a competent and experienced contractor is the prudent course of action, such as anytime the load-bearing structure of the house is involved. Here, you can get into a lot of trouble fast if you don't know exactly what you are doing. Be very cautious when dealing with the house structure.

In much of this book the focus is on materials. Knowing what products and materials to use for a particular job is essential to doing it well. Retail salespeople have an interest in recommending what they have in stock, not what they don't carry. You must acquire a good knowledge of the extensive range of products and materials available in order to make the wisest and most cost-effective choices for your repairs and improvements.

This book didn't just happen. Thanks are due to a lot of companies and associations who helped, including American Olean Tile Company, American Plywood Association, Andersen Corporation, Armstrong World Industries, Inc., Bruce Hardwood Floors, Chelsea Decorative Metal Co., Formica Corporation, Georgia Pacific, Louisiana Pacific, Masonite Corporation, Murphy Door and Bed Company, Owens-Corning Fiberglas, Stanley Hardware, Tile Council of America, Western Wood Moulding and Millwork Producers, and W. F. Norman Corporation.

People helped too. Special thanks to my wife, Carolyn, for her invaluable help in planning and writing the book, and her help in producing all of the finished artwork that illustrates the book. And thanks also are due to the editors at Meredith Press and Roundtable Press for putting it all together.

1 Tools and Workbench

To do your own home carpentry—repairs, remodeling, or renovation, building shelves and cabinets—you need some tools. You have probably been accumulating tools over the years and already have many of the items in the basic set listed here. (These are carpentry tools; tools for gypsum board and plaster repairs and ceramic tile work are listed in the chapters dealing with those subjects.) A workbench is also a necessity. This chapter has plans for an easy-to-build, inexpensive bench. A starting set of tools and the workbench are shown in Figure 1–1.

COMPILE A BASIC TOOL KIT

Buy quality tools, and buy name brands, not some strange brand on sale at a ridiculously low price. Be sure that each tool feels comfortable in your hand.

Claw hammer. Buy a 14- or 16-ounce hammer. A lighter hammer will take too many blows to drive a nail. A heavier hammer will wear you out and make driving small nails difficult. The claw can be either straight or curved. Straight is better for prying off molding and similar tasks; curved is better for pulling nails. Hammer handles are made of wood, fiberglass, or steel. One is as good as another, but be sure a nonwood handle has a rubber, leather, or other nonslip handgrip.

Screwdrivers. You will need a minimum of five—three flat-blade screwdrivers (3⁄16″, 1⁄4″, and 5⁄16″ wide) and two Phillips (crosshead) screwdrivers (#1 and #2). Get separate screwdrivers; don't go for a handle with insertable tips.

Electric hand drill. There are plug-in and cordless models. A cordless drill is handier for working in locations some distance from an electrical outlet, but a conventional plug-in drill runs faster and costs less. If you expect to be drilling holes in masonry (for anchors to hold screws, for example), you will need the higher speed of a plug-in drill. Your best choice is a drill with a 3⁄8″-capacity chuck, variable speed, and reversible rotation to remove screws easily. A basic set of bits should include bits of at least 1⁄16″, 1⁄8″, 3⁄16″, 1⁄4″, 5⁄16″, and 3⁄8″ diameter. Larger diameters are available on 1⁄4″ and 3⁄8″ shafts. Brad-point bits are best for wood as they will not skip when starting a hole, but they cannot be used in metal. For drilling metal, you must use twist drills. You should also

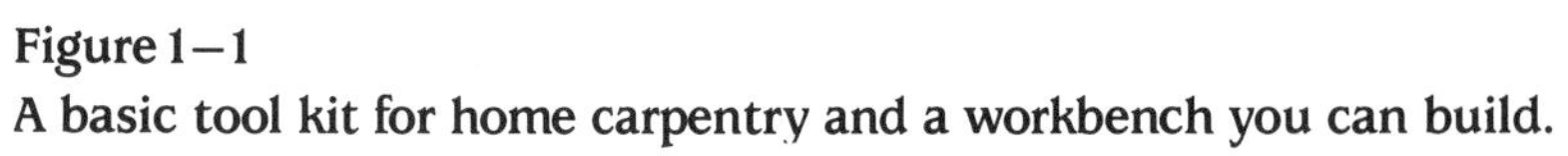

Figure 1–1
A basic tool kit for home carpentry and a workbench you can build.

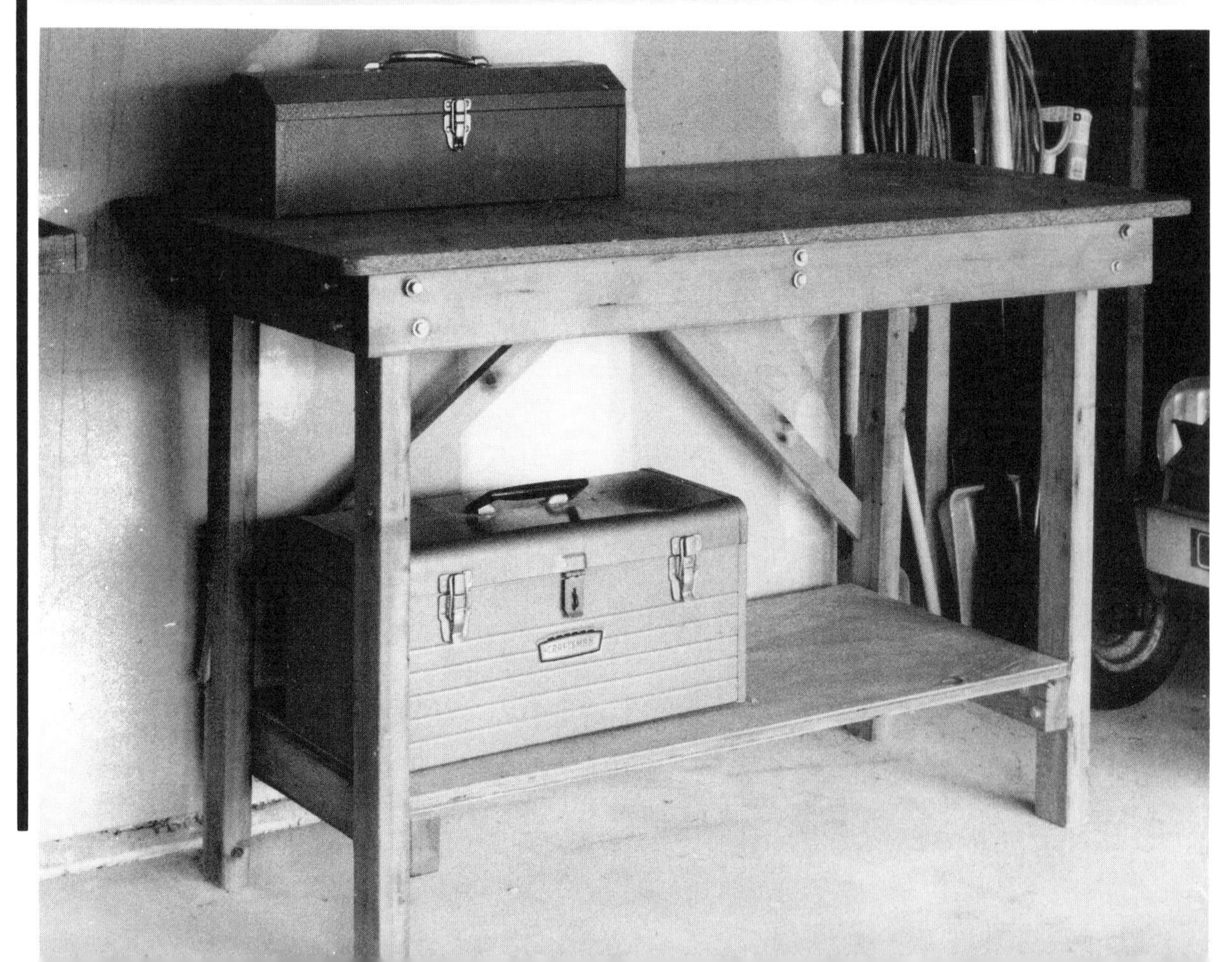

have a tapered (cone-point) countersink bit so you can set wood screws flush or below a wood surface. For masonry and gypsum board, get carbide-tipped bits in the required sizes.

Flexible steel measuring tape. Buy either a 10′ or 12′ tape. Longer tapes (25′, 50′, or 100′) are good for outdoor projects, but their bulky size is a nuisance indoors.

Saber saw. With the proper blade, you can cut just about any material with a saber saw. You can saw along straight or curved lines, and you can start holes in the middle of a piece. For best control of sawing, buy a variable-speed saw. Good-quality crosscut and rip handsaws are expensive, and you may want to put off purchase until you really need one or the other.

Straightedge. In all kinds of woodworking you need to draw straight lines. A wood yardstick is seldom straight. Get a steel or aluminum yardstick or a 4′ or 6′ metal straightedge.

Carpenter's square. Use this L-shaped steel measuring tool for marking wide boards for square crosscuts and for getting assembled parts of your construction perpendicular, or at right angles. The 24″ size is best.

Combination square. This tool consists of a 12″ blade and a handle that can be moved along the blade, locked at any point, or removed. It is used to draw square or 45° diagonal lines across small boards, to transfer lines from one surface of a piece of wood to another, and to check saw cuts to see if they are precisely square or mitered at a 45° angle. Most combination squares also have a built-in bubble level for checking vertical and horizontal alignment of work.

Portable circular saw. This versatile tool is the powered equivalent of a handsaw and a table saw. It's fast, accurate, and dangerous—a tool that must be used with care. A safety suggestion: clamp the work to your bench when sawing. Also, while learning to use the saw freehand, you can cut accurately by guiding the saw with a strip of wood or a commercial guide clamped to the work. Many special blades are available, for example, to rip or crosscut lumber, to cut plywood without splintering, or to cut metal or plastic laminate.

Wood chisels. Chisels are used for cutting mortises for hinges and tenons, as well as other recesses, and for carving wood to irregular shapes. You will need several widths—a set of six, ¼″, ⅜″, ½″, ¾″, 1″, and 1½″, will meet most needs. Buy chisels with plastic handles—the handles are virtually indestructible and can be driven with a metal-head hammer, although a plastic or wood mallet is better.

Plane. A wood plane in skilled hands is a good tool to finish-trim the long edge of a board or door. A 9″-long smoothing plane is a good choice for a start. However, a belt sander or router (see below) is a more versatile tool in that it can be used crossgrain and to edge plywood. A spokeshave is a handy substitute for small planing jobs.

Clamps. It is a great truth of woodworking that you never have enough clamps. They are made in many forms. C-clamps are the most versatile. As a start, choose 4″ and 5″ clamps—four of each as a minimum.

Pipe clamps are used for assembling furniture and gluing up boards for wide tabletops and similar purposes. You can buy sets of adjustable head- and tailpieces and use them with ¾″ pipe (be sure it's threaded on both ends). Buy extra lengths of threaded pipe and couplings to make longer clamps. If you don't want the heavy weight of pipe clamps, you can buy aluminum-alloy bar clamps. Their lighter weight makes them easier to handle, but you will pay a premium price for this convenience.

Wrench. You need a wrench to turn lag screws. A ⅞″-capacity adjustable end wrench is a good size if you must limit yourself to one wrench.

Prybar. Sometimes called a ripping bar. Buy the kind made from a flat bar rather than a crowbar made of hexagonal stock. It does a neater job removing molding.

Figure 1–2
Quarter-sheet small orbital pad sander (left) and belt sander.

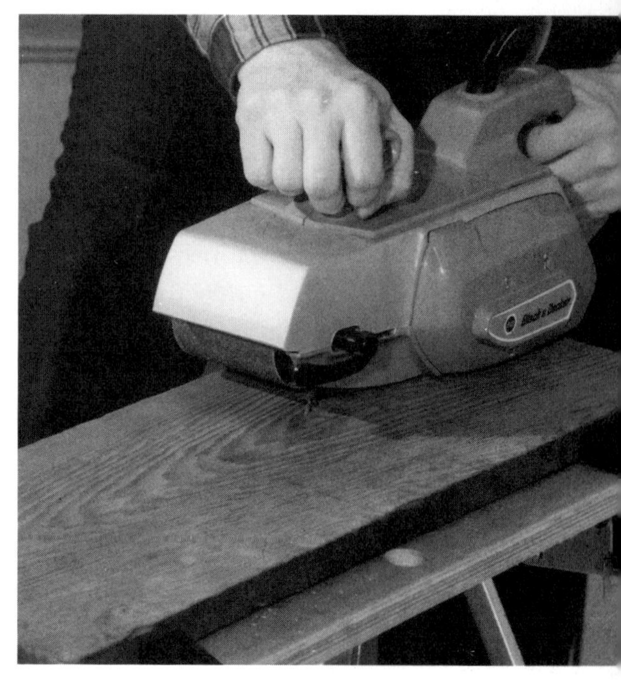

Pad sander. There are three kinds: the fast-sanding orbital, the smooth-sanding straight line, and the very small scratch-pattern orbital, which is both fast sanding and smooth finishing. Get the third kind; it is a worthwhile investment (Figure 1–2). All sandpaper removes wood by scratching it away, but the scratches made by this type of sander are so small that they can become invisible with fine-grit sandpaper.

Portable belt sander. You will become impatiently aware that a pad sander is really not very fast when it comes to heavy sanding. Use your pad sander for finishing, but get a belt sander for aggressive sanding. One with a sawdust-collecting bag makes cleanup much easier.

Vise. Half the job in bench woodworking is holding the work securely. One choice is a heavy-duty 3″ machinist's vise. Mount the vise on a square of ¾″ plywood so you can clamp it to the bench when you need it—and get it out of your way when you don't (Figure 1–3). Equip the jaws with wood faces so they won't mar your workpiece.

Figure 1–3
Vise mounted on scrap plywood for on-the-bench/out-of-the-way convenience.

Figure 1–4
Router and some of the many available bits.

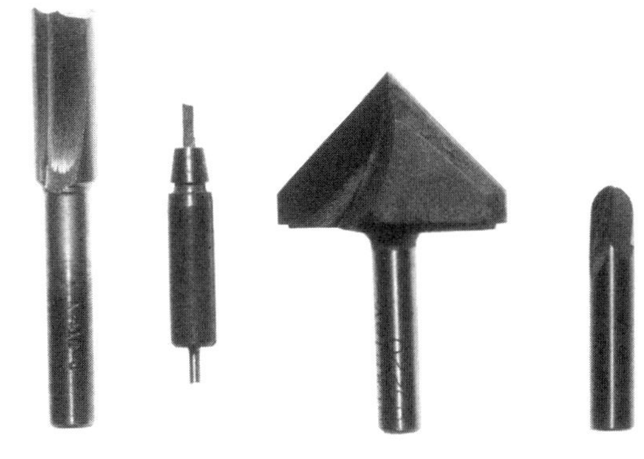

Router. A portable electric router is the most versatile tool you can have in your shop. With a router and the proper bits you can smooth and true up edges accurately; form dadoes, rabbets, slots; and make decorative edges (Figure 1–4).

Shop vacuum. Routers, saws, and particularly sanders generate a lot of sawdust. You can sweep and brush the mess from every surface after you finish for the day, or you can control the dust at its source by using a shop vacuum in the work area or even attached directly to a tool.

BUILD A BASIC WORKBENCH

Without a workbench you will waste a lot of time while fixing up, repairing, or building anything around the house. The kitchen table might seem like a good temporary substitute, but it has to be cleared every so often for meals and cooking; besides, no table is really sturdy enough to be a workbench. Inevitably your work surface is going to be marred, scratched, drilled, nailed, sawed into, and otherwise defaced. What you need is a bench constructed especially for carpentry work. The plans for an easy-to-build, low-cost workbench are shown in Figure 1–5; the parts are listed in Table 1–1.

Dimensions

The workbench top is 24″ × 48″. This is big enough for most jobs, and you can easily reach across it. At the same time the bench is small enough to fit into the end of your garage or corner of your basement. It also can be carried through doorways and up stairs, so you can take the workbench wherever you need it to work.

The top of the bench should be at a comfortable standup working height for you, 26″ to 31″, depending on your height. The top should be made of either ¾″ fir plywood or ¾″ particleboard (Chapter 2). Plywood is easier to cut and is lighter in weight; it can be covered with ¼″ hardboard to provide a smooth, durable work surface that can easily be replaced when necessary. Particleboard has solid edges and provides a harder work surface. The top should overhang on all four sides for clamping your work. Round the corners with a router or sander to avoid painful bumps as you work and prevent gouged walls when you move the bench.

For wallpapering and other large-surface jobs, you can extend the size of the top with an inexpensive flush door clamped to the bench top. To support the end of a long workpiece, make a sawhorse the same height as the bench.

The legs are cut from 2 × 3 lumber. However, the framing under the top—the rails—uses 2 × 4s. Here you need all the rigidity you can get to make the top a firm surface for your work.

The shelf provides storage; it also keeps the legs straight and takes the strain off the joints at the top of the legs. The shelf can be made of anything handy—¾″ plywood, particleboard, or pine shelving. The bottom of the stretchers should be 5″ up the leg so as not to be in the way of your feet. The diagonal braces from the center rail to the rear legs prevent the frame from racking.

Hardware

You can use nails, wood screws, gypsum board screws, lag screws, carriage bolts, or machine bolts to fasten pieces of wood together. Various fasteners are shown in Figure 1–6. Basic information about using fasteners is given below; more extensive information is in Chapter 3.

Nailing is the fastest way to assemble the frame of the bench, but there are two drawbacks

Table 1–1
WORKBENCH MATERIAL AND CUTTING LIST*

BUY		CUT
No.	Item	
1	2 × 3 (14′ long)	4 legs 28″–30¼″ long as desired
		2 stretchers 18″ long
1	2 × 4 (12′ long)	1 front rail 44″ long
		1 back rail 44″ long
		2 end rails 18″ long
		1 center rail 18″ long
1	¾″ × 48″ × 48″	1 top 24″ × 48″
	A-C fir plywood	1 shelf 18″ × 38″
	(½ panel)	
1	1 × 2 (5′ long)	2 diagonal braces 30″ long
28	¼″ × 2¾″ lag screws	
28	washers to fit lag screws	
18	1½″ #10 flat-head wood screws	

*Drill bits needed: ¼″, 3/16″, 5/32″, 3/32″, tapered countersink

Figure 1–5
Workbench parts. Adding a hardboard overlay to the top will give a smooth, durable work surface that can easily be replaced when necessary.

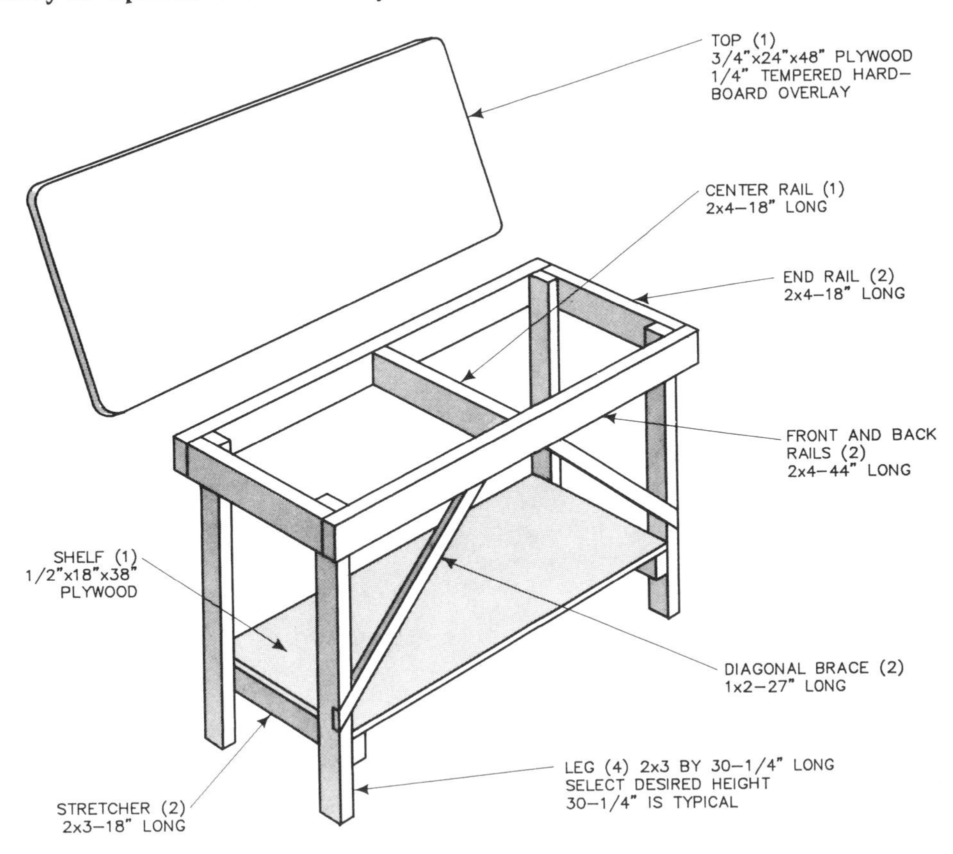

to using nails: (1) There is a good chance you will split at least some of the wood, and if you do, the result will be a weak joint. (2) Even if you avoid splitting any wood, the joints will be weaker than if you made them with threaded fasteners. (If you do choose to nail the bench together, use rosined box nails instead of common nails—see Chapter 3. They are thinner and cause fewer splits.) For the best results, you should use threaded fasteners such as lag screws or bolts.

Lag screws are the best fasteners to use for your workbench. However, you must drill two different-size holes for them—a clearance hole for the shank of the screw and a smaller pilot hole in the other wood piece for the threaded part of the screw. It is very important that this pilot hole be the correct size. If it is too small, the threads of the screw will mash up the wood and possibly cause splits instead of cutting clean, deep threads. If the pilot hole is too big, the threads cut will be shallow and weak, and the screw will strip out when you tighten it, which makes the fastener almost useless. General rules for placement and a method of aligning holes for assembly with lag screws are shown in Figure 1–7.

You must use a washer under the head of the lag screw. The washer keeps the screw head from chewing up the wood when you use a wrench to drive the lag screw into the wood, and it spreads the holding power of the head over more wood area, giving a tighter joint. This is especially important in softwoods, as even a washer can crush the wood. Positioning threaded fasteners for a strong joint is shown in Figure 1–8.

If you use *carriage bolts*, you must drill a hole the same size as the shank diameter through

Figure 1–6
Threaded fasteners. A machine or stove bolt (not shown) has a round head and a shaft threaded all the way to the flat underside of the head.

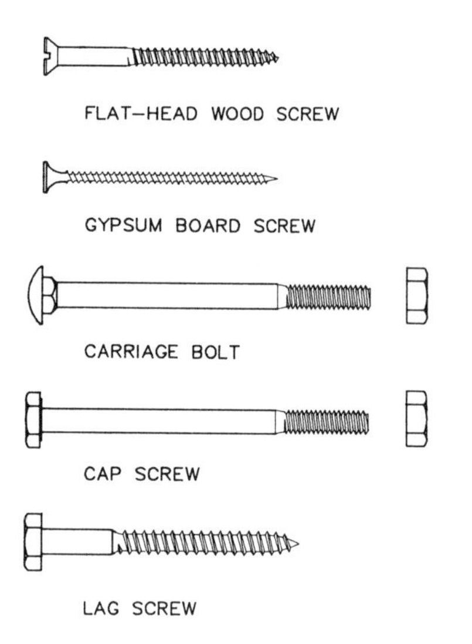

both pieces. Carriage bolts have two advantages over lag screws: (1) You don't depend on threads cut in wood to hold the joint together. (2) The smooth oval head makes a neater appearance. Under the head of a carriage bolt, the shank has a square section, which prevents the bolt from turning when you tighten the nut. You need a washer under the nut, but you cannot put a washer under the head, so carriage bolt heads tend to pull into softwoods easily.

Machine bolts (also called stove bolts) are another alternative. Because washers can be used under the heads of machine bolts, they can be wrenched tighter than carriage bolts. Machine bolts should be used to replace carriage bolts if the wood under the carriage bolt head allows the squared end of the shank to turn.

Workbench Assembly

Figure 1–9 shows how the workbench goes together. Begin with the workbench frame. After all parts of 2× lumber are cut to length, the frame goes together as follows: (1).Assemble rails, stretchers, and legs to form the ends. Note that the end rails are outside the legs at the top, and the stretchers that support the shelf are on the inside of the legs. (2) Connect the ends with front and back rails. (3) Fit the center rail between the front and back rails. (4) Miter the ends of the two crossed diagonal braces to fit against the center rail and notch the back legs to receive the braces (Figure 1–10). You can fasten the braces with nails or—better—wood screws; lag screws are not needed.

Figure 1–7
Locating fasteners in wood for strong, split-free joints. To avoid splitting, it is important to stagger holes so they do not cut through the same grain lines in either piece of wood.

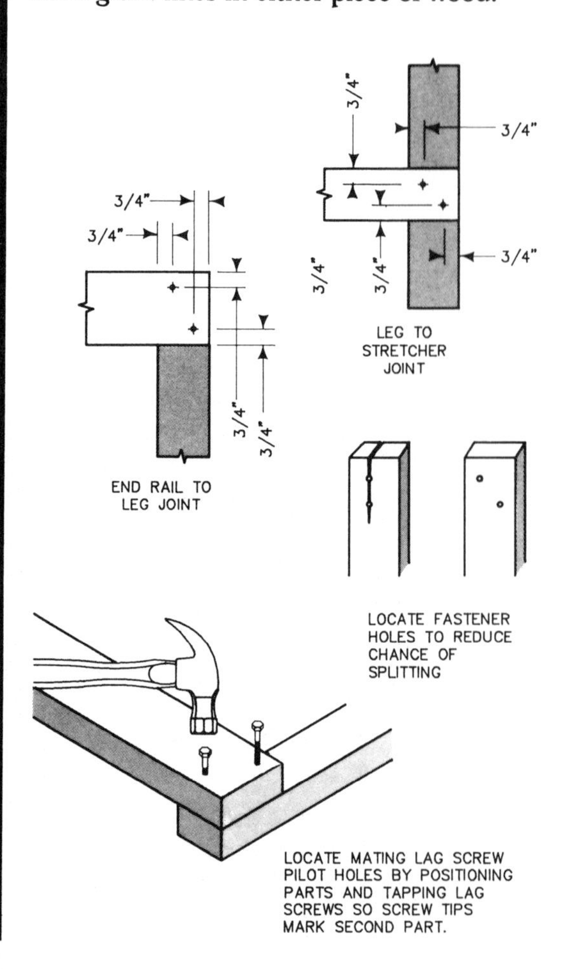

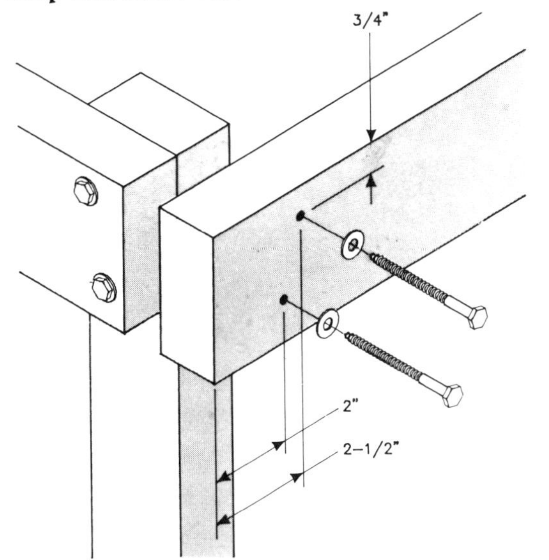

Figure 1–8
Lag screws should be 2¾" long for this joint, but you may have to settle for 2½". Use paraffin or a candle to lubricate screw threads. Soap will cause rust.

If you use a sheet of hardboard for a finished work surface, secure it to a plywood top with ¾" finish brads, or to a particleboard top with flathead screws (predrill screw holes). Drive brad heads below the surface with a nailset, or countersink screw heads.

Finish the bench top with varnish on both sides. It helps you keep the bench clean. If you also wax the top, most glue dribbles won't stick to it. Reverse the top when it gets shabby from use, or replace the hardboard surface sheet.

If you would rather buy a bench than build one, the Black & Decker Workmate shown in Figure 1–11 is recommended. This versatile workbench is made in three sizes. Each has a split top that functions as a clamp or vise, and folds for carrying and storage.

Figure 1–9
Assembling the frame. Lag screws are run into the side grain of the leg, not into the rail end grain. Screw threads do not hold well in softwood end grain.

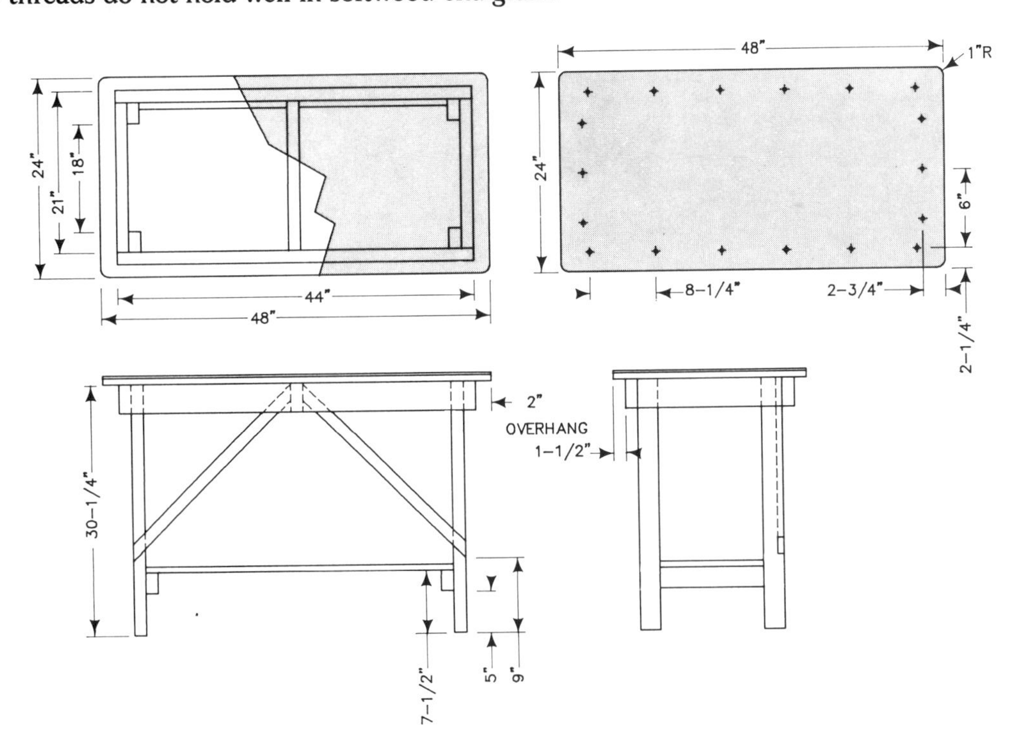

Figure 1–10
Diagonal braces are important. Don't omit them as they are necessary to prevent racking. Miter the upper ends of the braces to fit against the 2 × 3 center rail. Fit the lower ends into cutouts made in the legs. Fasten the brace with nails or screws and glue.

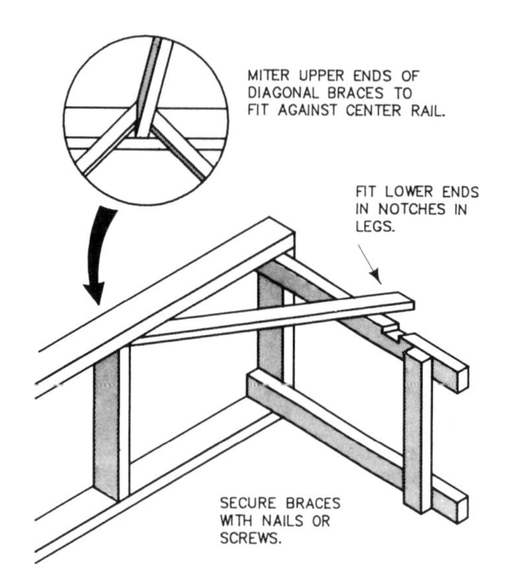
MITER UPPER ENDS OF DIAGONAL BRACES TO FIT AGAINST CENTER RAIL.
FIT LOWER ENDS IN NOTCHES IN LEGS.
SECURE BRACES WITH NAILS OR SCREWS.

Figure 1–11
A Black & Decker Workmate portable workbench.

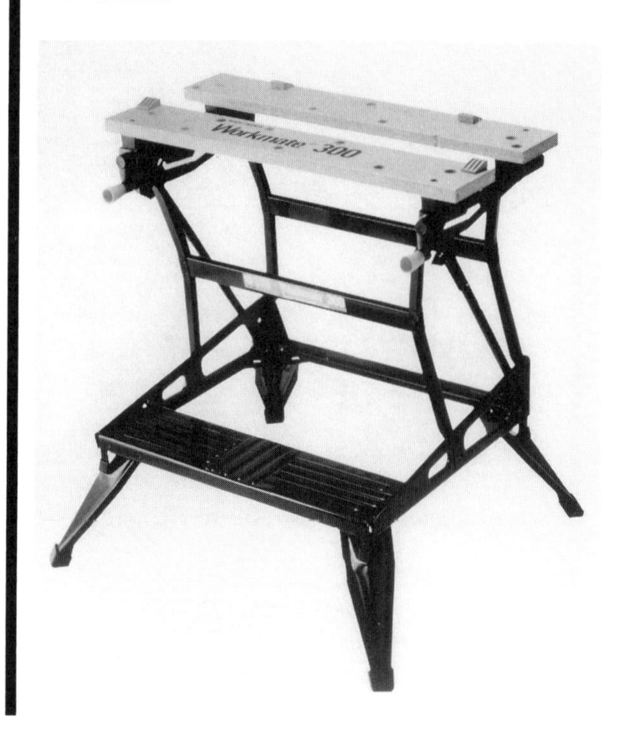
Workmate 300

2 Buying Lumber and Plywood

Buying carpentry materials can be confusing. Here's what you have to know about the way softwood and hardwood lumber and plywood are classified and graded so you can buy the right materials for your projects. There are also some tips on how to get the most for your money.

SOFTWOOD

Softwood lumber grading standards have been set by the government. The actual grading is done by inspectors of such lumber associations as the Western Wood Products Association (WWPA), Northeastern Lumber Manufacturers Association (NELMA), and Redwood Inspection Service (RIS). (See Figure 2–1.)

For grading purposes, all softwood lumber is divided into two groups: green lumber and dry lumber that has been seasoned or dried to a moisture content of 19 percent or less. All softwood lumber is further divided into three use classes: yard lumber for ordinary construction and building purposes, structural lumber for use where high stress will be encountered, and factory and shop lumber for manufacturing into molding, siding, and furniture.

The stock that you will find in lumberyards and home centers will be almost exclusively dry, yard lumber, classified as Board, Dimension, and Timber.

Board Lumber

A board is defined as a piece of lumber that is thin and of considerable length and width with respect to the thickness. Softwood boards are

Figure 2–1
Typical grade marks for board lumber:

A. Western Wood Products Association (WWPA). The lumber is Idaho White Pine (IWP) inspected under WWPA supervision, but graded under IWP rules. The grade is Sterling, equivalent to No. 2 Common.

B. West Coast Lumber Inspection Bureau (WCLB). Construction is their grade equivalent to No. 2 Common. The species is Douglas fir.

SPIB KD STUD 7

C. Southern Pine Inspection Bureau (SPIB). The select lumber grade is Stud, kiln-dried to a moisture content of 15% or less. The lumber is from mill number 7.

D. Redwood Inspection Service (RIS). The lumber is redwood and the grade is Clear Heart and may appear on seasoned or unseasoned rough lumber.

made in several standard thicknesses. Widths of softwood boards are also standard measures. (See Table 2–1.)

Grading. Softwood boards are sorted into select or finish grades (the good stuff) and common grades (with all the knots). They are graded after they are surface-planed, with the inspector looking at the good side. The board is graded as a whole board, with the idea that it will be used that way, not cut into shorter lengths. In the common grades, it is not the number of knots that determines the grade, but the size of the knots and whether they are sound, loose, or missing, plus items such as warp, wane, splits, rot, and other defects. In select grades, both the number and the size of knots are limited.

While manufacturing defects are also a factor, lumber grades are determined primarily by natural characteristics of a log that appear in the sawn lumber. These include the characteristics defined below and shown in Figure 2–2.

Bow: end-to-end curvature of wide face.
Cup: side-to-side curvature of wide face.
Twist: deviation from flat plane of four faces.
Crook: curvature of narrow face of board, measured end to end.
Wane: presence of bark or lack of wood on edge or corner of board.
Split: separation of wood, usually at end of the lumber, due to tearing apart of wood cells.
Knots: varieties including (A) sound, watertight, intergrown through two wide faces; (B) sound, encased, fixed round knot through two wide faces; (C) knothole through two wide faces; (D) edge knothole; (E) sound, watertight, tight intergrown spike through two faces; (F) edge knot; (G) intergrown round knot through four faces.
Shake: lengthwise separation of the wood. (Shakes usually occur between or through annual growth rings.)

What each board grade is called depends on what part of the country the tree came from. Western woods are graded mostly under WWPA rules. Lumber from the northeastern part of the country comes under NELMA grading rules.

Table 2–1
DRY SOFTWOOD LUMBER DIMENSIONS

BOARDS					DIMENSION LUMBER AND TIMBER	
THICKNESS			WIDTH		THICKNESS AND WIDTH	
Nominal		Actual	Nominal	Actual	Nominal	Actual
4/4*	1"	3/4"†	2"	1 1/2"	2"	1 1/2"
5/4	1 1/4"	1 5/32"	3"	2 1/2"	3"	2 1/2"
6/4	1 1/2"	1 13/32"	4"	3 1/2"	4"	3 1/2"
7/4	1 3/4"	1 19/32"	5"	4 1/2"	5"	4 1/2"
8/4	2"	1 13/16"	6"	5 1/2"	6"	5 1/2"
9/4	2 1/4"	2 3/32"	7"	6 1/2"	7"	6 1/2"
10/4	2 1/2"	2 3/8"	8" and wider, 3/4" less than nominal dimension		8" and wider, 3/4" less than nominal dimension	
11/4	2 3/4"	2 9/16"				
12/4	3"	2 3/4"				
16/4	4"	3 3/4"				

*"Quarter" sizes, often used for rough-cut boards not milled to finish sizes. The numbers indicate the thickness in number of quarter-inches, and are called "four-quarter," "five-quarter," and so on.
†Redwood is sometimes sold in 11/16" thickness rather than 3/4".

Figure 2–2
Wood grading characteristics.

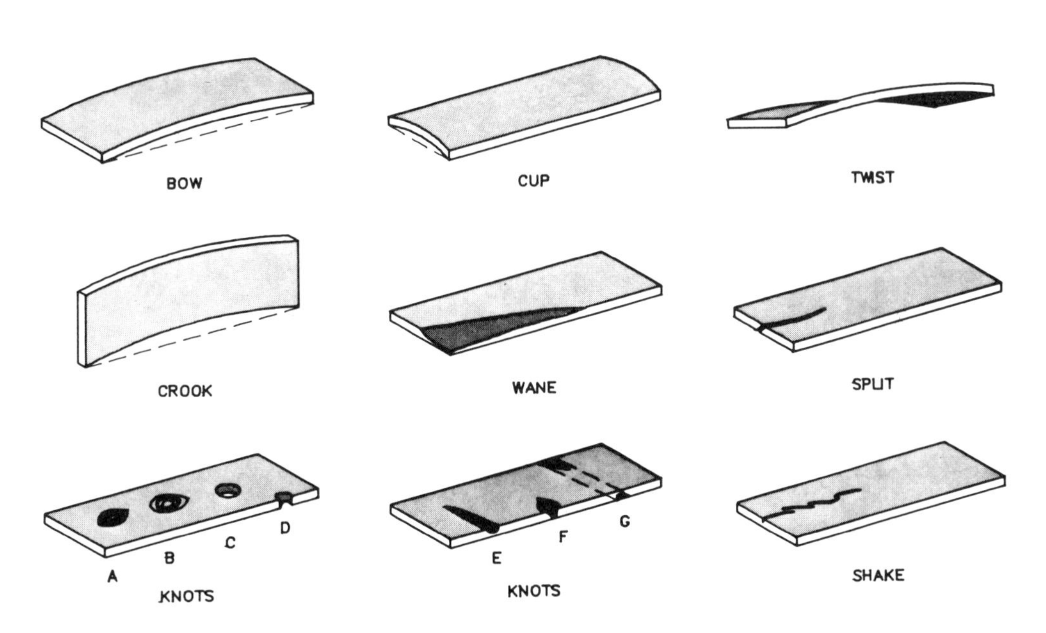

Idaho white pine, although graded under WWPA rules, carries its own grade designations.

Each board grade describes the type, size, and number of defects permitted in the worst board in that grade. Most boards in the grade will be better, all the way to just below the next higher grade. However, grading is done when a board is just out of the planer. If it has been improperly stored, it can possibly have a lot more warp, for example, than what was allowable at inspection.

Table 2–2 (pp. 16–17) gives typical grade names for softwood lumber. Table 2–3 (pp. 18–21) summarizes softwood dimension lumber and timber grade requirements, described below.

Dimension Lumber

Dimension lumber is so called because it is supplied in nominal 2″, 3″, and 4″ thicknesses of various standard nominal widths. It is used for framing—joists, studs, rafters, planks, light posts, and the like. Dimension lumber is divided into several classes, three of which—Light Framing, Studs, and Joists and Planks—are generally available in retail lumberyards.

Light Framing. Light framing dimension lumber is sorted into several grades. Construction, the top grade, is for general framing. Although the grade has good appearance, it is graded primarily for strength and serviceability. Standard grade, also for general framing, is almost as good, but bigger knots, knotholes, and other defects are permitted. At the bottom end are the Utility and Economy grades, which should be avoided.

Studs. There is just one grade in the National Grading Rules for this kind of lumber: Stud. It is suitable for all stud use, including load-bearing walls. Some suppliers also have an Economy Stud grade, in which, like Economy Light Framing, almost anything goes in the line of defects, as long as the piece of wood is there and you can nail into the ends.

Joists and Planks. This class of dimension lumber is intended to be used full length to support floors, roofs, and similar structures. Grades are: Select Structural, No. 1, No. 2, No. 3, and Economy; grade quality requirements are very similar to those for Light Framing except for allowable knots.

Timbers

Timbers have a minimum thickness of 5″ and are produced in several classes and grades.

Beams and Stringers. These are 5″ or more in thickness and more than 2″ wider than they are thick, making the minimum dimensions in this class 5″ × 8″. There are four grades; in descending order of strength and appearance they are: Select Structural, No. 1, No. 2 (No. 1 Mining), and No. 3 (No. 2 Mining).

Posts and Timbers. These posts are 5″ × 5″ and larger, but the width cannot be more than 2″ greater than the thickness. These are graded Select Structural (with good appearance), No. 1, and No. 2.

HARDWOOD

Unlike softwood lumber, which is graded looking at the piece as a whole, hardwood is graded with the assumption that it will be cut into pieces to make furniture parts. Softwood is inspected after the piece has been planed smooth; hardwood is graded in the rough. And the inspector doesn't care about the defects; he is interested only in the amount of clear wood between them. The grading is done under the rules of the National Hardwood Lumber Association (NHLA).

The highest hardwood cutting grade is called Firsts and the next grade Seconds. In a shipment, boards are nearly always combined in a single grade called Firsts and Seconds (FAS). Other grades, in descending order, are: Selects, No. 1 Common, No. 2 Common, Sound Wormy, No. 3A Common, and No. 3B Common. You are not likely to encounter the lower grades in a home supply center or a lumberyard that caters to do-it-yourselfers.

Hardwood grading is extremely complex, with all manner of exceptions to be taken into account

LUMBER AND PLYWOOD BUYING TIPS

- When working on a big home improvement project you may not be able to estimate materials to the last foot at the very beginning. So, buy about 80 to 90 percent of what you think you need, depending on how sure you are. By the time you get near the end and have changed the design a few times, you will be able to figure out exactly what additional material you need to complete the job.
- Avoid Economy grades of dimension lumber. If the wood was any good, it wouldn't be in that grade.
- Figure your board and dimension lumber needs in standard-length pieces rather than the total number of feet. On paper, figure out the combinations and quantities of pieces you can get from each standard piece ordered. You will be surprised how much money you can save by reducing waste with a little advance planning.
- Buy dimension lumber in pieces as long as you can handle. Long pieces in lumberyards tend to have less warp than short pieces.
- Yards will generally cut lumber for you so you can get it into your car without charge (if not, find another yard), but most will charge you extra to cut pieces to the exact dimension.
- Buy A-C exterior fir plywood in preference to A-D interior even for interior jobs. The better-quality back and core plys make it worth the small extra cost.
- For built-ins, buy birch veneer plywood rather than A-C fir plywood. While the cost will be appreciably more per panel than A-C exterior fir, birch ply can be stained or painted without all the surface preparation necessary with fir plywood. You will save time and work, and get a much better looking finished job.
- Lumberyards rightfully do not appreciate your sorting through a pile of lumber looking for the best pieces, and many do not permit it. But you don't have to take trash no one else wants either, or pieces that have defects that make them useless for what you want to build. You have a right to expect that each piece you buy meets the standards for the grade that you are being charged for.
- If you need top-quality veneer plywood in less than full-panel quantities, check local cabinet and millwork shops for leftovers from jobs. They can also be a good source for small quantities of hardwood.
- If you need a select grade of pine, but will be cutting to small pieces, buy common lumber and cut around the knots. The savings can be considerable. It all comes from the same trees, but inspect the boards carefully for cup.

for different wood species. Hardwood is sold by the board foot, not trimmed to standard lengths and widths like softwood dimension lumber. You will not get 100 percent usable wood when lumber is bought by the board foot, even from the best FAS grade. It is sold in random widths and lengths, rough or surfaced. If surfaced on opposite wide sides—designated S2S (surfaced two sides)—the thickness will be as specified; if in the rough, the thickness may vary.

PLYWOOD

Plywood is made of an odd number of thin layers of wood (called veneers) that are sandwiched together and glued up with the grain of the veneers alternating at right angles. Softwood plywood—usually called fir plywood—is manufactured under U.S. Product Standard PS-1. Most plywood is marked with a grade trademark of the American Plywood Association (Figure 2–3, page 22). This trademark tells you everything you need to know about the plywood, as explained in Figure 2–4. The APA performance standards encompass such other panel products as composites, waferboard, structural particleboard, and oriented-strand board.

Exposure durability. Fir plywood is made in four exposure durability classifications: Exterior, Exposure 1, Exposure 2, and Interior. There are many grades within each classification.

Table 2–2
SOFTWOOD LUMBER GRADING REQUIREMENTS

Use	Description; Limits of Defects	Northeastern Lumber NELMA	Western Lumber WWPA	Idaho White Pine IWP
Lumber of exceptional quality and appearance. Many pieces absolutely clear.	Two sound, tight pin knots (½" max.) or slight traces of pitch, or a very small pocket, or equivalent combination. Very light torn grain, skip, and cup. Wane on reverse side on an occasional piece.	—	B and Better 1 & 2 Clear Superior Finish	Supreme
Lumber of fine appearance for high-quality interior trim and cabinet work.	Two small, sound, tight knots (¾" max.) or light pitch, or a small pitch streak, or two very small pockets, or equivalent combination. Very light torn grain, cup, and crook. Light skip. Wane on reverse side on an occasional piece.	C and Better Select[1] C Select[2]	C Select	Choice
	Four small fixed knots (¾" max.) or four small pockets, or medium pitch, or equivalent combination. Medium stain, scattered light torn grain, light crook, short split on end, wane on reverse side.	D Select	D Select	Quality
Best-quality knotty lumber with all sound tight knots.	Sound, tight knots and smooth red knots (2¼" max.). Very light torn grain and cup. Very short splits, one each end. Light crook and pitch. Two small dry pockets. Wane on reverse side on an occasional piece.	Finish[1] and 1 Common[2]	1 Common	Colonial
Knotty lumber for exposed paneling and shelving and exterior house trim.	Sound and tight red knots (3" max.), sound and tight black knots (1⅜" max.). Light torn grain, cup, shake. Medium-light crook, short splits (one each end), three small dry pockets, medium wane, some firm heart pith, 12 scattered pinholes.	Premium[1] and 2 Common[2]	2 Common	Sterling
Knotty lumber for use where appearance and strength are important in shelving, paneling, siding, and fencing.	Sound and tight red knots (3½" max.), unsound knots, loose knots, knotholes (1½" max.). Medium torn grain and crook, cup. Light-to-medium shake. Some unsound wood, heavy pitch, split up to ⅙ of length. Only one knot in a board can be maximum size.	Standard[1] 3 Common	3 Common	Standard
Knotty lumber for general construction: subfloors, roof and wall sheathing, concrete forms, and crates.	Fixed, firm, tight knots up to ⅔ width, 3" loose knots and knotholes. Heavy torn grain, cup, wane. Large pockets. Heavy streaks and patches of massed pitch over ½ area. Medium to heavy shake full length, split ⅓ length. Pinholes and small holes unlimited.	4 Common	4 Common	Utility
Lowest grade knotty lumber.	Large knots, very large holes, unsound wood, massed pitch, heavy splits, shake, wane in any degree or combination. But in some pieces may be only slightly below No. 4.	5 Common	5 Common	Industrial

[1]Grading based on 1" × 8" × 12' piece. Defects in other size pieces proportional. Grading described based primarily on WWPA rules. Rules for other lumber may vary, particularly on allowable knots.

[2]Eastern white pine, Norway spruce.

West Coast Lumber WCLB	Redwood RLB	Western Red Cedar WCLB & WWPA	Southern Pine SPIB
——	Clear All Heart	Clear Heart	B & B Finish (B and Better)
C and Better Finish D Finish	—— ——	A B	C Finish D Finish
Select Merchantable	Select Heart Select	Select Merchantable	No. 2 Board
Construction	——	Construction	No. 2 Board
Standard	Construction Heart Construction Common	Standard	No. 3 Board
Utility	Merchantable	Utility	No. 4 Board
Economy	Economy	Economy	——

Table 2–3
SOFTWOOD DIMENSION LUMBER AND TIMBER GRADING REQUIREMENTS

	DIMENSION LUMBER			
	Light Framing (2″ to 4″ thick, 2″ to 4″ wide, 6′ and longer)			
Features	**Construction**	**Standard**	**Utility**	**Economy**
TYPE OF KNOT PERMITTED	Sound, firm, encased, and pith. Must be tight.	Any kind	Any kind	ALL LUMBER CHARACTERISTICS ALLOWED EXCEPT BROKEN OR SLABBY ENDS. LARGE KNOTS AND KNOTHOLES, UNSOUND WOOD, HEAVY SHAKE, SPLITS AND WANE OR ANY COMBINATION PERMITTED. PIECES 9′ AND SHORTER MUST BE USABLE FULL LENGTH
SOUND KNOTS, SIZE PERMITTED — WIDTH OF PIECE 2″	¾″	1″	1¼″	
3″	1¼″	1½″	2″	
4″	1½″	2″	2½″	
5″	—	—	—	
6″	—	—	—	
8″	—	—	—	
10″	—	—	—	
12″	—	—		
14″	—	—	—	
SPIKE KNOTS ACROSS WIDE FACE	¼ of cross section	⅓ of cross section	½ of cross section	
LOOSE KNOTS & KNOTHOLES, SIZE PERMITTED — WIDTH OF PIECE 2″	⅝″	¾″	1″	
3″	¾″	1″	1¼″	
4″	1″	1¼″	1½″	
5″	—	—	—	
6″	—	—	—	
8″	—	—	—	
10″	—	—	—	
12″	—	—	—	
14″	—	—	—	
KNOTHOLE SPACING	1 per 3 lin. ft.	1 per 2 lin. ft.	1 per lin. ft.	
SLOPE OF GRAIN	1 in 6	1 in 4	1 in 4	
SHAKE	Several heart shakes up to 2′ long, none through	½ thickness at ends; elsewhere longer, some through	Surface shakes at ends, same as split through; ⅓ length	
SPLITS (MAX.)	Length equal to width of piece	Length equal to 1½ times width of piece	⅙ length of piece	
WANE Typical, many exceptions	¼ thickness; ¼ width	⅓ thickness; ⅓ width	½ thickness; ½ length	
WARP Includes bow, crook, cup, twist, or any combination	½ of medium	Light	Medium	
UNSOUND WOOD • Spots and streaks only • Must not destroy nailing edge	Not permitted		⅓ of cross section	

DIMENSION LUMBER (cont'd.)			
Studs (2" to 4" thick, 2" to 6" wide, 10' and shorter)		**Joists and Planks** (2" to 4" thick, 5" and wider, 6' and longer)	
Stud	**Economy Stud**	**Select Structural**	**No. 1**
Any kind, well-spaced	Any kind	Sound, firm, encased, pith; tight and well-spaced	Sound, firm, encased, pith; tight and well-spaced
¾" 1¼" 1¾" 2¼"–3" 2¾"–3¾" — — — —	Knots up to ¾ of cross section permitted — — — — —	— — — 1"–1½" 1⅛"–1⅞" 1½"–2¼" 1⅞"–2⅝" 2¼"–3" 2⅜"–3¼"	— — — 1¼"–1⅞" 1½"–2¼" 2"–2¾" 2½"–3¼" 3"–3¾" 3⅛"–4"
N.A.	N.A.	N.A.	N.A.
¾" 1¼" 1½" 1¾" 2" — — — — 1 per lin. ft.	Knotholes up to ¾ of cross section permitted — — — — —	— — — ⅞" 1" 1¼" 1¼" 1¼" 1¼" 1 per 4 lin. ft.	— — — 1⅛" 1¼" 1½" 1½" 1½" 1½" 1 per 3 lin. ft.
1 in 4	—	1 in 12	1 in 10
At end, same as split if through; elsewhere ⅓ length	Not limited	½ thickness at ends; elsewhere 2' long, none through	½ thickness at ends; elsewhere 2' long, none through
Length equal to twice width of piece	¼ length of piece	Length equal to width of piece	Length equal to width of piece
⅓ thickness; ½ width; length unlimited	¼ thickness; ¾ face; length unlimited	¼ thickness; ¼ width; up to ½ length	¼ thickness; ¼ width; up to ½ of length
½ of medium	Crook and twist 1" max. in 8' stud	½ of medium	½ of medium
⅓ of cross section		Not permitted	Not permitted

(cont'd.)

Table 2–3
SOFTWOOD DIMENSION LUMBER AND TIMBER GRADING REQUIREMENTS (cont'd.)

	DIMENSION LUMBER (cont'd.)		
	Joists and Planks (2″ to 4″ thick, 5″ and wider, 6′ and longer)		
Features	**No. 2**	**No. 3**	**Economy**
TYPE OF KNOT PERMITTED	Any kind, well-spaced	Any kind, well-spaced	ALL LUMBER CHARACTERISTICS ALLOWED EXCEPT BROKEN OR SLABBY ENDS. LARGE KNOTS AND KNOTHOLES, UNSOUND WOOD, HEAVY SHAKE, SPLITS, WANE OR ANY COMBINATION PERMITTED. PIECES 9′ AND SHORTER MUST BE USABLE FULL LENGTH
SOUND KNOTS, SIZE PERMITTED — WIDTH OF PIECE 2″	—	—	
3″	—	—	
4″	—	—	
5″	1⅝″–2⅜″	2¼″–3″	
6″	1⅞″–2⅞″	2¾″–3¾″	
8″	2½″–3½″	3½″–4½″	
10″	3¼″–4¼″	4½″–5½″	
12″	3¾″–4¾″	5½″–6½″	
14″	4⅛″–5¼″	6″–7″	
SPIKE KNOTS ACROSS WIDE FACE	N.A.	N.A.	
LOOSE KNOTS & KNOTHOLES, SIZE PERMITTED — WIDTH OF PIECE 2″	—	—	
3″	—	—	
4″	—	—	
5″	1⅜″	1⅞″	
6″	1½″	2″	
8″	2″	2½″	
10″	2½″	3″	
12″	3″	3½″	
14″	3½″	4″	
KNOTHOLE SPACING	1 per 2 lin. ft.	1 per lin. ft.	
SLOPE OF GRAIN	1 in 8	1 in 4	
SHAKE	½ thickness at ends; elsewhere 3′ or longer, some through	Surface shakes permitted	
SPLITS (MAX.)	Length equal to 1½ times width of piece	⅙ length	
WANE Typical, many exceptions	⅓ thickness; ⅓ width; up to ⅔ length	½ thickness; ½ width; up to ⅞ length	
WARP Includes bow, crook, cup, twist, or any combination	Light	Medium	
UNSOUND WOOD • Spots and streaks only • Must not destroy nailing edge	Not permitted in thicknesses over 2″	⅓ of cross section	

TIMBERS				
Appearance (2″ to 4″ thick, 2″ and wider, 6′ and longer)	**Beams and Stringers** (5″ and thicker: width more than 2″ greater than thickness, 6′ and longer)		**Posts and Timbers** (5″ × 5″ and larger: width not more than 2″ greater than thickness, 6′ and longer)	
Appearance	**Select Structural**	**No. 1**	**Select Structural**	**No. 1**
Sound, tight, well-spaced	Sound, tight, well-spaced	Sound, tight, well-spaced	Sound, tight, well-spaced	Sound, tight, well-spaced
½″	—	—	—	—
¾″	—	—	—	—
1″	—	—	—	—
1¼″	1¼″–2″	1⅞″–3″	1″	1½″
1½″	1½″–2½″	2¼″–3¾″	1¼″	1⅞″
2″	1¾″–3″	2½″–4½″	1⅝″	2½″
2½″	2″–3¼″	3″–4¾″	2″	3⅛″
3″	2¼″–3½″	3¼″–5″	2⅜″	3¾″
3⅛″	2⅜″–3¾″	3½″–5½″	2½″	4″
N.A.	N.A.	N.A.	N.A.	N.A.
None permitted	None permitted	None permitted	None permitted	None permitted
1 in 10	1 in 15	1 in 11	1 in 12	1 in 10
Not specified	⅙ thickness	⅙ thickness	⅓ thickness on one end	⅓ thickness on one end
Length equal to width of piece	Length equal to ½ width of piece	Length equal to width of piece	Length equal to ¾ thickness	Length equal to width of piece
1/12 thickness; 1/12 width; ⅙ length	⅛ any face	¼ any face	⅛ any face	¼ any face
Very light	—	—	—	—
Not permitted	—	—	—	—

Figure 2–3
Plywood sheathing. Plywood is made in many grades to meet specific structural, environmental, and appearance requirements. (Photo courtesy American Plywood Association, © Arcon, Inc.)

Exterior panels have a fully waterproof bond and are designed for applications subject to permanent exposure to the weather or to moisture. Exposure 1 panels also have a fully waterproof bond, but are rated only for short-term weather exposure installation where high-moisture conditions will be encountered. Exposure 2 panels are made with a water-resistant glue and are rated for short-term exposure to moisture. Interior panels are made with glue that is not waterproof; they must be installed without exposure to weather or moisture and be protected from such exposure in service.

Grades. Plywood in each classification is made in many grades. The grade defines a panel in terms of the grade of the face and back veneers (A-C, A-D, etc.); it also defines the appearance (and quality) of the plywood veneers used in the core of the panel. Plywood grades also have names that indicate their primary application or suitability—Sheathing, Underlayment, and so on.

The face veneer grades, from best to worst, are N, A, B, C, C-plugged, and D.

Figure 2–4
The American Plywood Association (APA) trademark contains all the information needed to identify panel characteristics.

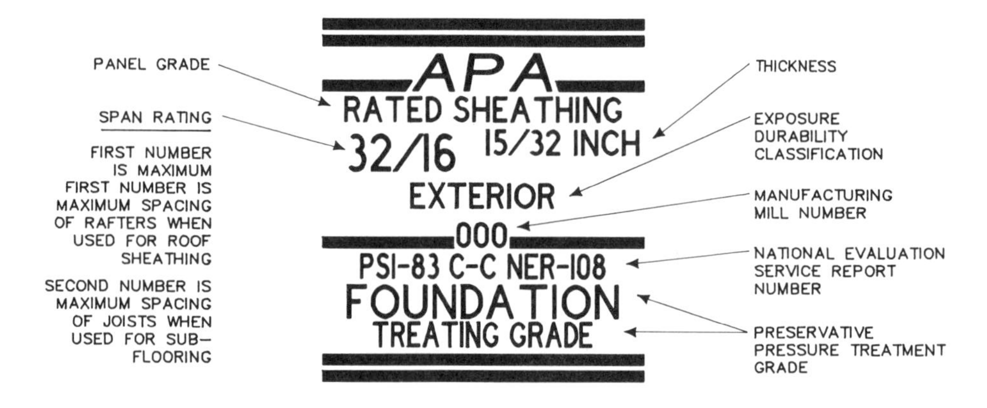

N Smooth surface, natural finish veneer. No open defects, and not more than six repairs (with wood) allowed per 4×8 panel, made to run with the grain and well matched for grain pattern and color. Rarely found in lumberyards.

A Smooth, paintable. Not more than 18 neatly made repairs per 4×8 panel permitted.

B Solid surface veneer. Shims, circular repair plugs, and tight knots up to 1″ across permitted. Minor splits permitted.

C-Plugged Improved grade C veneer with splits limited to ⅛″ in width and knotholes and borer holes limited to ¼″×½″. Synthetic compound repairs permitted.

C Tight knots up to 1½″ and knotholes up to 1″ across permitted, with some knotholes to 1½″ within specified limits. Synthetic or wood repairs. Minor splits and stitching permitted.

D Knots and knotholes up to 2½″ across the grain permitted, with some up to 3″ permitted within specified limits. Limited splits allowed. Grade D veneers are not allowed in Exterior panels.

Panels with grade A, B, or N faces are sanded smooth on both sides in manufacture. Panels with C, C-plugged, and D face panels are touch-sanded: high spots are sanded off to get uniform panel thickness.

The letter grades spell out veneer quality. Plywood is also divided into *Appearance* and *Engineered* grades. The common A-D interior and A-C exterior are appearance grades. They are used where how they look is the important consideration.

Engineered (performance) grades are used for sheathing, subflooring, underlayment, and decking where strength and solid integrity are more important than appearance and surface finish.

Span ratings. These ratings are the maximum recommended center-to-center support spacing for the panel in floor, wall, and roof construction. In the APA Rated Sheathing marking shown in Figure 2–5, the span rating is the large figure 32/16. The first number gives the maximum support spacing when the panel is used for roof decking (sheathing) with the long dimension of the panel running across three or more rafters. The second number gives the spacing for subflooring with the long dimension of the panel running across three or more joists. Thus, a panel marked 32/16 can go across rafters spaced up to 32″ on center and floor joists up to 16″ on center.

Guide to APA Performance-Rated Plywood

The following plywood listing contains those grades most likely to be used in residence interior renovation.

Sheathing (Figure 2–5). A plywood, particleboard, waferboard, oriented-strand, or composite panel designed for subflooring, wall sheathing, and roof sheathing (decking); also used for a broad range of other construction and do-it-yourself applications.

Sturd-I-Floor (Figure 2–6). A plywood, particleboard, waferboard, oriented-strand, or composite panel designed to be used as a combination structural subfloor and underlayment. The panel possesses high resistance to concentrated and impact loads and is sanded to provide a smooth surface for laying pad-and-carpet or resilient flooring. It is made with square or tongue-and-groove edges.

A-C Exterior (Figure 2–7). General utility plywood for use outdoors or indoors. One good side (A surface). Use for built-ins, shelves.

A-D Interior (Figure 2–8). General utility plywood for use indoors. One good side. Use for built-ins, shelves. D surface difficult to repair adequately for painting.

Underlayment (Figure 2–9). For use over structural subfloor. Provides smooth surface for carpet and pad and possesses high resistance to concentrated and impact loads. Touch-sanded. For areas to be covered with resilient nontextile flooring, specify panels with "sanded face."

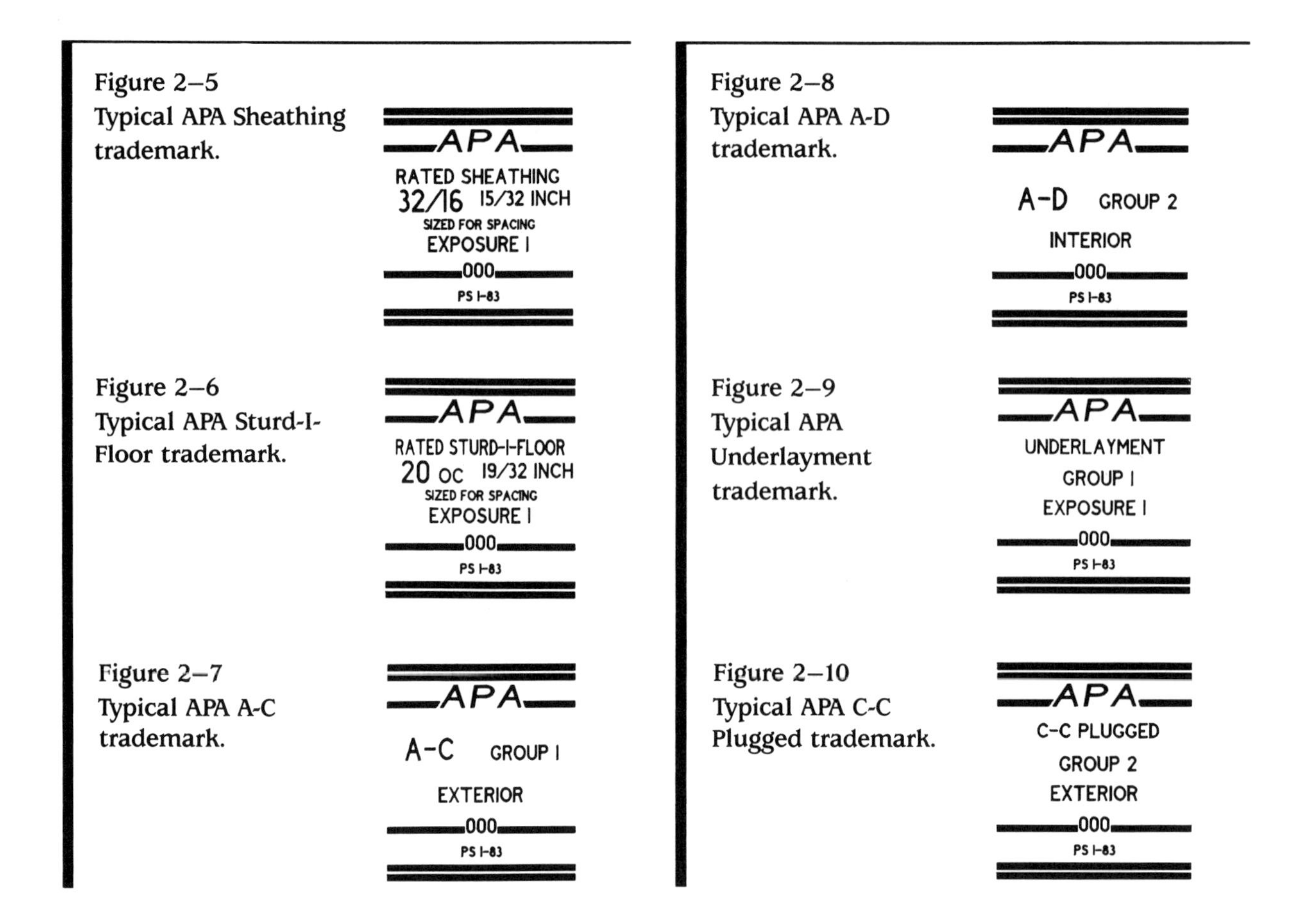

Figure 2–5 Typical APA Sheathing trademark.

Figure 2–6 Typical APA Sturd-I-Floor trademark.

Figure 2–7 Typical APA A-C trademark.

Figure 2–8 Typical APA A-D trademark.

Figure 2–9 Typical APA Underlayment trademark.

Figure 2–10 Typical APA C-C Plugged trademark.

C-C Plugged (Figure 2–10). For use as an underlayment over structural subfloor. Provides smooth surface for carpet and possesses high resistance to concentrated and impact loads. Touch-sanded.

Other Plywood

There are a number of kinds of plywood that are not covered by U.S. Product Standard PS 1–83 and are not performance-rated by the APA. Several are excellent for various home projects.

Lauan. This plywood is imported from Korea, Japan, and elsewhere, mostly as 3/16"- to 7/32"-thick underlayment. Lauan actually can be any of several wood species, some of which look like mahogany. Besides its use for underlayment, it is the best all-around inexpensive choice for anything needing 1/4" plywood, such as drawer bottoms and cabinet backs.

Veneer plywood. All plywood is made of veneer, but the term *veneer plywood* means plywood with something better than fir used for the face veneers, and usually for the interior layers too. This is the plywood to use for furniture or anything on which you plan to put a clear finish. The choice of veneers includes pine, birch and most other domestic hardwoods, redwood, mahogany, and even exotic imported woods.

Baltic birch. This high-quality birch plywood is imported from the Scandinavian countries and Russia. Panels are 3/8" and 3/4" thick and usually 60" square. It is excellent for doors, countertops, and similar uses.

3 Nails, Screws, and Bolts

Nails are the most common fasteners used in carpentry. They join wood parts faster, easier, and with less expense than screws, glue, and formed joints. Figure 3–1 shows some of the many kinds useful in home carpentry.

NAIL QUALITIES

There is no one kind of nail that is the best choice for all purposes. The size and shape of a nail's head determine what the nail can hold most efficiently and how it will look in place. To hold down asphalt shingles, especially in wind, a large-diameter, flat head is needed. If the nail must be inconspicuous or invisible, a very small head that can easily be driven below the surface, as on a finish nail, is needed. How well the nail will resist being pulled out depends on the shank diameter and to what extent the smooth shank has been squared, twisted, threaded, ringed, or grooved. The shape of the point affects both holding power and wood splitting. Nails to be driven into masonry (and some other nails) must be hardened to survive installation. Nails exposed to the elements or in a damp location must be corrosion resistant.

Holding power. A nail driven into wood resists being pulled straight out (withdrawal resistance) and resists being forced sideways in the wood (lateral resistance). Tests conducted by the U.S. Department of Agriculture Forest Products Laboratory have shown that for bright common nails withdrawal resistance varies directly with the diameter of the nail shaft and the depth of penetration, and that these nails hold better in dense (hard) than light (soft) woods. But large-diameter nails split the wood more than small ones, and hardwoods split more readily than softwoods. Nails will hold in dry or wet wood, but will lose as much as three-quarters of their grip if the wood alternates between moist and dry states after nails have been driven in.

A very effective way to increase withdrawal resistance, and at the same time reduce splitting, is to drill a pilot hole for the nail. The pilot hole should be slightly smaller than the nail. A pilot hole will also help you drive the nail straight.

The holding power of a nail can also be improved by deforming the smooth shank with some kind of a ridged pattern (see Figure 3–2). The Scotch nail is typical: the shank is square in section with angular serrations on all four surfaces. The equivalent round-shank nail has annu-

Figure 3–1
Nail types useful in home carpentry.

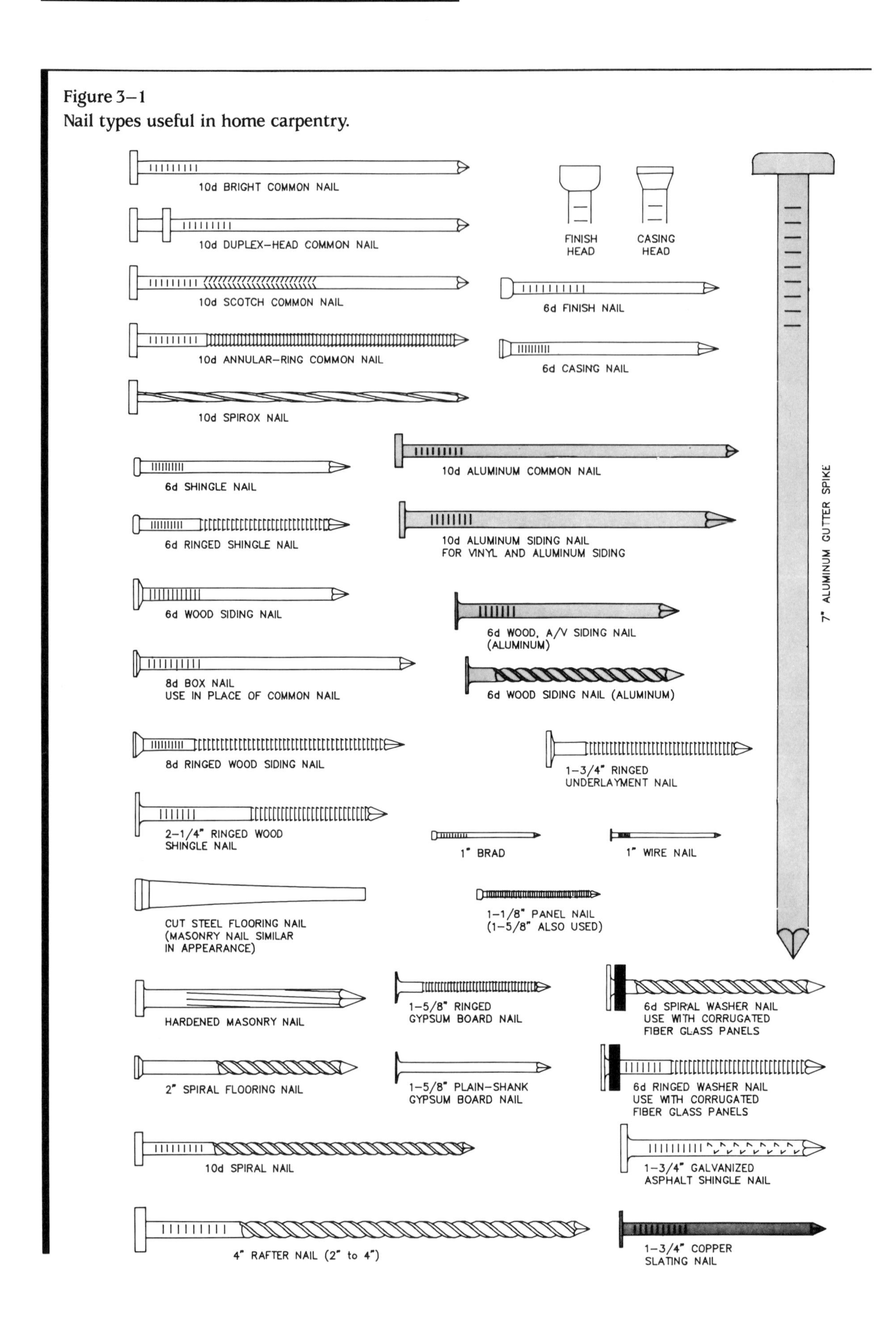

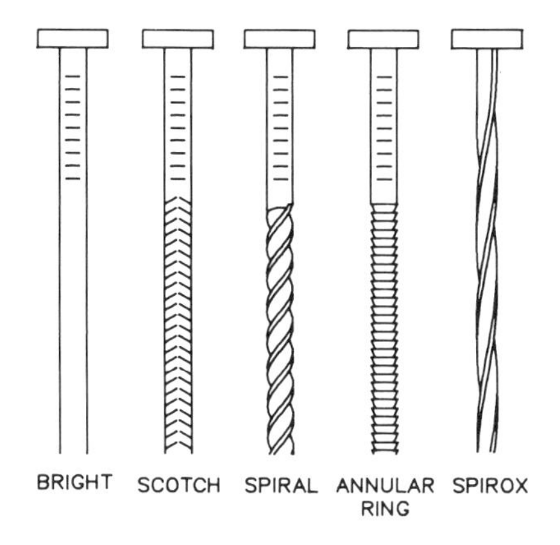

Figure 3–2
A smooth, round-shaft nail (the Bright), and four kinds of deformed-shank nails.

lar ringlike serrations. The holding power of a ridged- or barbed-shank nail is up to twice that of ordinary smooth-shank common nails. As the diagonal dimension of the square shank is the same diameter as that of the equivalent round-shank nail, you get more nails per pound. Scotch nails cost the same as the common hardware store variety, but only large suppliers carry them.

One big drawback to Scotch nails is that they can't be galvanized to prevent rusting. The Spirox spiral nail solves this problem. This twisted square-section nail can be galvanized and has a holding power roughly equivalent to a Scotch nail.

Corrosion resistance. Use galvanized nails or other corrosion-resistant nails any place you expect rust to be a problem. A galvanized nail has a coating of zinc. The zinc can be applied by electroplating, which results in a thin, smooth coating that gives only short-term rust protection. Or it can be applied by the hot-dip process—tumbling the nails in molten zinc—which produces a rough and sometimes sharp-surfaced nail, but one that will resist rust a long time. Only hot-dip galvanized nails should be used exposed to the weather outdoors.

Nails are also made of aluminum, copper, bronze, and stainless steel for corrosion resistance, but they are more expensive than galvanized nails.

NAIL TYPES

Nails are sized primarily by length from 1" to 6", but the sizes are traditionally given as "pennyweight," originally related to the cost of the nails. Pennyweight is abbreviated "d"; for example, a four-penny nail is designated 4d. Figure 3–3 shows the length of nails from 2d to 60d.

Common nails. These nails are general-purpose, heavy-duty nails for house framing and similar construction, and for rough work where the permanent visibility of the large nail head is not objectionable. Common nails are the kind shown in Figure 3–3.

Box nails. For most uses, you can substitute cement-coated box nails in the same lengths for common nails. A box nail has a head nearly as large as a common nail of the same length, but a smaller-diameter shank. The "cement" in the nail coating is a resin. As box nails have a smaller diameter than common nails, they are a lot less tiring to drive and are less likely to split wood. The resin coating increases friction, producing a higher initial withdrawal resistance in softwood,

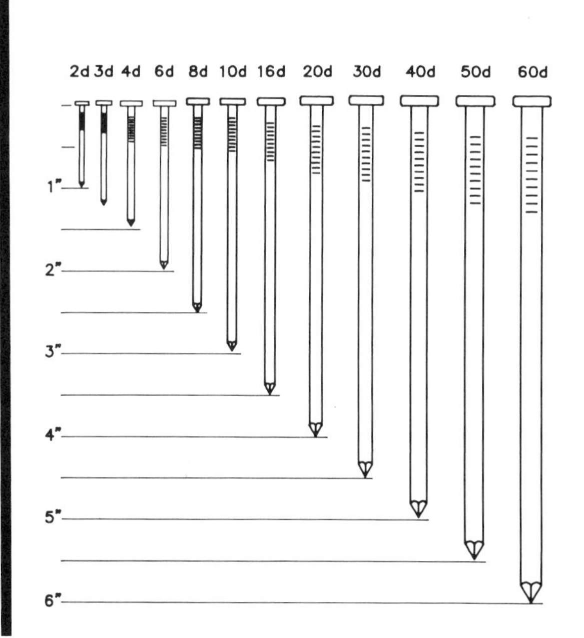

Figure 3–3
Sizes of common nails.

but the improvement decreases as the resin is absorbed in wood over time. In hardwood, the resin doesn't produce much of an advantage as it tends to get rubbed off going into the outside piece of wood.

Duplex nails. These nails have a regular flat head and a second "head"—actually a flange—about ½″ down the shaft. They hold better in temporary construction—anything you know you will dismantle—than common nails driven in only partway. Duplex nails are driven in as far as the lower head, the flange, and are removed using a claw hammer or prybar under the top head. They cost about the same as galvanized nails.

Finish nails. These nails are used in installing trim and in cabinetmaking. Finish nails are made of lighter-gauge wire than common or box nails of the same length and have much smaller heads, only about a third larger than the shank of the nail. The small heads are easily driven below the surface of the wood with a nailset so the nail hole can be filled and hidden. Finish nails are made from 2d (1″) up to 16d (3½″) and are also available galvanized and with Scotch shafts.

Smaller finish nails are called brads. They measure ½″ to 1½″ and are specified by length and wire gauge. Each length is made in several gauges, and some brad sizes are available galvanized. If a small nail of the same size range has a flat head, it is called a wire nail.

Gypsum board (drywall) nails. These nails are used to attach gypsum board to wood framing (see Chapter 12). They are made with either smooth or annular-ring shanks. The ringed variety should be used, as they hold better. With either variety, nail popping is a problem. If at all possible, use gypsum board screws instead.

Masonry nails. These heavy-shank, hardened nails are used to fasten furring or other rough wood to concrete and cement block walls. (When appearance is important, use screws driven into masonry anchors instead of nails.) The nail should be long enough to penetrate ¾″ to 1″ into the masonry; if shorter, the nail won't hold, and if longer, it may not go in all the way. Masonry nails must be driven with a hand-drilling hammer (it looks like a small one-hand sledge hammer). Tapping the nail with a claw hammer won't drive it. Masonry and wood flooring nails are shown in Figure 3–4.

Figure 3–4
Hardened masonry and flooring nails.

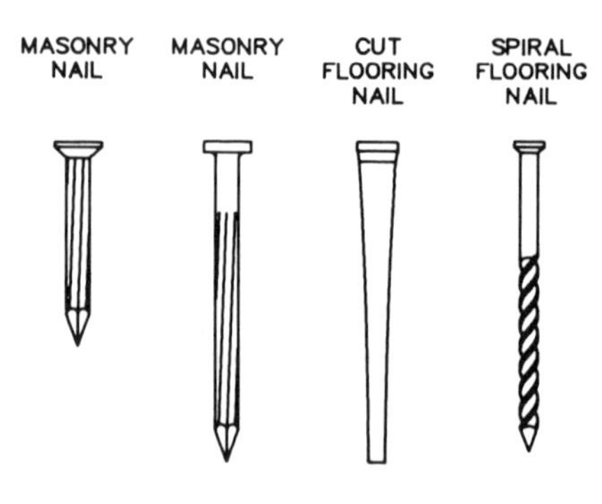

Flooring nails. Two kinds of nails are used for nailing strip flooring—hardened cut nails with very small heads and hardened spiral nails. Spiral nails have less tendency to split the wood, but cut nails stacked in strips are used in mallet-powered nailing machines. Block flooring can be nailed but is usually put down with a mastic adhesive. Some powered "nailers" now drive special flooring staples.

Spiral flooring nails have high holding power—about four times that of smooth-shank common nails of the same size. Similar, larger spiral nails include pallet nails (1½″ to 4″), rafter nails (2″ to 4″), and pole barn nails (2½″ to 6″).

Underlayment nails. These heavy-shanked annular ring nails with small flat heads are used to attach plywood subflooring to joists and plywood underlayment to subflooring.

Siding nails. Each type and form of siding material has specific nail and nailing pattern requirements. A wide variety of nails are used for fastening siding to a house, as shown in Figure 3–5. The nails must all be corrosion resistant, but there the similarity ends.

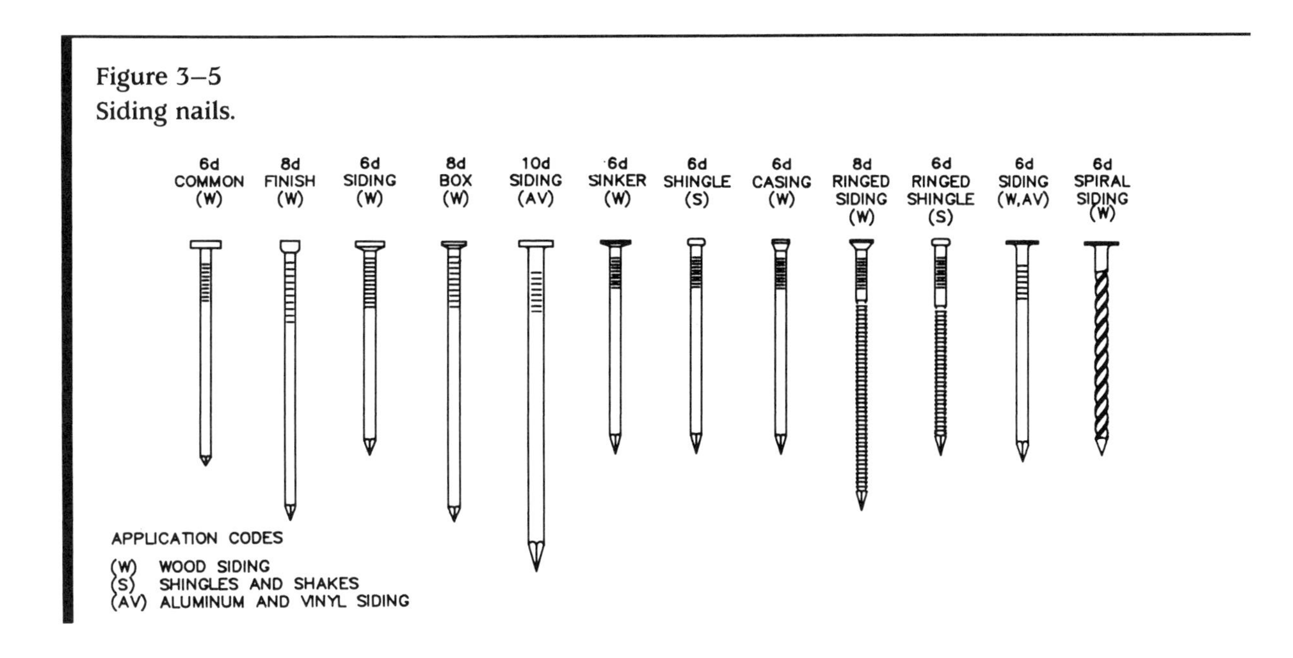

Figure 3–5
Siding nails.

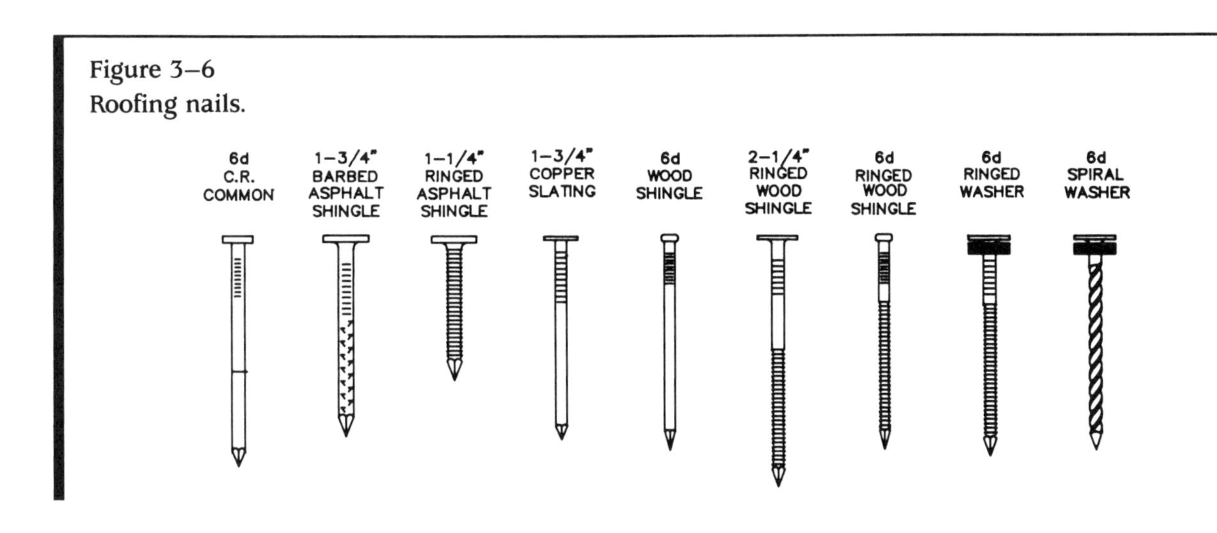

Figure 3–6
Roofing nails.

Aluminum siding requires a nail that is also aluminum, so there will be no corrosive dissimilar-metal reactions. The nail must have a big flat head because aluminum siding is hung loosely on the house, never fastened down tight; the siding will buckle if it is not free to expand and contract with temperature changes. Vinyl siding is hung the same way, with the same nails.

Wood lap siding is nailed with siding nails having sinker heads that can be driven flush with the wood surface or set below the surface and the hole puttied. Tongue-and-groove siding is nailed with finishing nails. Plywood siding is nailed with either ringed finish nails or ringed siding nails. Cedar shingles and shakes used for siding require one of the shingle nails shown in Figure 3–6 along with other kinds of roofing nails.

Roofing nails. Roofing usually means asphalt or fiberglass shingles. The nails for this roofing are hot-dip galvanized and have very wide heads. Cedar shingle and shake roofing require a different type of nail, as do slate and tile. (See Figure 3–6 for roofing nails.)

NAILING SCHEDULES

Selecting nail size, number of nails, and pattern of nailing in a joint is not something left to chance or casual choice. The proper fastening of framing members and sheathing provides the necessary rigidity and strength to the structure for it to withstand wind and other forces that would weaken it. The specifications for proper nailing are called a nailing schedule. The schedule given in Table 3–1 represents accepted practice in the home-building industry. However, some building codes may specify other nailing schedules. If you are going to be doing home remodeling or construction, be sure to check your local code carefully.

Table 3–1
NAIL SIZES AND SCHEDULES FOR RESIDENTIAL ROUGH FRAMING CARPENTRY

Joining	Nailing Method	Number of Nails	Nail Size[1]	Placement
Header to joist	End-nail	3	16d	Each joist
Joist to sill or girder	Toenail	3	10d or 8d	Each joist
Header and stringer joist to sill	Toenail		10d	16″ on center
Bridging to joist	Toenail	2	8d	Each end
Ledger strip to girder (2 × 2 or 2 × 3 ledger)	Face-nail	3	16d	At each joist
Subfloor, boards:				
1 × 6 and smaller	Face-nail	2	8d	To each joist
1 × 8	Face-nail	3	8d	To each joist
Subfloor, ¾″ plywood:				
At edges	Face-nail		8d	6″ on center
At intermediate joists	Face-nail		8d	10″ on center
Subfloor (2 × 6 T&G) to joist or girder	Blind-nail or face-nail	2	16d	Each bearing
Soleplate to stud (horizontal assembly)	End-nail	2	16d	At each end
Top plate to stud	End-nail	2	16d	At each end
Stud to soleplate	Toenail	4	8d	At each end
Soleplate to joist or blocking	Face-nail		16d	16″ on center
Doubled studs	Face-nail (stagger)		16d	24″ on center
End stud of intersecting wall to exterior wall stud	Face-nail		16d	16″ on center
Upper top plate to lower top plate	Face-nail		16d	16″ on center
Upper top plate, laps and intersections	Face-nail	2	16d	Each end
Continuous header, two pieces, along each edge	Face-nail		16d	16″ on center
Ceiling joist to top wall plates	Toenail	3	8d	Each joist
Ceiling joist laps at partition	Face-nail	3	16d	Each joist
Ceiling joist to parallel rafters	Face-nail	3	16d	Each joining
Rafter to top plate	Toenail	3	8d	Each joining
Rafter to ceiling joist	Toenail	5	10d	Each joining
Rafter to valley or hip rafter	Toenail	3	10d	Each joining
Ridge board to rafter	End-nail	3	10d	Each rafter
Rafter to rafter	Face-nail	3	8d	Rafter to rafter
(over girder)	Toenail	4	8d	Rafter to girder (2 each rafter)

Nailing Tips

Nails driven cross-grain near the end of the wood are very likely to cause the wood to split (see Figure 3–7). Position the nail away from the end—up to two-thirds of the thickness of the other piece in the joint. Drilling pilot holes will also help. This is a good place to use box nails instead of common nails.

In either an end- or cross-grain nailed joint, nails will resist withdrawal loads better if they are driven in at diagonally opposite directions rather than straight in.

When nailing close to the end of a piece be sure the nails are not aligned along the grain of the wood. That will most surely cause splitting. Stagger the nail positions.

Toenailing is a way to join wood framing when

Joining	Nailing Method	Number of Nails	Nail Size[1]	Placement
Rafter to rafter	Toenail	4	8d	Rafter to ridge
Through ridge board	Edge-nail	1	10d	Top of rafter to ridge
Collar beam to rafter:				
2" member	Face-nail	2	12d	Each end
1" member	Face-nail	3	8d	Each end
1" diagonal let-in brace to each stud and plate	Face-nail	2	8d	Each crossing
Built-up corner studs:				
Stud to stud	Face-nail		16d	24" on center
Studs to blocking	Face-nail	2	10d	Each side
Intersecting stud to corner studs	Face-nail		16d	24" on center
Built-up girders and	Face-nail	2	20d	At top and bottom
beams, three or		2	20d	32" staggered
more members		2	20d	at each splice
Wall sheathing:				
1 × 8 or less	Face-nail	2	8d	At each stud
Wider than 1 × 8	Face-nail	3	8d	At each stud
Wall sheathing, vertically applied plywood:				
⅜"	Face-nail		6d	6" edge
½" and over	Face-nail		8d	12" edge
Wall sheathing, vertically applied fiberboard, homasote, foamboard:				
½"	Face-nail		1½" roofing	3" edge 6" intermediate
1"	Face-nail		2" roofing	3" edge 6" intermediate
Roof sheathing boards 4", 6", 8" width	Face-nail	2	8d	At each rafter
Roof sheathing, plywood:				
⅜"	Face-nail		6d	6" edge 12" intermediate
½" and over	Face-nail		8d	6" edge 12" intermediate

[1]Box nails may be substituted for common nails.

Figure 3–7
Nailing at the ends of stock.

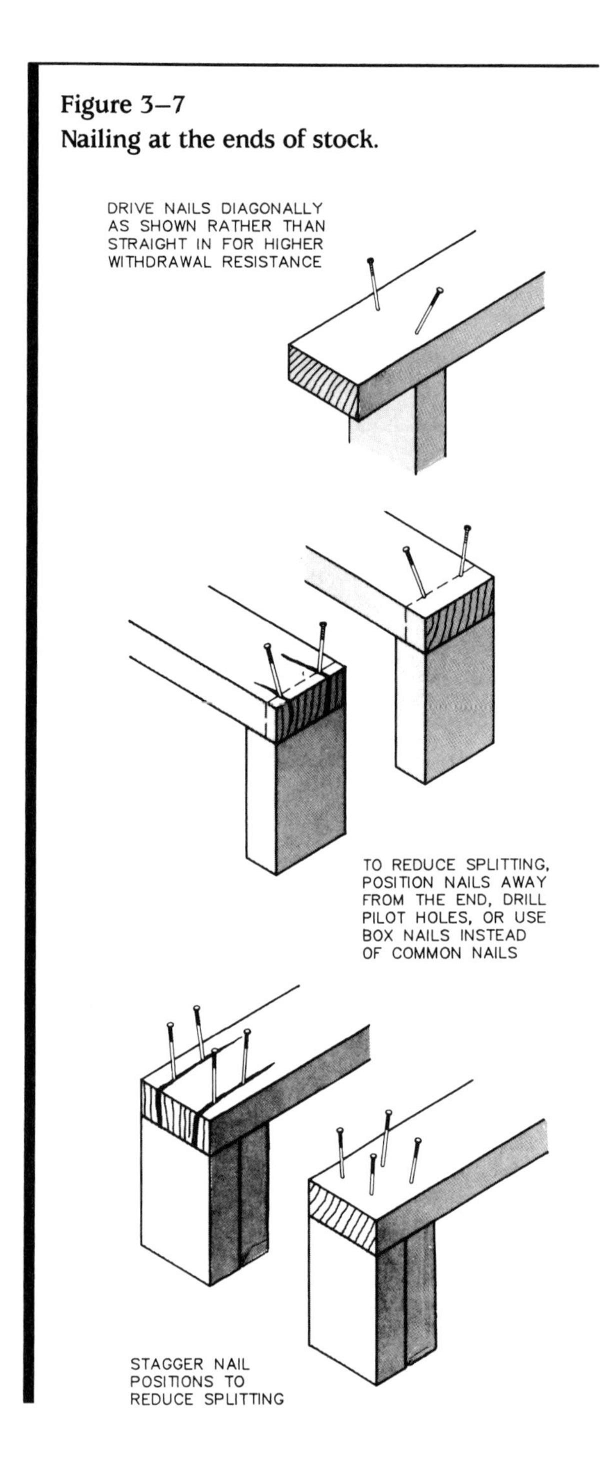

you can't get at the face of either part to drive nails. In toenailing, nails are driven diagonally through the attached piece into the main piece. Figure 3–8 shows a toenailed joist-to-plate joining on top (a structural wood fastener could also be used). On the bottom the figure shows nail positioning for maximum lateral and uplift strength in a toenailed joint. Toenailing takes practice, but results in joints that are stronger than face-nailed joints. The general rules for toenailing are: (1) the largest nails that won't cause splitting should be used, (2) one-third of the length of the nail should be in the atttached piece, (3) nails should be at a 30-degree angle with the attached member, (4) the full length of the nail should be in the wood, but you should avoid mutilating the wood with your hammer. Tests of stud-to-plate joints have shown that four 8d common nails make a stronger joint than one with two 16d common nails, and if you can drive them without splitting the wood, 10d nails make stronger joints than 8d nails. If you drill pilot holes slightly smaller than the nail shank diameter, you will have a strong, neat joint with nails going in at a correct angle and not splitting the wood.

DRIVE NAILS DIAGONALLY AS SHOWN RATHER THAN STRAIGHT IN FOR HIGHER WITHDRAWAL RESISTANCE
TO REDUCE SPLITTING, POSITION NAILS AWAY FROM THE END, DRILL PILOT HOLES, OR USE BOX NAILS INSTEAD OF COMMON NAILS
STAGGER NAIL POSITIONS TO REDUCE SPLITTING

Toenailing techniques for wall studs are shown in Figure 3–9. Brace the attached member with your foot offset to the nail side of the desired final position and drive the first nail. Then brace the member with your foot from the other side and

Figure 3–8
Toenailing.

JOIST–TO–PLATE JOINT
JOIST
PLATE
30°
USE FOUR 8d BOX OR COMMON NAILS. 10d NAILS HOLD BETTER, BUT ARE MORE LIKELY TO SPLIT WOOD.
1/3 NAIL LENGTH

Figure 3–9
Toenailing technique.

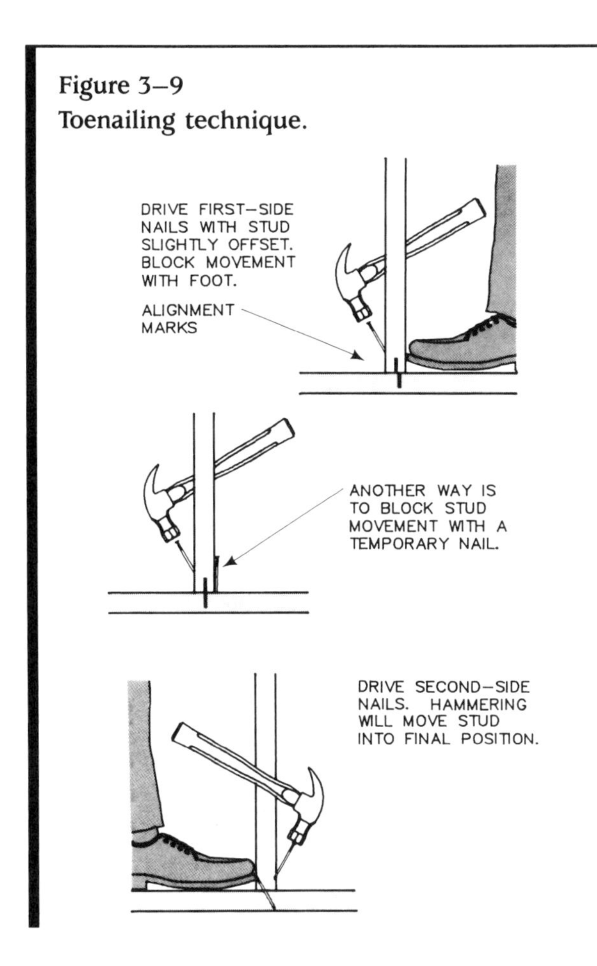
DRIVE FIRST–SIDE NAILS WITH STUD SLIGHTLY OFFSET. BLOCK MOVEMENT WITH FOOT.
ALIGNMENT MARKS
ANOTHER WAY IS TO BLOCK STUD MOVEMENT WITH A TEMPORARY NAIL.
DRIVE SECOND–SIDE NAILS. HAMMERING WILL MOVE STUD INTO FINAL POSITION.

drive the first nail on the other side, allowing the attached member to be moved into final position with the hammering. A toenailed stud-to-plate joint should always have three or four nails. The stud can twist if only two nails are used.

Another way to align the attached member for toenailing is to drive a temporary nail to hold the piece in position for the first nail. This is handy when you can't get your foot into position.

SCREWS AND BOLTS

A screw can be run into a threaded hole in one of the pieces being assembled, or it can cut its own threads as it goes in. A screw is tightened or loosened by turning the head. A bolt is used with a nut. It is run through drilled clearance holes and usually is tightened or loosened by turning the nut, not the head. Bolts require nuts, screws do not; anything that can be used with or without a nut is a screw.

Most threaded fasteners are made from standard steels. Even in dry locations, steel fasteners corrode and lose strength, and require protection with paint or other coatings. Moisture in wood also causes corrosion, and corroding screws can discolor wood. Corrosion-resistant aluminum, brass, stainless steel, and bronze are also available. Threaded fasteners useful in home carpentry are shown in Figure 3–10.

A joint assembled with screws is more rigid than one assembled with nails. The joint cannot be worked as readily as a nailed joint, and the joint can be disassembled. With glue added, both nailed and screwed joints are equally rigid; it is the glue in this case that produces the rigidity.

Figure 3–10
Threaded fasteners useful in home carpentry.

2" #10 FLAT–HEAD WOOD SCREW
1" #8 PAN HEAD TAPPING SCREW
1-5/8" GYPSUM BOARD SCREW
2" #8 ALL–PURPOSE SCREW
1/4"x 2-1/4" HEX–HEAD LAG SCREW
3" #10 DOWEL SCREW
1/4" x 3" HANGER BOLT

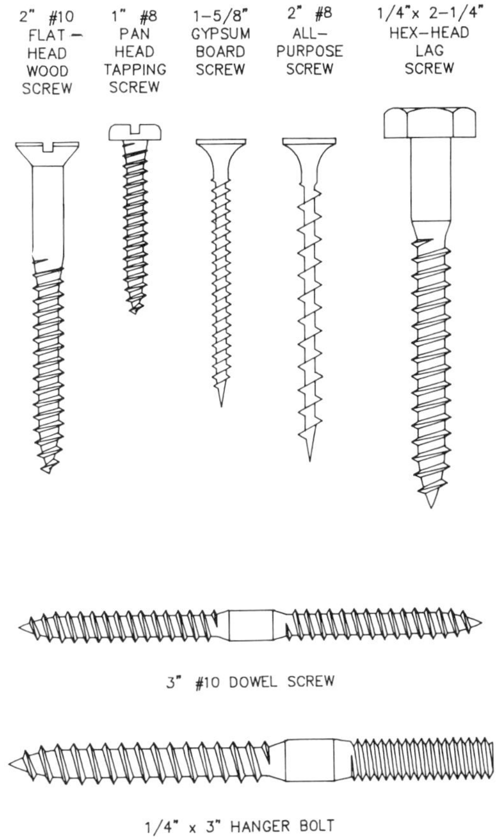

Wood screws. Screws have flat, round, or oval heads. The underside of flat and oval heads is slanted or tapered to the shank; the underside of a round head is flat. The threaded portion of the shank tapers gradually over most of its length, then tapers sharply to the tip. The unthreaded portion of the shank is not tapered. Flat-head screws are used if the screw head is to be flush with the surface of the wood or driven below the surface in a countersunk hole. Round-head wood screws are used, normally with washers, when the top piece being joined is thin wood, metal, plastic, or another material that cannot be countersunk for flat-head screws, or when countersinking is objectionable. Oval-head wood screws are used when the screw head is to show decoratively or when the screw is to be used with a cup washer.

Heads may be straight-slotted, Phillips (cross-slotted), or square recessed. A flat-blade screwdriver must fit the slot tightly, or driving will be difficult, with gouged wood or burred screw heads the result. The Phillips head with its crossed slots is the most widely available alternative to the slot. Using the right size screwdriver is important. It is also important to discard a Phillips screwdriver when the tip becomes worn or deformed.

Figure 3–11
Using screws to join wood. Use a tapered countersink bit to drill a chamfered hole for flat-head screws.

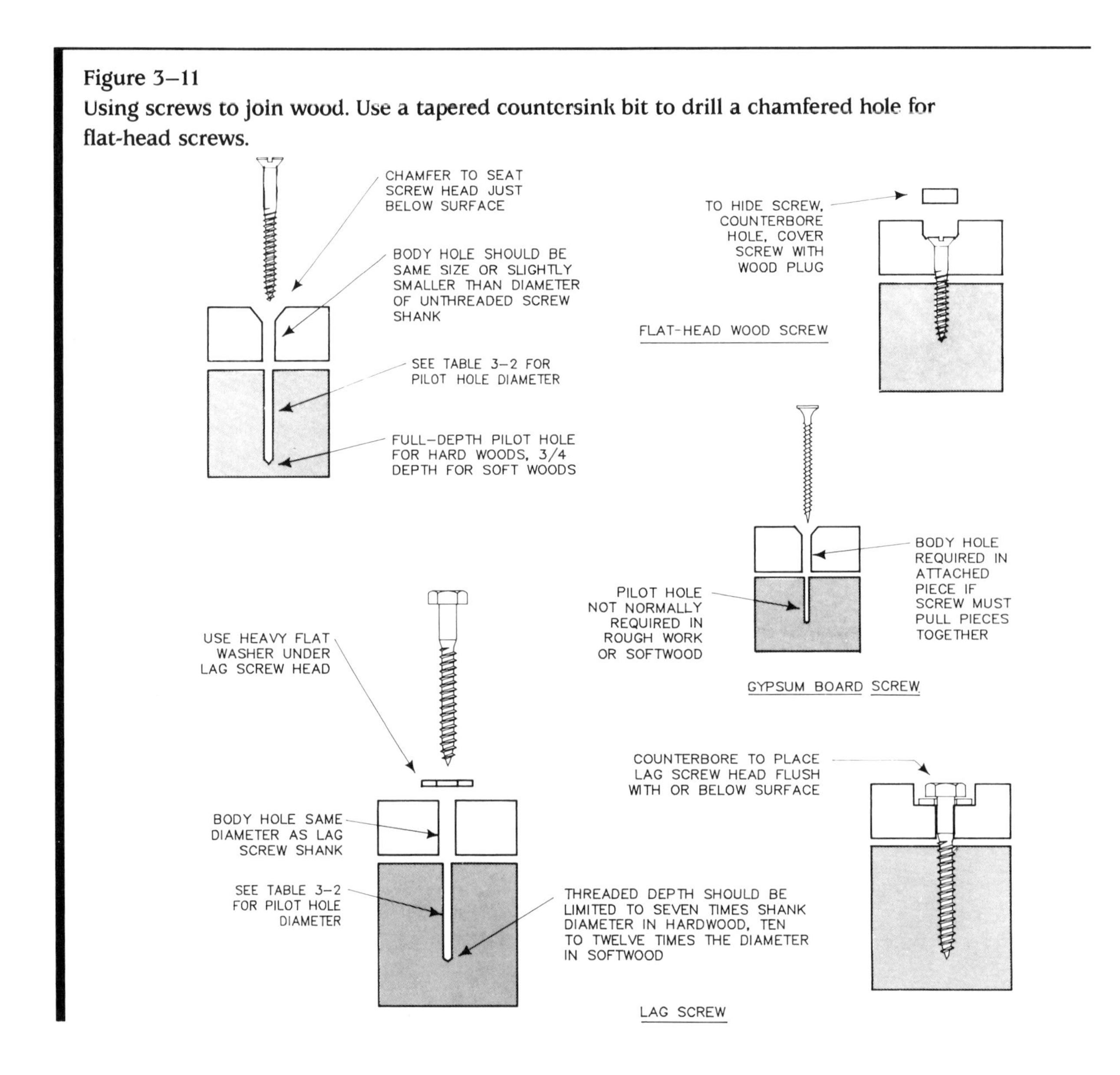

CHAMFER TO SEAT SCREW HEAD JUST BELOW SURFACE

BODY HOLE SHOULD BE SAME SIZE OR SLIGHTLY SMALLER THAN DIAMETER OF UNTHREADED SCREW SHANK

SEE TABLE 3–2 FOR PILOT HOLE DIAMETER

FULL-DEPTH PILOT HOLE FOR HARD WOODS, 3/4 DEPTH FOR SOFT WOODS

TO HIDE SCREW, COUNTERBORE HOLE, COVER SCREW WITH WOOD PLUG

FLAT-HEAD WOOD SCREW

PILOT HOLE NOT NORMALLY REQUIRED IN ROUGH WORK OR SOFTWOOD

BODY HOLE REQUIRED IN ATTACHED PIECE IF SCREW MUST PULL PIECES TOGETHER

GYPSUM BOARD SCREW

USE HEAVY FLAT WASHER UNDER LAG SCREW HEAD

BODY HOLE SAME DIAMETER AS LAG SCREW SHANK

SEE TABLE 3–2 FOR PILOT HOLE DIAMETER

COUNTERBORE TO PLACE LAG SCREW HEAD FLUSH WITH OR BELOW SURFACE

THREADED DEPTH SHOULD BE LIMITED TO SEVEN TIMES SHANK DIAMETER IN HARDWOOD, TEN TO TWELVE TIMES THE DIAMETER IN SOFTWOOD

LAG SCREW

Square recesses have a slight taper that matches the taper of the square-recess screwdriver, resulting in its being able to hold the screw for starting.

Wood screws are sized by length in inches and by body diameter in gauge numbers from 0 to 24. (The shank of a #24 screw is about ⅜".) Dimensions are standard.

Pilot holes are essential for all screws used with wood, with the possible exception of gypsum board screws (see Figure 3–11). To get maximum holding power, the hole must be the correct size. For wood screws, the pilot hole for the unthreaded portion of the shank should be slightly less or the same diameter as the shank. The pilot hole for the threads should be about 70 percent of the root diameter of the threads in softwood and 90 percent in hardwood. In hardwood the pilot hole should be full depth, about three-quarters of the thread length in softwood. See Table 3–2 for pilot hole dimensions for screws in wood.

Wood screws hold better in hardwood than they do in softwood. Withdrawal resistance also depends on the depth of threaded penetration and the diameter of the screw. Increasing the length of the screw to get greater holding power can be carried only so far. Seven times the diameter of the shank in hardwood and ten times in softwood appear to be the limit for threaded penetration; more than that usually results in the screw being twisted off as you try to drive it in.

Driving a screw of any kind in wood is easier if you lubricate the threads first. Use wax or paraffin. Soap is not good because the water in the soap (even if it feels dry) will rust the threads. Wax lubrication has little effect on the holding power of the screw.

Gypsum board screws. These screws are designed for attaching gypsum board panels to wood framing or metal channels, but are now widely used for other purposes. These hardened screws have Phillips bugle heads (similar to flat heads), sharp self-drilling pyramid points, and are threaded the entire length of the shank. In gypsum board installation, the screws are driven without pilot holes, using a power screwdriver (called a screwgun) that drives the screw just below the surface and automatically disengages before the gypsum board panel is damaged by overdriving. They can be driven with a variable-speed electric drill equipped with a hex-shank Phillips bit; inexpensive adapters are available to prevent overdriving.

In rough carpentry, gypsum board screws can be used without pilot holes and will cause no more splitting than nails of the same diameter. The screws will pull pieces more tightly together if a pilot hole is drilled in the outside piece. The head will partially strip in the top piece as the screw penetrates into the lower piece, pulling the pieces together. There will also be less likelihood of the wood splitting, and the driving will be a lot easier. In finish work, the outside piece should also be drilled to countersink the head below the surface.

All-purpose screws. These screws look like gypsum board screws, but have thicker shanks (several sizes), coarser threads, and are corrosion resistant. Drill pilot holes and countersink as for gypsum board screws.

Self-tapping screws (sheet-metal screws). These screws are made in several thread and point configurations. Tapping screws sold in hardware stores and home centers have type AB threads, which hold in wood as well as wood screws do. Both flat heads and pan heads, slotted and Phillips, are available. Tapping screws are superior to wood screws for attaching thin materials. All hinges should be attached with tapping screws because they are threaded the full length of the shank; the unthreaded portion of a wood screw shank contributes little or nothing to holding power in the wood.

Lag screws. These screws are available in larger sizes than wood screws, ranging from ¼" to 1" shank diameter. Heads are either square or hexagonal, and are driven with a wrench, making them easier to drive and remove than slotted wood screws of comparable size. Pilot holes are essential for maximum withdrawal resistance and to eliminate splitting (see Table 3–2).

Withdrawal resistance of a lag screw from end grain is about 75 percent of the side-grain resistance for the same piece of wood. As the ultimate

Table 3–2
THREADED FASTENER PILOT AND BODY HOLE SIZES, GIVEN IN DRILL BIT GAUGE NOS.

	WOOD SCREW SIZE				
	#6	#8	#10	#12	#14
Softwood					
Body hole	#31	#26	#18	#10	#3
Pilot hole	#52	#48	#43	#38	#33
Useful threaded depth	1⅜"	1⅝"	1⅞"	2⅛"	2⅜"
Hardwood					
Body hole	#29	#19	#11	#3	C
Pilot hole	#45	#39	#32	#29	#26
Useful threaded depth	1"	1⅛"	1¼"	1½"	1⅝"

	LAG SCREW SIZE (diameter)			
	¼"	5/16"	⅜"	½"
Body Hole Size	¼"	5/16"	⅜"	½"
Cedar, White Pine				
Pilot hole (min.)	#38	#30	#25	#7
Useful threaded depth	3"	3¾"	4½"	6"
Fir, Yellow Pine				
Pilot hole (min.)	#25	#13	#1	5/16"
Useful threaded depth	2½"	3⅛"	3⅝"	5"
Hardwood (all)				
Pilot hole (min.)	#19	#6	¼"	11/32"
Useful threaded depth	1¾"	2⅛"	2⅝"	3½"

withdrawal resistance is dependent on the tensile strength of the root diameter of the screw, screws should be selected to penetrate only seven times the shank diameter in hardwood and ten to twelve times the diameter in softwood. Greater penetration risks destroying the screw when wrenching. Washers should always be used under the heads of lag screws regardless of the material under the head.

Particleboard screws. Because of the composition of particleboard, fiberboard, and waferboard, ordinary wood screws, tapping screws, and gypsum board screws do not hold well. Particleboard screws have coarse-pitched threads angled to prevent gouging as the screw is driven in. The special screws can be driven without pilot holes, and inserted and withdrawn several times without loss of holding power. For a neat surface after screw insertion, however, the hole should be countersunk for a flat screw head. If wood screws are used, they should not be larger than #8 and the unthreaded shank must have an equal-diameter pilot hole to prevent chipping.

MACHINE SCREWS AND BOLTS

Bolts (Figure 3–12) are used to assemble wood framing side-grain to side-grain and to attach metal members to wood side grain. Bolt holes should be drilled the same size as the bolt body diameter or slightly smaller. Ideally, the bolt should have to be lightly hammered into the hole. To prevent bolt heads and nuts from crushing the wood, washers must be used under the head and nut to spread the stress over a larger bearing area.

Figure 3–12
Bolt configurations.

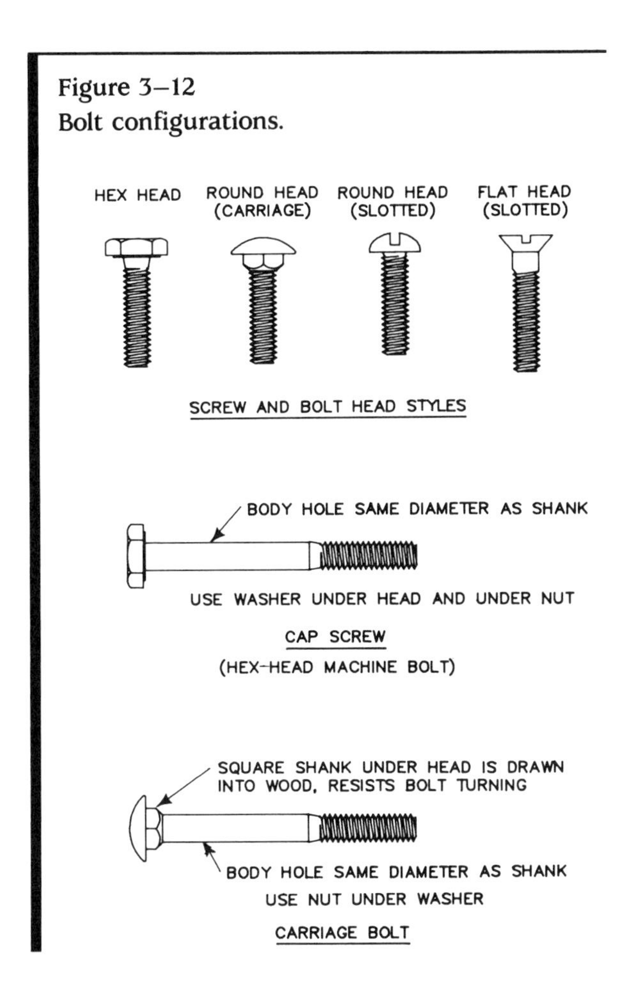

Through bolts alone should not be depended on to produce a rigid wood-to-wood joint. The pieces of wood should be joined in such a way that they cannot move perpendicular to the axis of the bolt as long as the bolt is compressing the pieces together. This can be accomplished by using a joint in which one piece is notched to receive the other, such as a half-lap, cross-lap, or bridle joint. Or, nail short cleats (scabs) on one piece that butt against each side of the other piece.

Carriage bolts. These fasteners have button heads—smooth, round tops—without any recess for holding or turning. Most have a square neck on the shank just under the head, but some are made with fins or ribs. They may be threaded partway up the shank or, more likely today, threaded all the way to the head. When used in wood-to-wood or wood-to-metal assembly, pilot holes the diameter of the shank or slightly smaller are required. The bolt is driven into wood; the squared section keeps the bolt from turning in the wood as the nut is torqued. A flat washer should always be used under the nut of a carriage bolt. The carriage bolt head may be seated on the surface or set into a counterbored hole for a flush surface.

Carriage bolts are also useful in metal-to-metal and metal-to-wood assembly. The square locking holes are usually made in the metal, but they can be in the wood.

Stove bolts. These bolts are made with slotted flat or round heads. They may be threaded the full length of the shank or only partway. In woodworking, stove bolts are used with nuts, or with special fasteners such as threaded inserts. Flat washers should be used under round heads and under nuts. If one of the joined members is wood, lock washers are not useful.

Cap screws (hex-head bolts). Holding a screw head with a screwdriver while you tighten the nut doesn't always work well. The advantage of using cap screws is that their hex heads can be held with a wrench as you torque the nut or vice versa.

Nuts. There are many kinds of nuts; hex and square nuts are the types most commonly used. Square nuts are normally used only for light-duty applications.

4 Interior Doors

A hinged door swings on hinges fastened to one side of the opening and latches on the other side. This chapter tells you how to install interior hinged doors; Chapter 5 deals with installing exterior hinged doors. Chapter 6 tells you how to correct common problems with hinged doors. Sliding doors and pocket doors are installed in completely different manners and are covered in Chapter 7.

Interior door styles should be selected to complement the decorative style of the house. While inexpensive mahogany flush doors are most commonly used today, paneled doors are far more attractive in traditional styles of decoration.

TYPES OF DOORS

Residence doors are made in a wide range of styles and standard sizes (see Table 4–1). There are two basic door types—panel and flush. Interior doors are usually 1⅜″ thick (exterior doors are 1¾″ thick).

Standard *panel doors* consist of solid wood frames into which multiple panels are inserted. Figure 4–1 shows a few of the many panel and light (glass) combinations available. The panels are usually, but not always, thinner wood. Interior doors in the patterns shown usually have all-wood panels, except French doors, which have panes of glass called lights in place of wood panels. The vertical frame members of a door are called stiles; the horizontal members, rails. Paneled doors are used for both interior and exterior doors when a traditional style is desired. Exterior doors often have glass lights in place of wood panels in the upper portion of the door. Interior panel doors are also made of pressed steel or molded plastic.

Flush doors are flat and plain on both sides. There are two kinds of flush doors—hollow core and solid core. Hollow-core doors have a frame around the edge with reinforcement where the latch is to be installed, as shown in Figure 4–2. An interior latticework of corrugated strips supports the thin plywood covering glued to the frame on both sides.

Solid-core flush doors have the interior space between the narrow stiles and rails filled with either blocks of wood planed to exact thickness or, more likely, a single piece of particleboard as thick as the frame. Thin plywood is glued to each side. Solid-core flush doors are more secure than paneled doors because they have no thinner por-

Figure 4–1
Panel door patterns.

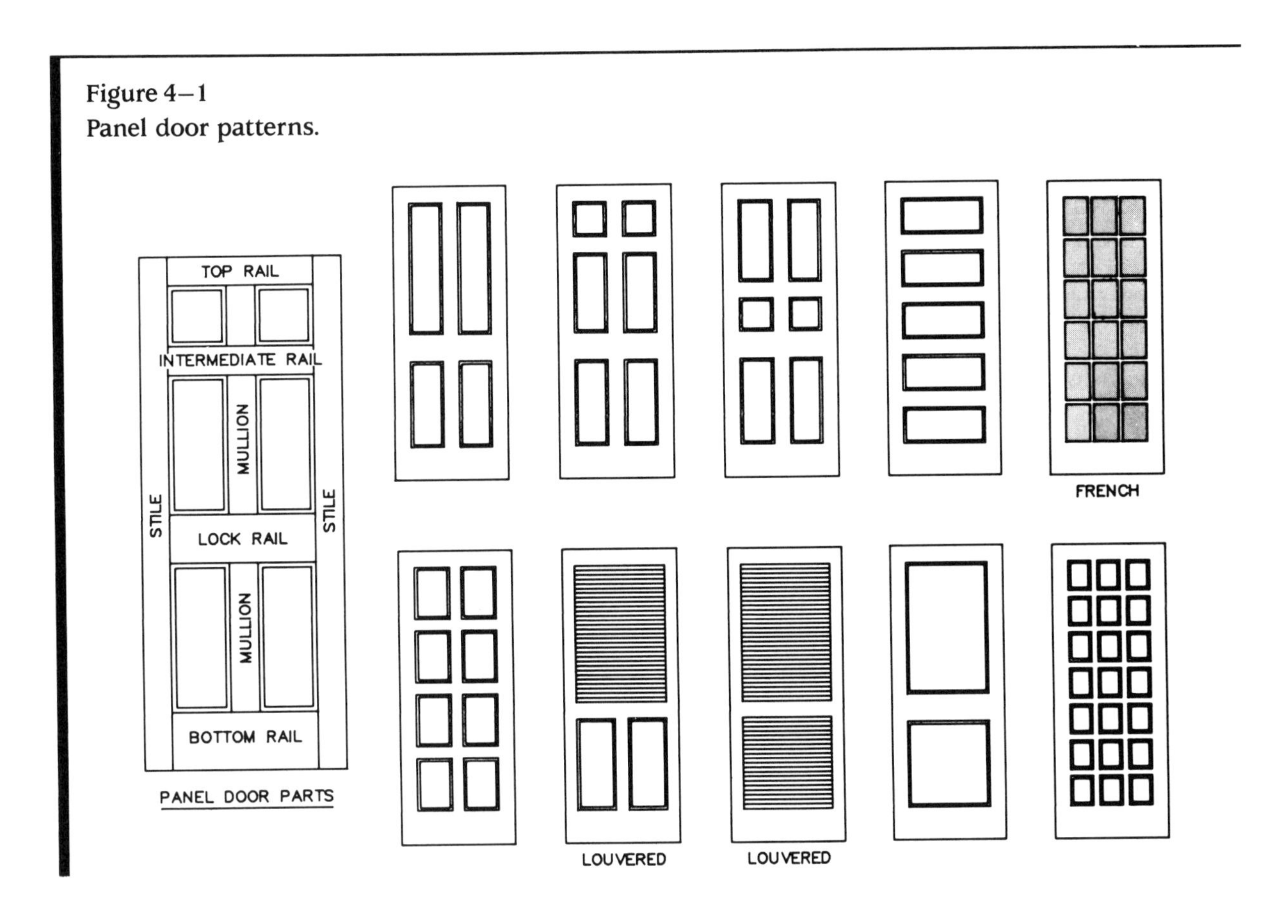

Table 4–1
STANDARD INTERIOR AND EXTERIOR DOOR SIZES

DOOR WIDTHS	PANEL DOORS[1]			FLUSH DOORS[2]															
	1¾″ and 1⅜″			1¾″ Solid Core			1¾″ Hollow Core			1⅜″ Solid Core				1⅜″ Hollow Core					
	6′8″ (80″)	7′0″ (84″)	6′8″ (80″)	6′8″ (80″)	6′10″ (82″)	7′0″ (84″)	6′8″ (80″)	6′10″ (82″)	7′0″ (84″)	6′0″ (72″)	6′6″ (78″)	6′8″ (80″)	7′0″ (84″)	6′0″ (72″)	6′6″ (78″)	6′8″ (80″)	6′10″ (82″)	7′0″ (84″)
12″																X		
14″																X		
15″																X		
16″																X		
18″										X	X				X	X	X	X
20″														X	X	X	X	X
22″														X	X	X	X	X
24″			X							X	X	X		X	X	X	X	X
26″														X	X	X	X	X
28″			X	X	X					X	X	X	X	X	X	X	X	X
30″	X	X	X	X	X	X								X	X	X	X	X
32″	X	X	X	X	X	X	X	X	X					X	X	X	X	X
34″	X	X					X	X	X					X	X	X	X	X
36″	X	X		X		X	X	X	X					X	X	X	X	X
40″				X		X												

[1]Not all styles of panel doors are made in all sizes.
[2]Some 1¾″ solid-core flush doors are not exterior rated.

Figure 4–2
Flush door construction.

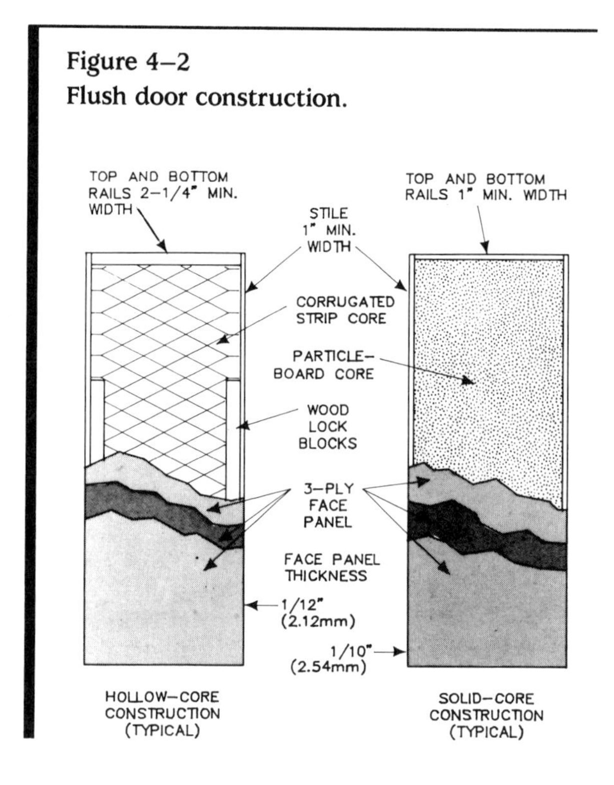

Figure 4–3
A plain flush mahogany door can be dressed up with applied moldings.

tions that might be easily broken through. Also, solid-core doors transmit less sound and permit less heat loss. Both types of flush doors can be decorated with molding, as shown in Figure 4–3.

A door is fitted into a frame called a jamb that forms the top and sides of the doorway. The side pieces are called side jambs, and the piece across the top is called the head jamb (Figure 4–4). It is, most of the time, far easier and more practical to make the jamb fit the door than the other way around. It is particularly important to fit the jamb to the door if you are installing a hollow-core flush door. The stiles and rails of a hollow-core door are so narrow that you cannot trim off much to adjust the width or height without seriously weakening the frame.

After assembly, the jamb is carefully centered, plumbed, and leveled in the oversize rough opening with shims and nails. The gap between the jamb and the 2× studs and header of the rough opening is covered with casing. Casing is not only a decorative wood trim, it also structurally reinforces the connection of the jamb to the wall framing.

To keep the door from swinging all the way through an interior jamb, a projecting strip of wood called a doorstop is nailed to the inside surfaces of the jambs, giving the door something to butt against when closed. (In an exterior jamb, for weather-tightness and security, the doorstop and the jamb are usually made in one piece, or the stop fits into a dado in the jamb.)

Whether you need to replace a hinged door or are hanging a new door in an opening, the task can be approached two ways. One way is to buy a door and either trim it to fit an existing jamb or install a new jamb the proper size for the door. Both are a lot of work. The easier alternative is to install a prehung door, one that comes already hinged into a jamb and drilled for handle and latch hardware (called a lockset or passage set). Hanging a prehung door is fast: you tip it into the rough opening, square it, shim it, and nail it in place.

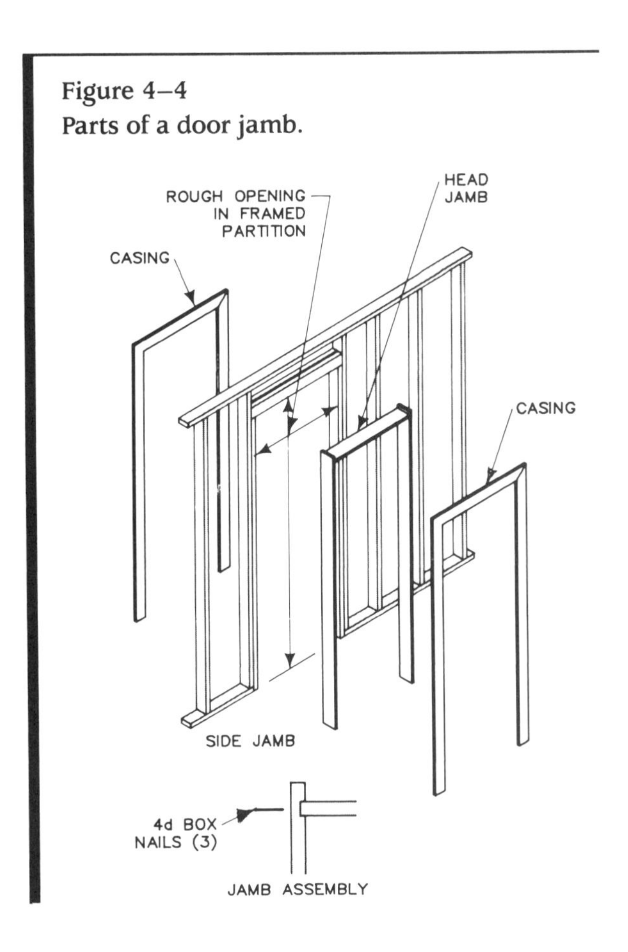

Figure 4–4
Parts of a door jamb.

HANGING A PREHUNG INTERIOR DOOR

Prehung doors are received with the door hinged in the jamb frame and bored for the lockset. The door is restrained in the jamb with a clip of some sort. There are two kinds: prehung door and jamb only, and split jamb. The first requires separately purchased and installed casing; the second has casing integral with the two halves of the split jamb. Installation is shown in Figure 4–5.

The jamb-only type is installed much as separate doors and jambs would be, except that you know the door will fit the jamb without trimming. You must get a jamb to fit your wall thickness, or trim down the next deeper jamb to size. The door and jamb unit is placed in the rough opening, shimmed to be perfectly vertical and horizontal, and nailing (from one side of the door) is started. Once the jamb is in position, the door can be freed of its locking clip and opened for completion of the installation.

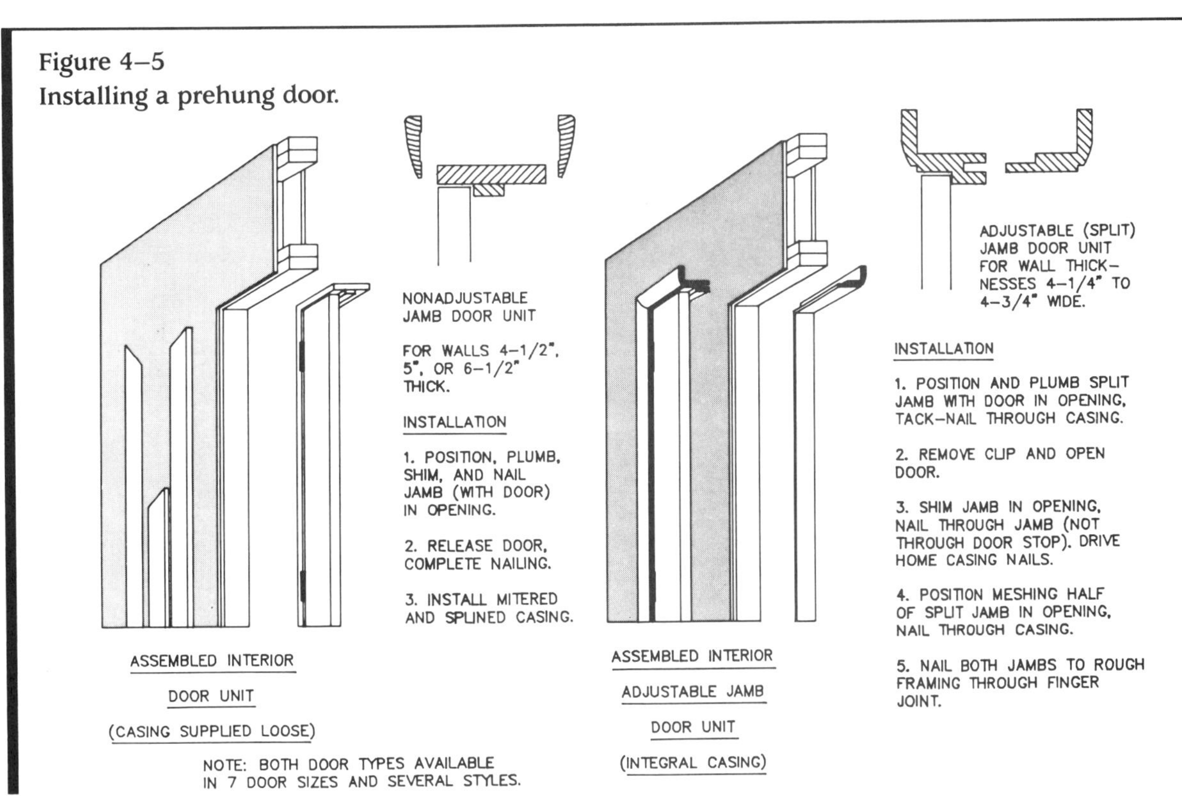

NONADJUSTABLE JAMB DOOR UNIT
FOR WALLS 4-1/2", 5", OR 6-1/2" THICK.
INSTALLATION
1. POSITION, PLUMB, SHIM, AND NAIL JAMB (WITH DOOR) IN OPENING.
2. RELEASE DOOR, COMPLETE NAILING.
3. INSTALL MITERED AND SPLINED CASING.
ASSEMBLED INTERIOR DOOR UNIT (CASING SUPPLIED LOOSE)
NOTE: BOTH DOOR TYPES AVAILABLE IN 7 DOOR SIZES AND SEVERAL STYLES.
ADJUSTABLE (SPLIT) JAMB DOOR UNIT FOR WALL THICKNESSES 4-1/4" TO 4-3/4" WIDE.
INSTALLATION
1. POSITION AND PLUMB SPLIT JAMB WITH DOOR IN OPENING, TACK-NAIL THROUGH CASING.
2. REMOVE CLIP AND OPEN DOOR.
3. SHIM JAMB IN OPENING, NAIL THROUGH JAMB (NOT THROUGH DOOR STOP). DRIVE HOME CASING NAILS.
4. POSITION MESHING HALF OF SPLIT JAMB IN OPENING, NAIL THROUGH CASING.
5. NAIL BOTH JAMBS TO ROUGH FRAMING THROUGH FINGER JOINT.
ASSEMBLED INTERIOR ADJUSTABLE JAMB DOOR UNIT (INTEGRAL CASING)

Figure 4–5
Installing a prehung door.

The split-jamb type can be installed in walls of various thicknesses. The door is hinged in the half of the jamb that includes the doorstop. This half is positioned in the rough opening and nails are started through the jamb into the rough framing. The second half is then placed in the opening from the other side with the jambs lapped. Nailing is completed with nails through the lap.

HANGING A STANDARD DOOR

Why would you go to the work of hanging a door that was not prehung? There are several reasons: The opening requires a nonstandard jamb size. You want a style or quality of door not available prehung. You want to use an existing period-style jamb and casing.

Hanging a hinged door is not a difficult undertaking, but it must be done carefully and the task will take some time. Go at it slowly and methodically, step by step. Make sure each step is completed properly before starting the next.

First, prepare the new opening or measure an existing opening. A rough opening should be 2½″ to 3″ wider than one of the stock door dimensions and about 2½″ taller, as shown in Figure 4–6. These figures are for a reasonably square rough opening. If the rough opening is out of square, you will need more width allowance. Misalignment of the rough header is seldom a source of trouble; the problem is usually the cripple studs that run beneath the header and the floor at each side. If they are more than ¼″ off vertical, you must compensate for that in choosing a door width.

Once you have measured the opening and determined how square it is, pick your door size from the table of stock sizes. It is better to shim the rough opening on one or both sides with 1× lumber than to cut down a too-wide door, particularly with a hollow-core flush door, as there can be as little as 1″ of solid stile at each side. Panel doors can be shortened up to 2″; flush doors not more than 1″, sometimes not at all.

Wood for the jambs should be clear or select pine, or clear redwood. Jambs should be 1/16″ wider than the thickness of the wall, measured

Figure 4–6
Standard door dimensions, and required jamb and rough opening dimensions.

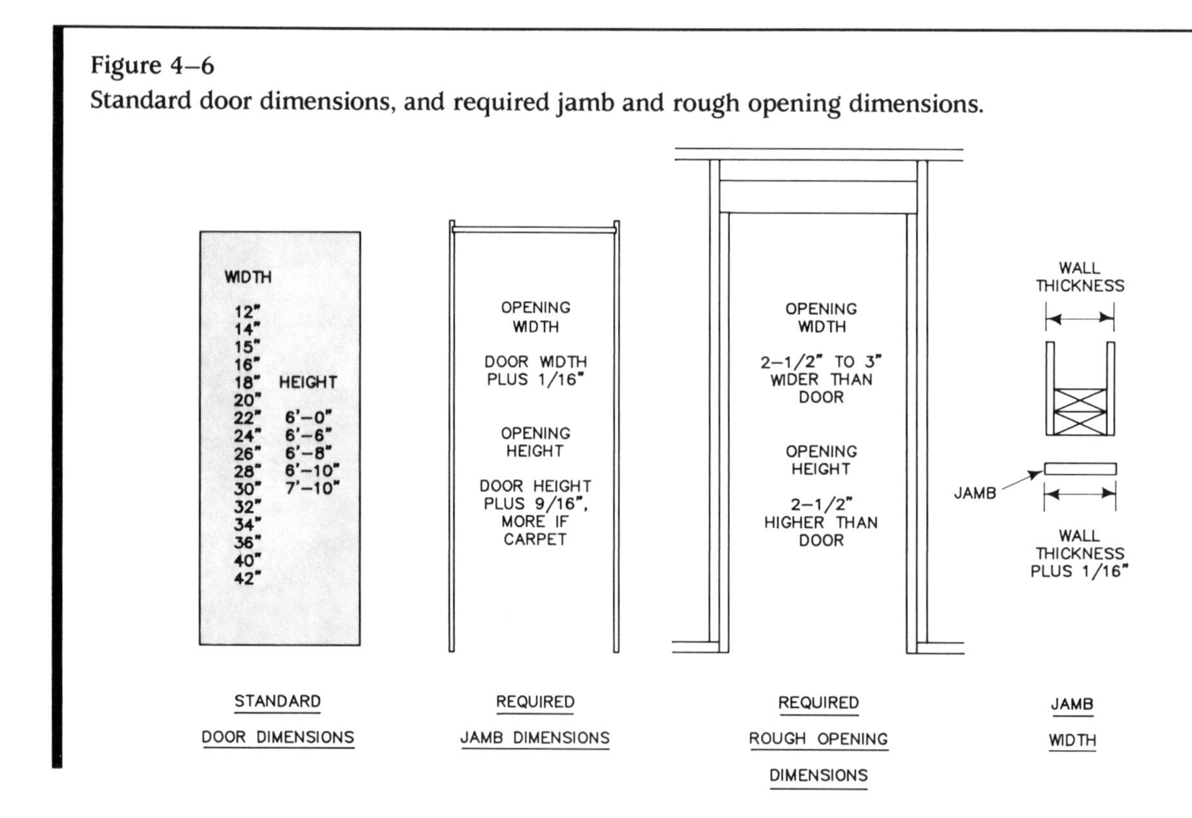

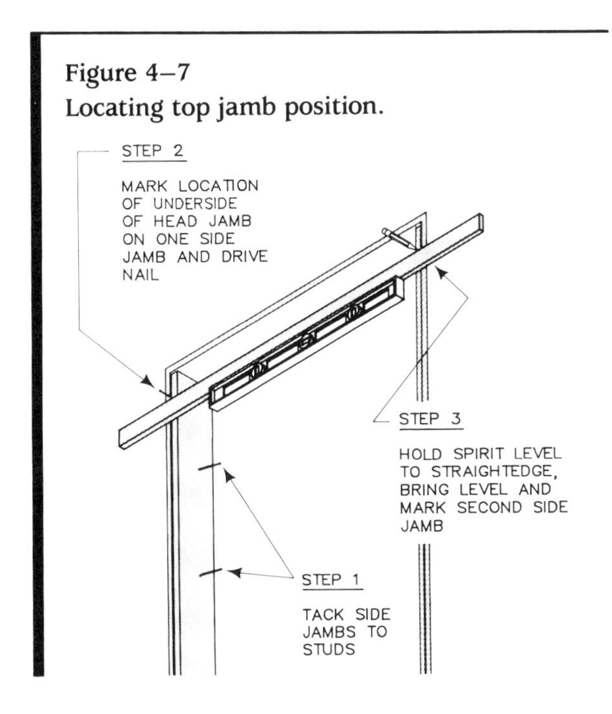

Figure 4–7
Locating top jamb position.

between finish wall surfaces. This is to make sure the casings will lie flat against the jambs in spite of slight wall unevenness. Cut the side jambs to length and rip to width. Tack the side jambs to the studs with partially driven nails and roughly shim them to vertical.

Next, mark the location on the side jambs for the dado that will receive the top jamb. Don't simply measure up from the floor unless you have checked the floor and found it to be level. With a straightedge and a level, locate the height of the underside of the top jamb. Figure 4–7 shows this procedure. Allow 1/16″ additional clearance to the top of the door for a painted door (see below) and whatever clearance you want at the bottom (1/2″ minimum). If the floor slopes from one side of the doorway to the other, make your height measurement from the low side of the floor. Now rout the dadoes for the top jamb and cut the top jamb to length (door width plus depth of the two dadoes plus 1/8″ clearance). Rip the jamb to width. Now you must choose hinges and prepare for installing them.

Selecting Butt Hinges

Although there is a wide variety of hinges available, the basic butt hinge is probably used more in residence carpentry than all other types combined. There are also many types and grades of butt hinges, and selecting the right one for a door is important if the door is going to operate properly and stay hung. Until recently, specialized hinges were available only to industrial purchasers; now, many are available in home centers and hardware stores.

Some hinge types are made for left-hand or right-hand doors and must be used only that way. The "hand" of a hinge is usually restricted by having leaves made for specific door or jamb configurations coupled with loose pins, which must be inserted only from the top. The "hand" of a door is defined in Figure 4–8.

Butt hinges are made in three weight grades: utility, standard, and heavy. Cheap utility-weight hinges are not adequate as replacement hinges for the doors in your home. Standard-weight hinges in residence door sizes are surprisingly inexpensive—they often cost less than smaller cabinet hinges. Standard-weight hinges of appropriate size are suitable for all residence doors unless the doors are unusually heavy, in which case you should substitute heavyweight hinges

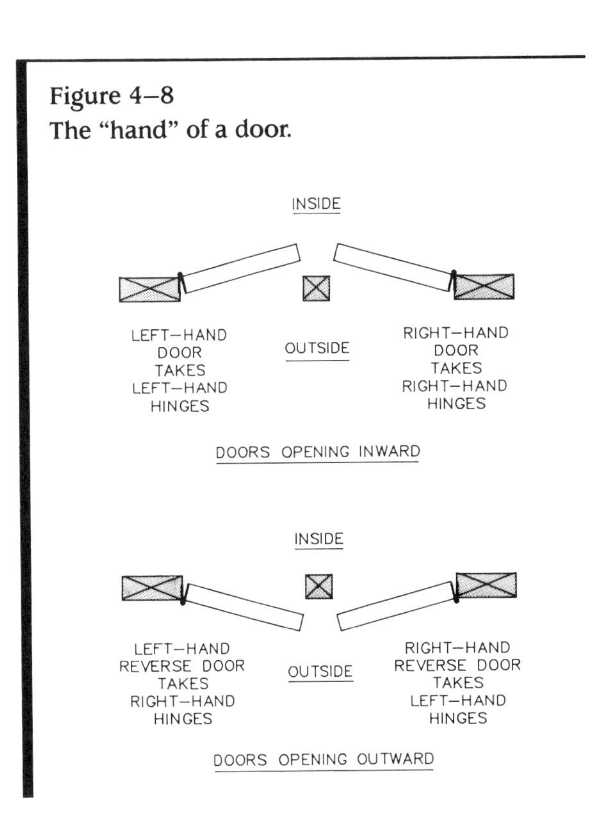

Figure 4–8
The "hand" of a door.

of the same size. (Heavyweight hinges are used in commercial buildings.)

The hinge size must be appropriate for both the width and the thickness of the door. The required hinge height is determined from the combination of door thickness and width as shown in Table 4–2.

Determining the necessary hinge width is more complicated. You have to take into account door thickness, inset, backset, and clearance required. *Clearance* is how far the flat surface of a flush door or the frame of a paneled door must be out from the flat surface of the wall beside the jamb when the door is opened 180 degrees (see the detail in Figure 4–9). *Inset* is the distance from the knuckle edge of the hinge leaf to the near door edge. *Backset* is the distance from the straight edge of the leaf to the far edge of the thickness.

Factors entering into establishing the clearance requirement include the thickness of the jamb casing, any raised molding or trim on the face of the door, and the baseboard, chair rail, or other protrusions on the wall. If the door can only open about 90° because the wall on the hinge side of the doorway is perpendicular to the door, then the clearance requirement is zero. Determine the hinge width as follows:

Hinge width = twice the door thickness minus the backset, plus the inset, plus the clearance.

Table 4–2
DETERMINING HINGE HEIGHT

Door Thickness	Door Width	Hinge Height (min.)
¾″ to 1⅛″	24″ (max.)	2½″
(cabinet doors)	36″ (max.)	3″
1⅜″ (most residence	32″ (max.)	3½″ to 4″
interior doors)	over 32″	4″ to 4½″
1¾″ (most residence	36″ (max.)	4½″
exterior doors)	48″ (max.)	5″
	over 48″	6″

Figure 4–9
Hinge locations and hinge terminology.

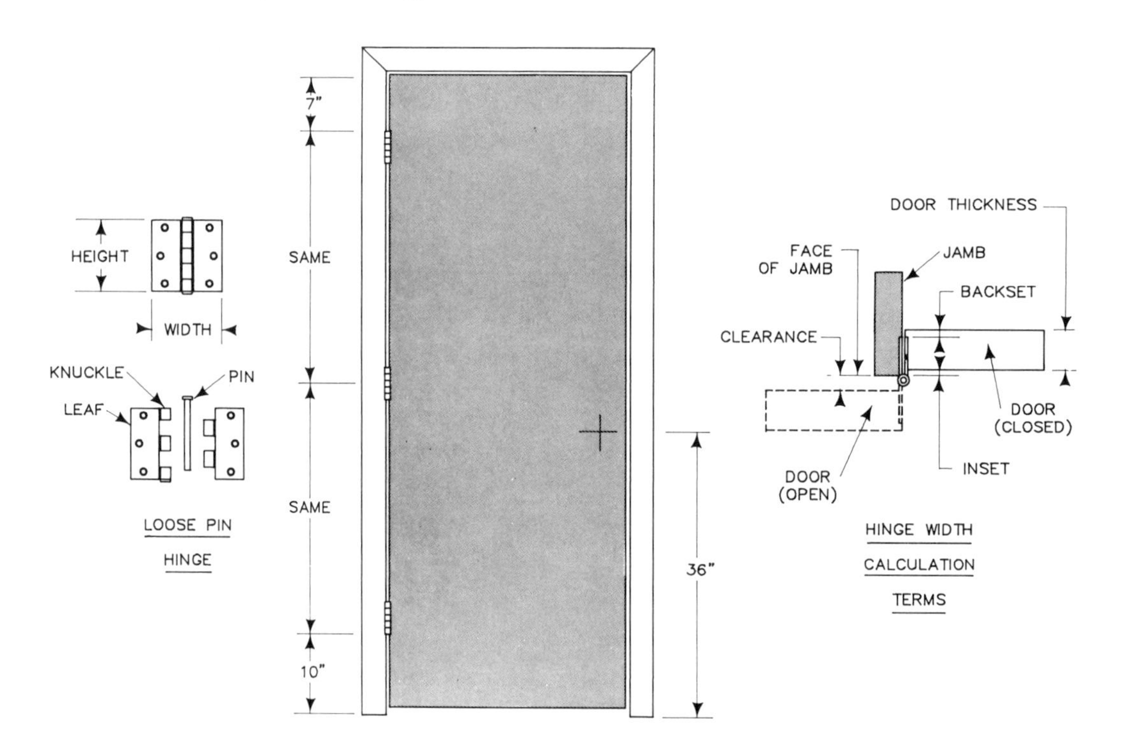

Backset seems to have no standard. On 1⅜″ lightweight doors ⅛″ to ¼″ should be used, with some attention paid to keeping the hinge screws somewhat centered in the edge of the door. Exterior 1¾″ doors should have a backset of ¼″, and thicker doors ⅜″.

Forget the widespread builder practice of using only two hinges on interior doors. You should always use three hinges on any doors up to 90″ high. (The only exception is the Dutch door, which has two hinges per section, spaced as far apart as practical.) Taller doors require four hinges. Position the hinges the same as on the rest of the doors in your house. Generally, the top hinge should be down 5″ to 7″ from the jamb, and the bottom hinge up 8″ to 10″ from the floor. Center the middle hinge between the other two hinges, as shown in Figure 4–9. Even lightweight interior doors should be hung with three 3½″ × 3½″ butt hinges, or match existing hinges.

In any door installation, it is the top hinge that takes most of the stress and abuse. As it loosens, the center hinge will take up the load and prolong the life of the installation. Wood doors do warp. The center hinge helps control this warping and reduces the hinge binding caused by warping.

Mortising for Hinges

To avoid a wide gap at the hinge edge of the door, hinge leaves should fit into mortises cut in the jamb and the door. The functional purpose of hinge mortises is to take lateral door weight off the sides of the screws and put it on the edges of the hinges. Mortising techniques are shown in Figure 4–10. It is a lot easier to cut mortises for the hinges with the jamb lying on your bench than it is with it nailed in the rough opening.

If you are skillful with a chisel, use it to chop out the hinge mortises. They can also be cut with a router, using a jig as shown in the figure. The dimensions given are for a 1⅜″ door, 3½″ × 3½″ hinges, ¼″ router bit, and 7/16″ guide bushing. Inserting the 3/16″ scrap for each cut prevents tearing the wood at the end of the mortise.

Try the jig on scrap first to get the depth of the cut set precisely. The hinge leaf should set flush with the surface of the wood. Sand the jamb parts while they are on your bench, too. Tack the jamb together with 4d finishing nails and see how it fits in the rough opening. Check to be sure the door fits in the jamb.

The required clearance between the door and the jamb depends on the finish you plan to use. One-sixteenth inch as shown (use a dime to check) is adequate for a stain and varnish finish; if you plan to use paint, leave a little more clearance (a nickel) for eventual paint buildup.

Take the door down, and replace the finishing nails holding the jamb parts together with 6d box nails. Box nails are better than common nails because, being thinner, they do not split the wood as readily and they will hold just as well.

Installing the Jamb

To install the jamb in the rough opening, wedge it in place with shimming shingles, sold in bundles for door and window jamb shimming (see Figure 4–11). Always use two shingles at a time, one from each side, unless the cripple stud is twisted.

But first cut a spreader stick exactly the width of the jamb opening at the top. Place the spreader stick on the floor between the jambs to keep them the correct distance apart. (Or, you can nail a stick across the jamb near the bottom for the same purpose.) Drive pairs of wedges top, bottom, and center between one jamb and the cripple stud. Tack them in place with nails driven partway in.

Check with a level to make sure the side jamb is plumb; then check with a straightedge—such as the side of the door itself—to make sure it is flat. Now repeat the shimming and leveling for the other side jamb. When you are satisfied that the jamb is correctly placed, drive the nails in through the shingles, then cut off the protruding excess. It is not necessary to nail the head jamb up into the rough header.

If you have built the jamb accurately and installed it carefully, you should not have to do any door trimming, except at the bottom to compensate for an uneven floor and possibly some beveling on the latch side (see below). However, if you are installing a door in an existing jamb, you may have some trimming to do, so we will proceed on that basis.

Figure 4–10
Forming hinge mortises. This can be done with chisels or a router, using a shop-made template.

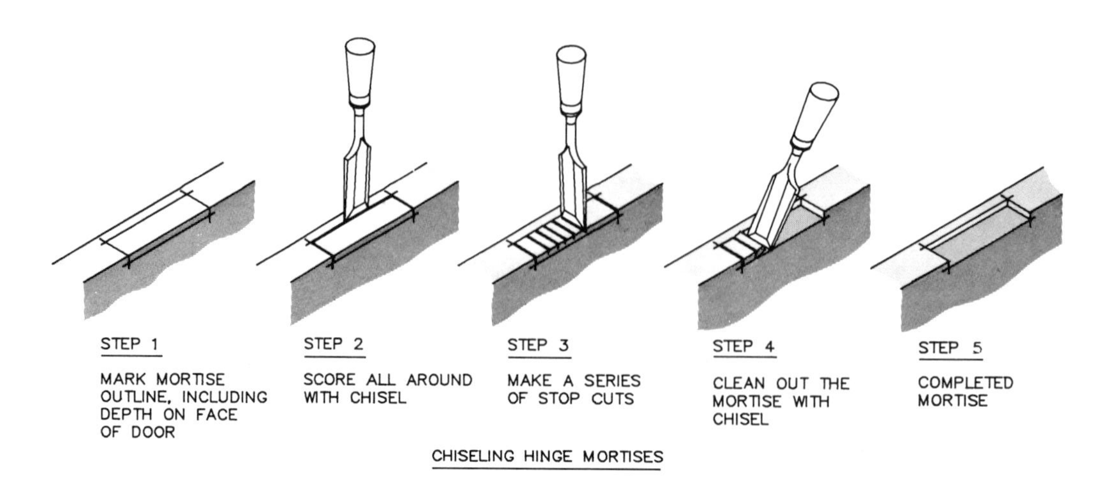
STEP 1
MARK MORTISE OUTLINE, INCLUDING DEPTH ON FACE OF DOOR
STEP 2
SCORE ALL AROUND WITH CHISEL
STEP 3
MAKE A SERIES OF STOP CUTS
STEP 4
CLEAN OUT THE MORTISE WITH CHISEL
STEP 5
COMPLETED MORTISE
CHISELING HINGE MORTISES

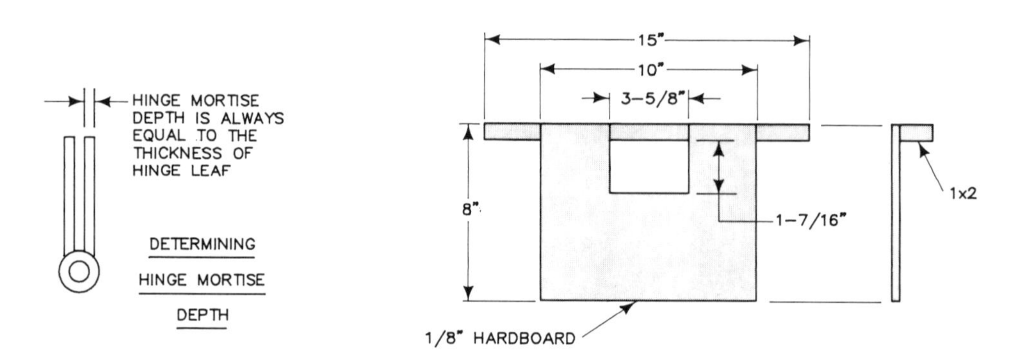
HINGE MORTISE DEPTH IS ALWAYS EQUAL TO THE THICKNESS OF HINGE LEAF
DETERMINING HINGE MORTISE DEPTH
15"
10"
3-5/8"
8"
1-7/16"
1x2
1/8" HARDBOARD

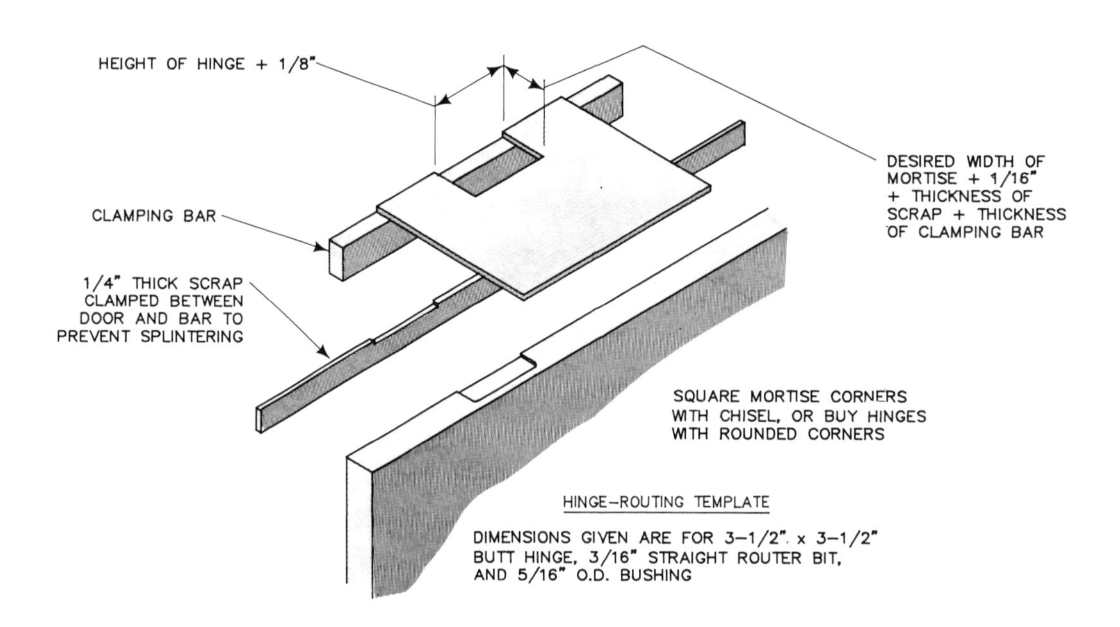
HEIGHT OF HINGE + 1/8"
DESIRED WIDTH OF MORTISE + 1/16" + THICKNESS OF SCRAP + THICKNESS OF CLAMPING BAR
CLAMPING BAR
1/4" THICK SCRAP CLAMPED BETWEEN DOOR AND BAR TO PREVENT SPLINTERING
SQUARE MORTISE CORNERS WITH CHISEL, OR BUY HINGES WITH ROUNDED CORNERS
HINGE-ROUTING TEMPLATE
DIMENSIONS GIVEN ARE FOR 3-1/2" x 3-1/2" BUTT HINGE, 3/16" STRAIGHT ROUTER BIT, AND 5/16" O.D. BUSHING

Figure 4–11
Shimming side jambs using shingle wedges.

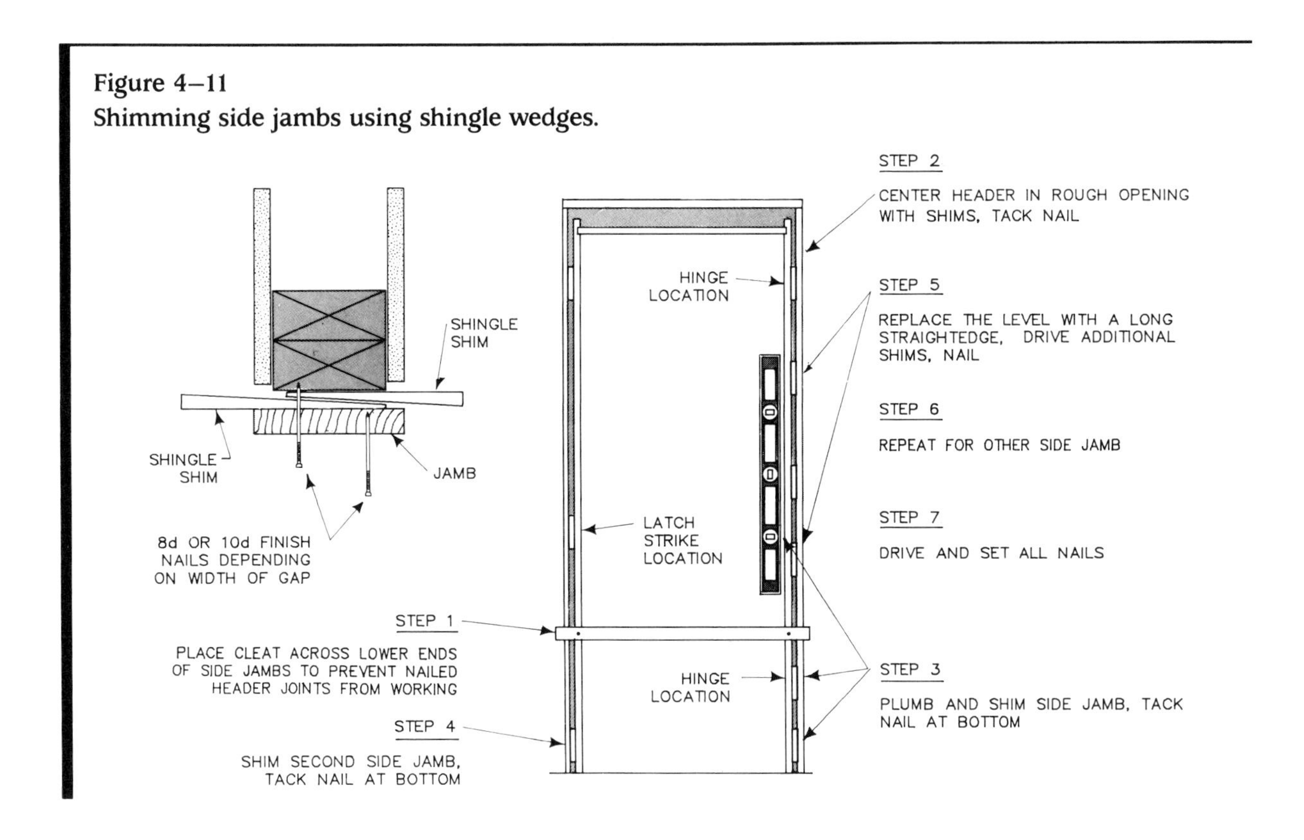
SHINGLE SHIM
SHINGLE SHIM
JAMB
8d OR 10d FINISH NAILS DEPENDING ON WIDTH OF GAP
STEP 1
PLACE CLEAT ACROSS LOWER ENDS OF SIDE JAMBS TO PREVENT NAILED HEADER JOINTS FROM WORKING
STEP 4
SHIM SECOND SIDE JAMB, TACK NAIL AT BOTTOM
HINGE LOCATION
LATCH STRIKE LOCATION
HINGE LOCATION
STEP 2
CENTER HEADER IN ROUGH OPENING WITH SHIMS, TACK NAIL
STEP 5
REPLACE THE LEVEL WITH A LONG STRAIGHTEDGE, DRIVE ADDITIONAL SHIMS, NAIL
STEP 6
REPEAT FOR OTHER SIDE JAMB
STEP 7
DRIVE AND SET ALL NAILS
STEP 3
PLUMB AND SHIM SIDE JAMB, TACK NAIL AT BOTTOM

Fitting the Door

Start by wedging the door tightly against the butt (hinge) jamb and up against the head jamb. If the door doesn't fit tightly against the butt jamb top to bottom, plane or belt-sand the door until it does. Now trim the latch side of the door so that with the door wedged against the butt jamb there is an even ⅛″ clearance top to bottom.

Now wedge the door up against the head jamb. Mark the top of the door and trim for a tight fit against the jamb all the way across. Last, put the door in the opening with a 4d finish nail between the top and the head jamb, mark the bottom of the door for the desired clearance, and trim off the excess.

Mortising for hinges. With the door still wedged in the jamb, and the 4d finish nail between the top of the door and the head jamb, transfer hinge locations from the jamb to the door and chisel or rout the hinge mortises.

Drilling for passage set. Most interior passage sets—also called passage locks or locksets—have locking latches for privacy. Modern passage locksets mount in the same hole combination as key-in-knob locksets, although the hole sizes and sometimes the backset may differ. Mortise locks for interior doors mount the same as their large exterior door counterparts. Small interior surface-mount (or rim) latches are also available for renovating and restoring. Instructions for locating the holes for the passage set come with it. The large hole in the side of the door can be made with a portable drill and a hole saw. If you have a brace, you can also make the holes with an expansion bit. Either way, work from both sides.

First, mark the side and edge holes (Figure 4–12, step 1). Then drill from one side until the pilot drill or bit point breaks through; turn the door over and complete the hole from the other side. After the hole is made through the side of the door, drill the hole for the latch at the center of the door edge (step 2). You can use a spade bit for this. Insert the latch in the hole, and mark around the edge of the door for the mortise. Chisel out this mortise so the latch seats flush (step 3).

Figure 4–12
Drilling the door for a passage set (lockset).

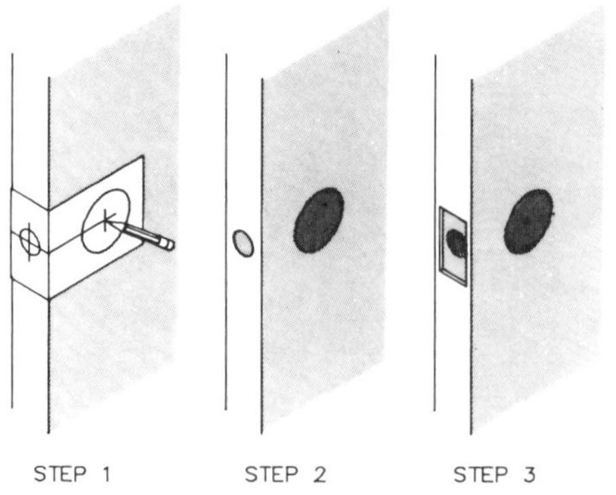

STEP 1 STEP 2 STEP 3

Hinging the door. Now attach the butt hinges to the door and jamb, but don't drive the screws quite all the way. Hang the door, first from the top hinge, then the middle and bottom hinges. After dropping the pins in partway, check how the door swings and whether the clearance between the jambs and the door is even and about 1/16" wide on both sides. (If the clearance on the hinge side is not satisfactory, see Chapter 6.)

With the clearance set correctly on the hinge side, the edge of the door should just barely miss the lock-side jamb as it closes. If there is interference, the edge of the door can be beveled slightly (Figure 4–13), but under no circumstances should more than 1/16" be taken from the closing side, because the latch will not work properly. (*Note:* The latch edge of an exterior door is normally beveled: interior 1 3/8" doors are usually not beveled.) A sander is usually better for this job than a plane. A good way to support the door on edge is to clamp it in a Workmate, as shown in Figure 4–14.

Figure 4–14
Sanding a door edge. Note method of supporting door.

Figure 4–13
Bevel for latch edge of door. This is seldom necessary on a 1 3/8" interior door, but is usual on an exterior door.

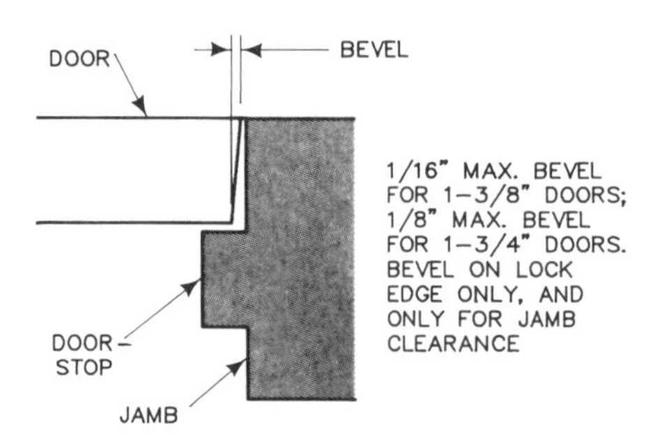

Installing the doorstop. You can buy doorstop molding in several profiles. Most are 1/2" thick and about 1 1/4" wide. Cut the doorstop for the side and top jambs to length and miter the ends for the joints at the top (see Figure 4–15).

Start with the butt-side doorstop. With the door closed, tack the stop in place, leaving 1/32" clearance to the door, checked all along the door. (You must allow for some paint buildup.) Now set your

Figure 4–15
Fitting and nailing doorstop.

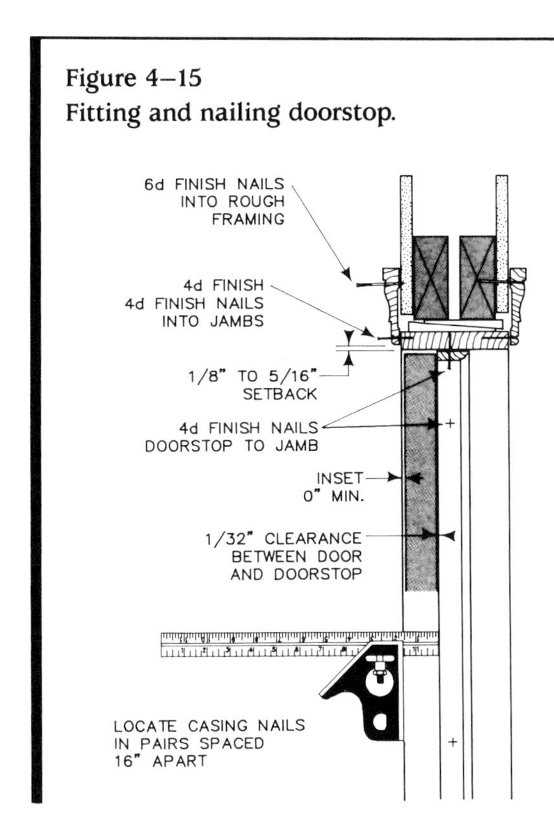

combination square to match this setback, and use that to tack up the latch-side doorstop at the same depth. Check to see that the door can close properly with 1⁄32" clearance on the latch side, then nail both doorstops tight and add the doorstop to the head jamb.

Installing a mortise lock. If improvement for you is a step back to Early American (not too early) or Victorian or turn-of-the-century decoration, you can add a lot to your rooms by getting rid of brass and chrome tubular locksets and going back to the old-time mortise locks on interior doors. (From a security standpoint, the mortise lock offers more than a key-in-knob lockset on your front door too.) Old-style interior mortise locksets can be installed in any door authentically, even with the old-style bit key, and with a wide choice of brass, glass, and porcelain knobs.

Unhinge the door and move it to your workbench to remove the tubular lockset. Cut plugs and glue them into the large holes (Figure 4–16, step 1). If you are going to drill out the new mortise on a drill press or with a commercial drill guide, or if you are going to chisel all the way, the plugs need go into the door only ½" or so. However, if you are going to drill out the mortise with a spade bit, the plugs should meet in the center so you won't have an unbalanced breakthrough with the bit. As an aid in starting drilling, put a lattice-wood plug in the edge mortise too.

After making the mortise, locate and drill holes for the knob spindle and keyholes (steps 2 and 3). If you want to be able to lock the door from one side only, put a keyhole only on that side. Use the knob and spindle inserted in the lock to accurately locate the rosettes. You will also have to replace the strike plate.

Figure 4–16
Replacing a tubular passage set with an old-style mortise lockset.

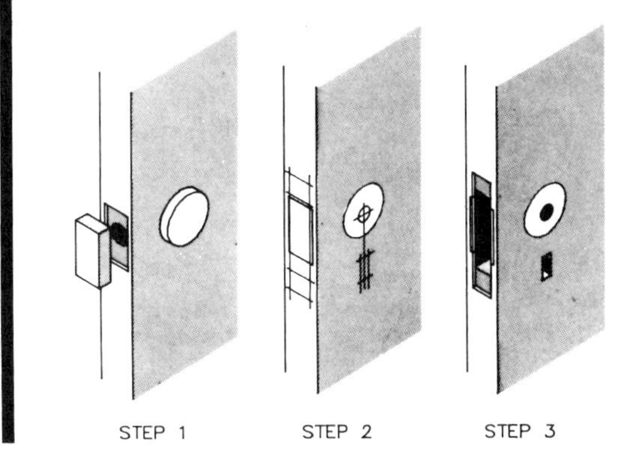

Installing the Door Casing

The casing is the final trim of the doorway, and its decorative style should complement that of the door. Various profiles and installation details are shown in Figure 4–17. Modern styles are simple, narrow, and unadorned—for example, a "ranch" casing, appropriate for flush doors. Colonial casing is more in keeping with paneled doors and a traditionally styled house. Victorian houses used very ornate carved casings.

You cannot simply cut the casing to length,

miter the ends, and nail it in place. That would work if the walls were flat, but they usually are not, particularly wet-plastered walls. There are two ways to put up casing: you can cut and trim on the wall until the corner joints go together neatly, or you can assemble the casing flat on your workbench. (You don't need a huge bench—only the upper end has to be clamped down.)

The corner miter joints can be assembled accurately, glued, and reinforced with a long screw diagonally through the top casing into the side casing at each miter. The casing can then be nailed to the jamb and the wall without problems of joint alignment. Gaping cracks behind the casing can be filled with drywall taping compound, patching plaster, or plaster crack filler.

Figure 4–17
Casing patterns and installation.

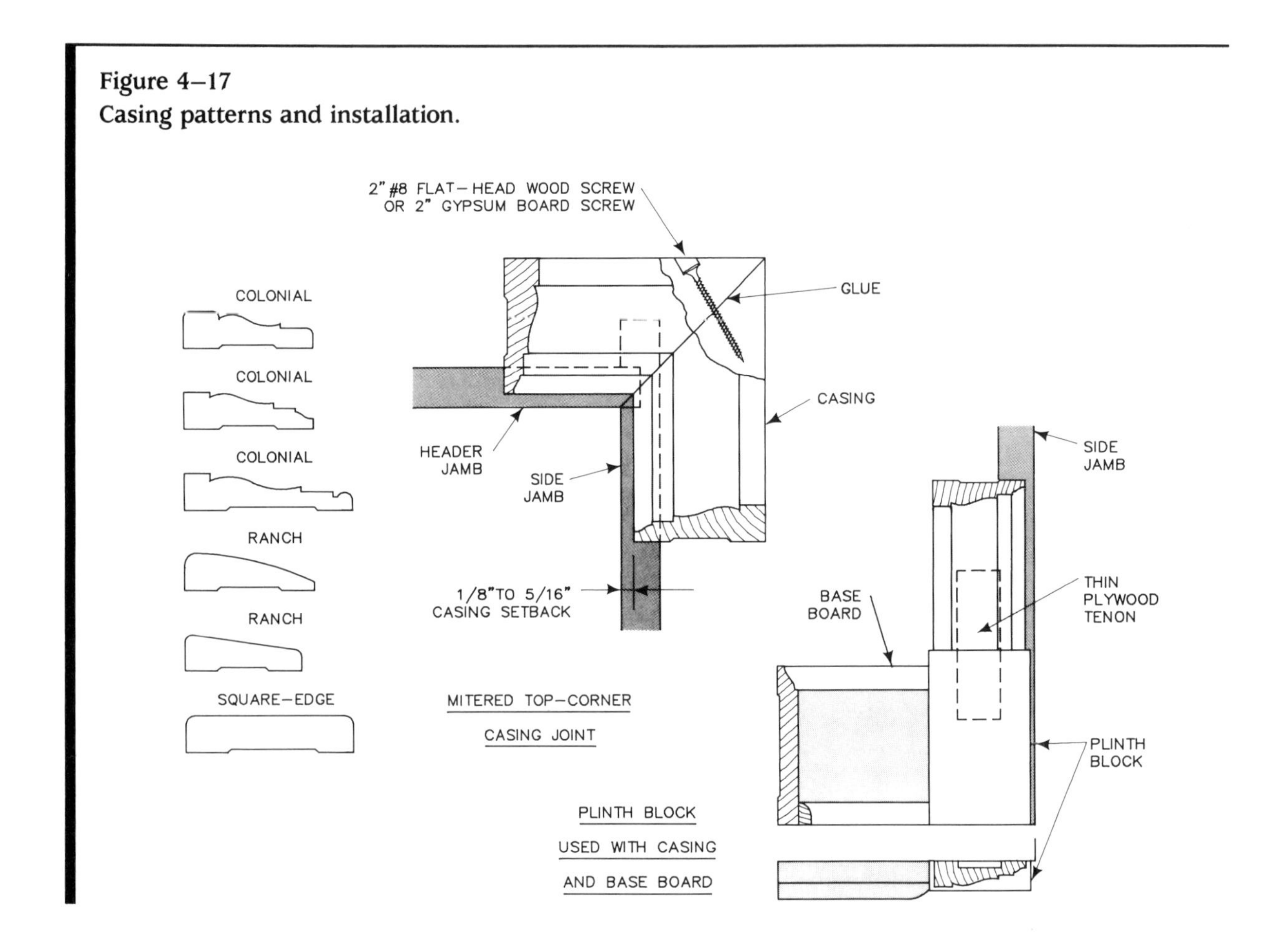

5 Exterior Doors

Exterior doors are better made today than they were only a few years ago. The most significant improvement is in weatherstripping. Gone are the spring metal strips, and in their place are weatherstripping systems that make the doors virtually airtight. This eliminates a major source of heat loss in the home. Wood, the traditional door material, has been joined by doors made of molded hardboard or with steel and plastic skins over foam-insulation cores. This kind of construction reduces the warping and swelling that can cause seasonal sticking in the opening, and it improves the R-value—the resistance to heat transmission—of the door. These new doors come prehung, making them easier to install. All in all, today there are good reasons to replace an old or thermally inefficient entry door. Figure 5–1 shows some of the standard doors that are hung on the job, as well as prehung entry doors available today.

Hanging an entry door is different from hanging an interior door, and more involved. First, the door will be larger and heavier, and you may want some assistance moving it around. If you are hanging a door that isn't prehung, the fit must be accurate for the door to be weathertight and mechanically secure. It will also require weatherstripping.

SELECTING A DOOR

You have several alternatives. You can go traditional and buy the door and jamb separately. This gives you the widest possible choice of doors, and may be necessary if you are replacing a deteriorated jamb and door in a nonstandard size opening. Jambs can be purchased assembled, or you can buy milled door jamb (usually pine or fir) and threshold (oak or yellow pine) pieces by the foot, and build the finish jamb to fit your choice of entry door.

INSTALLING A PREHUNG DOOR

A prehung door is today's popular alternative to the traditional on-the-job hung door. Unless you are an experienced carpenter, it's the way to go when selecting a replacement exterior door. The range of available styles is also large. There are two kinds of prehung doors: those hinged into a jamb and those hinged into a steel frame that fits into an existing jamb. The difference is important.

Most prehung entry door sets are a combina-

Figure 5–1
Three examples of new entry doors you can buy and install.

tion of door, jamb, threshold, and exterior casing designed to be installed in place of the old door and jamb, or in a new rough opening, as shown in Figure 5–2. Some of these doors are supplied without exterior casing, but otherwise they are the same.

To install, the old door is unhinged, and all hardware, interior and exterior casing, and the jambs and threshold are first removed (see Figure 5–3). With the backside of the brick mold or exterior casing caulked, the replacement door and jamb are tipped into the rough opening from the outside of the house, shimmed and centered in the opening, and the exterior molding nailed to the house frame all around. The retaining clips can now be removed and the door swung open to allow the jamb to be attached to the studs. The interior casing is then replaced and the lockset installed.

A second variety of prehung door is hung in an

Figure 5–2
Installing a prehung door.

Figure 5–3
Prehung door installation details.

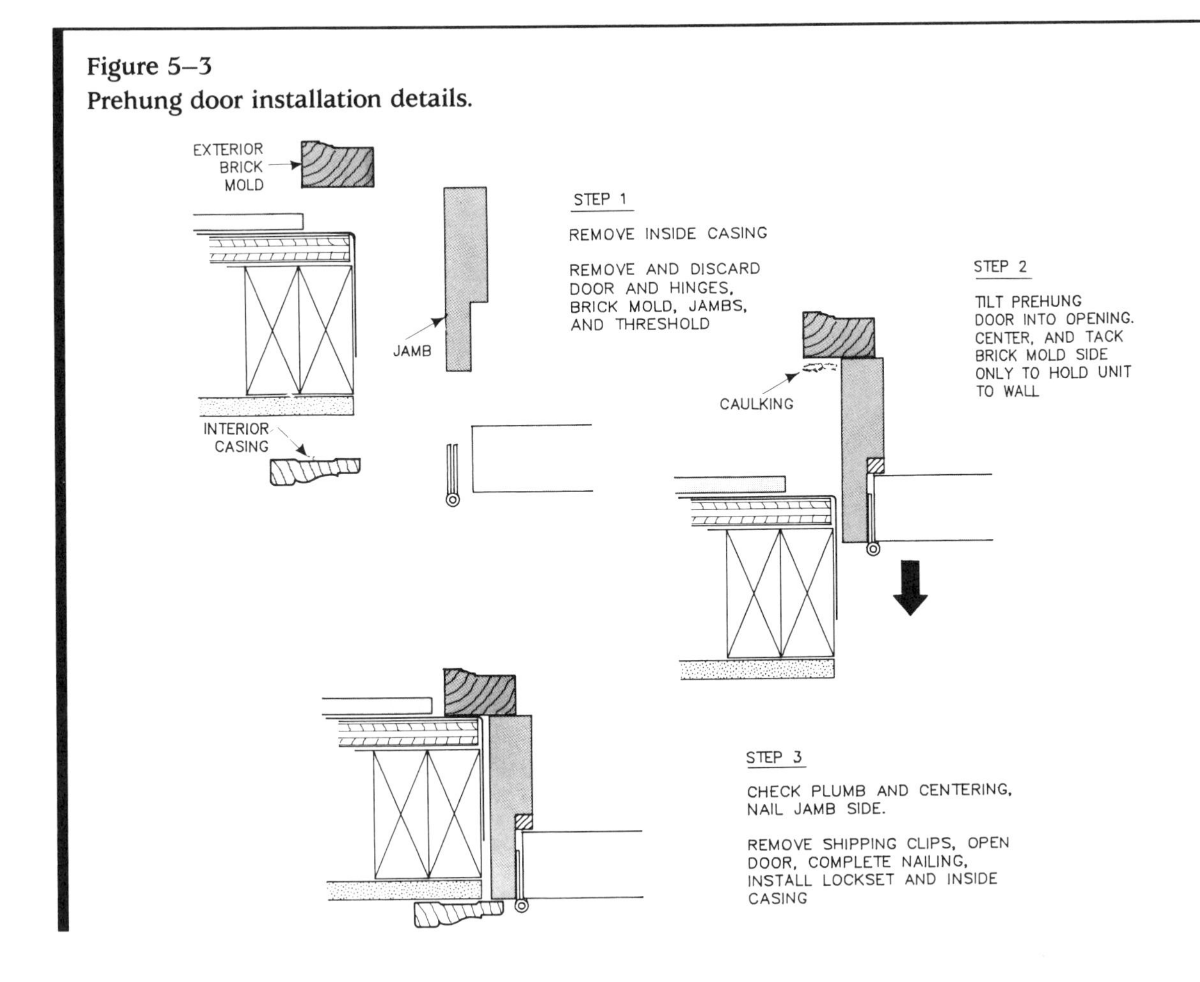
EXTERIOR BRICK MOLD
JAMB
INTERIOR CASING
STEP 1
REMOVE INSIDE CASING
REMOVE AND DISCARD DOOR AND HINGES, BRICK MOLD, JAMBS, AND THRESHOLD
STEP 2
TILT PREHUNG DOOR INTO OPENING. CENTER, AND TACK BRICK MOLD SIDE ONLY TO HOLD UNIT TO WALL
CAULKING
STEP 3
CHECK PLUMB AND CENTERING, NAIL JAMB SIDE.
REMOVE SHIPPING CLIPS, OPEN DOOR, COMPLETE NAILING, INSTALL LOCKSET AND INSIDE CASING

Figure 5–4
Measuring door opening for steel-framed replacement door.

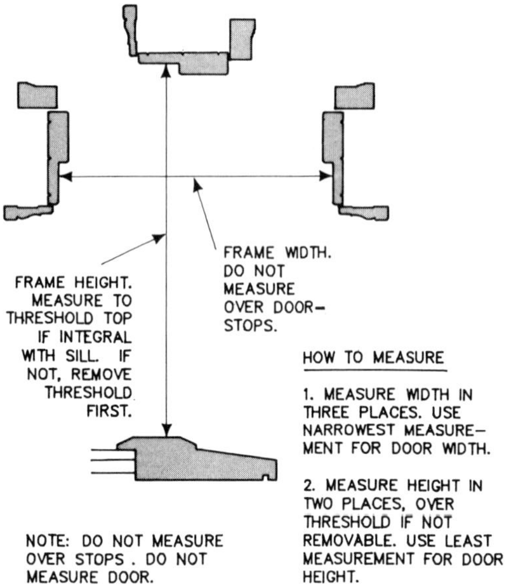

L-shaped steel frame that fits into an existing door jamb in place of the old door. All you have to do is unhinge your old door and remove the interior casing and all door hardware attached to the jamb. It works because the new door is slightly smaller than the old door.

Fit is critical. Measure the door opening width and height between side jambs and head jamb to sill (not threshold, unless it can't be removed). Do not measure doorstop-to-doorstop and do not measure your door (see Figure 5–4).

To install, tip the new door frame replacement unit into place in the opening from the inside of the house. Center and plumb the unit in the jamb and nail the flange of the new door frame to the old jambs. When that is done, remove the retainers that hold the new door and frame together, and complete the installation with security screws. Complete the installation by nailing new wood stops on top of the old ones and replacing the casing over the steel replacement door flanges.

Figure 5–5
Installing a replacement door.

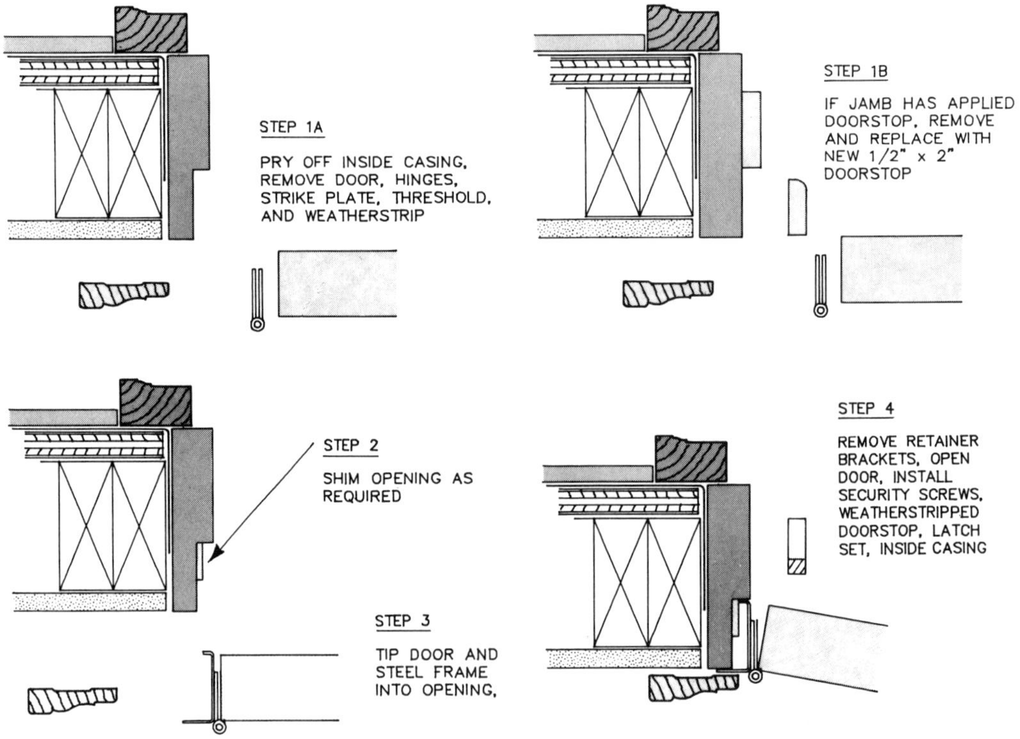

HANGING A STANDARD DOOR

The procedure is generally the same as that for an interior door. An exterior jamb is shimmed into the rough opening in the same manner as an interior jamb, except that it is nailed in more securely (see Figure 5–5).

Interior doors often are hung on two hinges (although they really should be hung on three). Exterior doors are considerably heavier and usually wider; they must be hung with three hinges. The minimum size for these hinges is 4″ × 4″. The center hinge helps prevent warp and supports some of the dead weight of the door. Do not use inexpensive lightweight hinges.

Figure 5–6
Dutch door.

The lock side of an exterior door is beveled, not only to maintain tight edge clearance, but also to get better sealing action from the weatherstripping. The bottom edge of the door usually requires additional shaping to get the desired weather seal and drip protection, depending on the method of sealing used.

Once the old door has been taken out, the technique of hanging and fitting the replacement door, except for beveling the edge, is exactly the same as for an interior door, as described in detail in Chapter 4. The outside edges of an exterior door are not square, but are beveled on the closing side for both closing clearance and a wedging fit against weatherstripping.

For something different, consider a Dutch door, one divided into two separately hinged sections; a typical installation is shown in Figure 5–6. A Dutch door not only makes an extremely attractive front door for almost any style house, it is practical. Traditionally, Dutch doors were used to keep small children and pets either in or out, without shutting the door completely.

CHOOSING ENTRY DOOR LOCKS

The familiar "builder's grade" key-in-knob lockset is shiny, attractive, and cheap—and so easy to break loose that burglars seldom waste time picking it. Old-time mortise locks are not much of a barrier either. It makes sense to replace a weak exterior door lock with something more secure, or back it up with a second lock—or do both. The backup, or secondary lock should be either a cylinder deadbolt lock or a rim deadlock.

Security aside, there are other reasons for replacing door locks. If you are renovating a house in Early American or Victorian style, you would want to replace a key-in-knob lockset with a mortise lock or a box lock while retaining the cylinder lock for security. Inside the house, for appearance, tubular passage locksets can be replaced with interior mortise locks with bit keys.

Key-in-knob locks and mortise locks are primary locks. They combine key locking with a doorknob and latch—so you can open and close the door, and latch it closed without locking it.

Key-in-knob locksets require two holes for installation: a large—usually 2⅛″ diameter—hole through the door and a smaller intersecting hole in the edge of the door. Hole location and size are fairly standard today, so inserting a replacement, even with a different brand, is seldom a problem. But you should always check whether you can return an unused lockset that doesn't fit the old holes.

A weakness of the key-in-knob lockset is the location of the keying cylinder in the outside handle. If the handle is knocked off—not hard to do with a hammer or brick—the door can be unlocked with a screwdriver.

Mortise locks are more secure. A knob-operated spring-loaded latch is backed up with a cylinder lock deadbolt. Except for the key cylinder, which is mounted flush with the outside of the door, the works of the mortise lock are enclosed in a steel case that is inserted into the door in an opening hollowed out from the edge of the door. There is more work involved in installing a mortise lock than any other kind, and mortise locks are expensive.

As already noted, the most expedient way to improve exterior door security is to install a secondary lock, either a cylinder deadbolt lock or a rim lock, sometimes called a night lock.

A cylinder deadbolt requires the same intersecting mounting holes as a key-in-knob lockset, but the latch bolt hole in the edge of the door may be larger. Keyed operation on both sides offers better security, but a turn button on the interior side may get you out of the house faster in an emergency. If you have a door with glass lights, or one with thin, easily kicked out panels (and most can be kicked out very easily), you should have a key on both sides of the door.

The body of a rim lock (Figure 5–7) is installed on the inside surface of the door, and the strike—the part the lock connects to—is on the inside surface of the door jamb. This kind of mounting may not be particularly attractive, but it offers good security. You must bore a hole through the door to house the cylinder of the lock so it can be opened with a key from the outside. The lock can also be installed as a night lock with no access from the outside. Modern security rim locks may have a horizontal bolt much the same as a cylinder deadbolt lock, or they may have a vertical bolt that is secured in the strike the same way a hinge pin holds the leaves of a hinge together. A vertical bolt rim lock provides superior security; it is almost impossible to force entry without destroying the jamb or the lock.

Figure 5–7
Rim lock.

If you are doing renovation or restoration work, you might want a reproduction brass rim lock called a box lock, which combines the function of latch and lock, just as mortise or key-in-knob locksets do. Some box locks offer a modern keyed cylinder on the outside and a historically appropriate bit or so-called skeleton key inside.

If you want to go high-tech, you could install an electronic lock actuated by depressing buttons in a preset combination. This kind of lock might have a high fumble-factor on dark stormy nights when you want to get into the house in a hurry. A better solution is a Key 'N Keyless electronic lock with deadbolt, as shown in Figure 5–8. Installation requires two sets of large and small holes in the door and two holes in the jamb. The lock can be operated by entering a four-digit code or with a conventional key, handy if you forget the code or the batteries go dead. An alarm sounds inside the house whenever the lock is tampered with.

Figure 5–8
Electronic lock with deadbolt. Lock operates with four-digit combination or conventional key. Tampering sounds alarm inside house. (Photo courtesy Schlage Lock Company)

Replacing an old mortise lock with a key-in-knob lockset is made quite simple with a modernization kit consisting of a latch plate and trim escutcheon plates. The kits come in more than one size. For easiest conversion, pick one with a latch plate that matches the height of the old mortise lock latch plate. Using the same-size latch plate will result in the new door knob being slightly higher or lower than the old one, depending on the design of the old mortise lock. If this isn't acceptable, use a larger size kit.

Remove the mortise lock knobs, plates, and strike. Pull the mortise lock out of the door (Figure 5–9). Cut a wood plug to fill in the door mortise and glue it in place to maintain door strength. Place the new latch plate in the mortise and mark the centerline on the door. This will locate the new knob height. Use the kit template and locate and drill holes. Attach the lockset (Figure 5–10).

INSTALLING LOCKS

Installing any of these types of locks in a new door is simply a matter of following the instructions packed with the lock—carefully. Replacing one of these locks with another of the same kind is a problem only when the new lock requires a different-size hole through the door or a different amount of backset from the outside door edge. Changing the size of the latch bolt hole in the edge of the door is a lesser problem. However, changing from one kind of lock to another, such as from a mortise lock to a rim lock, requires some carpentry.

Figure 5–9
To remove an old mortise lock, take off the knobs and escutcheon plates, pull out the knob shaft, and take out the screws above and below the latch in the door edge. Then pry the mortised unit out.

Reinforced Strike

The strike plate that comes with most locksets leaves a lot to be desired from a security standpoint, even when installed in an adequate jamb. Far better is a reinforced strike for a deadbolt, which consists of a strike box and strike plate (see Figure 5–11). The plate is extra-long, and may have holes for additional screws to spread the load over more of the length of the jamb. Additional strength can be obtained by using screws long enough to anchor into the rough framing behind the jamb. The strike box is attached to the mortised jamb with long screws through the box bottom into the rough 2× framing. When used with a lock with a 1″ or longer deadbolt, this provides greatly increased security.

Figure 5–10
Fill the mortise in the door edge as necessary and drill clearance holes for a new lockset. A new escutcheon plate the same size or larger than the old one will cover unused holes in the door face.

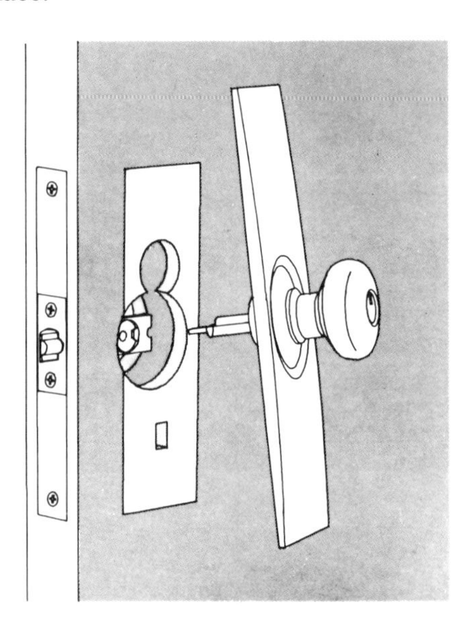

Figure 5–11
Installing a reinforced strike.

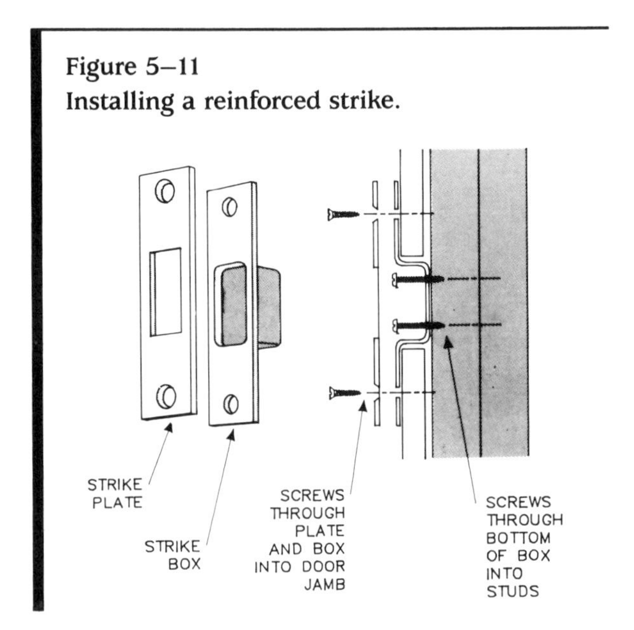

6 Problems with Hinged Doors

The most common problem with a hinged door is its sticking closed. A door may stick for many reasons. High humidity may cause the wood to swell. Hinge screws may be loose, or the hinges may be improperly mortised. The house foundation could be settling, or the partition containing the door sagging.

DETERMINING THE CAUSE OF STICKING

Before grabbing a plane or belt sander to take some wood off the door, it is a good idea to identify the source of the problem, and that is usually not difficult to do. Most of the time, door sticking can be corrected without a lot of wood removal.

Open the door, take hold of the knobs on both sides, and try to lift the door up. If you can feel any movement, there is a loose hinge—most likely, the top hinge. Doors seldom stick everywhere. Close the door and find out exactly where along its edges it is in contact with the door frame. An easy way to do this is to run a piece of stiff paper along the crack between the door and the jamb. A dollar bill is a good thickness. If the latch edge of the door rubs the side jamb near the top, the top hinge is probably the source of the difficulty. If the sticking is near the bottom on the latch edge, the bottom hinge is the most likely culprit. Figure 6–1 shows common door-sticking symptoms.

If the hinges are not loose, and the door rubs against the top of the jamb, the problem may be either the house foundation settling or a sagging partition. Check the latch bolt in the strike plate opening. It should be centered. Examine the strike for signs of wear. Is the bolt still aligned with the wear, or has it shifted up or down? If the bolt is not centered, or not lined up with the wear, or rubs on the top or bottom edge of the strike plate opening, or won't go into the opening at all, a foundation or partition problem may be the cause of the sticking.

If it is summertime, and it is hot and humid, the problem may be wood expansion due to the

Figure 6–1
Door sticking symptoms.

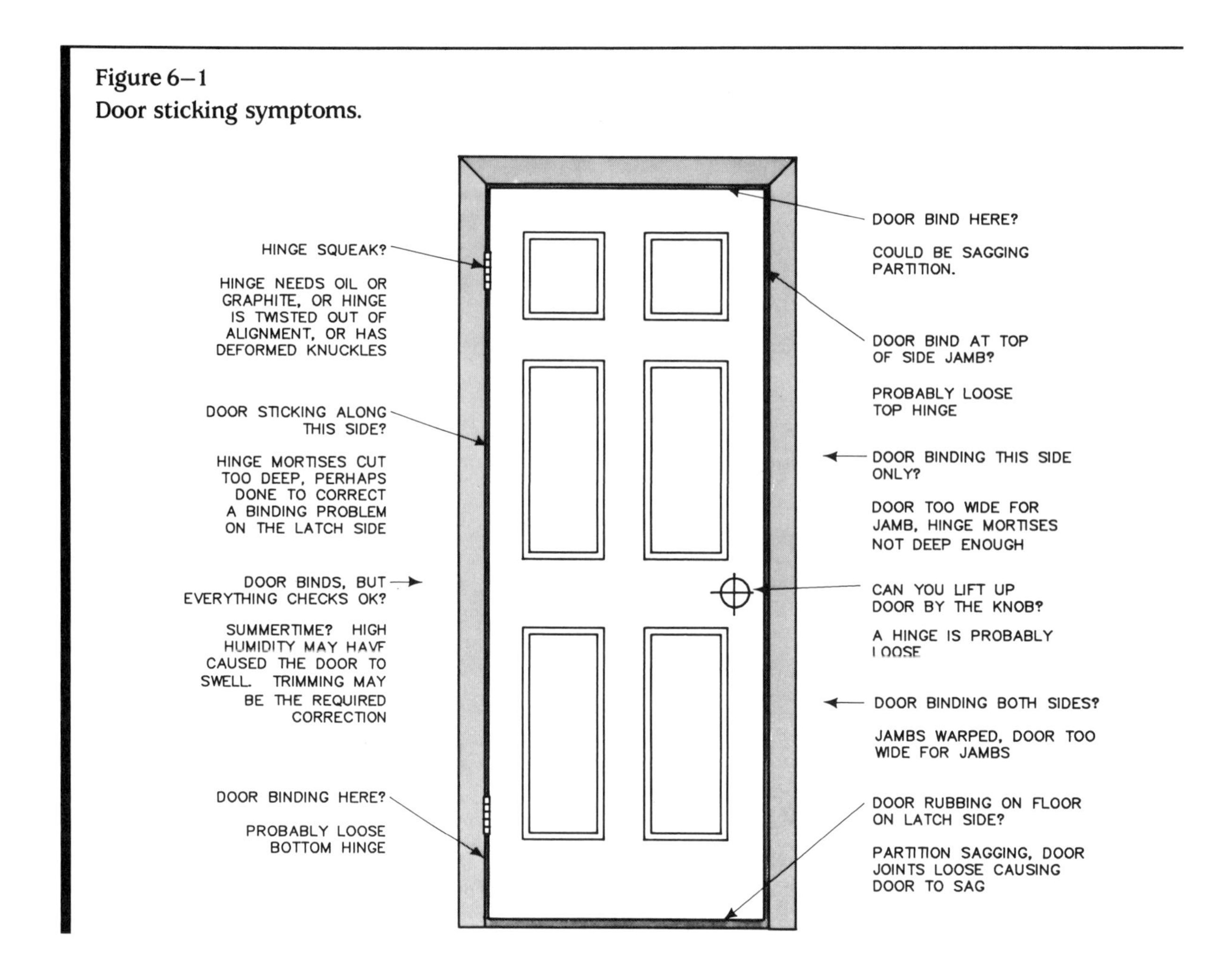
HINGE SQUEAK?
HINGE NEEDS OIL OR GRAPHITE, OR HINGE IS TWISTED OUT OF ALIGNMENT, OR HAS DEFORMED KNUCKLES
DOOR STICKING ALONG THIS SIDE?
HINGE MORTISES CUT TOO DEEP, PERHAPS DONE TO CORRECT A BINDING PROBLEM ON THE LATCH SIDE
DOOR BINDS, BUT EVERYTHING CHECKS OK?
SUMMERTIME? HIGH HUMIDITY MAY HAVE CAUSED THE DOOR TO SWELL. TRIMMING MAY BE THE REQUIRED CORRECTION
DOOR BINDING HERE?
PROBABLY LOOSE BOTTOM HINGE
DOOR BIND HERE?
COULD BE SAGGING PARTITION.
DOOR BIND AT TOP OF SIDE JAMB?
PROBABLY LOOSE TOP HINGE
DOOR BINDING THIS SIDE ONLY?
DOOR TOO WIDE FOR JAMB, HINGE MORTISES NOT DEEP ENOUGH
CAN YOU LIFT UP DOOR BY THE KNOB?
A HINGE IS PROBABLY LOOSE
DOOR BINDING BOTH SIDES?
JAMBS WARPED, DOOR TOO WIDE FOR JAMBS
DOOR RUBBING ON FLOOR ON LATCH SIDE?
PARTITION SAGGING, DOOR JOINTS LOOSE CAUSING DOOR TO SAG

high humidity. This is often the cause for sticking along the whole side of the door. If swelling affects the hinge edge, the door will bind as you close it, and may not close far enough for the latch to engage the strike. However, do some more checking before taking corrective action.

So far, it has been assumed that the door was correctly hung in the first place. Close the door and inspect the gaps between the door and the jamb. Start on the hinge side of the door. The gap should be 1/16″, and it should be the same from top to bottom.

Now check the gap on the latch side. The gap should be 1/16″ to 1/8″ and, again, the same from top to bottom. There should be a 1/8″ gap at the top, and it should be the same from side to side. The gap at the bottom should be at least 1/2″. You can check the narrow gaps with a slip of cardboard the proper thickness, and the wide gap with a piece of scrap wood of appropriate size.

SOLVING HINGE PROBLEMS

If you found the top hinge to be loose, try turning in the screws. If the screw threads are stripped in the wood, remove them (not all at once), apply white glue (not epoxy) to toothpicks, stuff them in the holes, and run the screws back in right away. This works sometimes, but you won't know until the hinge has been in use again for a while (Figure 6–2).

Another way is to install new screws of the same size diameter, but significantly longer. Simply increasing the size—from a #10 screw to a #12, for example—is not a good idea because the larger heads won't seat flush in the countersunk holes in the hinges, and will probably interfere with the hinge closing. A slight increase in length is no solution either, because the screw will only come out the back of the finish jamb, which is only ¾" thick for interior doorways. You need replacement screws long enough to go through the finish jamb, bridge the gap, and thread into the 2 × 4 or 2 × 3 rough jamb inside the wall. For that you'll need flat-head wood screws or flat-head tapping screws 2" to 2½" long. Drill pilot holes in the rough jamb, and lubricate the screw threads with wax.

SOLVING HINGE MORTISE PROBLEMS

The gap between the door and jamb on the hinge side should be 1/16", top to bottom. If the door is rubbing against the jamb on the lockset side and the hinge side gap is not 1/16", the mortises must be corrected. Check the gap right at the hinge. If less than 1/16" at a hinge, shims can be inserted in the mortise, under the hinge leaf.

Remove the screws from either the door or jamb (whichever mortise appears to be deeper) from that hinge only. Cut pieces of cardboard to correct the gap, insert them between the hinge leaf and the mortise, punch screw holes, and drive in the screws. You may have to use longer screws.

A hinge leaf set in a mortise that is too deep can itself be the cause of loose screws. Every time the door is closed, the flexing hinge leaf will put excessive force on the screws, loosening them. A mortise of correct depth puts the hinge leaf surface just flush with the surface of the door or jamb.

To correct a door-to-jamb gap that is wider than 1/16", the hinge mortise in the door or the jamb has to be chiseled deeper. To do this, drive out the hinge pins and take down the door. It is easier to chisel out the door mortise, and preferable, because there's more wood to cut into. Remove the hinge leaf and draw a line on the side of the door or edge of the jamb to mark the bottom of the mortise, then carefully chisel down to that depth.

Figure 6–2
Hinge problems and cures.

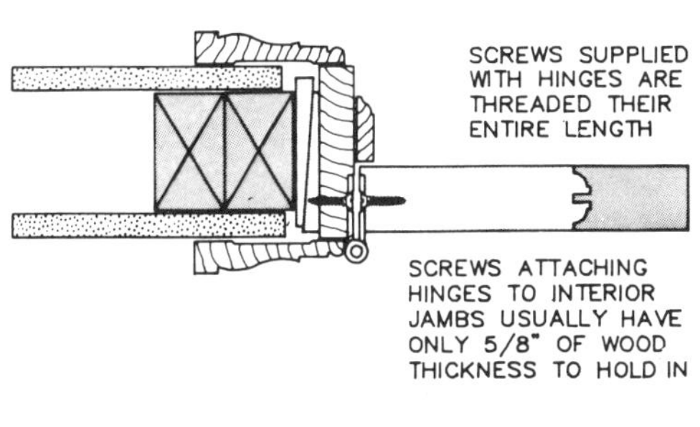

SCREWS SUPPLIED WITH HINGES ARE THREADED THEIR ENTIRE LENGTH
SCREWS ATTACHING HINGES TO INTERIOR JAMBS USUALLY HAVE ONLY 5/8" OF WOOD THICKNESS TO HOLD IN

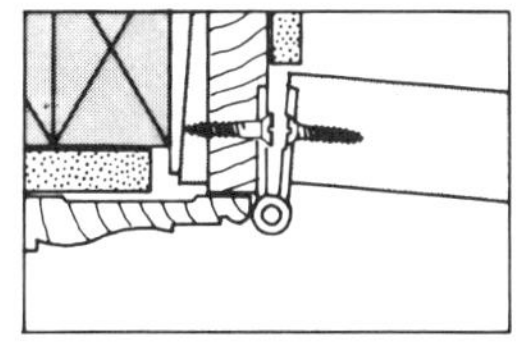

REPLACING A STRIPPED HINGE SCREW WITH A LARGER DIAMETER SCREW WORKS, BUT A COMMON WOOD SCREW AS SHOWN HAS FEW THREADS IN CONTACT WITH THE JAMB

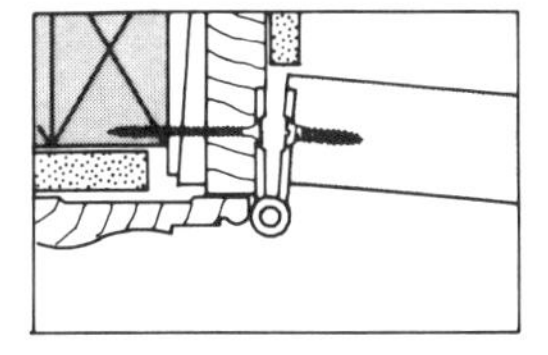

IF POSSIBLE, USE A LONG GYPSUM BOARD SCREW OR AN ALL-WEATHER SCREW TO REPLACE A STRIPPED-OUT HINGE SCREW
BE SURE THE SCREW IS DRIVEN INTO THE ROUGH FRAMING

FIXING A SAGGING PARTITION

A doorway is a weak spot in the rigidity of an interior partition, and any deformation will be concentrated there. If the top edge of the door rubs against the overhead jamb on the latch side, suspect that this is the problem (Figure 6–3).

Look for open miters at the corners of the casing. Also look for cracks in the wall near the top corners of the doorway. If you find either, the partition is sagging, or has sagged. You can either jack up the partition (see Chapter 11) or plane off the top of the door. It is best to fix the structural problem, or at least be sure it is not getting any worse, before you trim the top of the door. Otherwise the problem may recur. Unless you are a whiz with a plane, the best way to trim down the top edge is with a portable belt sander. Take off just enough to get the door operating.

Sticking of the bottom edge of an interior door against the threshold is not a common problem because the gap there is usually ½" or more (it may be less with an exterior door). If you encounter this condition, check for a sagging partition, or for swelling or other trouble in the threshold.

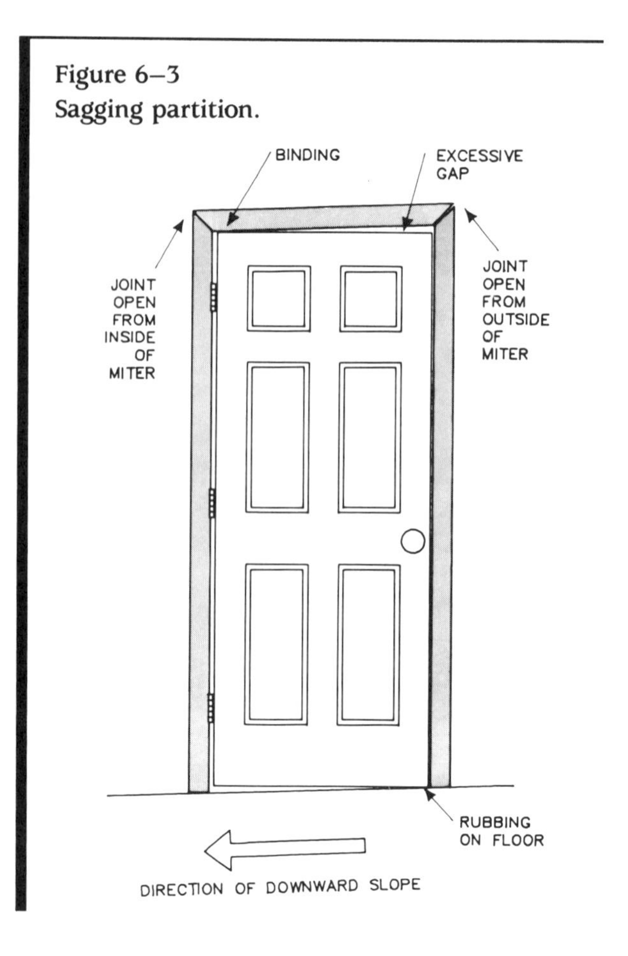

Figure 6–3
Sagging partition.

REPAIRING A SWOLLEN OR OVERSIZE DOOR

If the hinges are not loose, and the gap on the hinge side is 1/16" from top to bottom, but the door rubs on the other side or the top (and you have checked for settling foundation or sagging partition), the door must be trimmed.

But wait. If the door sticks in the middle of a hot, humid summer and you trim it, when winter and the dried-out heating season is upon you, the gap where you trimmed will be big—maybe bigger than you would like. A little sticking in the summer, as long as you can open and close the door, might best be tolerated.

If the sticking or apparent swelling seems to be uniform along one or more edges, the problem may actually be a built-up thickness of paint. In that case, take down the door, remove the hardware, and strip the edges carefully so as not to chip the paint on the flat surfaces. Repaint with one primer and one finish coat and replace the hardware.

If the door is sticking at the top or bottom of the latch side, check to see if the edge is beveled. The latch edge of exterior doors should always be beveled; interior doors usually are not. If an interior door edge is square, a slight bevel will usually eliminate the sticking. Unless the sticking is occurring near the latch, you can plane the latch side.

If the sticking is at the latch, you can plane the edge only if you can set the latch plate deeper into the edge of the door. You may be able to do this if the holes for the knobs are oversize, which allows the knobs to be repositioned.

Otherwise, you must take down the door. Chisel the hinge mortises a little deeper if you are lucky to find the gap on that side more than 1/16". If you are not so lucky, plane or sand the hinge side of the door, cut the hinge mortises deeper so they remain the correct depth, replace the hinges, and rehang the door.

Setting up the door for trimming the long edge can be a problem. A Workmate will hold the door nicely, or back the end of the door into a corner of the room, or wedge it in a doorway—or find an assistant.

LATCH BOLT

A latch bolt that doesn't go into the strike plate is usually a symptom of another problem, and not the problem itself. Correcting loose hinges, gaps, and settling or sagging will usually clear the problem. But if all else fails, relocate the strike plate, or remove it and file the top or bottom of the opening as required.

7 Folding and Sliding Doors

Folding doors and sliding doors are two alternatives to conventional side-hinged doors that you should consider in any remodeling project. While hinged doors have advantages—they give complete privacy when closed, they can be latched, locked, and weatherstripped—they tend to be in the way when open. They either stick out in the room, or they cover wall space that you might want to use for other purposes. Sliding and folding doors don't have these disadvantages.

TYPES OF FOLDING AND SLIDING DOORS

Bypass door. This is the most familiar type of sliding door, where two or more doors slide in front of or behind one another to provide access to the space beyond. A bypass door is the easiest type to install, and it is often used for bedroom closets. Two doors can cover a wide opening for a closet, but only half the width is open at a time. When open, the doors take up no space in the room. For extra-wide closets—as wide as the end of a room, for example—three or more doors may be used.

Pocket door. Shown in Figure 7–1, this is a sliding door in the grand style. Nothing is more elegant in interior doors than a pair of pocket doors gliding noiselessly out of sight into the wall—with or without the butler announcing dinner. A pocket door allows the entire width of a doorway to be open; however, installing this kind of door requires tearing out the partition to double the doorway width in order to accommodate the pocket.

Surface door. A third type of sliding door is the surface door. Instead of sliding into a pocket, it moves along the surface of the wall, hung from a boxed-in overhead track. It requires wall space equal to its width, but often can slide behind a bookcase or a large appliance, such as a refrigerator. Or it can be boxed in a shallow compartment constructed on the wall surface.

Bifold doors. Making more decorative closet doors than bypass doors, bifold doors provide almost full access to the closet when open. These

Figure 7–1
Pocket door. A conventional hinged door between the kitchen and dining room was replaced by a pocket door.

wood or plastic folding doors are made in many styles, ranging from a single pair to multiple doors that can convert the entire end or width of a room to an immense storage closet. Bifold doors can also be used as room doors or in any interior opening where complete privacy is not a requirement. (In normal installation there is a gap between the pivot side door edge and the jamb when the door is open or closed.)

Hardware kits. Bypass doors and pocket doors are sold separately from the hardware kits required for their installation. Bifold doors and their hardware kits can be purchased as separate items, or as a complete package of doors plus hardware. For a surface sliding door, the required track and hangers are purchased separately.

BYPASS DOOR INSTALLATION

Two-door hardware kits are made for openings of 4′, 5′, 6′, and 8′. You can cut the track for other widths. If you have a wider opening using three

Figure 7–2
Bypass sliding door parts and method of hanging.

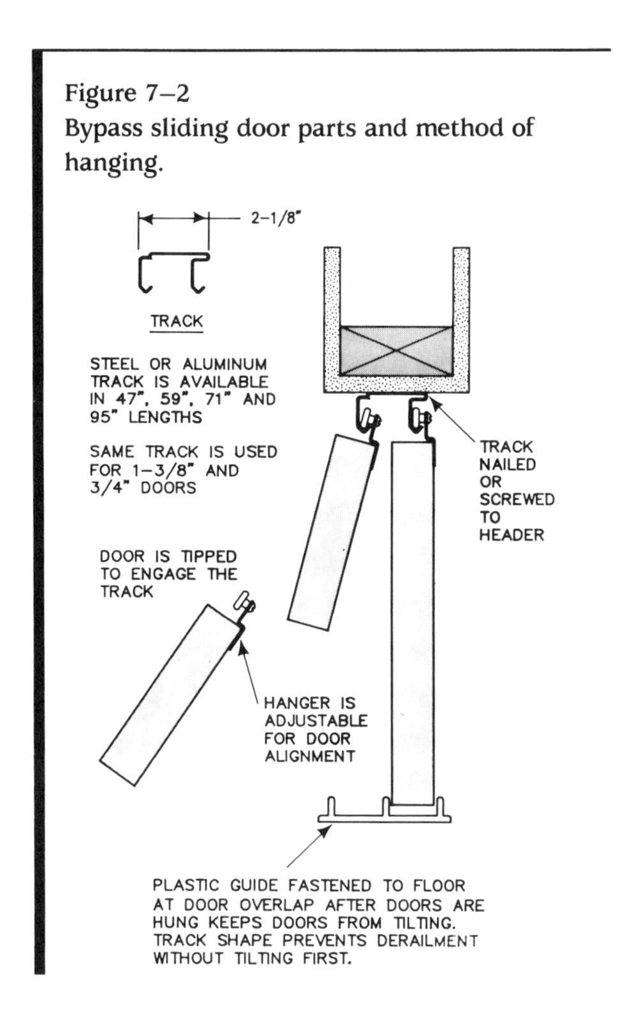
2-1/8"
TRACK
STEEL OR ALUMINUM TRACK IS AVAILABLE IN 47", 59", 71" AND 95" LENGTHS
SAME TRACK IS USED FOR 1-3/8" AND 3/4" DOORS
DOOR IS TIPPED TO ENGAGE THE TRACK
TRACK NAILED OR SCREWED TO HEADER
HANGER IS ADJUSTABLE FOR DOOR ALIGNMENT
PLASTIC GUIDE FASTENED TO FLOOR AT DOOR OVERLAP AFTER DOORS ARE HUNG KEEPS DOORS FROM TILTING. TRACK SHAPE PREVENTS DERAILMENT WITHOUT TILTING FIRST.

Figure 7–3
Bypass sliding door overlap.

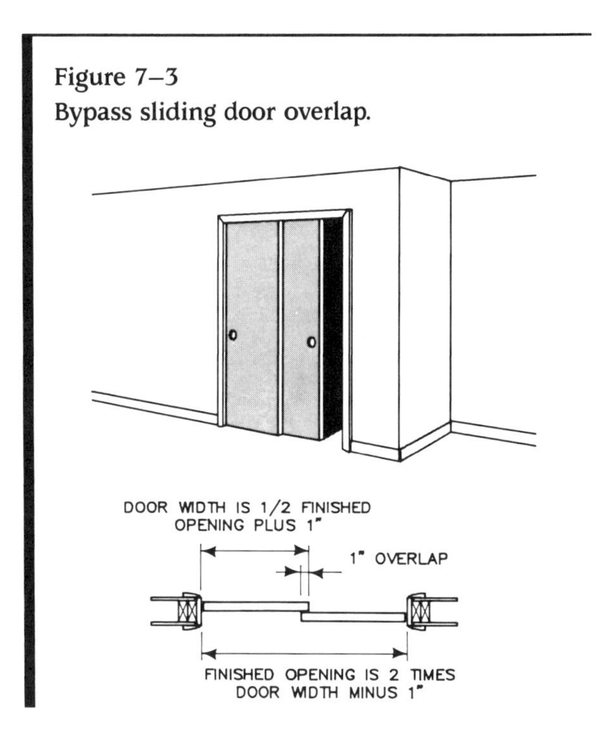
DOOR WIDTH IS 1/2 FINISHED OPENING PLUS 1"
1" OVERLAP
FINISHED OPENING IS 2 TIMES DOOR WIDTH MINUS 1"

or more doors, the tracks are installed in sections. The doors used should not weigh more than 60 pounds each.

The hardware for a pair of bypass sliding doors consists of a steel or aluminum track, four hangers, and a floor guide (Figure 7–2). Standard 1⅜"-thick, hollow-core flush doors are normally used in two-door installation. Doors of other thicknesses can be used, but they may result in interference if they are thicker, or a wide gap between them if they are thinner. The combined width of two doors must be 1" more than the finished width of the opening. This will produce 1" of overlap when the doors are closed. For multiple-door installations, add 1" for each additional door. The combined width of three doors must be 2" wider than the opening, four doors must be 3" wider, and so on (see Figure 7–3).

The opening for the bypass sliding doors can be bare plaster or gypsum board, or framed with wood jambs and casing (Figure 7–4). If hinged doors are being replaced, the doorstop molding must be removed, and hinge and strike mortises plugged and finished over.

The position of the track in the wall thickness is determined by the thickness of the doors and whether you want the outer door flush with the wall or casing, or inset. Manufacturer's instructions give the information needed. The screws supplied with the track usually are long enough for installation in a wood jamb. For openings finished in plaster or gypsum board, they should be replaced with screws long enough to penetrate the rough header.

The track must be level, which may require shimming. The doors are suspended from the track with hangers attached to their upper corners. Each hanger has an adjustment cam with which the doors can be plumbed and set to the same height. Each door is hung on the track by tilting the top inward and lifting up so the rollers on the hangers can engage the track. When the door swings to its vertical position, the hanger wheels cannot hop out of the tracks. The floor guide, when installed, keeps the doors from tilting.

Figure 7–4
Bypass sliding door track location options.

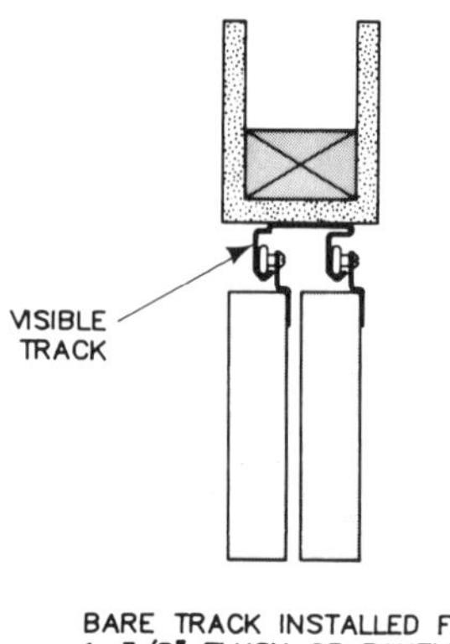
VISIBLE TRACK
BARE TRACK INSTALLED FOR 1–3/8" FLUSH OR PANELED DOORS IN GYPSUM BOARD OPENING.

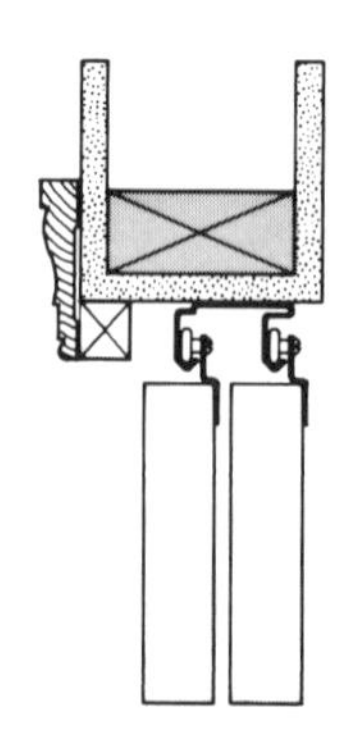
TRACK INSTALLED IN GYPSUM BOARD OPENING HIDDEN WITH CASING. BACK CASING AS SHOWN. PROVIDE CLEARANCE FOR TIPPING DOOR DURING INSTALLATION.

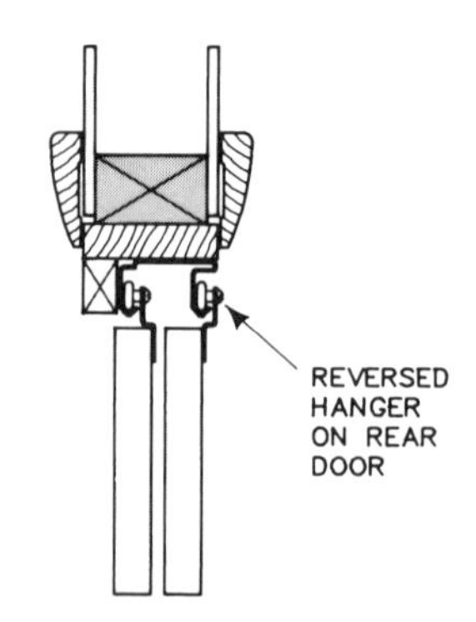
REVERSED HANGER ON REAR DOOR
3/4" DOORS INSTALLED IN 2x3 FRAMED OPENING WITH PLYWOOD PANELED WALLS

SURFACE SLIDING DOOR INSTALLATION

This is a real do-it-yourself project. For a single-door installation (Figure 7–5), nail or screw a 2 × 2 across the opening and extend it an equal distance along the wall at one side of the opening. Fasten a sliding door track to the underside of the 2 × 2 for its entire length. Use a 1 × 4, wide molding, or other 1 × trim to box in the 2 × 2 and conceal the track. A floor guide keeps the door from tilting. You can leave the door visible, but it will look better boxed into a shallow compartment or a false wall, or hidden by furniture or an appliance.

Figure 7–5
Surface sliding door installed to run between the wall and refrigerator.

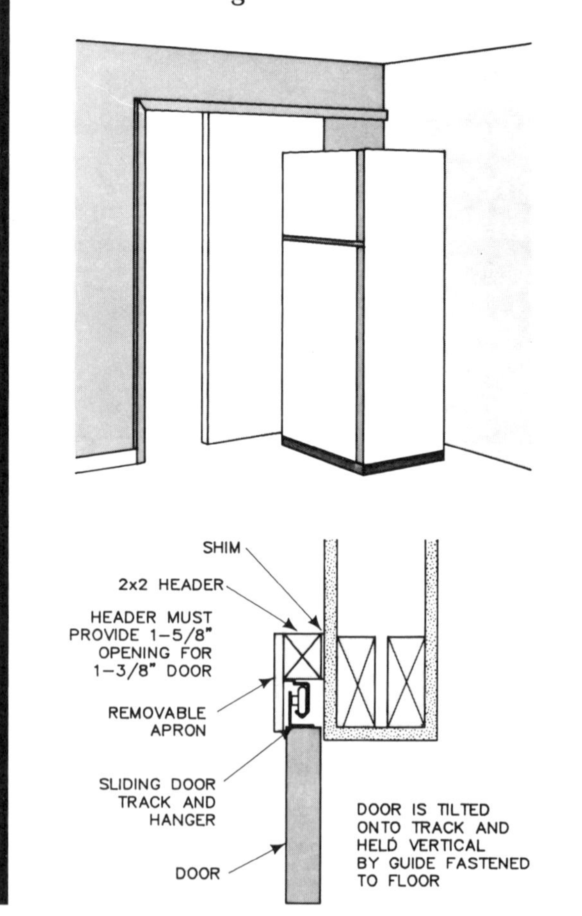
SHIM
2x2 HEADER
HEADER MUST PROVIDE 1–5/8" OPENING FOR 1–3/8" DOOR
REMOVABLE APRON
SLIDING DOOR TRACK AND HANGER
DOOR
DOOR IS TILTED ONTO TRACK AND HELD VERTICAL BY GUIDE FASTENED TO FLOOR

FOLDING OR BIFOLD DOOR INSTALLATION

You can install bifold doors in almost any opening in your home. The wood or plastic doors are made in many styles, including paneled, flush, and louvered. They come in two heights, one to fit in standard 6′8″ doorways, and one for an 8′

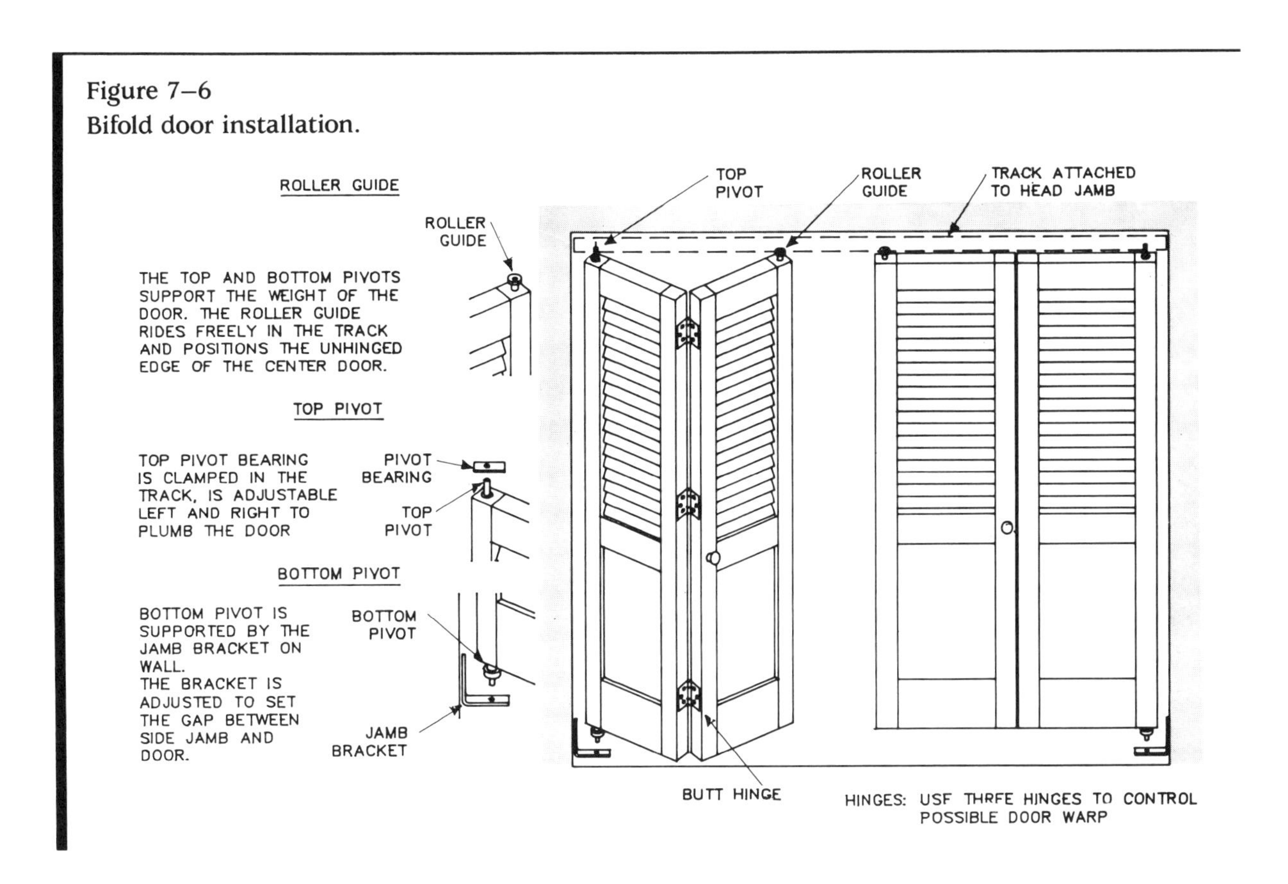
ROLLER GUIDE
ROLLER GUIDE
THE TOP AND BOTTOM PIVOTS SUPPORT THE WEIGHT OF THE DOOR. THE ROLLER GUIDE RIDES FREELY IN THE TRACK AND POSITIONS THE UNHINGED EDGE OF THE CENTER DOOR.
TOP PIVOT
TOP PIVOT BEARING IS CLAMPED IN THE TRACK, IS ADJUSTABLE LEFT AND RIGHT TO PLUMB THE DOOR
PIVOT BEARING
TOP PIVOT
BOTTOM PIVOT
BOTTOM PIVOT IS SUPPORTED BY THE JAMB BRACKET ON WALL.
THE BRACKET IS ADJUSTED TO SET THE GAP BETWEEN SIDE JAMB AND DOOR.
BOTTOM PIVOT
JAMB BRACKET
TOP PIVOT
ROLLER GUIDE
TRACK ATTACHED TO HEAD JAMB
BUTT HINGE
HINGES: USE THREE HINGES TO CONTROL POSSIBLE DOOR WARP

Figure 7–6
Bifold door installation.

floor-to-ceiling height. Wood doors can be trimmed slightly for a good fit; most plastic doors cannot be trimmed. (If trimming for width is required, all wood doors must be trimmed equally.) Bifold doors are hinged to side jambs with pivot hinges; a roller guide attached to the outer top corner of the free door rides in an overhead track to support the weight of the door and guide it (Figure 7–6).

The maximum width of each section is 24″. Thus a single two-door unit attached at one side can be used in an opening up to 4′ wide. Two of these units, one at each side, can be used in openings up to 8′ wide, or the two two-door units can be hinged together to fold to one side. Two such four-door units can span up to 16′ of opening. However, 24″ doors will extend almost two feet into the room when open. Multiple pairs of narrower doors hinged together usually are a better way to span a wide opening attractively.

While there are differences in the design of the mounting hardware for folding doors, the basic principle is the same (Figure 7–7). A metal overhead track is leveled and fastened to the header

Figure 7–7
Bifold door track location options.

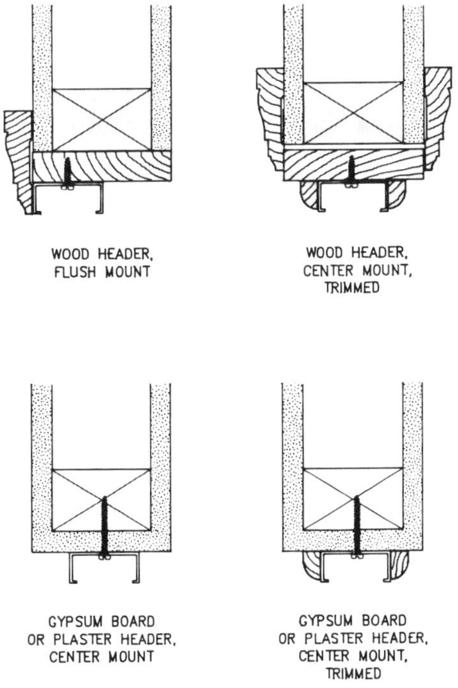

WOOD HEADER, FLUSH MOUNT

WOOD HEADER, CENTER MOUNT, TRIMMED

GYPSUM BOARD OR PLASTER HEADER, CENTER MOUNT

GYPSUM BOARD OR PLASTER HEADER, CENTER MOUNT, TRIMMED

or overhead jamb with screws. A roller-guide is attached to the top of the free end of the hinged pair of doors. This guide slides in the track, both guiding and supporting the doors. The edge of one door is secured at one side of the opening by pivot pins inserted in the top and bottom rails of the door. The pins engage brackets fastened to the jamb or plates in the top jamb/header and the floor.

POCKET DOOR INSTALLATION

A pocket door is a lot of work to install, but the results can be worth the effort. It is one door that will never be in the way when it is open, for it simply disappears into the wall. In fact, you will need a pull ring set into the edge of the door to pull it out of the pocket. Traditional pocket door installation details are shown in Figure 7–8.

Pocket door hardware kits will accommodate doors from 24″ to 36″ wide. Two kits are required for a pair of doors that move into pockets on either side of an opening. Any door without raised moldings up to 1¾″ thick and 80″ high can be installed in a pocket, including even the hinged door originally hanging in the opening. There are two kinds of kits: track and hanger only, and track, hangers, and steel split studs to support any kind of dry or wet-plaster wall construction.

To install a pocket door you will need a wall at one side equal to the existing doorway width plus 1″, as shown in Figure 7–9. Pocket door hardware is designed for use in partitions framed with 2 × 4s. With a little ingenuity you can install a pocket door in a 2 × 3 studded wall, but it leaves you with a somewhat flimsy wall.

Installation begins with removing the partition where the pocket is to go. Everything except the top plate comes out. In Figure 7–10 the plaster finish walls have been removed to expose the framing. Installation will be less involved if there is no electrical wiring in the wall, but if there is the wiring has to be removed (see Chapter 11). A baseboard radiator simply has to be carefully worked around.

If the partition is load-bearing (see Chapter 11),

Figure 7–8
Traditional pocket door construction. The removable piece in the opposite jamb allows the door to enter that side so its normally closed end becomes accessible.

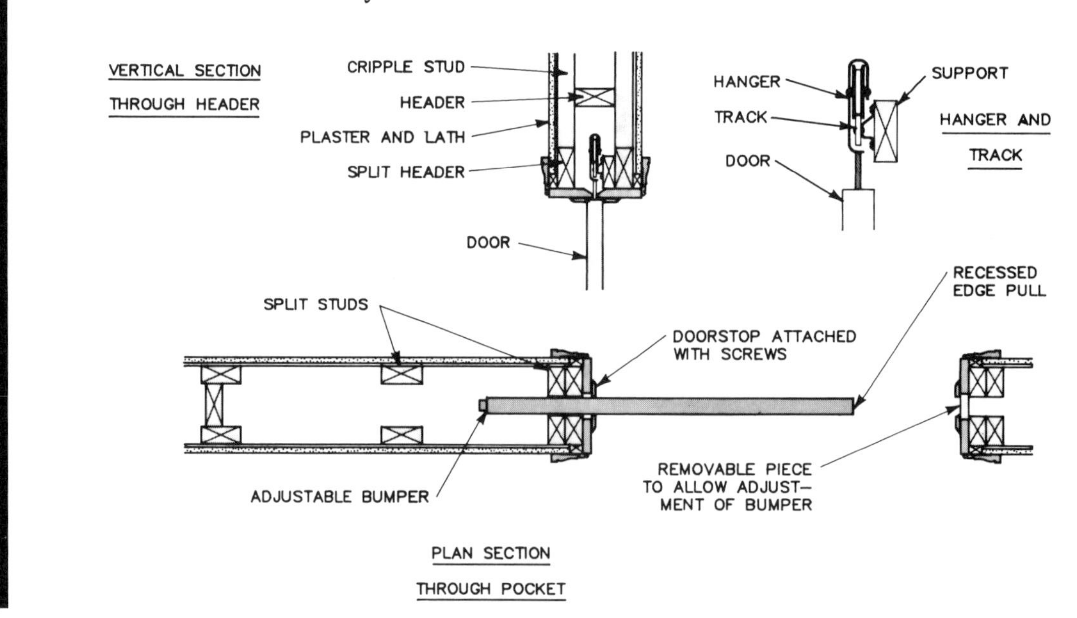

Figure 7–9
Rough opening dimensions for a single-door pocket.

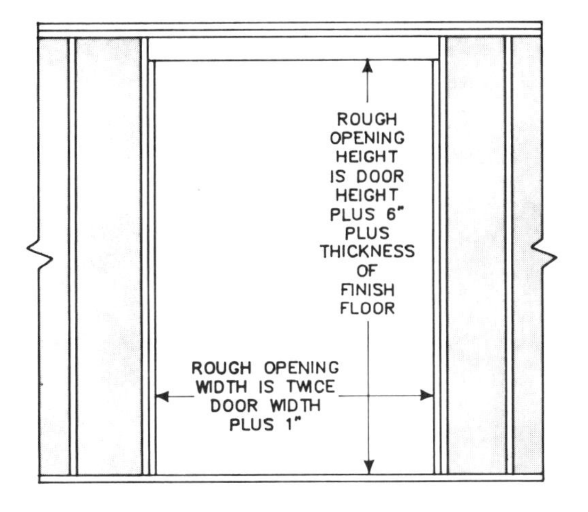

Figure 7–10
Old wall covering and studs must be removed to install a single door and pocket.

Figure 7–11
Typical header construction. Split header configurations vary with brands of pocket door kits.

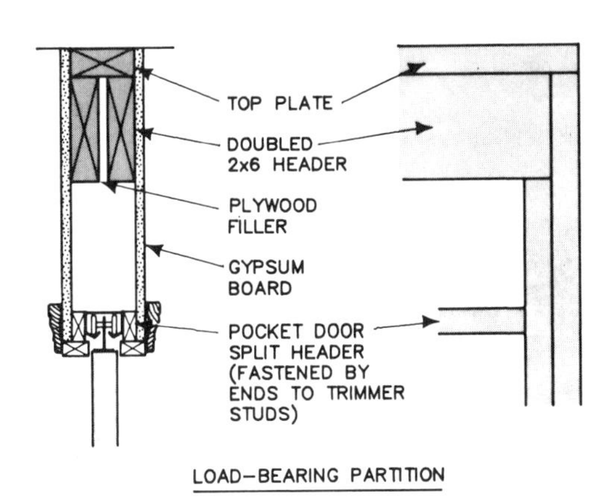

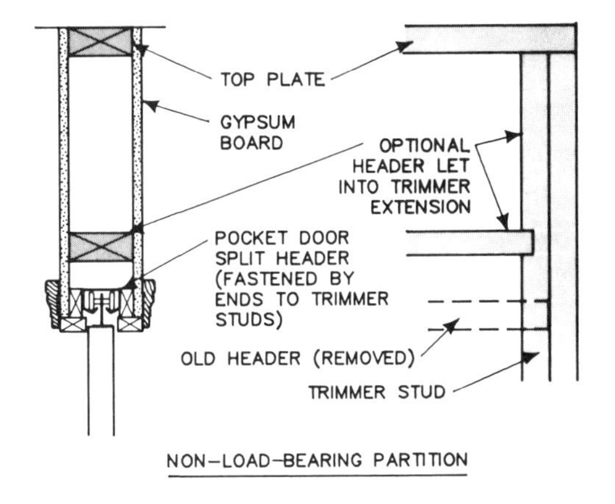

the ceiling must be independently braced while the partition is being demolished and a load-bearing header across the doorway and pocket is installed. Unless you have some house carpentry experience, cutting into a load-bearing partition is a job for an experienced contractor.

Header construction for a pocket door in a non-load-bearing partition is shown in Figure 7–11. Attach the steel track to the header and special split steel studs and jambs between the header and the floor. These split studs contain wood inserts for attaching gypsum board and trim. Use gypsum board screws and gypsum board adhesive instead of nails. Figure 7–12 shows the split studs, mounting plates, and backing for replace-

Figure 7–12
Steel and wood split studs form pocket for door.

ment gypsum board along the floor at the sides of the pocket.

The completely finished door is hung from the header before the opening is trimmed, as part of the door never comes out. The door can be plumbed and the height adjusted any time after the trim is installed, because the hangers remain accessible. When adding the trim at the sides of the doors, use screws; if the door hops the track or a hanger comes apart (it happens), you will have to remove the trim to rerail the door.

8 Window Installation

New windows can work wonders for any aging house, or they can subtly change the character of a contemporary house. This chapter tells you how to install new windows. Chapter 9 deals with the repair and adjustment of existing windows.

TYPES OF WINDOWS

All windows consist of two parts: the window frame that is attached to the wall of the building and the sash or sashes in which the glass panes, called lights, are mounted. The sashes may be fixed or movable. Windows are classified by the way in which the sash or sashes move in the frame and by the frame shapes. The most common types are shown in Figures 8–1 and 8–2; some special shapes are shown in Figure 8–3.

Double-hung window. This window has two sashes that slide vertically in tracks (channels) in the frame (Figure 8–4). This is the conventional all-purpose window installed in more than half the homes of this country. Air entry is easily controlled, but only half the window area can be opened at one time. Double-hung windows fit in with almost all periods of architecture. Various means are used to counterbalance the sashes. In old windows, cast-iron sash weights enclosed in pockets are attached to the sashes with rope or chain running over pulleys. Modern windows use torsion or coiled springs, or controlled friction on the edges of the sash. The glass area in each sash may be a single large pane or may be divided into a number of lights—either by a removable grille over the large pane or by a framework of dividers (called muntins) that hold smaller panes. The number of lights in each sash is usually the same, and the window is described according to the number in each. For example, Figure 8–4 shows an eight-over-eight double-hung window.

Sliding window. This type of window has two sashes that slide horizontally, right or left, in a common frame (Figure 8–5). The window can be positioned high on a wall, providing improved ventilation, lighting, and privacy compared with a double-hung window, and allowing more freedom in placing furniture in front of the window. Only half of the window area can be opened at a time. The sashes do not require counterbalances and usually run on nylon rollers.

Figure 8–1
Standard window types. The intersection of dashed lines points to the latch side. A casement window is hinged at either side. An awning window is hinged at the top to open outward. A hopper window is hinged at the bottom and usually opens inward.

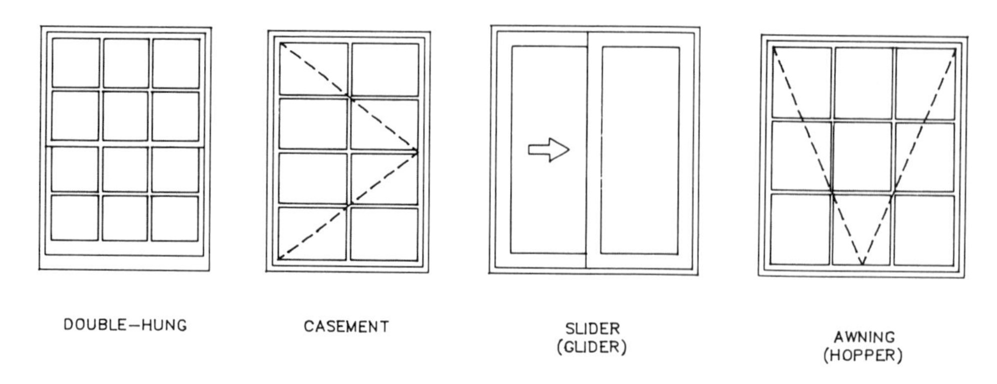

Figure 8–2
Stock special shapes. These can be installed as fixed windows, or hinged or pivoted so they can be opened. If a roof window can be opened, the down-slope end can be raised.

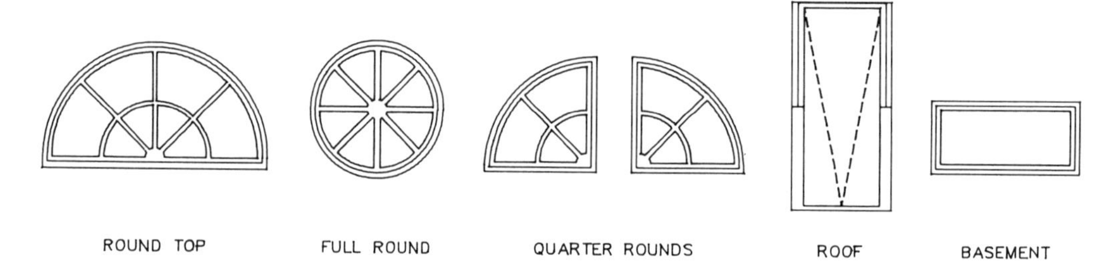

Figure 8–3
Other special shapes. These are usually custom-built to size and installed as fixed windows.

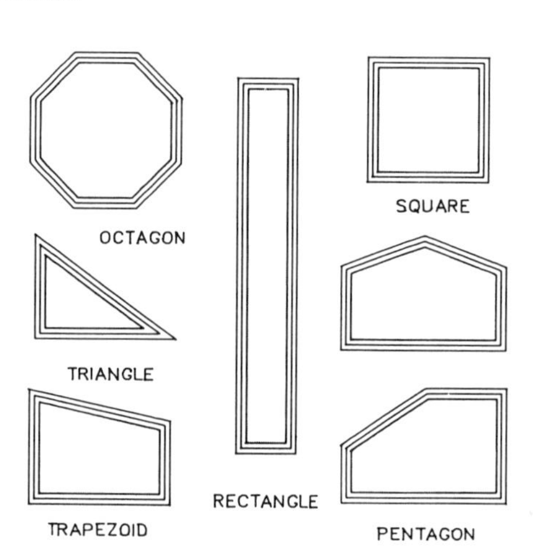

Casement window. This window has a sash hinged at the side, which is opened and closed by a crank. The window opening may have one full-width sash or two half-width sashes, as shown in Figure 8–6. This style of window is the most weathertight. The window can be completely opened, not just halfway, as for a double-hung or sliding window. In multiple casement window units all sashes may open, or one or more may be fixed, and the fixed sash may be wider than the operable sashes. Casement windows are usually made today with a single light, which for appearance may be divided into smaller rectangular or diamond shapes by removable grille inserts. Older windows may have a framework of muntins to hold smaller lights. Screens must be installed on the inside of the window. Double glazing is the only practical

Figure 8–4
Vinyl-encased wood double-hung window parts and construction.

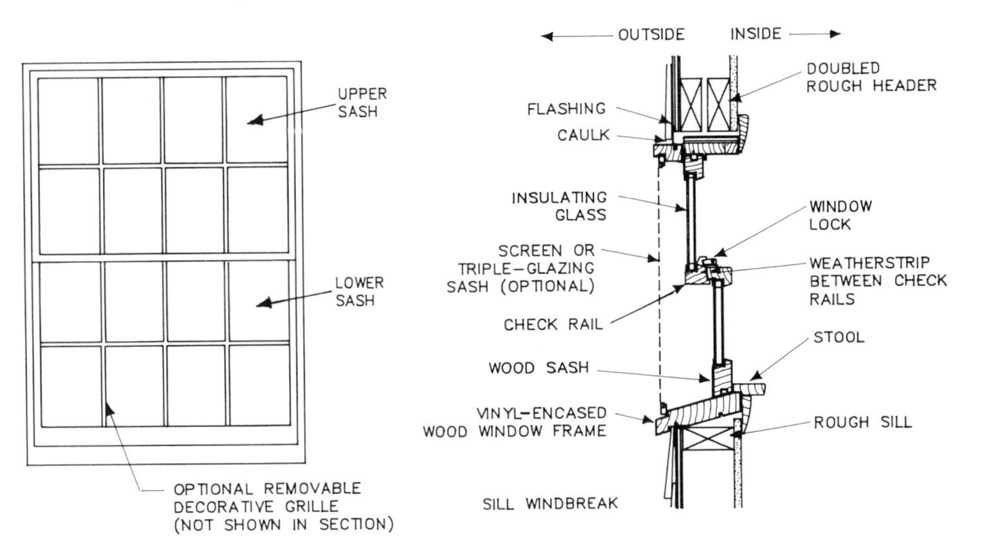

Figure 8–5
Wood sliding window construction.

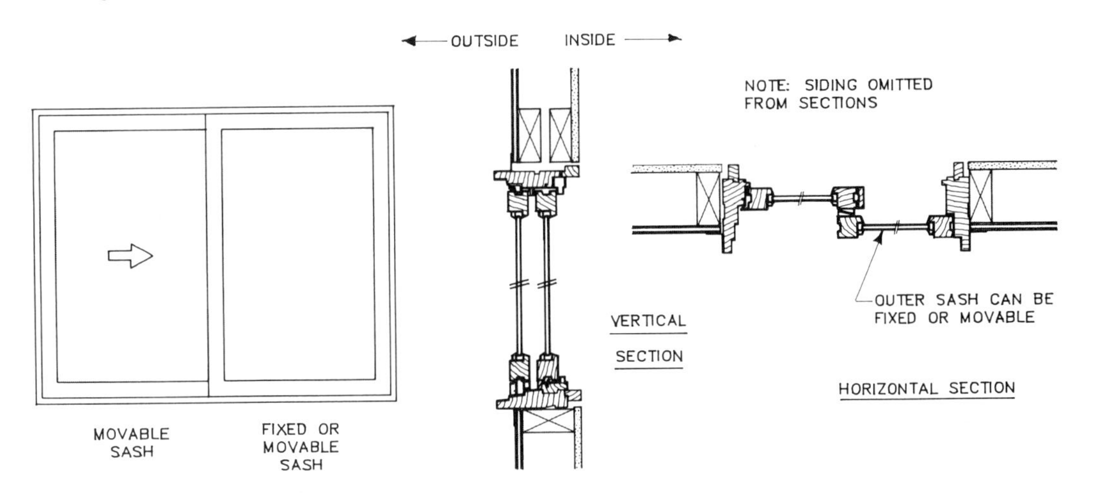

means of obtaining storm window weather protection if the window is to be opened.

Awning windows. An awning window is like a casement window, but with the sash hinged at the top instead of the side. To open, the bottom swings out and up. A similar design, the hopper window, is hinged at the bottom. The top swings down and in when it opens. Awning windows are effective in controlling ventilation and can be opened in the rain. A popular style of window consists of a fixed sash above, for an unobstructed view, with an awning sash below for easily controlled natural ventilation.

Jalousie. This window resembles a venetian blind. Horizontal slats of glass 3″ deep are secured by their ends in metal holders pivoted in

Figure 8–6
Casement window parts and construction.

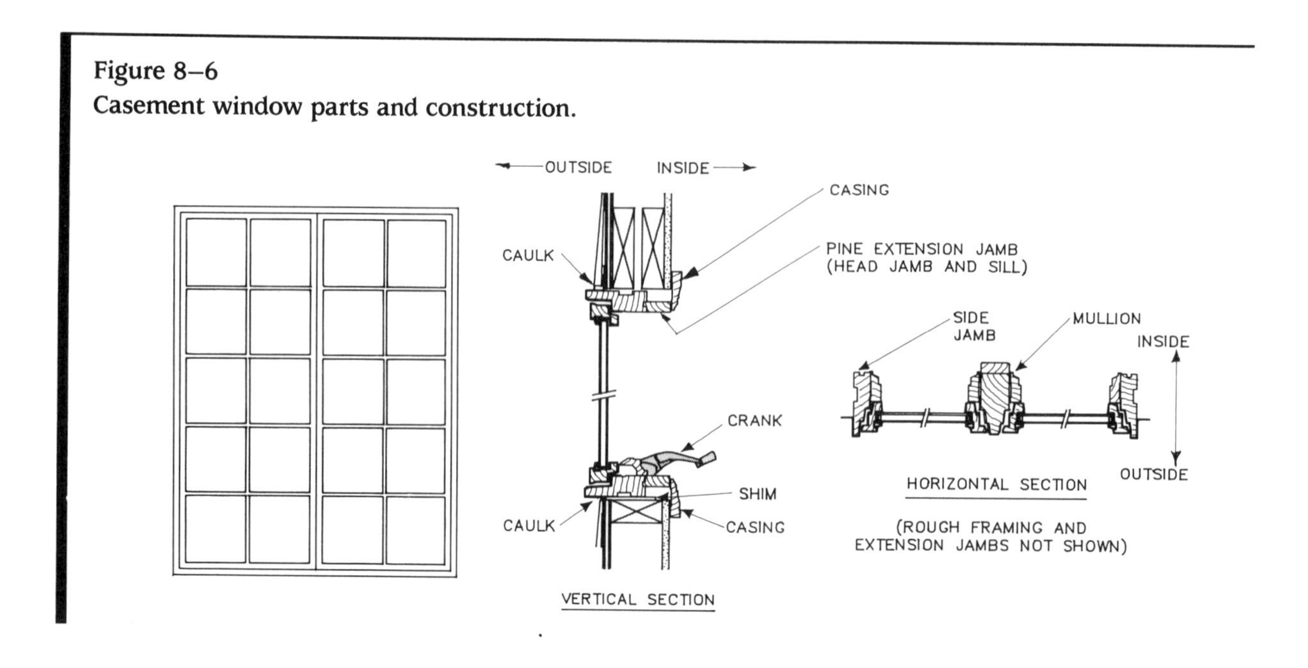

Figure 8–7
Bow window.

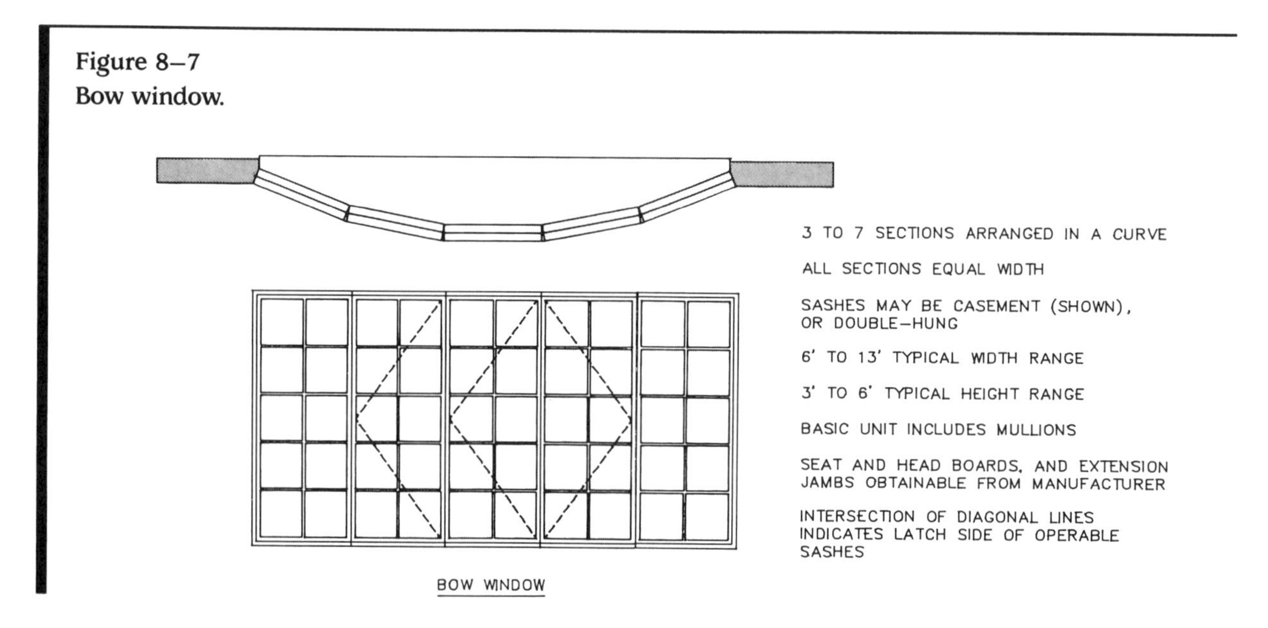

frames at the sides of the window frame. The bottom edge of each slat overlaps the top of the slat below by about ½". The frames are linked together and open simultaneously, controlled by a crank. Jalousie windows are not weathertight and are generally used only in warm areas. They provide superior ventilation: the glass slats can be positioned to direct air up or down and can regulate the amount of air admitted.

3 TO 7 SECTIONS ARRANGED IN A CURVE
ALL SECTIONS EQUAL WIDTH
SASHES MAY BE CASEMENT (SHOWN), OR DOUBLE-HUNG
6' TO 13' TYPICAL WIDTH RANGE
3' TO 6' TYPICAL HEIGHT RANGE
BASIC UNIT INCLUDES MULLIONS
SEAT AND HEAD BOARDS, AND EXTENSION JAMBS OBTAINABLE FROM MANUFACTURER
INTERSECTION OF DIAGONAL LINES INDICATES LATCH SIDE OF OPERABLE SASHES
BOW WINDOW

Bow and bay windows. In a bow window (Figure 8–7), several units are arranged to curve outward from a wall; in a bay window (Figure 8–8), a center unit parallels the house wall with an angled unit at each side. (If the side units are angled 90°, the bay window becomes a box window.) The floor can be extended out with the window, or a window seat may be built. Both bays and bows can be purchased as complete win-

Figure 8–8
Bay window.

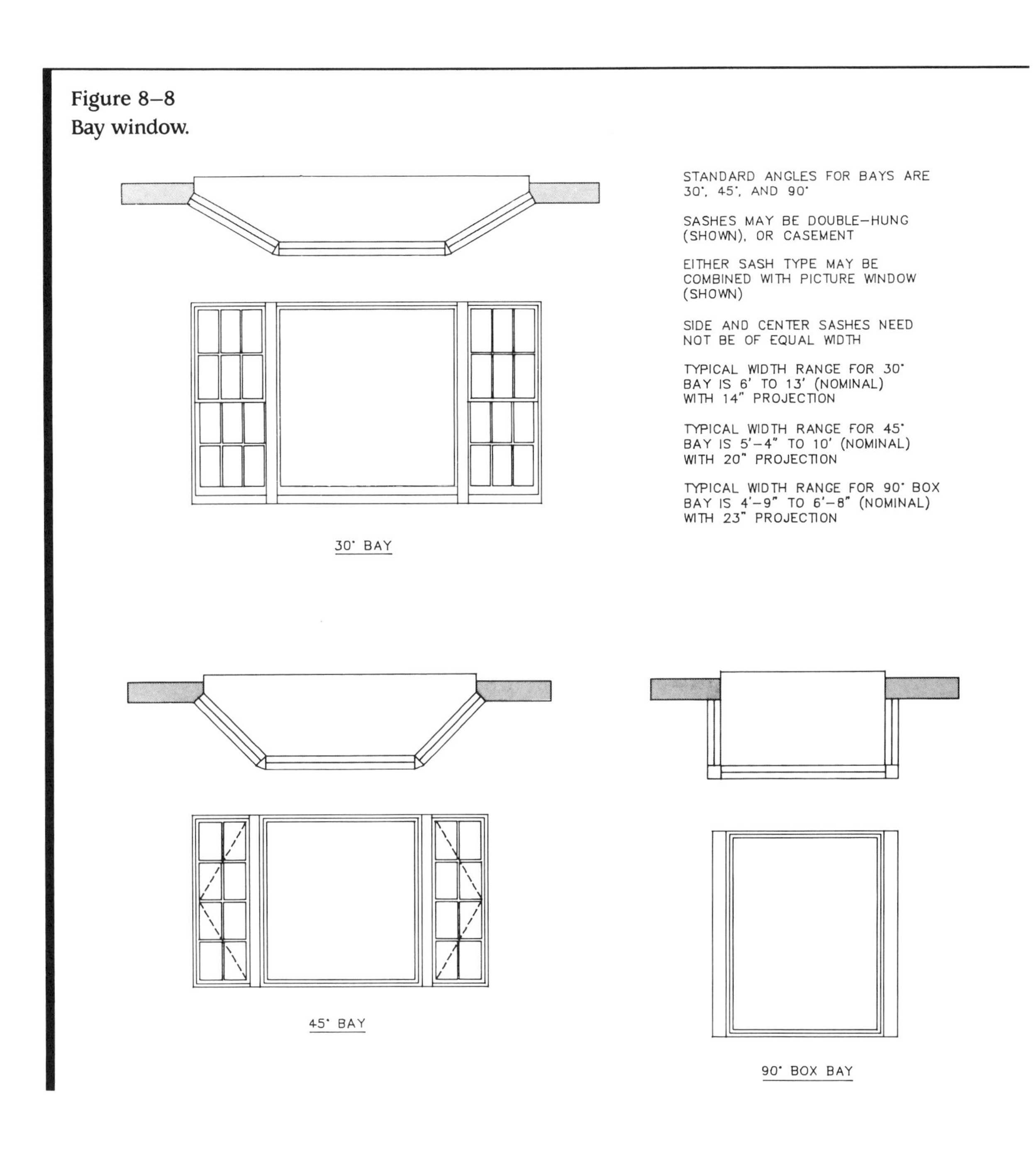

dows or assembled from individual windows—casement, double-hung, or fixed-sash—to fit a nonstandard opening. Bows and bays are very attractive windows. They are appropriate for Georgian and Colonial-style houses, but can be used anywhere. Bow windows are usually made with fixed glazing—the glass is sometimes mounted directly in the frame without separate sashes.

STANDARD ANGLES FOR BAYS ARE 30°, 45°, AND 90°
SASHES MAY BE DOUBLE-HUNG (SHOWN), OR CASEMENT
EITHER SASH TYPE MAY BE COMBINED WITH PICTURE WINDOW (SHOWN)
SIDE AND CENTER SASHES NEED NOT BE OF EQUAL WIDTH
TYPICAL WIDTH RANGE FOR 30° BAY IS 6' TO 13' (NOMINAL) WITH 14" PROJECTION
TYPICAL WIDTH RANGE FOR 45° BAY IS 5'-4" TO 10' (NOMINAL) WITH 20" PROJECTION
TYPICAL WIDTH RANGE FOR 90° BOX BAY IS 4'-9" TO 6'-8" (NOMINAL) WITH 23" PROJECTION
30° BAY
45° BAY
90° BOX BAY

WINDOW MATERIALS AND CONSTRUCTION

The traditional window frame and sash material is wood, and quality windows are still being made of all wood. Steel frames have also been around a long time, principally for casement windows, but they present problems of rust and repair. Preservation of old steel frames is difficult at best.

Wood as a window frame and sash material requires periodic painting, and even with regular maintenance is liable to rot and require replacement or extensive repair. Aluminum windows were introduced as an improvement that would eliminate the rotting and the need for maintenance.

But aluminum windows have their own problems. While wooden window frames provide a small amount of thermal insulation, aluminum is one of the best conductors of heat and cold, with the result that aluminum windows are cold to the touch and are usually dripping wet in winter from moisture condensation on their inside surfaces. Ice buildup is not unusual. Thermal breaks—nonmetallic material placed between the inside and outside halves of the aluminum frames—can reduce the heat conduction and condensation.

Figure 8–9
Vinyl window construction.

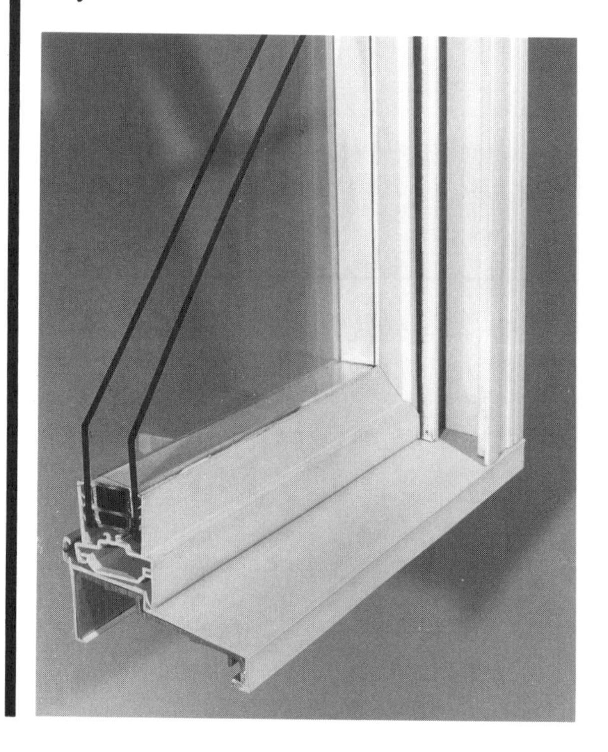

The next window material to arrive was vinyl, initially in the storm window and replacement window market. Frame and sash extrusions are easily cut and mitered for assembly into windows that are custom-dimensioned for each existing rough opening (Figure 8–9). However, even with careful extrusion design, many vinyl windows do not have the solid feel of wood windows.

Vinyl and wood have been combined in a construction that provides the low-maintenance, no-painting features of vinyl and the traditional solid window construction of wood. The wood frame and sash are clad in a thin vinyl casing.

Prime windows are windows manufactured in wide ranges of standard sizes for use in new construction or as replacement windows if a reasonable fit in the rough opening can be obtained. An aluminum or vinyl replacement window is custom-assembled from stock extrusions to fit an existing rough opening. A new replacement window should always provide a good visual match to old windows remaining on the house.

Windows are a major source of winter heat loss, however well made. The loss is through the glass and frames (conduction loss) and also through the cracks between the sashes and where the sashes fit into the frame (infiltration loss). Weatherstripping can be used to minimize the infiltration loss by blocking the cracks. Double glazing—adding a second pane of glass separated from the initial pane by dead air—will reduce the conduction loss. Double glazing can be achieved by mounting both panes in the same sash or by adding a second sash system—a storm window. A separate storm sash has the advantage of also decreasing the infiltration loss, but it is an inconvenience when you want to open the window. And unless it is a double-hung unit, it usually must be taken down and stored in the temperate part of the year.

REPLACING A WINDOW

Replacing a window is a project that is easy to put off. It often seems to be less work to caulk, paint, wedge, and ignore the rot and warp in the old window one more time. But eventually you must face the fact that besides all their maintenance problems, rattling, and sticking sashes, old windows run up the heating and air-conditioning bills.

Installing a new window is a task that must be done right. If the window frame isn't square, the sashes will bind. If care isn't taken with caulking and fastening the window frame to the house, the new window will be no more watertight than the old one—maybe even worse.

While most of the installation can be done from the inside, there is some outside work. For that, you will need a ladder or ladders, or scaffolding, to get you safely into position to work on the window.

The first step is to find out exactly what size window is required. Rip off the *inside* casing at the sides and top, and the apron under the stool. If the stool is a separate piece from the sill, remove it too. (See the window component labels in Figure 8–4.) This will expose the rough opening and the window frame so you can take measurements. To determine what size window you should buy, measure both the rough opening and the existing window as follows.

If you are replacing only one window, you will want the glass size of the new window to match the glass size of the rest of the windows on that side of the house. Glass size is given per sash, ignoring any muntins dividing the glass area. You cannot order a window, prime or replacement, by glass size alone; you must also specify frame dimensions.

The unit, or frame, dimensions are the height and width to the outside edges of the window unit, excluding any flanges or attached outside casing (Figure 8–10). The rough opening is measured between the surfaces of the studs, and between the header and sill in the wall framing. Several measurements should be taken and the narrowest and shortest used.

If the window sashes are counterbalanced by weights on ropes or chains at each side of the window frame, measure the width over the side jambs, not over the weight pockets if they are integral with the window. The area between the window jambs and the rough opening must be framed in.

If the window sashes are counterbalanced by visible springs and not by weights, measure the width of the rough opening at its narrowest point and the height at its minimum to determine the maximum-size window you should order. The ordered unit should be ½″ narrower than the rough opening and ¾″ shorter.

When you go to order or buy your new window, take all the measurements with you—rough opening, frame, and glass. Be sure you and the

Figure 8–10
Parts of a double-hung window and dimensioning conventions.

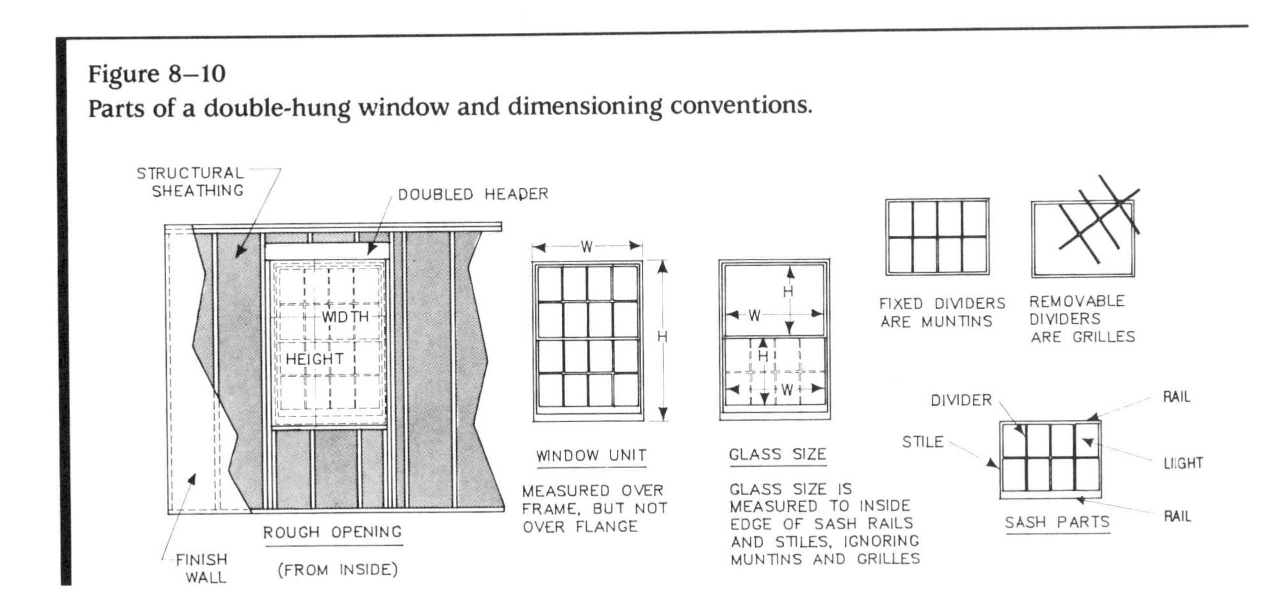

dealer understand each other. A wrong-size window can be a real problem. Some stock windows come with mounting flanges or with the outside casing already mounted. While either makes installation in a framed wall simpler, installation in a masonry wall may be more difficult. If your window is to go into a masonry wall opening, take along a sketch of how the old window was attached.

Prime window manufacturers do not all offer the same range of stock sizes. If you can't find the exact size you want, try another retailer and brand. You can order wood windows in custom sizes from millwork shops, but be sure that you know what the price will be when you place the order.

Replacement aluminum and vinyl windows are ordered in custom sizes. Don't look for the cheapest windows that you can find. Select quality.

Installation

Although all the accompanying illustrations show double-hung windows, the installation information is equally applicable to casement, awning, and fixed windows.

Trimming the opening. If the window you removed had weights, you will have to install 2 × 4 trimmers to fill the voids where the weights formerly hung. Block them out to provide the cor-

Figure 8–11
The old, tall windows in this house were candidates for replacement. (Photo courtesy Andersen Corporation)

Figure 8–12
Throughout the house the openings were resized to take modern replacement units with rigid-vinyl sheath covering the sill and frame and with traditional wooden sashes. (Photo courtesy Andersen Corporation)

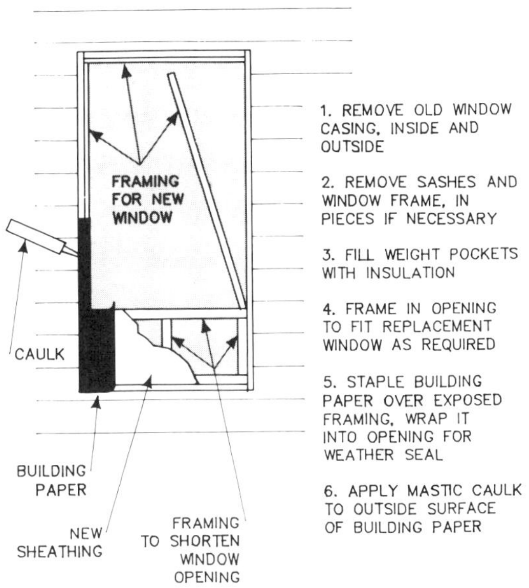

rect opening size for your replacement window with ½″ clearance.

Changing the rough opening size. If you want to change the size of the window, parts of the opening can be framed in. Before you do this, consider how you will match the existing siding. As shown in Figures 8–11 and 8–12, a tall window typical of an older house can be replaced by a shorter window by adding new sill framing. The method of reframing the opening is explained in Figure 8–13.

Installing a flanged window. A flanged window is easiest to install. After a trial fit in the opening, put a bead of caulk on the back side of the flange and press the window in place (Figure 8–14). Wedge the window and tack-nail one corner. Plumb the window, then complete the nailing. Follow the same procedure for a window with outside casing attached.

Figure 8–13
Resizing a window opening to accept a new unit.

FRAMING FOR NEW WINDOW
CAULK
BUILDING PAPER
NEW SHEATHING
FRAMING TO SHORTEN WINDOW OPENING
1. REMOVE OLD WINDOW CASING, INSIDE AND OUTSIDE
2. REMOVE SASHES AND WINDOW FRAME, IN PIECES IF NECESSARY
3. FILL WEIGHT POCKETS WITH INSULATION
4. FRAME IN OPENING TO FIT REPLACEMENT WINDOW AS REQUIRED
5. STAPLE BUILDING PAPER OVER EXPOSED FRAMING. WRAP IT INTO OPENING FOR WEATHER SEAL
6. APPLY MASTIC CAULK TO OUTSIDE SURFACE OF BUILDING PAPER

Figure 8–14
Installing a flanged window unit.

1. NAIL WINDOW FLANGE AT ONE UPPER CORNER
2. PLUMB WINDOW WITH CARPENTER'S LEVEL AND TACK-NAIL OPPOSITE LOWER CORNER. RECHECK PLUMB BEFORE COMPLETING NAILING

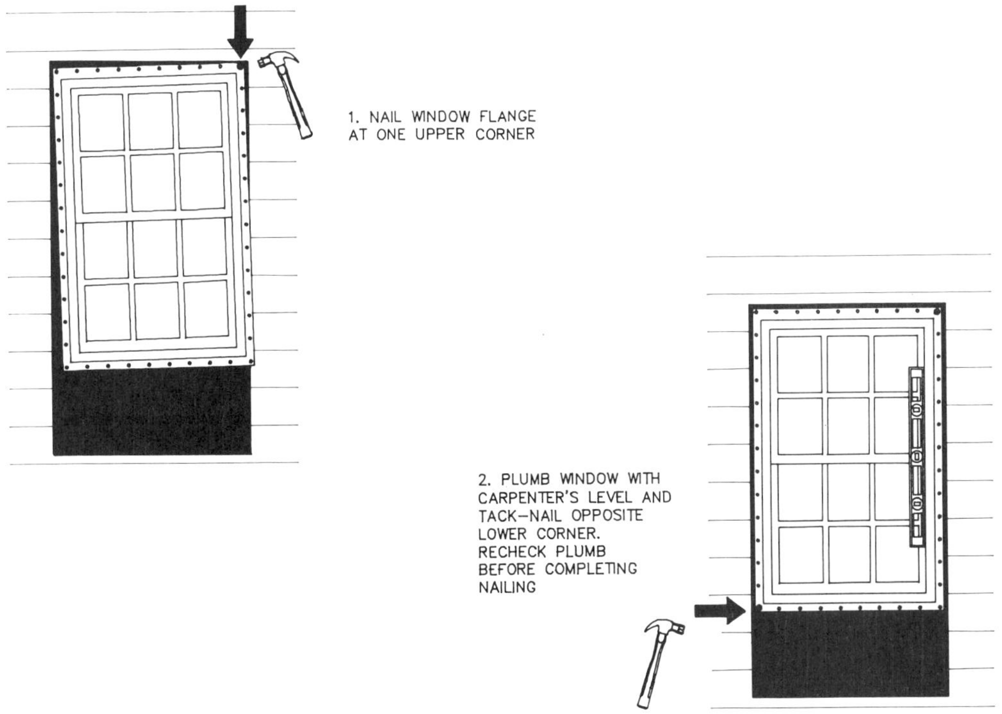

Trimming the window. First, install the flashing and drip cap molding supplied with the window unit. Follow the directions with the unit and see Figure 8–15. Next, install casing at the sides. Third, apply matching siding below the window if the opening has been resized to take a smaller window. Finally, caulk all seams between the window trim and the siding at the top, bottom, and sides.

Figure 8–15
Trimming an installed replacement window.

INSTALL FLASHING AND DRIP CAP MOLDING ABOVE WINDOW
INSTALL CASING TO CLOSE GAP BETWEEN WINDOW AND SIDING
CAULK ALL SEAMS THOROUGHLY
APPLY MATCHING SIDING BELOW WINDOW

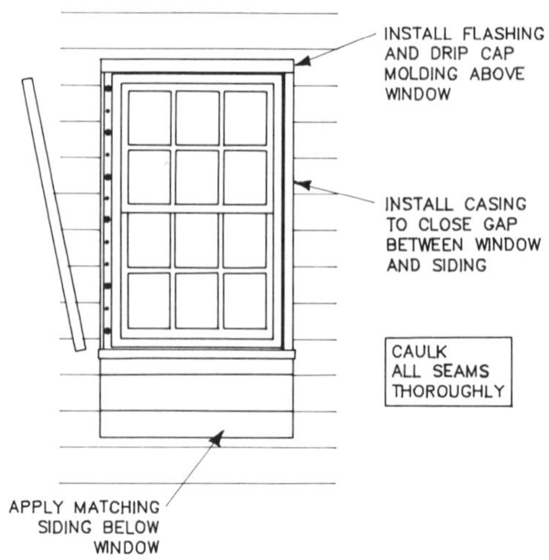

9 Double-Hung Window Repair

Many things can go wrong with a wood double-hung window as it ages. Wood swells and warps, causing sashes to bind. Or old, dry wood shrinks, and loose sashes rattle in the wind. Conscientious painting through the years can leave sashes glued tight shut. Sash cords and chains break; sash pulleys jam.

Should you repair or replace a troublesome window? Unless the wood is hopelessly rotted, all these conditions can be corrected without replacing the window. This chapter tells you how.

FIXING A STUCK SASH

Perhaps the most common problem is a sash stuck firmly shut. There are many possible causes: warping, swelling, paint, dirt in the channels, too much weatherstripping—someone in the past may even have nailed or screwed the sashes together. Figure 9–1 identifies and locates possible problem areas; Figure 9–2 shows the parts of a wood double-hung window.

First, be sure the sashes are not locked together by a nail or screw. Work a putty knife blade between the sticking sash and the parting and stop strips. Go all around both sides of the sash and be sure the blade gets all the way to the surface of the frame—tap the handle of the putty knife with a hammer if necessary. The sash may have been painted shut a long time, or a locking screw or nail may have been covered by paint decades ago. If you find a nail or screw, cut it with a hacksaw blade inserted between the sashes.

If the sash is still stuck, try tapping around the edge of its frame with a hammer. Use a block of wood to protect the sash from hammer tracks. The vibration may get the sash loose. It may also crack the glass.

If that doesn't work try a prybar. Apply force only at the sides of the sash, working the sides alternately. Insert the prybar between two putty

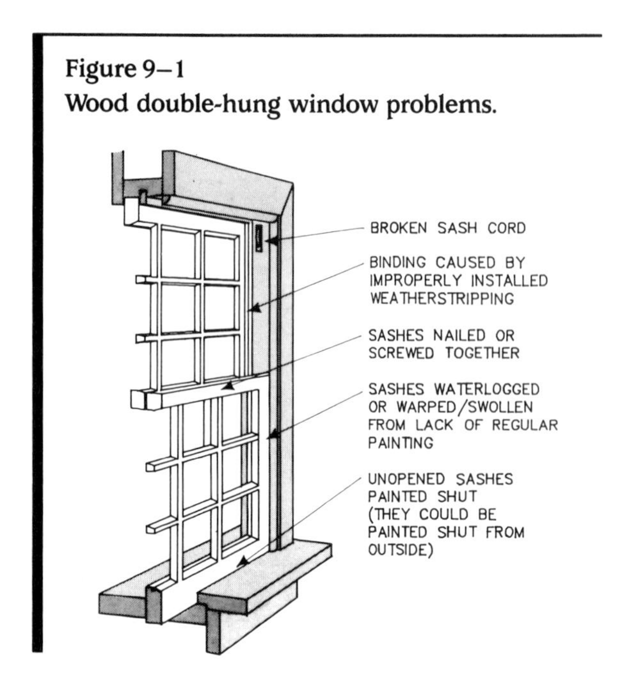

Figure 9–1
Wood double-hung window problems.

knife blades to protect the wood. Work near the top and bottom of each side first. Force applied at the center of the sash can break the rail or the glass.

If nothing else works, pry off the inside stop molding to get the lower sash out; or pry off the inside stop, remove the lower sash, and pry off the parting bead to get the top sash out. To remove the stop, insert a putty knife blade between it and the frame and pry. It is normally held in place only by coats of paint and small finishing nails. Getting the parting bead out can be more difficult as it is normally seated in a dado. Slit down the sides with a sharp blade, then try prying with a putty knife and pulling with a pair of pliers. You may have to settle for getting it out in pieces, then replacing it with new wood. Unless a sash is stuck fully open or fully closed, you usu-

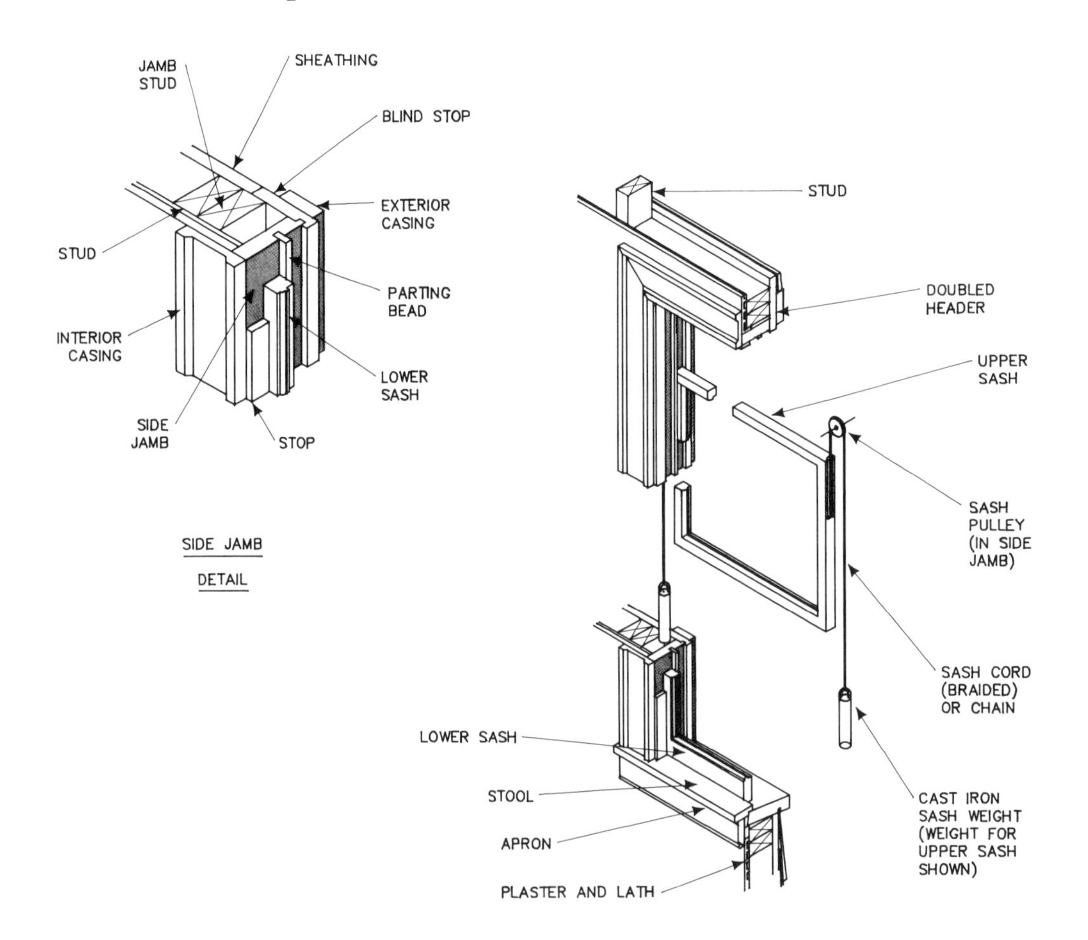

Figure 9–2
Parts of a wood double-hung window.

ally need to remove the stop and parting bead from only one side of the window.

Once the sashes are out you can scrape and sand encrusted paint. Sand or plane off excess wood if the sash frame has warped or swelled, but be careful not to remove too much wood or the sash will rattle when you put it back. You may find that all you have to do to eliminate the sticking is to reposition the inside stop. Use new nails.

If the sash or frame wood has rotted, trim the rot back to solid wood and rebuild the sash with epoxy filler made for the purpose.

If you have cleaned the paint off the sashes, repaint them while they are out of the frame so as to protect all surfaces. Also scrape or sand and repaint the frame and tracks before you reassemble the window. Board up the window opening for the time it takes for the paint to dry thoroughly.

If too much weatherstripping is causing the sash to bind, replace it with a thinner type, or sand or move the stop strips to allow more clearance.

REPLACING SASH CORDS AND WEIGHTS

There are several solutions to the problem of broken sash cords or chains. If you want to continue using the weights to counterbalance the sash you must be able to get at the weights. The stiles usually have removable cover plates for that purpose near the bottom, fastened with screws top and bottom (Figure 9–3). But they may be painted or nailed shut, or the carpenter may have omitted the cover plates (they took shortcuts in those days too).

If you can see the outline of the cover plate, you ought to be able to get it off—if no other way, in pieces. Use a utility knife to break the paint seal. If the cover plate has been nailed from the side, cut the nails with a hacksaw blade. If there are no cover plates to provide access to the weight pockets, you have the choice of chopping through or counterbalancing your sashes some other way.

Before you replace cords or chains, check the pulleys (Figure 9–4). If they don't turn freely, remove, clean, and lubricate them. The pulleys are usually held by screws accessible from the

Figure 9–3
Sash weight pocket covers may have rabbeted ends, as shown, or miter-cut ends.

LOCATE WEIGHT POCKET ACCESS COVER, REMOVE RETAINING SCREW AND TAKE OFF COVER

Figure 9–4
Sash weight pulleys are located at the top of the jamb on each side of the window.

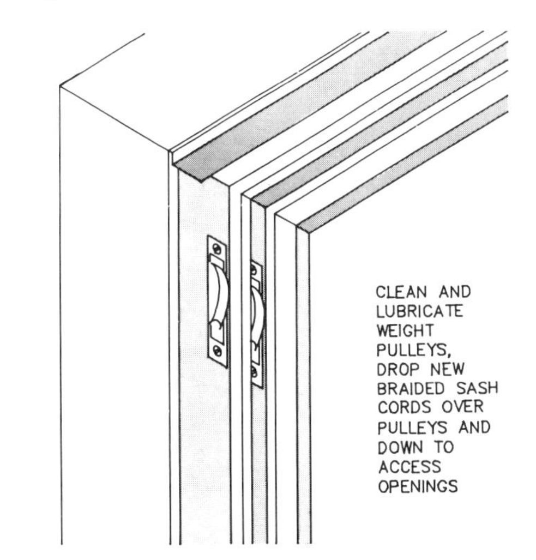
CLEAN AND LUBRICATE WEIGHT PULLEYS, DROP NEW BRAIDED SASH CORDS OVER PULLEYS AND DOWN TO ACCESS OPENINGS

face of the frame. Scrape away the old paint to find them.

Figure 9–5 shows how to install a new sash cord. The procedure is the same with a sash chain, but the chain can be secured to the sash with a screw inserted through one link. Getting the new cord or chain the right length is important. If the cord is too long, the weights will hit bottom and not hold a lower sash fully open. If it is too short, the sash cannot close completely.

There are alternatives to sash weights, as shown in Figure 9–6. One type of sash balance acts like a tape measure and is mounted in the hole where the pulley was mounted. The tape is pulled down and attached to the side of the sash. Tension is adjustable. One unit is required at each side.

A torsion-spring sash balance operates by winding a helically coiled spring when the sash is lowered. The upper end of each balance is fastened to the window frame stile. The lower end is attached to the bottom of the sash on each side. As the sash is pulled down, a spiral strip causes the lower end of the helical spring to turn, winding it up. It is this torsion that provides the counterbalancing action. To use this type of balance, the sides of the sash must be recessed (ploughed or routed) for clearance.

Figure 9–5
Replacing a sash cord. The procedure is the same with a sash chain.

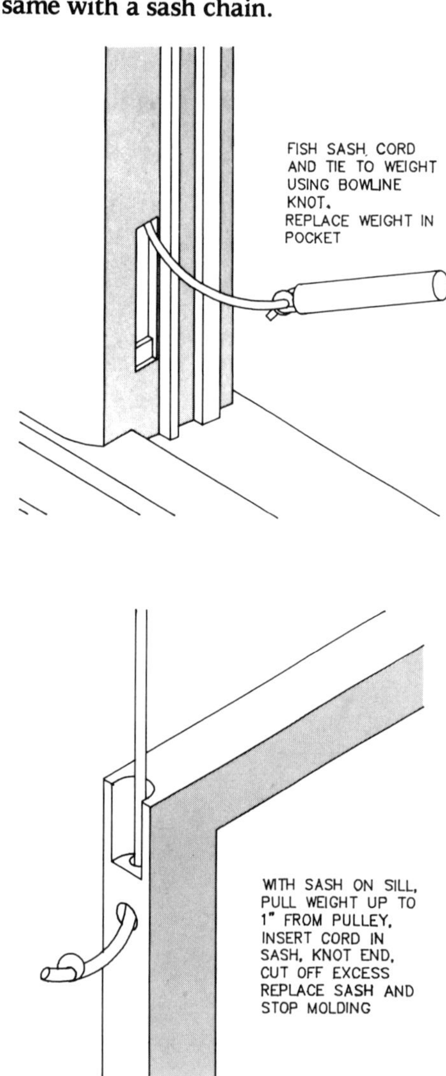
FISH SASH CORD
AND TIE TO WEIGHT
USING BOWLINE
KNOT.
REPLACE WEIGHT IN
POCKET
WITH SASH ON SILL,
PULL WEIGHT UP TO
1" FROM PULLEY,
INSERT CORD IN
SASH, KNOT END,
CUT OFF EXCESS
REPLACE SASH AND
STOP MOLDING

FIXING A LOOSE SASH

A sash loose in its channels rattles, it lets rain and cold drafts into the house, and it may not hold a position up or down unless locked or blocked. In the winter, when the problem is most noticeable, shimming the channels to eliminate the looseness can lead to trouble because you may have a hopelessly bound-up window in the summer. Fortunately, there is a quick fix available: sash holders (Figure 9–7). These are formed metal strips that are placed between the side of the sash and the stile. They operate by wedging the sash against the other stile. One type fastens to the center or meeting rail of each sash; another type fastens to the jamb, as shown.

A second way to deal with loose sashes is to mount them in patented side channels manufactured by Quaker City Co. (Figure 9–8). These readily available replacement channels offer a good way to replace sash cords and weatherstrip the sides of the sashes in one shot, but they work only for sashes 1⅜" thick separated by a ½" parting bead. These dimensions are standard on today's wood windows, but check your window dimensions before you buy—particularly if you have an old house.

Channel construction and installation are

shown in Figure 9–9. To install the channel, you must first remove both sashes. Temporarily remove the inside stop. If the lower sash is held by weights, cut the cords or chains and lower the weights. If you can open the pockets, remove the weights. After the lower sash is out, permanently remove the parting bead, then remove the upper sash (and its weights, if accessible).

Figure 9–6
Sash weight alternatives.

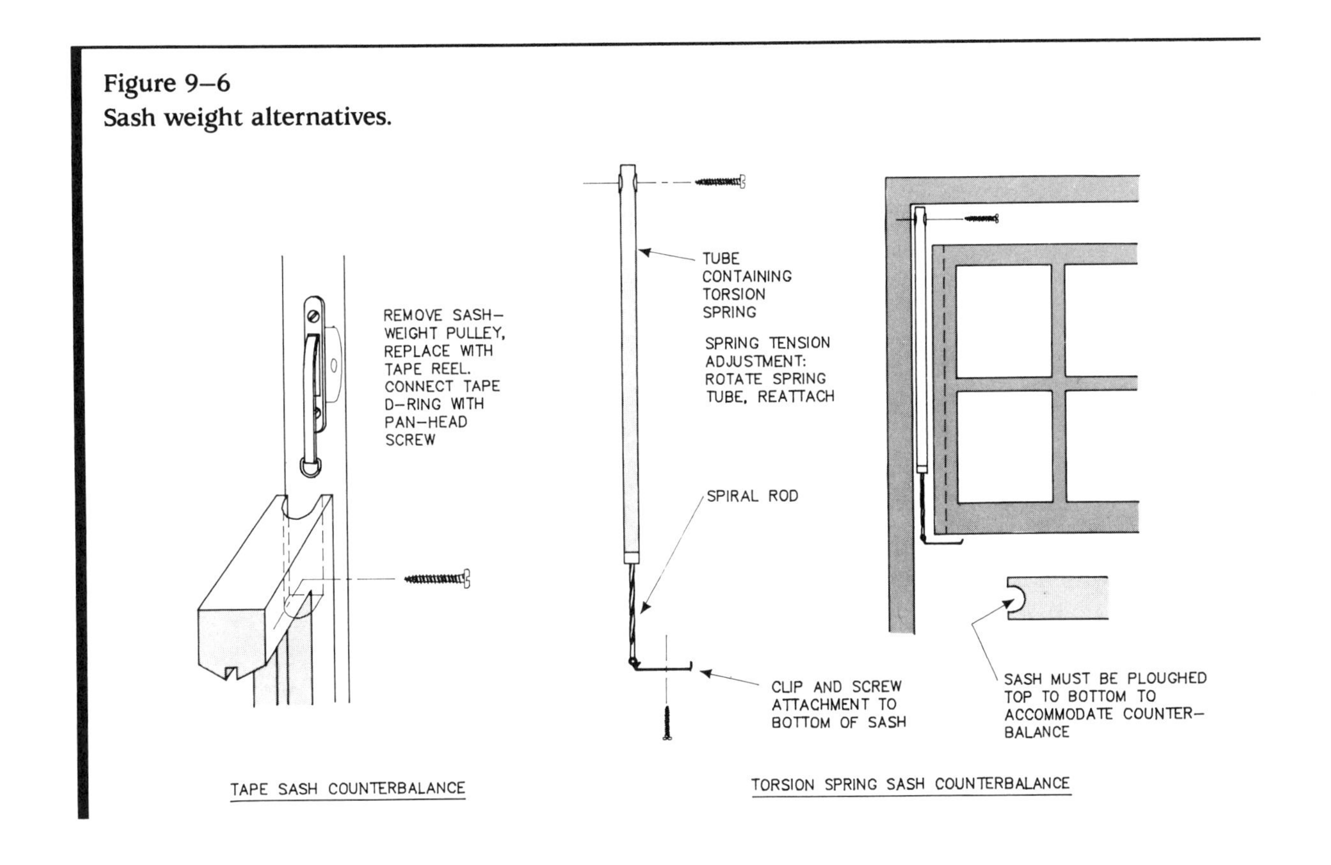
REMOVE SASH-WEIGHT PULLEY, REPLACE WITH TAPE REEL. CONNECT TAPE D-RING WITH PAN-HEAD SCREW
TUBE CONTAINING TORSION SPRING
SPRING TENSION ADJUSTMENT: ROTATE SPRING TUBE, REATTACH
SPIRAL ROD
CLIP AND SCREW ATTACHMENT TO BOTTOM OF SASH
SASH MUST BE PLOUGHED TOP TO BOTTOM TO ACCOMMODATE COUNTER-BALANCE
TAPE SASH COUNTERBALANCE
TORSION SPRING SASH COUNTERBALANCE

Figure 9–7
Spring-clip sash holders. Sash holders are a quick and easy way to correct loose, rattling sashes.

SASH-MOUNTED SASH HOLDER
INSERT BETWEEN SASH AND SIDE JAMB, SECURE TO SASH WITH SCREW
JAMB-MOUNTED SASH HOLDER
PRONGS POSITION STAINLESS-STEEL HOLDER, USE SCREWS TO SECURE

Figure 9–8
Replacement side channels can eliminate a number of window problems.

Remove the pulleys, because they will be in the way. Weight pockets are a major source of heat loss around old windows. If the pockets are open below, replace the cover plates, then pour in vermiculite through the pulley openings. Finally, stuff the top with scrap glass or mineral fiber insulation. Scrape off any excess paint in the channel area and trial-fit the replacement channels. You may have to trim the ends of the top parting bead for clearance. If the channels are too long, trim from the slanted bottom only. The amount of trimming you can do is limited. Follow the instructions closely.

The sashes and channels are installed together as shown in Figure 9–10. Stand the sashes on a chair, place the channels on either side, then pick up and tip the sash and channels into the window, bottom first. Align the channels vertically, check their operation, and nail them top and bottom. Replace the inside stop to finish the job. Be careful not to nail them too tight.

Sash holders and Quaker City replacement channels are widely available at home centers, lumberyards, and hardware stores. The torsion

Figure 9–9
Construction and installation of replacement side channels.

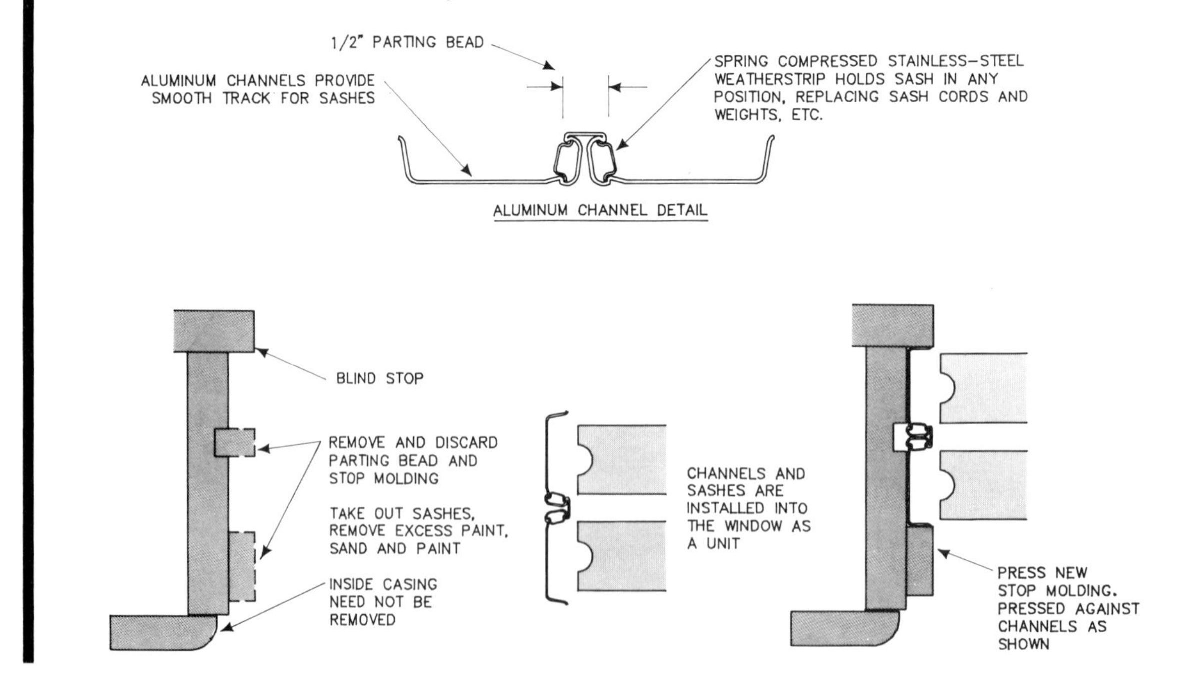

Figure 9–10
Sashes are inserted in the channels and everything is tipped into position at the same time.

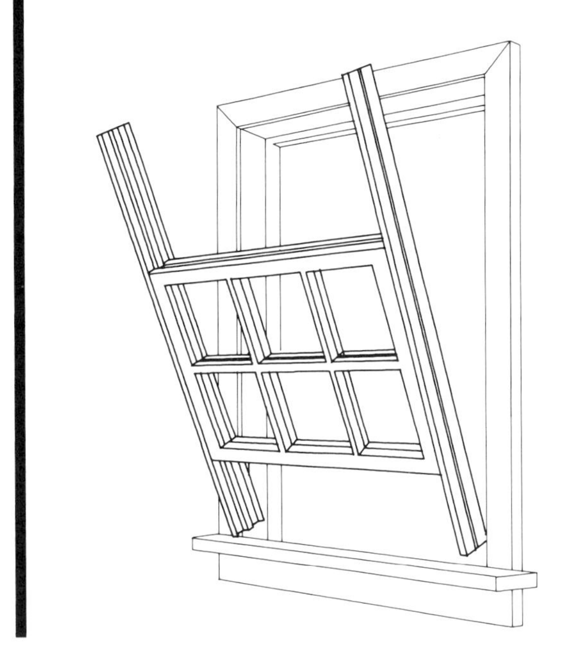

(Spinex) counterbalance and the "tape measure" type sash balance may require some hunting. Sash weights may be available at millwork shops, where they are probably wondering what to do with them. Sash cords should be braided, not twisted: a twisted cord will unwind.

10 Interior Trim

Interior trim consists primarily of moldings around windows, doors, walls, and ceilings. Moldings, more than any other detail of construction, determine the style of your home (Figure 10–1). The unique beauty of Early American and Victorian houses is attributable as much to the moldings used as to any other feature. Examples of how Victorian moldings can be reproduced today using stock moldings are shown in Figures 10–2 and 10–3.

In remodeling, as in new construction, wood moldings also serve a second function: they cover and finish off the seams, gaps, rough edges, and misfits inherent in most construction.

Until around 1850, moldings were made using wood molding planes—a carpenter might have thirty different patterned planes in his tool chest. Moldings were cut by hand on the job. Then, moldings began to be mass-produced and the carpenter's handcrafted profiles were replaced by standardized patterns.

Today, the Western Wood Moulding and Millwork Producers—a trade association—identifies stock moldings by "WM" number in a published pattern book. Each number identifies both the

Figure 10–1
Oak cornice assembled on wall using stock crown and dentil moldings combined with square-edged stock.

Figure 10–2
Victorian oak dentil crown molding made from purchased parts.

size and shape of a molding profile. Examples are shown in Figure 10–4. The profiles are all drawn to the same scale.

The widest variety of moldings is obtainable in unfinished pine and oak. Many common stock moldings are also available in lauan (Philippine mahogany), and in prefinished pine, or plastic. Moldings in any wood can be made up for you in cabinet and architectural woodworking shops, but they will be expensive.

BUYING MOLDING

When estimating your needs for a room, don't just get the total footage; instead, record the

Figure 10–3
Home-built cornice and doorway pediment combining stock moldings and shop-made framing and decoration.

Figure 10–4
Examples of the wide variety of stock moldings available.

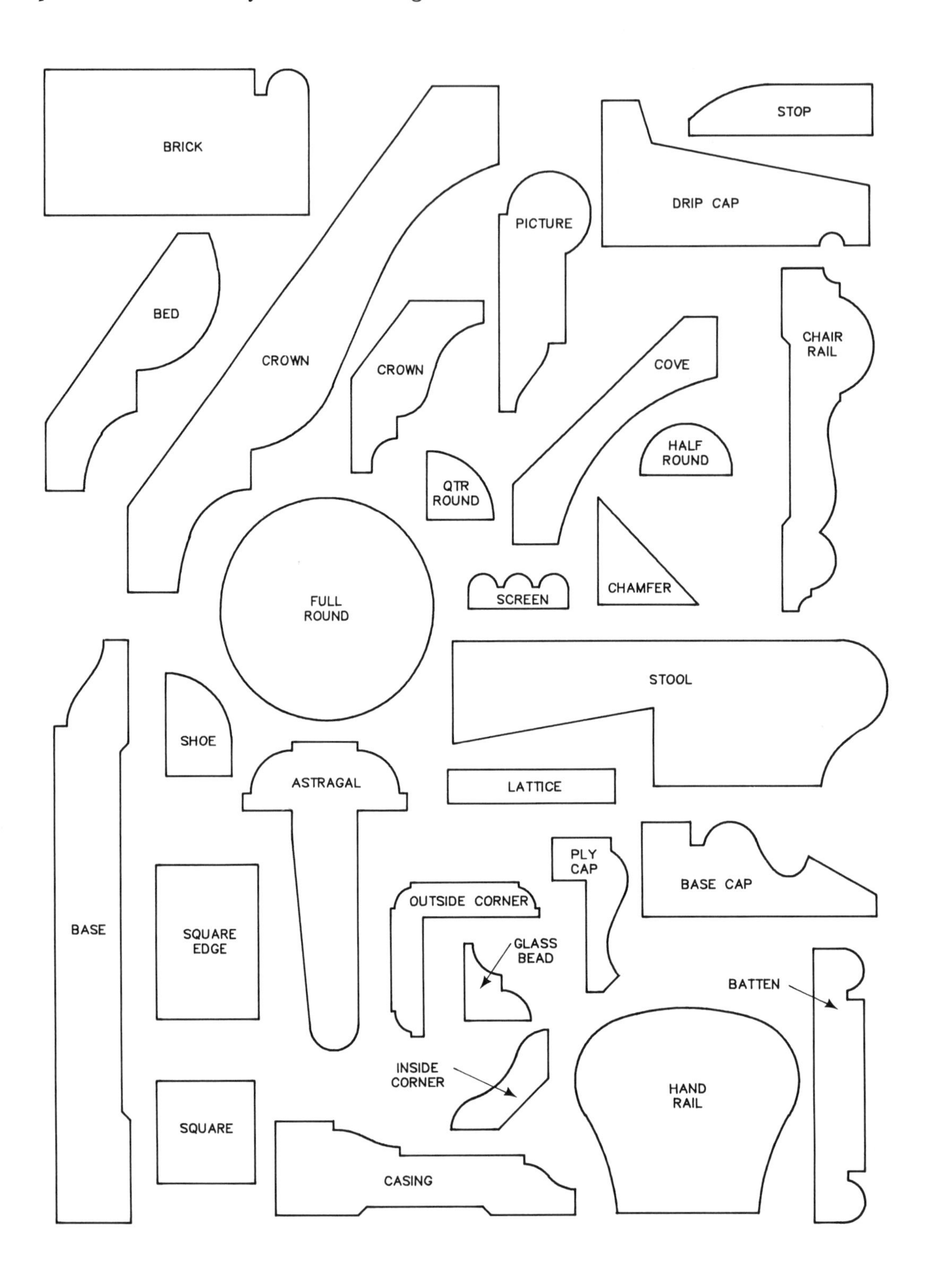

individual lengths of the pieces you will need. Whenever possible, you want to do each side of the room with a continuous piece. (There are times when this won't work; see "End-Joining Molding" later in this chapter.) When you buy molding, inspect every foot of it. Industry grading standards allow some defects. It may be possible to fill some defects before you paint, but for the cost of molding today you should select perfect or near-perfect pieces.

If you will be mitering outside corners, don't forget to allow extra length equivalent to the molding width for each miter cut. And don't calculate your lengths too closely—everybody makes mistakes mitering, and a foot long is far better than an inch short.

Molding size is described by thickness, width, and length, in that order. Thickness is the actual maximum thickness of the finished molding; width is the extreme width (see Figure 10–5). Length quoted is to the nearest full foot. If the molding is 8′6″ long, it goes as 8′. In lumberyards, moldings are usually described by width only and the length you need—2¼ Colonial casing by 10′, for example.

If you are planning to match existing molding, take along a sample so you can be sure the profiles of the new and the old are exactly the same. Although nominally the same pattern number, moldings from two different sources are not likely to match exactly. Figure 10–6 shows typical variations. Moldings that don't match cannot be joined or mitered neatly.

Figure 10–5
How molding dimensions are measured in standard practice.

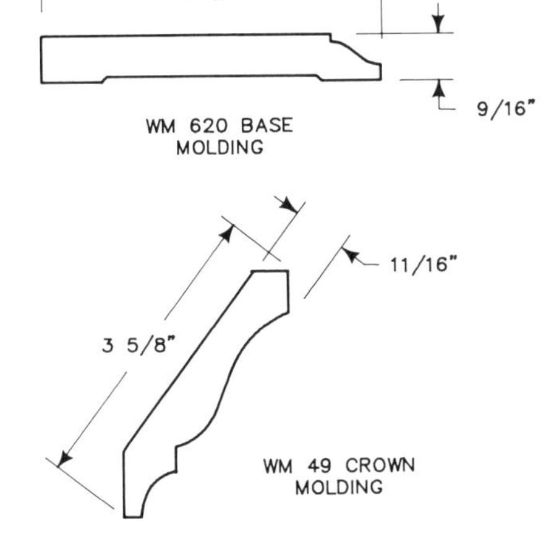

Figure 10–6
Differences in moldings nominally made to the same pattern number. The differences are not important unless you are joining moldings from different sources.

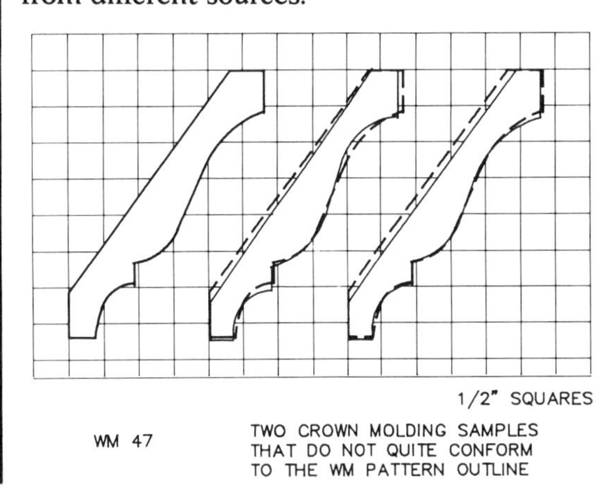

MOLDING PATTERNS

The molding patterns described and shown on the following pages are generally available anywhere in the United States, although probably no lumberyard in your area will carry all of them.

Ceiling Moldings

Crown, bed, and cove moldings. Many types and combinations of moldings are used to cover or decorate wall-to-ceiling joints. They give the room a finished look and provide a pleasant transition from one surface to the other. Crown, bed, and cove moldings are the most commonly used ceiling moldings. All three moldings are made in a wide range of sizes. Crown, bed, and cove moldings are always installed "sprung"—diagonally positioned to bridge the ceiling-to-wall joint, as shown in Figure 10–7. Sprung moldings can be put in neatly in spite of wall

Figure 10–7
Ceiling moldings. A variety of moldings can be used to hide or decorate the joint between walls and the ceiling.

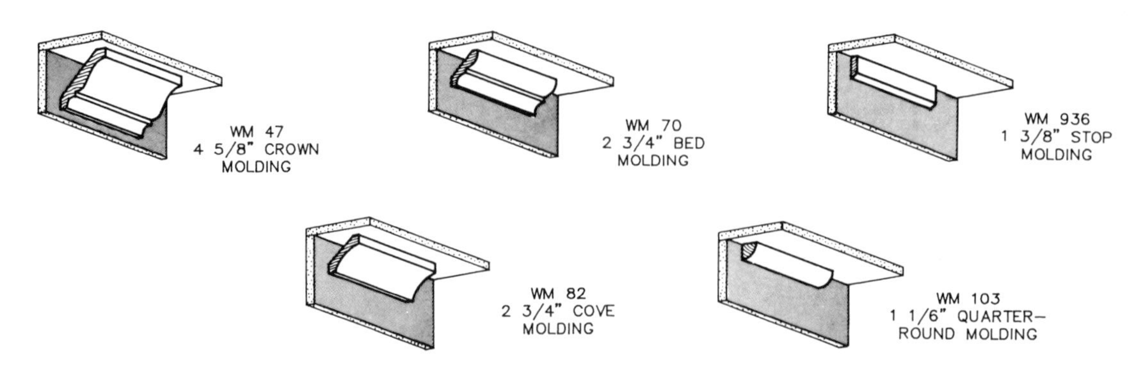

Figure 10–8
Crown, bed, and cove molding patterns.

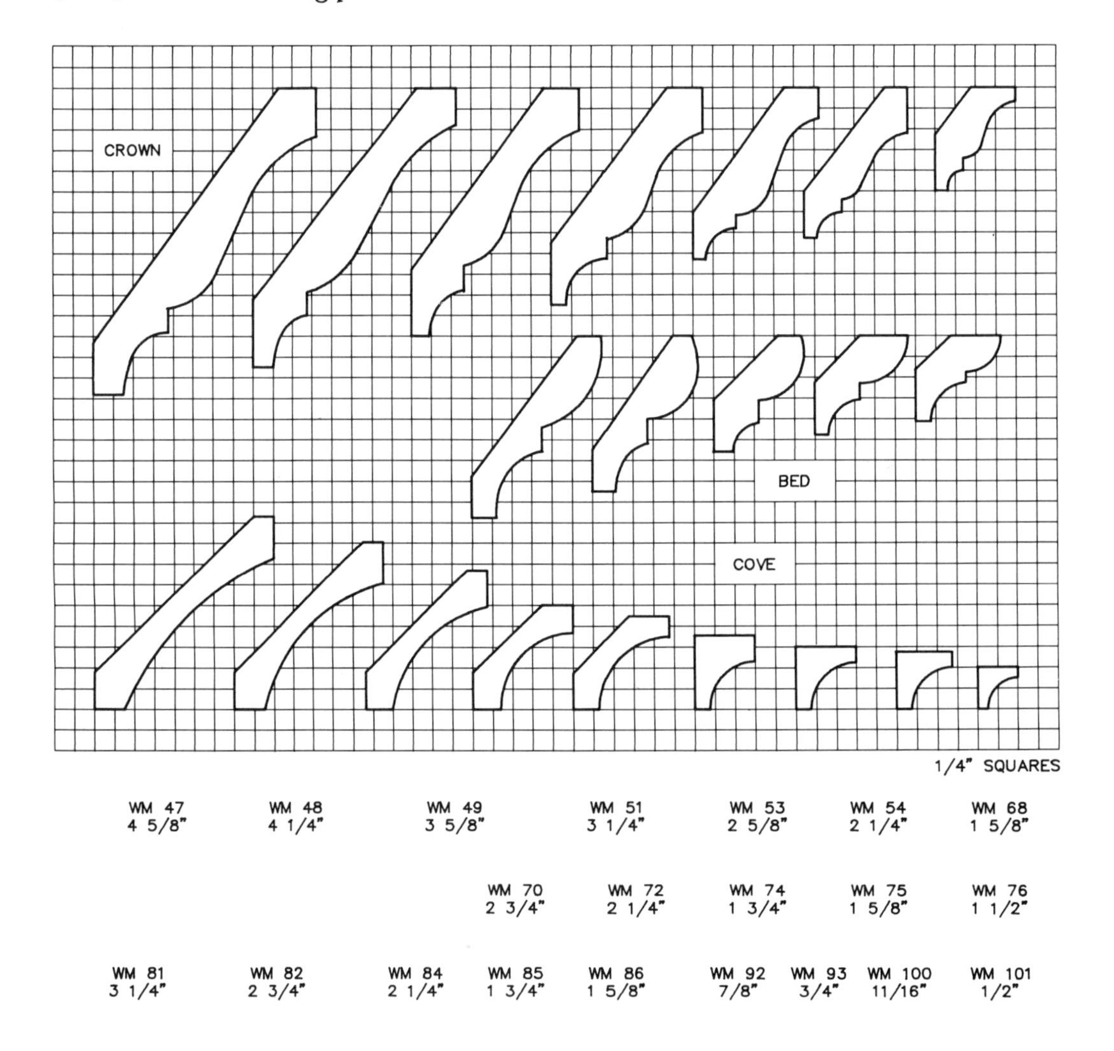

Figure 10–9
Cornices using combinations of stock moldings.

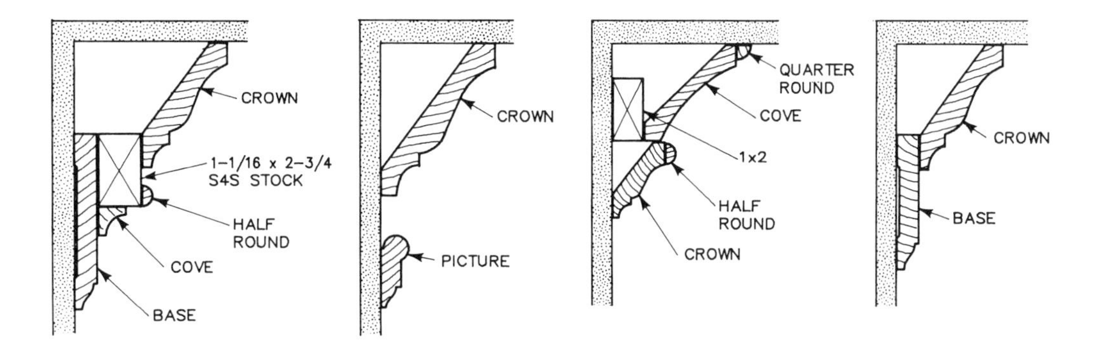

and ceiling irregularities. Patterns for crown, cove and bed moldings are illustrated in Figure 10–8.

Cornices. A cornice has been described as a crown molding that got out of hand. Simple and complex combinations of stock moldings can result in pleasing and ornate decoration at a wall/ceiling joint, at the top of any vertical part of the house (including partition paneling and screen paneling), or at the top of the exterior house wall.

Crown and stop, crown and Colonial base, crown, square edge, and base cap are but a few of the possible combinations (see Figure 10–9). Additional cornices can be made by combining stock moldings and shop-made decoration, as shown in Figure 10–10.

Chair Rails

One early function of moldings was to keep the backs of chairs from marring walls. Chair rails are an important detail in Colonial and traditional decor. Today they are usually just decorative, often providing a dividing trim between different wall coverings such as paint, wallpaper, and paneling. You can use available chair rail molding or make up more elaborate ones. Chair rails should be located 30" to 32" above the floor regardless of ceiling height (see Figure 10–11).

Baseboard

Baseboard commonly consists of three separate moldings: base molding, base shoe, and base cap (see Figure 10–12).

Base molding. The large section of a baseboard positioned tightly against the floor and extending up the wall is the base molding. It is nailed to the wall framing with 6d or 8d finish nails driven into the sole plate and studs as shown in Figure 10–13. Base molding is installed after the door casing has been nailed in place. It serves both as a decorative trim for the wall and to protect the wall from kicks, bumps, and scraping by brooms and vacuum cleaners. Often, base molding is used alone. Base molding is "backed out"—that is, the back of the molding is relieved (slightly cut out) to allow a tight fit against the wall at the top edge.

Base shoe. This molding resembles quarter-round molding except that its flat sides are not the same width. Base shoe is nailed at the bottom of base molding to conceal gaps between the base molding and an uneven floor. If wall-to-wall carpeting or a new finish floor is to be installed over an existing finish floor, the base shoe should be taken up first, then reinstalled to hide the edge of the new floor or covering.

Base cap. A baseboard can consist of only base and base shoe moldings. A more ornate base-

Figure 10–10
Construction details for the cornice shown in Figure 10–3.

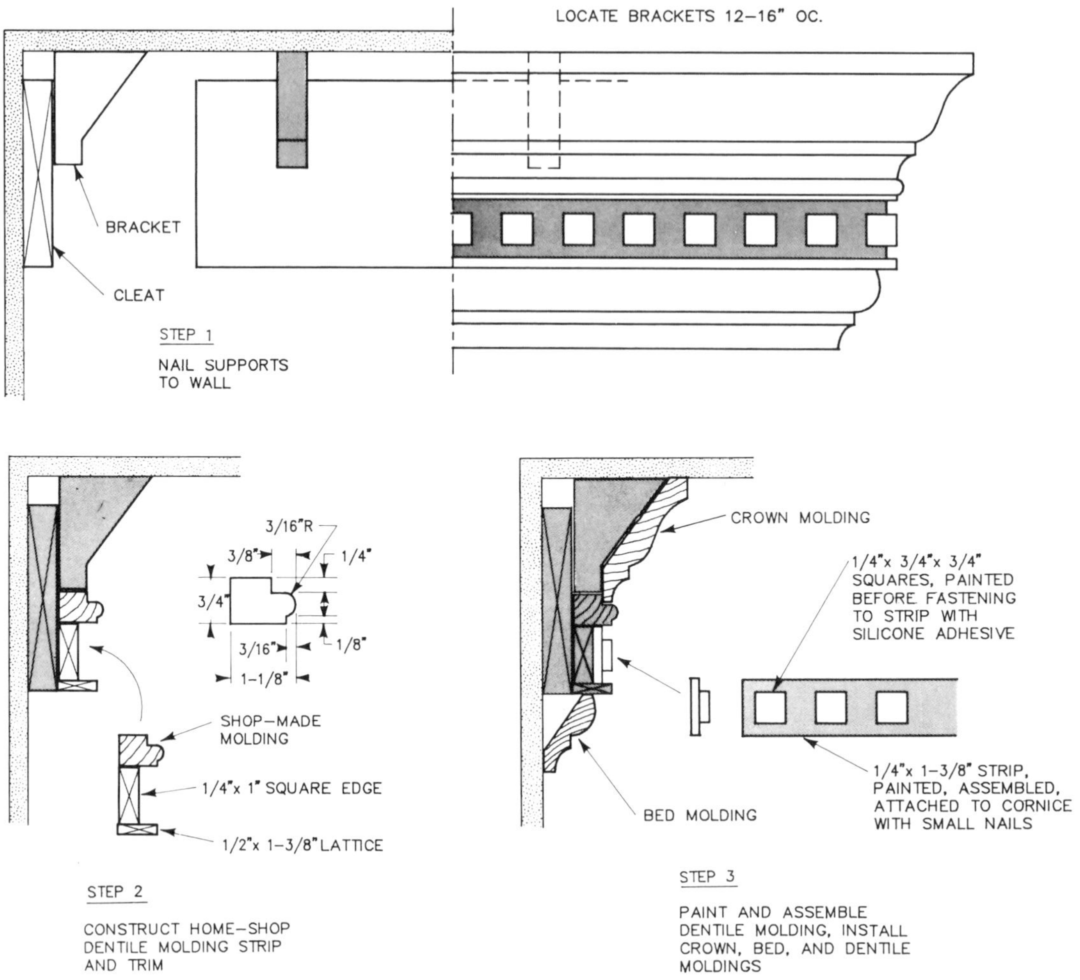

board can be made by substituting square-edged stock and a decorative base cap molding for the base molding. This type of baseboard, shown in the center of Figure 10–12, is recommended for any wall that is uneven, because the base cap can be bent to follow the wall's unevenness.

Casing

Casing is the wall trim installed around doors and windows, both to give the opening a decorative, finished look and to conceal and seal gaps between the door and window jambs and the

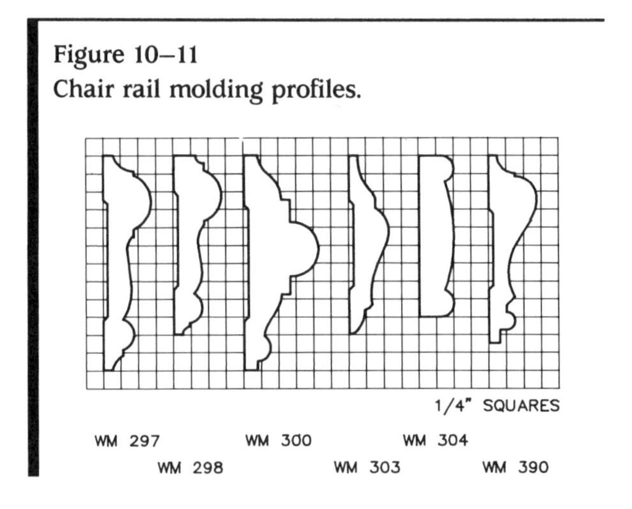

Figure 10–11
Chair rail molding profiles.

Figure 10–12
Base molding profiles, and use as a baseboard with base cap and shoe moldings.

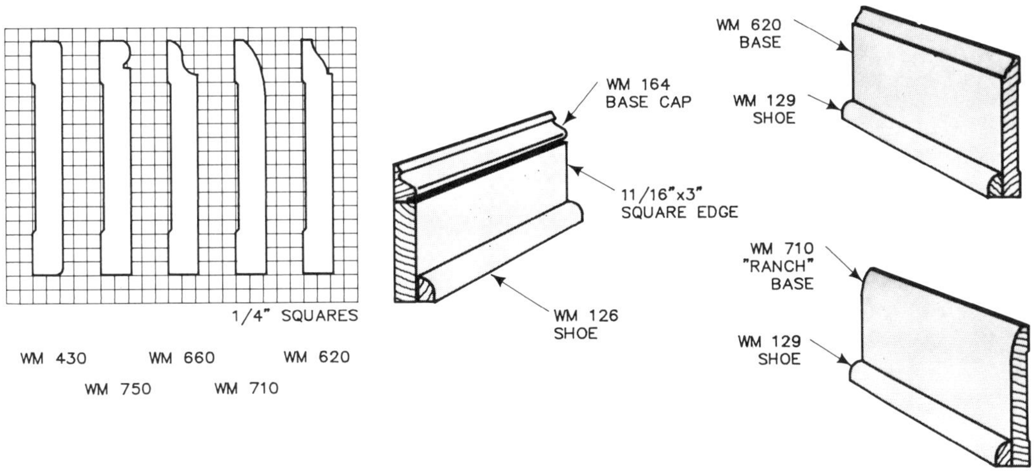

1/4" SQUARES
WM 430 WM 660 WM 620
WM 750 WM 710
WM 164 BASE CAP
11/16"x3" SQUARE EDGE
WM 126 SHOE
WM 620 BASE
WM 129 SHOE
WM 710 "RANCH" BASE
WM 129 SHOE

Figure 10–13
Baseboard installation.

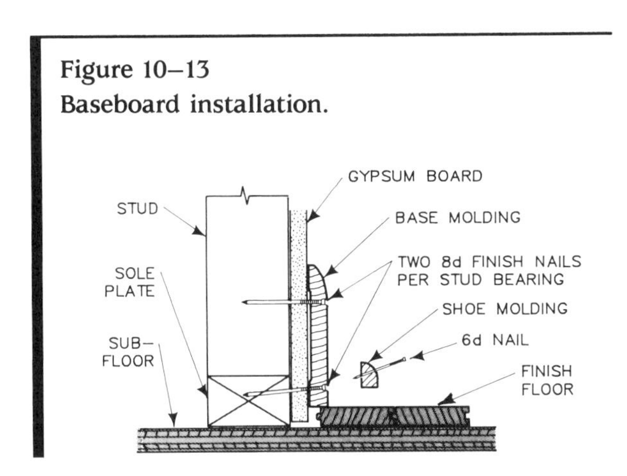

finish wall. Window casing can be either traditional or contemporary in style (Figure 10–14). In the contemporary style, casing is applied like a picture frame—mitered at all four corners. In a traditionally cased window, the bottom ends of the side casing rest on a protruding window sill (stool molding), with an additional piece of casing, called the apron, applied horizontally below the sill. In contemporary styles, casing molding is normally used alone, but it can be dressed up with a base cap or with a back bend molding, which is similar to ply cap molding (described below) but with a deeper rabbet. Casing is also relieved or backed out. It is made in many patterns, as shown in Figure 10–15.

Figure 10–14
Traditional and contemporary window and door casing.

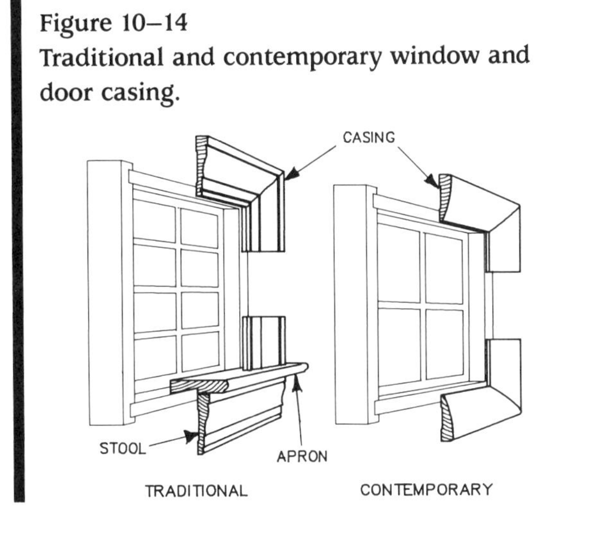

Figure 10–15
Casing molding profiles.

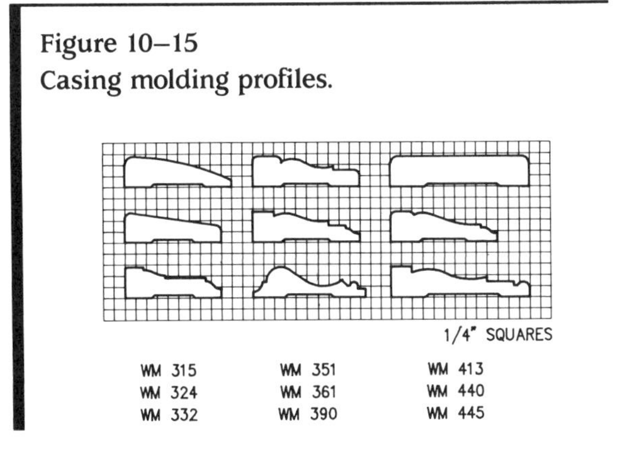

Other Moldings

Battens, panel strips, and mullion casings. Symmetrical moldings in this group are used to conceal joints between boards or panels and for decoration (see Figure 10–16).

Brick molding. These moldings (Figure 10–17) have a thick section for siding materials to butt against when used for exterior window and door casings. They can also be used in combination with other moldings for interior decoration.

Corner molding (outside and inside). These moldings (Figure 10–18) cover the corner joints of gypsum board and plywood paneling.

Drip cap. Applied over exterior door and window frames, drip cap (Figure 10–19) helps keep water from seeping under the siding and directs water away from the window glass or door surface. The molding can be used for an attractive contemporary interior door and window casing.

Panel molding. This molding (Figure 10–20) is a pattern originally assembled as a frame and applied to interior walls to obtain a paneled effect. Other moldings and combinations of moldings can also be used.

Figure 10–16
Batten, panel strip, and mullion casing molding profiles. These moldings have a wide variety of uses.

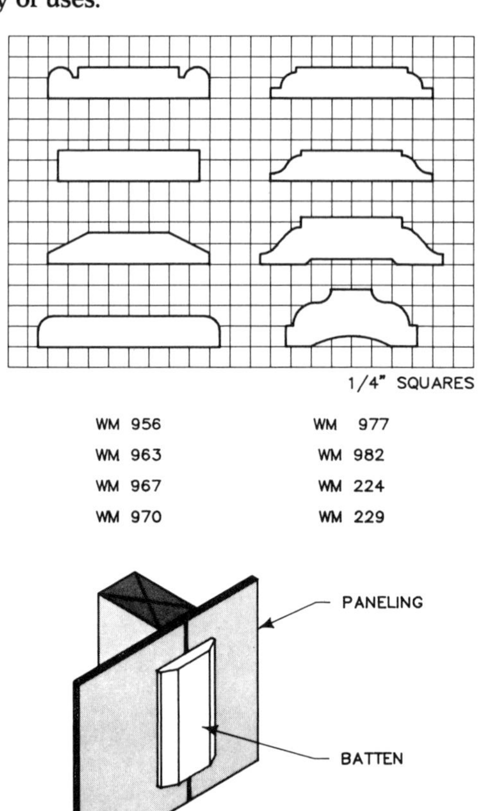

Figure 10–17
Brick molding profiles. Brick moldings are used as exterior casing around doors and windows with all kinds of siding (except cedar/pine shingles) in addition to brick exterior walls.

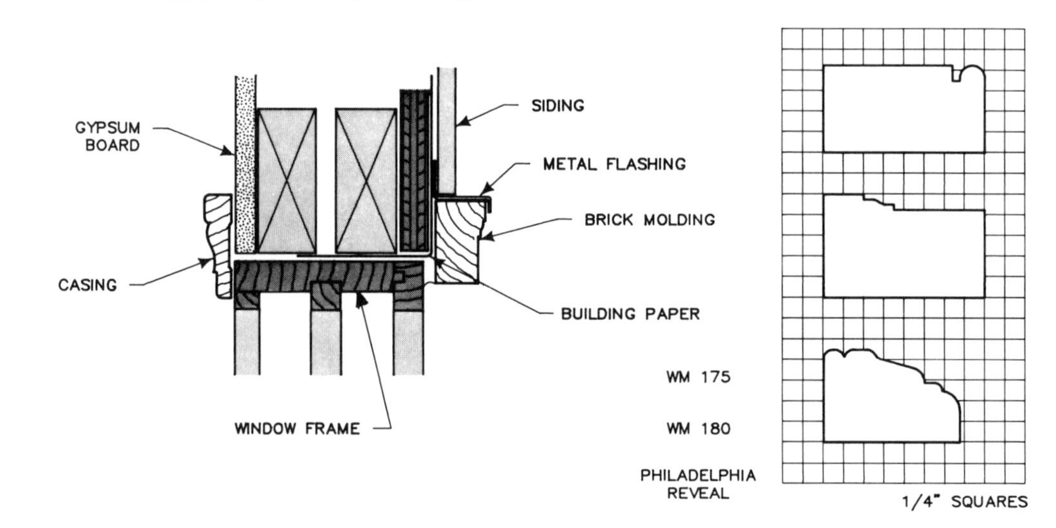

Figure 10–18
Corner molding profiles. Inside and outside corner moldings are used to trim interior paneled walls and gypsum board walls that have not been jointed. Outside corner moldings are also used to provide physical protection to papered or painted walls.

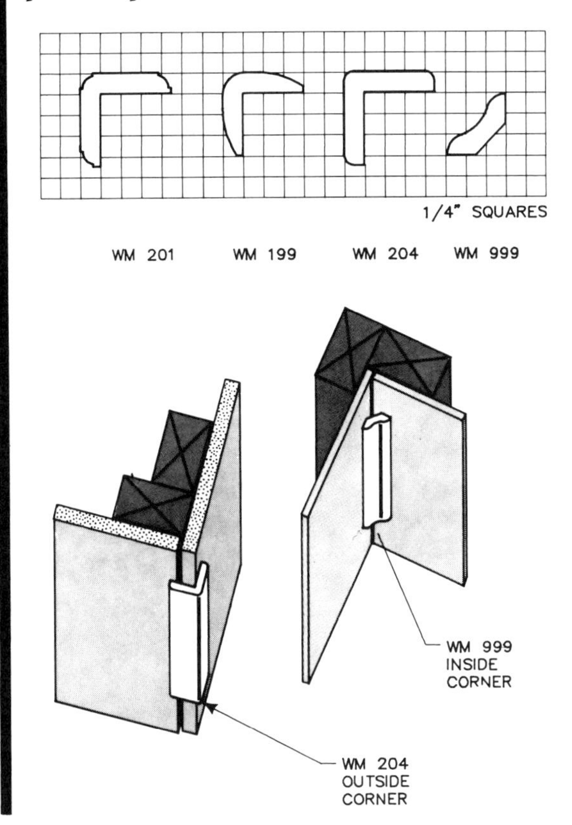

Ply cap, wainscot cap, and lip molding. These moldings (Figure 10–21) trim the top of plywood or solid wood wainscot walls, hiding edges or plank ends. They are also used with other moldings in making chair rails.

Stool. These moldings, used as interior window sills, are made in several forms, as shown in Figure 10–22.

Stop. This molding (Figure 10–23) is nailed to the finish side and head door jambs to prevent a

Figure 10–20
Panel molding profiles.

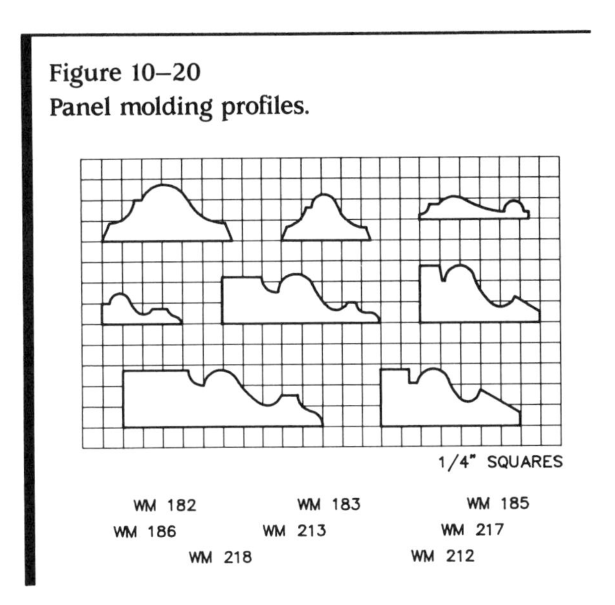

Figure 10–19
Drip cap molding profiles and application. These exterior moldings are used in combination with metal flashing above doors and windows to prevent infiltration of rain and dripping water.

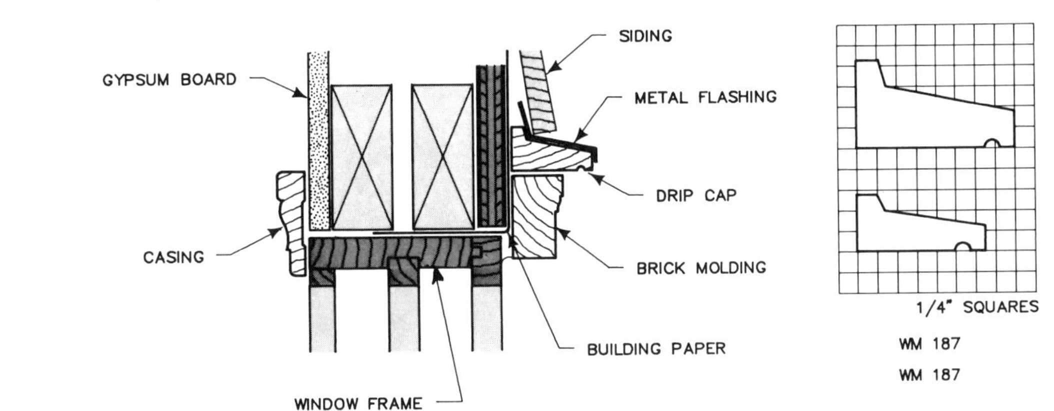

Figure 10–21
Ply cap, wainscot cap, and lip molding profiles and applications.

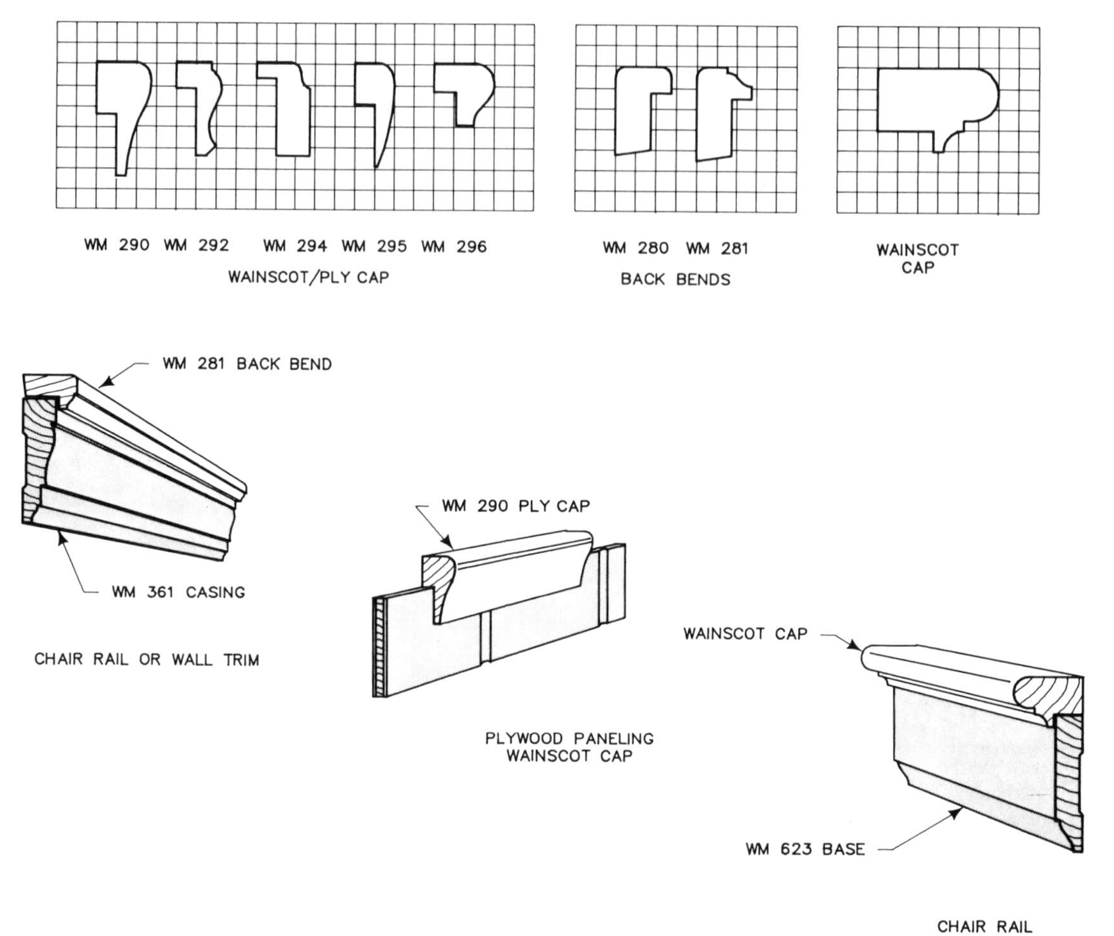

Figure 10–22
Rabbeted stool molding profiles and application.

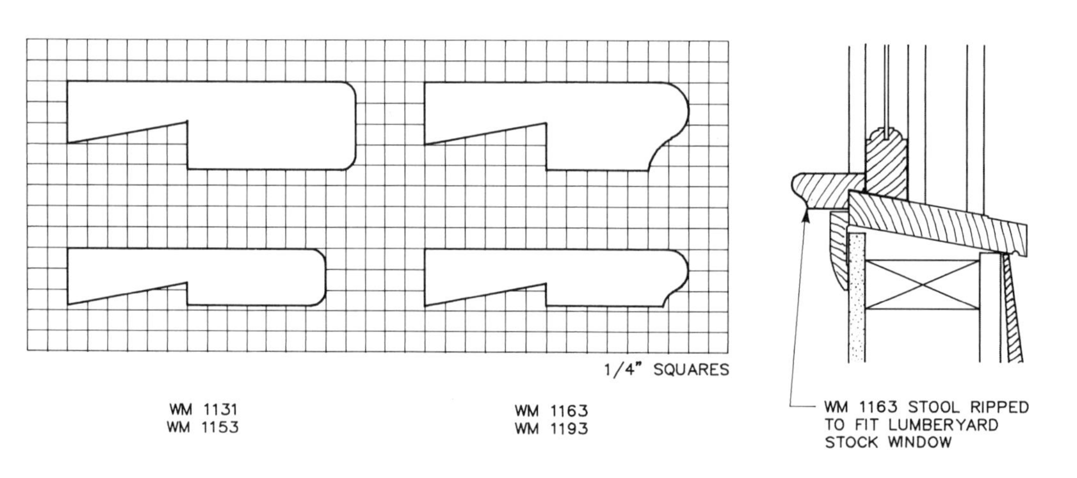

Figure 10–23
Stop molding profiles.

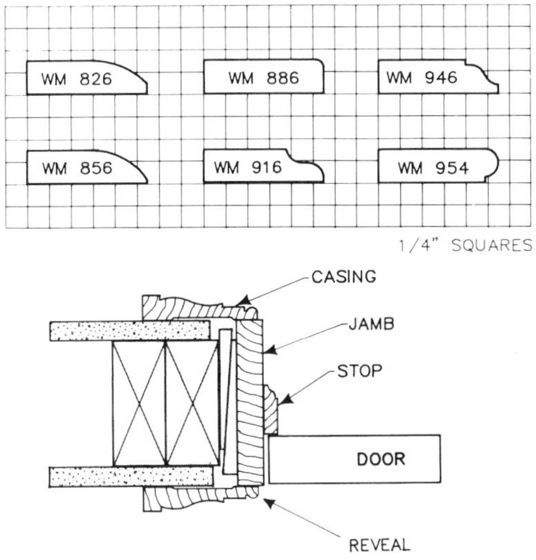

WM 826
WM 886
WM 946
WM 856
WM 916
WM 954
1/4" SQUARES
CASING
JAMB
STOP
DOOR
REVEAL

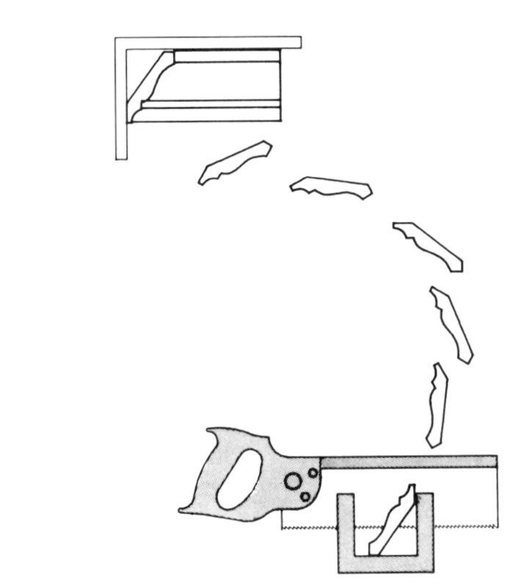

Figure 10–24
Mitering crown molding. Visualize how the molding is flipped over in the miter box.

closing door from swinging through the opening. In a double-hung window, stop molding holds the lower sash in place.

HOW TO MITER

Mitering is the basic operation in working with molding. How you do your mitering depends on what tools you have available.

As a minimum, you will need a miter box and a fine-toothed crosscut saw, or preferably a backsaw, which is designed to be used with a miter box. The most basic miter box is made from three pieces of hardwood lumber or plastic and has slots to guide the saw for 90° and left or right 45° cuts. Better-quality metal miter boxes can be adjusted to other angles; some have an integral fine-toothed blade.

Whether you cut a miter in a miter box or on a table or radial-arm saw, the saw blade tends to pull the molding sideways, producing an uneven cut. Either keep a good grip on the work or clamp it securely in place.

If you have a bench-mounted belt or disk sander, you can make extremely accurate miter joints by sawing the stock slightly overlength and then trimming it to the exact length and angle on the sander.

Mitering Crown Molding

This is not difficult once you visualize what you are doing. Crown molding is "sprung" at an angle when installed, and it must be placed in the miter box at the same angle when mitered. Think of the base of your miter box as the ceiling and the back of the box as the wall.

Place the molding in the box upside down, sloped at the same angle between the back (wall) and the base (ceiling) as it will be installed, but upside down (Figure 10–24). Put two blocks of wood in the miter box with the molding to wedge the molding at the same sprung angle for all pieces (Figure 10–25). The block will also help keep the molding from slipping sideways while you saw it.

End-Joining Molding

When a wall is longer than available molding, fit the mitered ends into the corners before cutting the moldings to butt together at a stud in the

Figure 10–25
Positioning crown molding in a miter box.

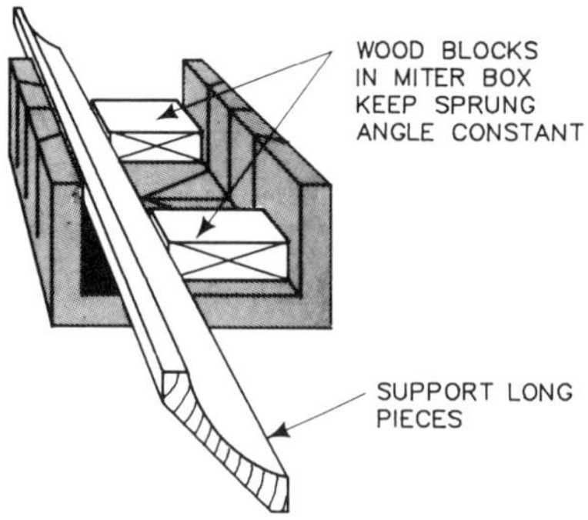

WOOD BLOCKS IN MITER BOX KEEP SPRUNG ANGLE CONSTANT

SUPPORT LONG PIECES

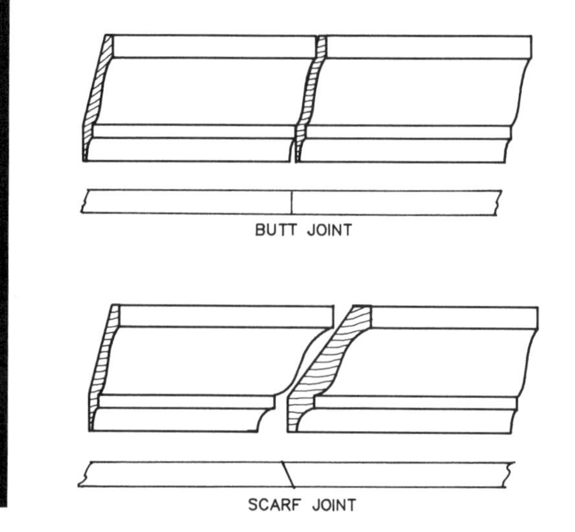

Figure 10–26
Butt and scarf joints. The angled scarf joint usually produces a less visible joint line, but is more difficult to make.

middle of the wall, because the corners are where the problems are. The joint can be butted square-cut, but it is usually better to make a scarf joint (see Figure 10–26). With the two pieces cut at an angle to overlap, that will make a less visible join.

NAILING

Small ceiling moldings can be randomly nailed because the nails will go into the wall top plate. (Be sure the nails are long enough.) Studs should be located before wide crown molding and cornices are installed; otherwise the nails are likely to be in gypsum board or plaster only and will not hold.

INSTALLING CASING

Casing nailed to door and window jambs is positioned ⅛″ to ¼″ short of the inner edge of the finish jamb. The exposed edge of the jamb is called the reveal (see Figure 10–23). Having a reveal makes fitting the casing easier, produces a better appearance, and keeps the casing clear of the door hinges. The window and door casings in a room should be the same style and size molding.

You cannot simply cut the casing to length, miter the ends, and nail it in place. That would work if the walls were flat, but they usually are not, particularly wet-plastered walls. There are two ways to put up casing: you can miter the pieces and trim on the wall until the corner joints go together neatly, or you can assemble the casing flat on your workbench. If the edge of the jamb and the wall surface are not flush, shim the outer edge of the casing pieces before assembly. (You don't need a huge bench—only the upper end of the casing has to be clamped.)

With bench assembly, the corner miter joints can be assembled accurately, glued, and reinforced with a long screw diagonally through the top casing into the side casing at each miter, as shown in Figure 10–27. The casing can then be nailed to the jamb and the wall without problems of joint alignment.

Door Casing

Position the assembled casing on the jamb, and start nailing the sides from the top and work down. Tack first into the trimmer studs with 8d finish nails, particularly if the casing stock is warped. Use 4d nails into the jambs. Pairs of nails should be not more than 16″ apart. Nail the head casing last. Fill any cracks between the casing and the wall.

Figure 10–27
Assembling casing miter joints.

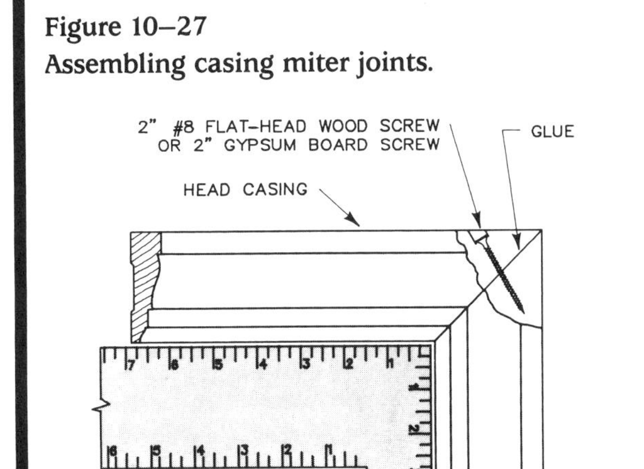

Window Casing

Assemble contemporary window casing (see Figure 10–14) on your bench and install it as a picture-frame unit. With traditional window casing, the rabbeted stool must be in place before fitting and installing the casing. The ends of the stool should be either mitered or shaped by hand to carry the molded edge around the ends. This is called a *return*. The apron should be similarly treated.

CUSTOMIZING FLUSH DOORS

The appearance of any flush door—exterior, interior cabinet, or garage—can be given a facelift with applied moldings, as shown in Figure 10–28.

If it is a new door, cut hinge mortises and bore holes for the lockset first. A used door should be

Figure 10–28
Customizing flush doors with molding.

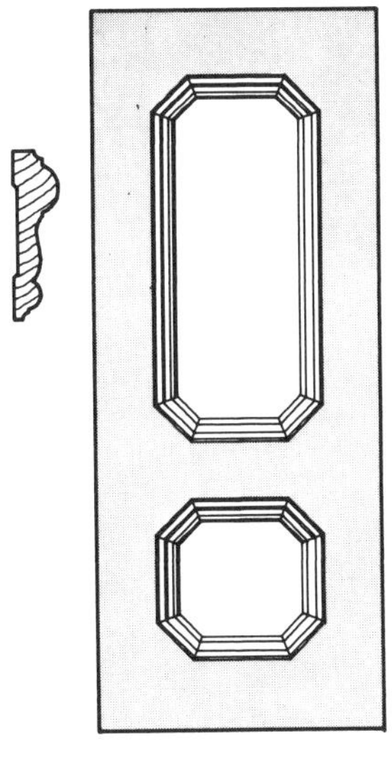

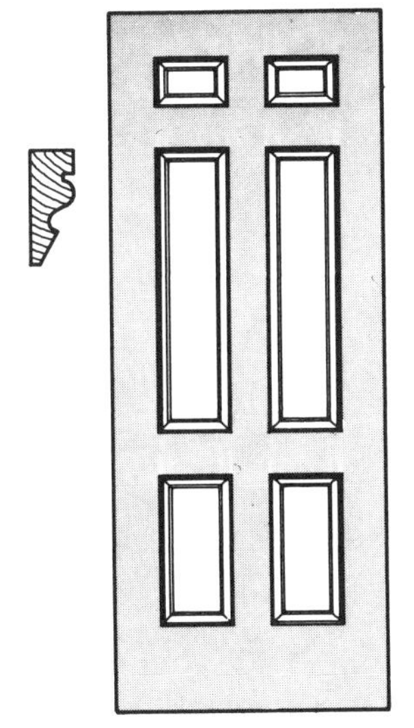

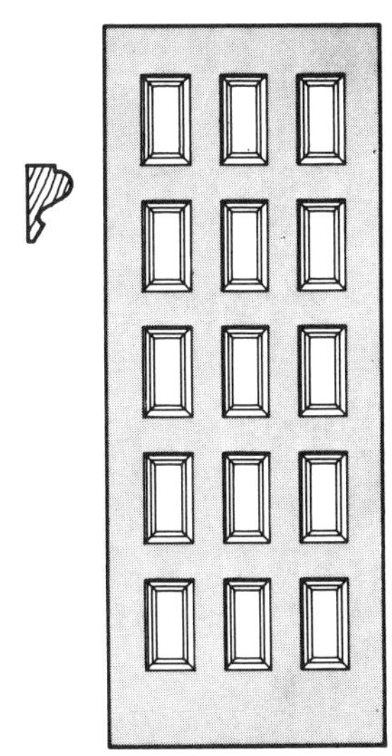

sanded to prepare the surface for glue. If the door is mahogany and you are going to paint it, fill and sand before you start applying molding. Prepared spackling paste makes a good filler for painted mahogany.

Moldings can be nailed to a solid-core door, but must be glued to a hollow-core door because the ⅛" or thinner plywood skin has little ability to hold nails of any kind. If you use white glue instead of contact cement, weight down the molding and use small nails to hold it in position while the glue dries. Moldings applied to the outside of exterior doors must be assembled and glued to the door with a waterproof glue such as resorcinol.

In general, when planning to add moldings to full-length doors, keep 4½" clear at the sides for knob clearance, and 6" at the top and 9" at the bottom, for appearance. Assemble the moldings before applying them to the door.

11 Interior Partitions

New houses today tend to have a rather open interior design—there are relatively few fully partitioned-off spaces and, except perhaps for an extra lavatory or two, almost no small rooms. Older houses, on the other hand, are full of partitions. These houses generally were built with a greater number of rooms, of a variety of sizes, than is the case today. Also, partitions were often added at a later time to divide one or more large rooms to accommodate a growing family. Or, a large multiple-room house may have been divided into apartments.

In either kind of house, a new owner may very well want to do some partition work. The most common projects are adding a partition to make two rooms of one, taking out a partition to create a larger space, opening up a doorway into a wide archway, and removing a closet to enlarge a room.

Closing in a furnace or laundry area so the remainder of the utility room can be used as a home shop or office or exercise room is an example of the need for a new partition. The room in Figure 11–1 has been all three. The furnace air supply is drawn down from the attic through a duct in the new partition. When woodworking is done, air filters in the louvered door keep sawdust out of the furnace. The room has a new acoustical drop ceiling (see Chapter 17) with built-in fluorescent lights. Because the shop was so noisy, it has been soundproofed (see Chapter 15). As another example, the bedroom shown in Figure 11–2 has been divided into a study and a bedroom.

Adding partitions seldom involves many construction problems; however, check your building code carefully regarding window and access requirements for interior rooms. It is taking out partitions that can be dangerous. The partition might be load-bearing—that is, it might support part of the house structure, not just its own dead weight. Removing a load-bearing partition can be structurally disastrous to your house and a risk to the safety of your family. Unless you are absolutely sure a partition you want to take out is non-load-bearing, hire an experienced contractor to do the job and to erect a replacement load-bearing structure to continue to carry the load.

Most often you run into the problem of taking out a partition when renovating an older house that has had several owners. But even in a new house, you may want to take out an under-the-

Figure 11–1
Before and after. This laundry/furnace utility room was converted to a laundry room/home shop by partitioning off the furnace.

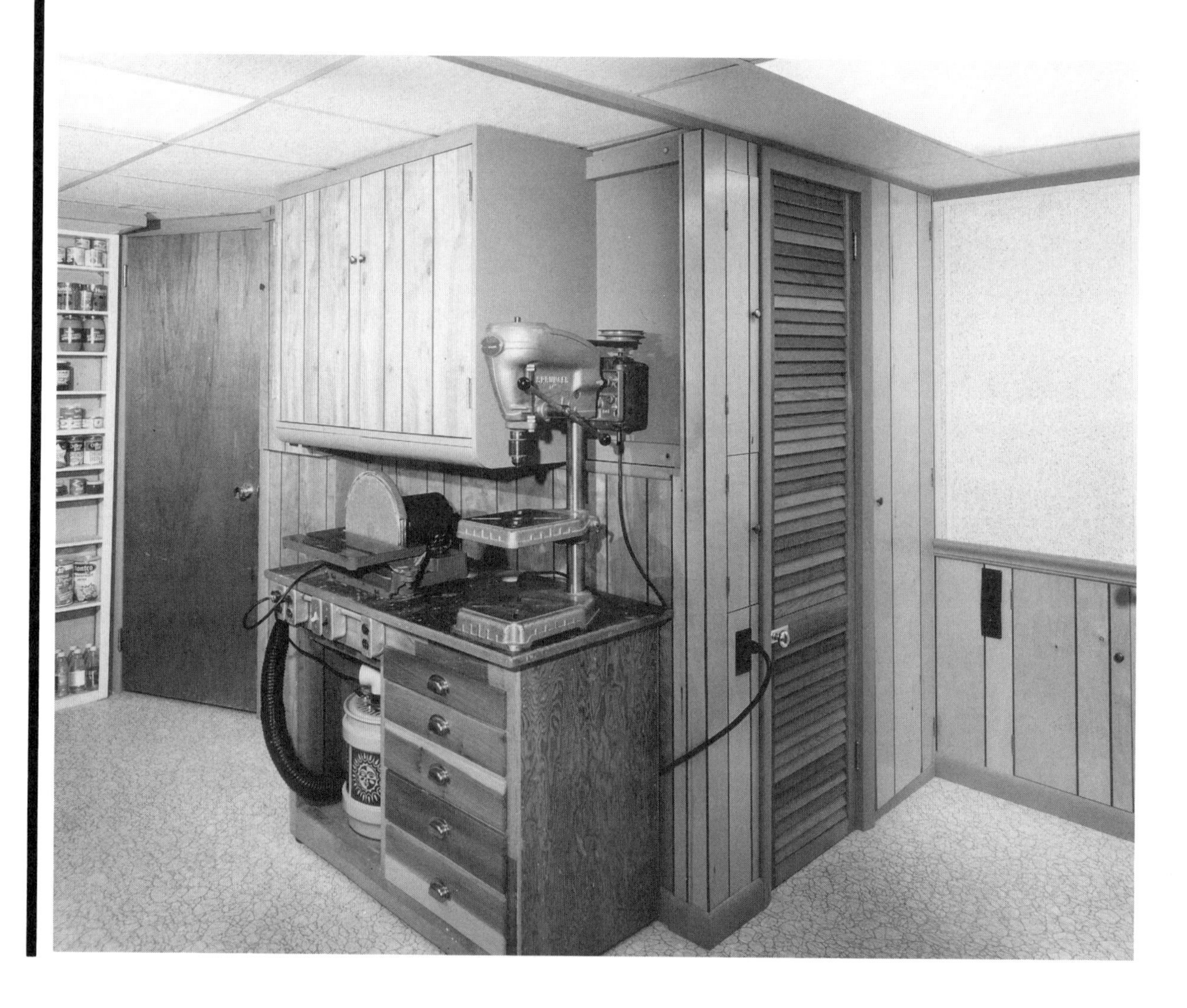

Figure 11–2
Before and after. A wood-panel partitioning converted a bedroom into a combination bedroom and study.

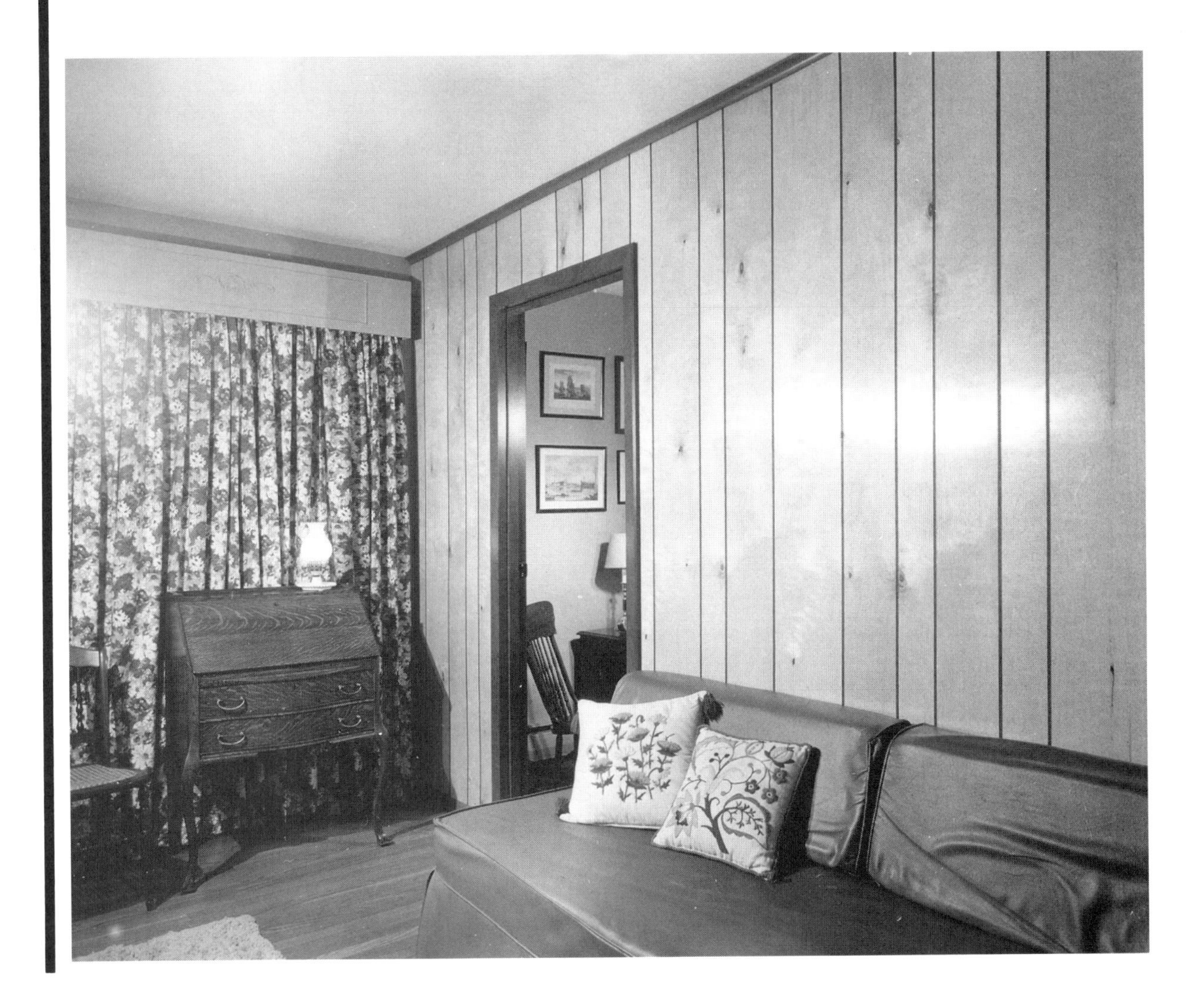

stairs closet to give yourself an alcove or to rearrange kitchen and dining room space. There are many reasons to take out a partition.

Choosing the right techniques for taking out the partition is not the problem. Literally bashing it down, if approached in the proper frame of mind and with common sense, can be downright fun, except for the mess. The problem is determining if it is structurally safe to remove the partition in the first place.

IDENTIFYING LOAD-BEARING AND NON-LOAD-BEARING PARTITIONS

We are dealing here only with interior partitions, not the outside walls of the house. As already indicated, there are two kinds of interior partitions: load-bearing and non-load-bearing. A load-bearing partition supports another part of the house—the ceiling, the next floor, a partition directly above it. Taking out a load-bearing partition will risk structural damage to the house. On the other hand, a non-load-bearing partition supports only its own weight. The problem is how to be very certain that the partition you plan to take out is non-load-bearing.

If you are not familiar with how a wood-frame house is structured, hire a contractor. Otherwise, proceed with a little detective work in your house. Generally, it is easier to determine that a partition is load-bearing than to be sure it is not.

Top-Floor Partitions

If the partition you plan to remove or modify is on the top floor of your house and can be seen from the attic, or is otherwise accessible from above, go up and look for the location of the top of the partition. You may have to pull back insulation to find it, or possibly pull up a floor board. But you don't necessarily have to see the top plate of the partition itself.

The partition is load-bearing under any one of the following conditions (see also Figure 11–3).

- The ends of joists rest on the partition.
- The centers of joists that run in one piece from one outside wall of the house to another rest on the partition.
- There is a partition or any kind of vertical framing directly above or on top of it.
- Even though the joists end some distance away, the partition is the last support under them.

The partition is non-load-bearing if it runs parallel to and between the joists above it. If the length of the partition runs directly under the length of a joist and the joist is supported at both ends beyond the partition, *and* there is no house structure resting on the joist, *and* the joist is not doubled or reinforced in any way, *and* the joist looks just like all the other joists beside it—then the partition is non-load-bearing.

If the partition runs crosswise to the joists, and begins and ends short of the joist ends, and both ends of the joists are supported, then the partition is *probably* not load-bearing.

If, where the top of the partition should be, you can see nothing except the back side of the ceiling from above, then the partition was probably added after the house was built, and the partition is *probably* not load-bearing.

First-Floor Partitions

For first-floor partitions, make an inspection from the basement or crawl space to determine what is under the partition. Locate the main girder of the house—either a steel I-beam or (typically) two or three 2 × 12s spiked together. If the girder or a wall of any kind runs directly under and parallel with the partition (Figure 11–4), the partition is load-bearing.

Figure 11–5 shows other clues to identify the bearing status of a partition. If the partition is not located above the main girder or a basement wall, as described above, but runs across joists that have been strengthened under it, the partition is load-bearing. "Strengthened" means that every joist or every other joist is doubled, or that

Figure 11–3
Load-bearing or non-load-bearing? These are the things to look for from an attic or other space above the partition.

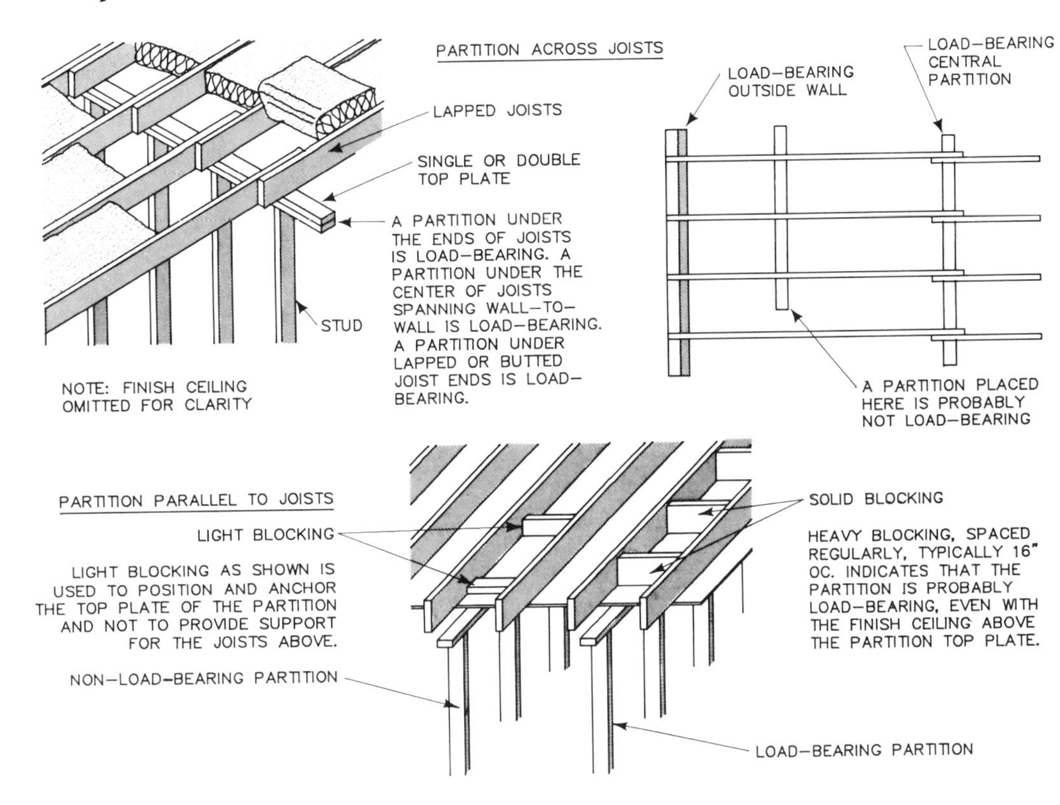
PARTITION ACROSS JOISTS
LAPPED JOISTS
SINGLE OR DOUBLE TOP PLATE
A PARTITION UNDER THE ENDS OF JOISTS IS LOAD–BEARING. A PARTITION UNDER THE CENTER OF JOISTS SPANNING WALL–TO–WALL IS LOAD–BEARING. A PARTITION UNDER LAPPED OR BUTTED JOIST ENDS IS LOAD–BEARING.
STUD
NOTE: FINISH CEILING OMITTED FOR CLARITY
LOAD–BEARING OUTSIDE WALL
LOAD–BEARING CENTRAL PARTITION
A PARTITION PLACED HERE IS PROBABLY NOT LOAD–BEARING
PARTITION PARALLEL TO JOISTS
LIGHT BLOCKING
LIGHT BLOCKING AS SHOWN IS USED TO POSITION AND ANCHOR THE TOP PLATE OF THE PARTITION AND NOT TO PROVIDE SUPPORT FOR THE JOISTS ABOVE.
NON–LOAD–BEARING PARTITION
SOLID BLOCKING
HEAVY BLOCKING, SPACED REGULARLY, TYPICALLY 16" OC. INDICATES THAT THE PARTITION IS PROBABLY LOAD–BEARING, EVEN WITH THE FINISH CEILING ABOVE THE PARTITION TOP PLATE.
LOAD–BEARING PARTITION

Figure 11–4
In the basement or crawl space, the main girder or a wall of any kind directly under the partition means the partition is load-bearing.

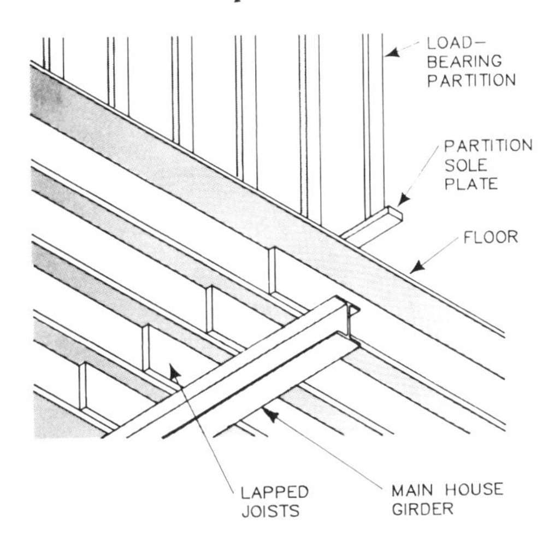
LOAD–BEARING PARTITION
PARTITION SOLE PLATE
FLOOR
LAPPED JOISTS
MAIN HOUSE GIRDER

the joists are closer together than 16″ on center.

If the partition runs parallel to the joists, either between joists or on top of a joist that has not been doubled, the partition is not load-bearing.

If the joist directly under the partition is doubled, or two joists have been placed close together with 2 × 4 blocking spaced every few feet between them, the partition is load-bearing.

Measure the thickness of the partition and determine how it is finished. Figure 11–6 gives thicknesses for the two standard sizes of stud framing. If you determine that the joist is framed with 2 × 4s in a house less than twenty-five years old, it is probably load-bearing. In an older house you don't know. If the partition is framed with 2 × 3s and the house is more than five years old and the partition is in the second story, the 2 × 3 framed partition might be load-bearing.

Figure 11–5
These are ways to identify partitions not located over the main beam. The primary clue is the amount of strengthening of the joists under the partition.

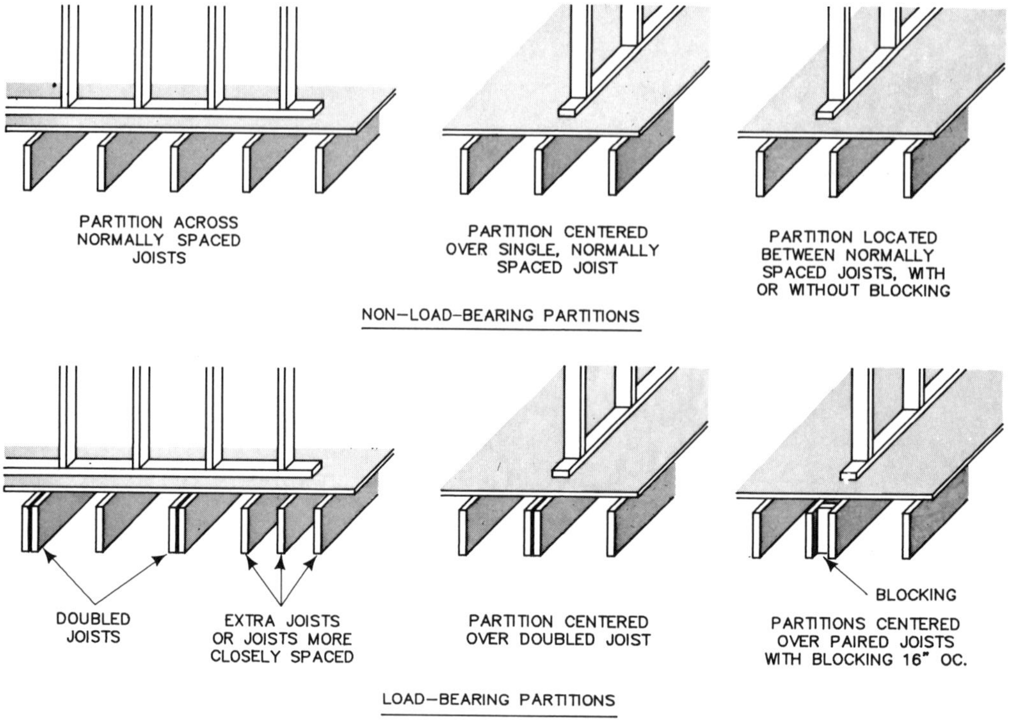

Other Indications

From all of this you can see it is a lot easier to tell if a partition is load-bearing than to establish that it is non-load-bearing. A partition added to a house after it was built would seem to be almost surely non-load-bearing, but I added a short partition in the basement of our house the year after it was built because the floor above was sagging and squeaking—its function was to be load-bearing.

Any partition under a bathroom should automatically be assumed to be load-bearing. A partition designed to be non-load-bearing may have had a load, such as a bathroom above, added later.

Be very cautious when thinking about taking out a partition. If you have any doubt whatsoever about its bearing status, get a second opinion. Accidentally taking out a load-bearing partition could cause you a lot of grief and expense.

TAKING OUT A NON-LOAD-BEARING PARTITION

Start by taking everything out of the rooms that you don't want covered by plaster dust, or just plain dust if the partition is paneled. Wear full face protection, and a respirator if plaster is involved.

Pry off all trim—ceiling molding, baseboards, corner molding. If there's a door, unhinge it and pry off the casing and jambs. Remove all outlet and switch plates. Identify the circuit and turn it off at the main power panel. (Figure 11–7 shows the framing inside a typical interior wall.)

From here, plywood paneling is easy—if it has been nailed and not glued. Find a seam, insert

Figure 11–6
The thickness of the partition may give a clue to load-bearing or non-load-bearing status, but not too reliably.

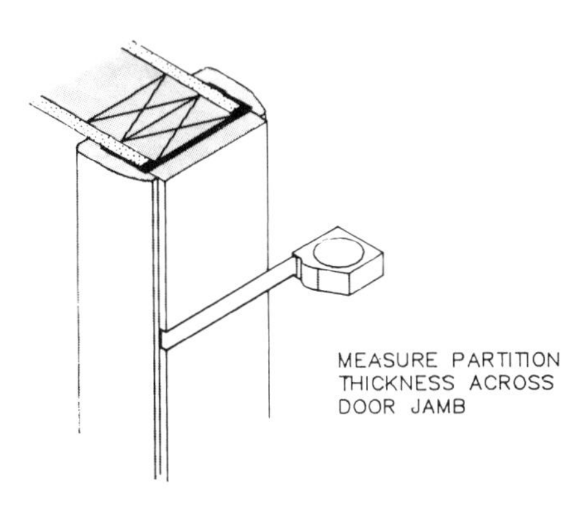

WALL FINISH	THICKNESS WITH 2x3 FRAMING	THICKNESS WITH 2x4 FRAMING
GYPSUM BOARD	3-1/4" - 3-1/2"	4-1/4" - 4-1/2"
PLASTER	3-3/4" - 4-1/4"	4-3/4" - 5-1/4"
PLYWOOD PANEL	2-7/8" - 3"	3-7/8" - 4"
SOLID WOOD	3-1/2" - 4"	4-1/2" - 5"
PANELING OVER GYPSUM BOARD	3-5/8" - 4"	5-5/8" - 5"

your ripping bar, and pry. Remove nails as you pop them. If the plywood has been glued, you will have to pry harder. Once the first side is off, the second side can be whacked loose from the rear.

Gypsum board is not much harder to remove, but it is messier. Locate the studs and score the gypsum board with a utility knife alongside the studs, then bash the board in between the studs. It will break out in fair-sized pieces. Remove the remains caught between the nails, then pull the nails.

How much trouble you will have removing plaster depends on whether it was troweled over wood lath, expanded metal lath, or gypsum lath. If the last, it won't be much harder than gypsum board to remove, but messier. For any plaster, don't waste time and blades scoring, just start bashing and prying.

If there is wiring inside the partition, you cannot just cut it off and tape the ends. (The circuit should have been turned off at the main power panel before you started to remove the wall covering.) Trace the wire back to the next box, and disconnect it. If you cannot get back to another box, you can tape off the wires inside a box

Figure 11–7
Interior partition framing and preliminary steps to removing a partition.

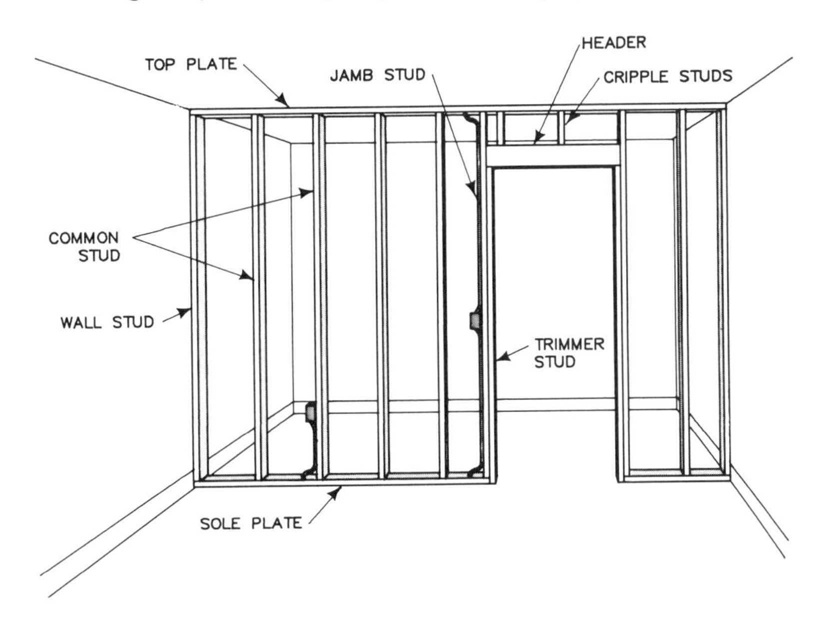

installed just for the purpose. The box must not be buried inside a wall; it must remain visible and accessible.

After the studs are bare, saw through the middle of each, and twist and pry the pieces loose from the sole plate and top plate as shown in Figure 11–8. Don't try to save the pieces for reuse, except perhaps to cut into short lengths for blocking. If you are not able to pry off the plates in one piece, you may have to use a chisel and splinter them out.

Should the top plate be removed? Yes, if you want an uninterrupted ceiling. No, if you are putting in an archway, or if you want to retain an appearance of two parts to the room, or if you don't feel a continuous ceiling between the two rooms is desirable.

BUILDING A NEW NON-LOAD-BEARING PARTITION

Plan to locate a new partition for the most useful space in both new areas and for easy attachment, if possible. If the partition and ceiling joists run in the same direction, locate the partition directly below a joist. Anchoring the top of the partition just to the gypsum board or plaster of the finish ceiling, even with expansion (or Molly) bolts, is not a good practice. However, if you have access to the ceiling joists from above, you can insert blocking between the joists, to which the partition top plate can be attached.

Measuring and Planning

Mark the intended position of the top of the partition on the ceiling by locating the end points and snapping a chalk line between them. Or, stretch a string taut between nails driven into the end points and trace along it, being careful not to push the string out of alignment. Use a string and plumb bob to mark the floor directly below the ends of the ceiling line, then use a chalk line to establish the floor line of the partition. Another way is to mark several points on the floor using a plumb bob and then use a long straightedge to draw a line connecting them.

Now measure the room where the partition is to go (see Figure 11–9). Don't assume anything is square, or that either the floor or the ceiling is level. Measure the height of the ceiling at each wall and at intervals across the room. These measurements will tell you how long to make the

Figure 11–8
Removing the framing of a non-load-bearing partition.

ROUGH FRAMING REMOVAL
1. COMPLETE THE REMOVAL OF THE FINISH WALL MATERIAL.
2. SAW THROUGH COMMON STUDS AS SHOWN.
3. TWIST STUDS FREE OF TOP AND SOLE PLATES
4. SAW THROUGH JAMB STUDS ABOVE HEADER, TAKE OUT ROUGH JAMB AS A UNIT.
5. PRY SOLE PLATE LOOSE, REMOVE NAILS, TWIST OUT FROM UNDER WALL STUD SIDEWAYS.
6. PRY WALL STUDS FROM WALL.
7. PRY TOP PLATE FROM JOISTS. IT MAY BE NECESSARY TO CUT THE TOP PLATE AT THE CENTER AND TAKE IT OUT IN TWO PIECES.

Figure 11–9
Measuring. At the partition location, measure the room width at floor and ceiling height; measure the ceiling height at several locations. The floor and ceiling may not be level or flat, and the walls may bulge or be out of plumb.

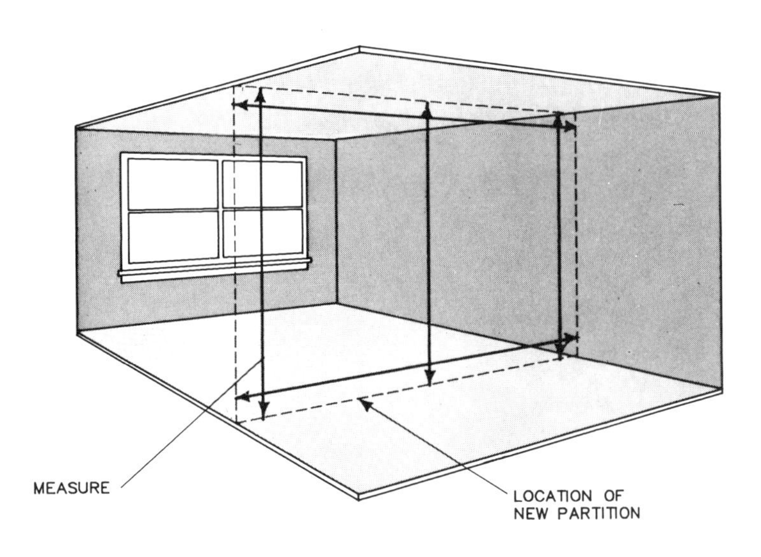

studs. Measure the width at the ceiling and at the floor.

Next, draw a plan for your partition frame. The easiest way to assemble it will be flat on the floor, to be tipped up into position, so don't plan to build the partition to fit exactly. You will need some leeway to turn the framework around and to raise it, and to compensate for bowed and out-of-square walls, ceiling, and floor. The width of the framing should be 1″ less than the room width if the partition is to be wall-to-wall, and ½″ less than the ceiling height, which may not be the same from one side of the room to the other.

The other way to build a partition is to nail the top and sole plates in place on the ceiling and floor, then nail the studs and other parts to them. It is usually easier to do the nailing at floor level than overhead, and face-nailing is a lot easier than toenailing. (If you were putting up a load-bearing partition, you might have to build it in position because the temporary bracing that supports the overhead load usually leaves no room to raise a preassembled framework.)

Framing

An interior non-load-bearing partition can be framed with 2 × 3s with a single header. Cut the top plate and sole plate to length and mark the stud locations, as shown in the lower portion of Figure 11–10. The first stud at one end goes flush with the end of the plate. Then measure 16″ from the outside end of the plate to find the *center* of the second stud position. All stud positions after that are 16″ on center until you get to the other end, where there may be less than 16″ to the end of the plate and the final stud. This procedure ensures that the edges of 4′-wide sheets of gypsum board or paneling will fall on the center of a stud for nailing. Note that Figure 11–10 shows layouts for a two-piece sole plate, cut out at the door opening, and for a single, continuous sole plate. The single sole plate provides rigidity when a wall frame built on the floor is raised and moved into position. However, the doorway portion of the sole plate must be cut away after the frame is erected, and that may risk damaging the

Figure 11–10
Lay out and mark stud positions on sole and top plates. When dimensioning the door opening, allow for finish jambs (see Chapter 4). To assemble the partition on the floor, follow the sequence indicated.

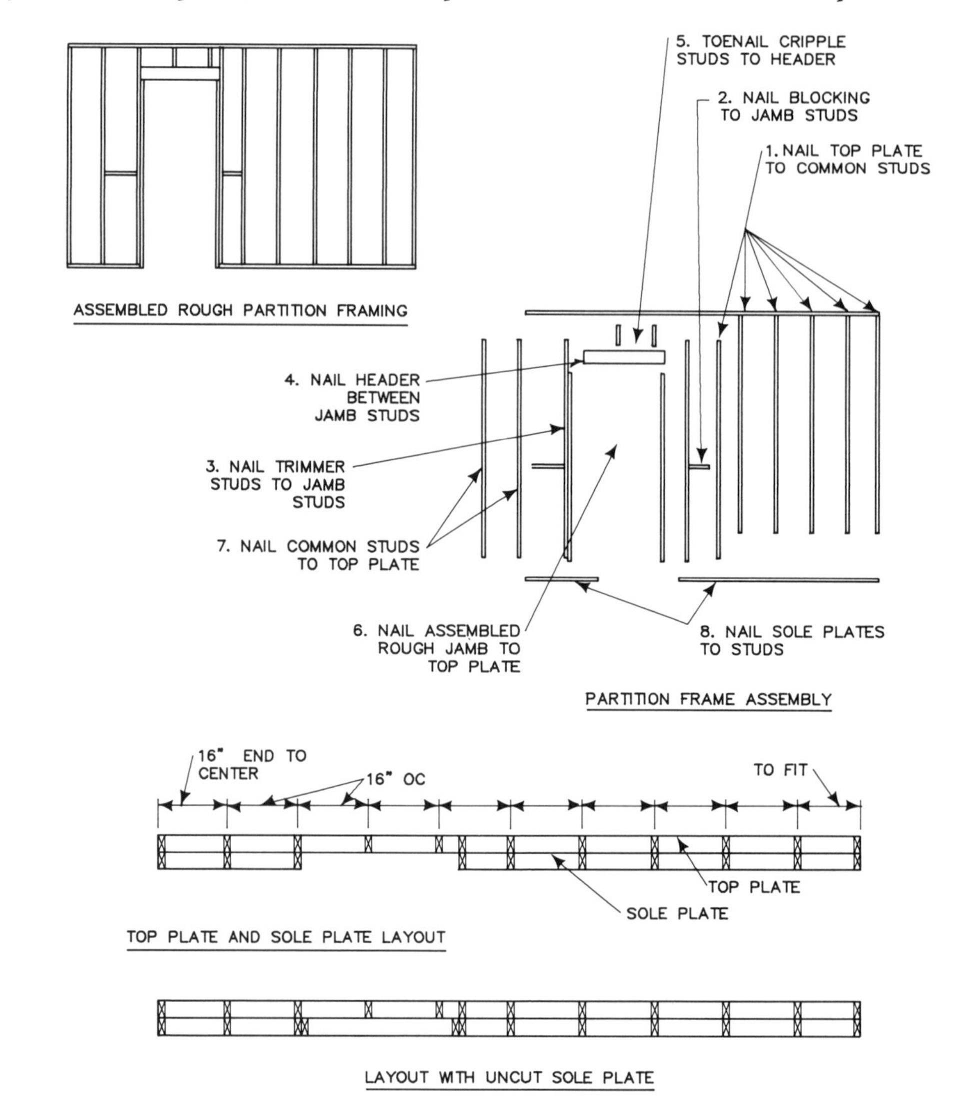

adjacent finish flooring. If you use a two-piece plate, stiffen the doorway opening as described below when you move the frame into place.

If you assemble the frame on the floor, follow the numbered sequence in Figure 11–10. In that case all studs can be cut to the same length and any variations in ceiling height taken care of by shims over the top plate. If you are building the frame in place, attach the plates in their floor and ceiling positions and measure each stud individually for a snug fit.

The final step in floor assembly is to nail the sole plate into the bottom ends of the studs (Figure 11–11). Note that short horizontal blocking is nailed in place to give rigidity to the studs on each side of the door opening. If you have made a framework with a cutout sole plate, attach a 2 × board across the opening near the bottom to hold the framework aligned when you move it. If the frame is more than 6–7′ wide on either side of the doorway, temporarily nail a 2 × 3 across the stud from a top corner to the diagonally opposite lower corner. Put one nail into each stud to keep the framework aligned until it is permanently fastened in position.

Raise the partition and (you may need a helper) brace it so it will stand free in approximate position while you shift it into the exact position for fastening (see Figure 11–12).

With the partition centered between the side walls, check the trimmer studs to see that the sides facing the opening are vertical, or can be pushed to vertical without allowing the top plate to move. If not, adjust the position of the top plate toward one side wall or the other. Steps in positioning the partitions are listed in Figure 11–13.

Securing the Framework

The partition must be fastened to the ceiling joists at the top and to the floor at the bottom. Fastening to floor joists is not as important as getting nails or lag screws into the joists overhead. Lag screws are preferable because they can be installed without hammering, which might pop nails in existing walls. Always fill a gap with a shim at any point where fastening is to be made. The partition should also be fastened to the side wall studs or the side wall covering (with expansion bolts) to prevent vibration.

Attachment details are shown in Figure 11–14. Drill a hole in the top plate near each end and drive a lag screw partway into the ceiling, enough to keep the framework upright but not to lift it off the floor. Adjust the partition so it is vertical. Lock it in place with several nails driven partway into the floor on either side of the sole plate. Place shims between the ends of the framework and the side walls so that the doorway trimmer studs are vertical. Drive shims in pairs from opposite sides of the partition to get a parallel wedging action. Fasten the ends to the side walls through the shims.

Now drill clearance holes in the sole plate and pilot holes in the floor for the lag screws and fasten the partition to the floor. There must be a lag screw on each side of the doorway opening and one at each end. Space intermediate screws about 36″ apart. Then pull out the nails that were temporarily holding the plate in position.

Next, drive opposing pairs of shims at several

Figure 11–11
Floor assembly. Nail the sole plate to the studs last. With a carpenter's square, check that each stud is perpendicular to the plate as you nail.

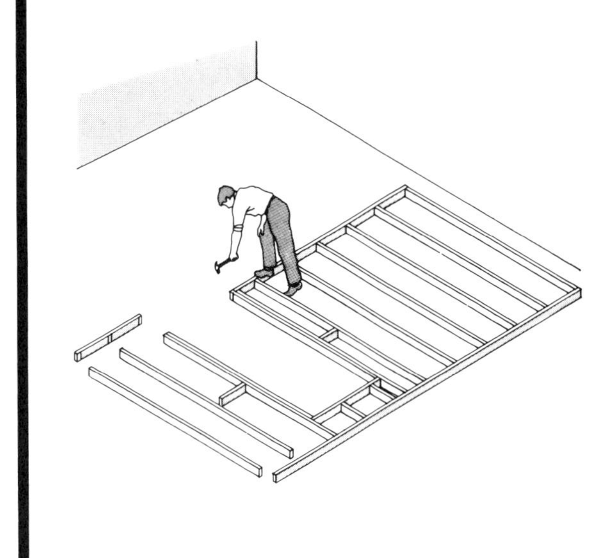

Figure 11–12
Erect the partition. Brace the partition for support while you move it into final position.

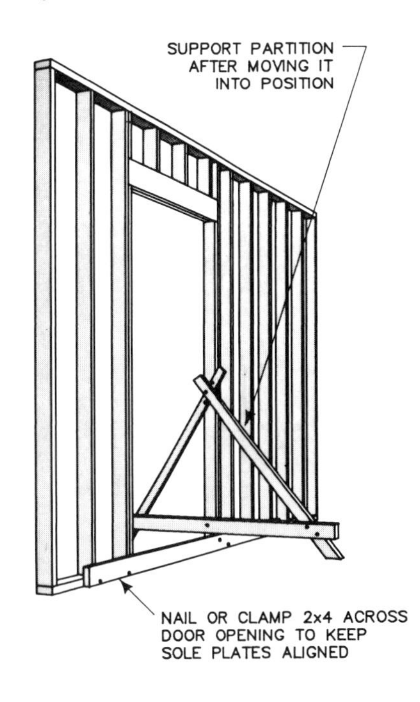
SUPPORT PARTITION
AFTER MOVING IT
INTO POSITION
NAIL OR CLAMP 2x4 ACROSS
DOOR OPENING TO KEEP
SOLE PLATES ALIGNED

Figure 11–13
Positioning and attaching the partition framework.

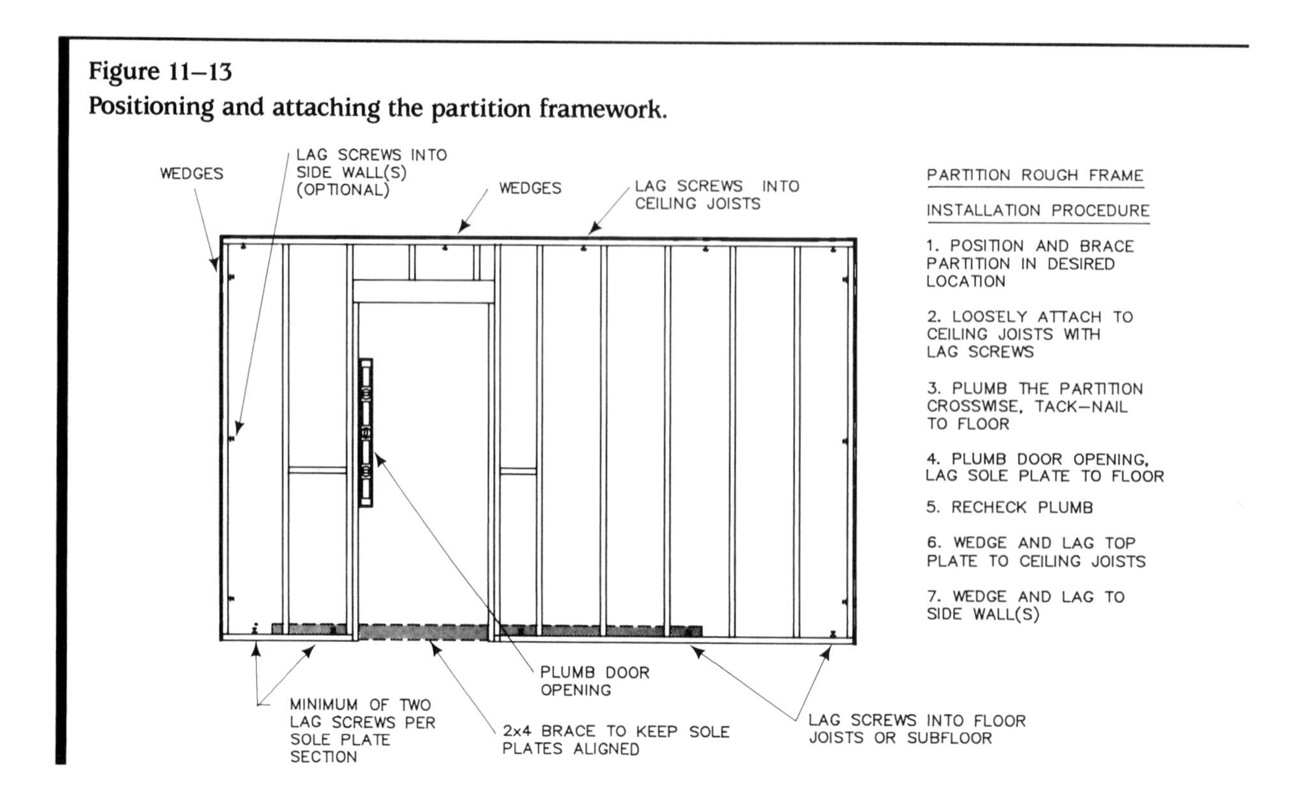
WEDGES
LAG SCREWS INTO SIDE WALL(S) (OPTIONAL)
WEDGES
LAG SCREWS INTO CEILING JOISTS
PARTITION ROUGH FRAME
INSTALLATION PROCEDURE
1. POSITION AND BRACE PARTITION IN DESIRED LOCATION
2. LOOSELY ATTACH TO CEILING JOISTS WITH LAG SCREWS
3. PLUMB THE PARTITION CROSSWISE, TACK–NAIL TO FLOOR
4. PLUMB DOOR OPENING, LAG SOLE PLATE TO FLOOR
5. RECHECK PLUMB
6. WEDGE AND LAG TOP PLATE TO CEILING JOISTS
7. WEDGE AND LAG TO SIDE WALL(S)
PLUMB DOOR OPENING
MINIMUM OF TWO LAG SCREWS PER SOLE PLATE SECTION
2x4 BRACE TO KEEP SOLE PLATES ALIGNED
LAG SCREWS INTO FLOOR JOISTS OR SUBFLOOR

points between the top plate and ceiling. Drill at these points and drive lag screws tightly into the ceiling and joist above. Be sure to fasten the framework at a point at each side of the doorway. Remove the partially driven lag screws at the ends, which you installed just after raising the framework. Insert shims as needed at these points and drive the lag screws all the way in.

Figure 11–15 shows the partition framework installed, ready to be covered with gypsum board or paneling (see Chapters 12 and 14). Figure 11–16 shows details for three partition-building problems: constructing corners, doorway headers, and freestanding partition ends.

Figure 11–14
Attachment details.

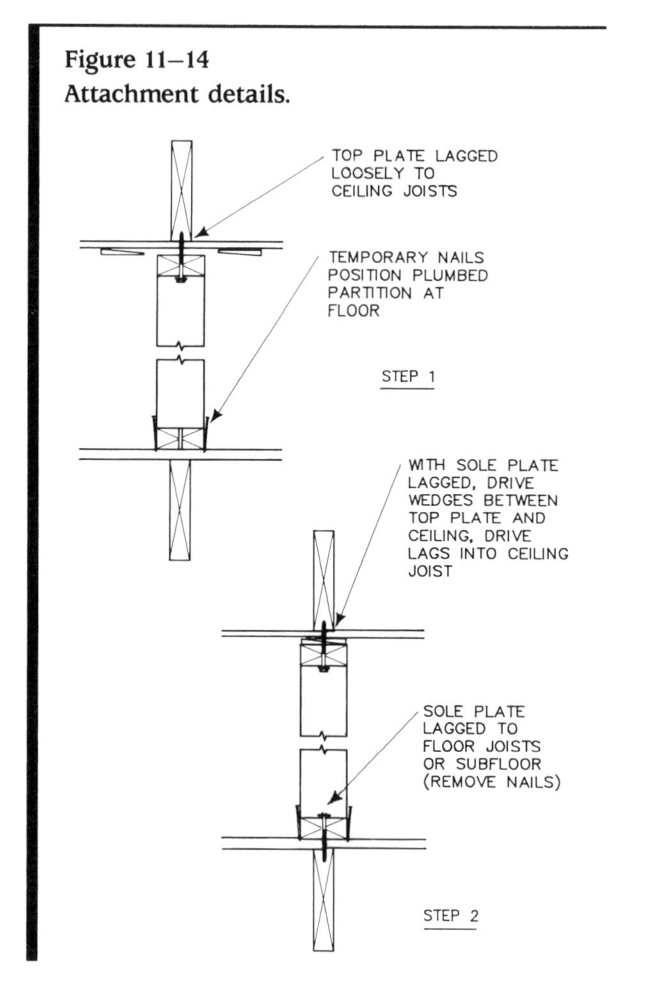
TOP PLATE LAGGED LOOSELY TO CEILING JOISTS
TEMPORARY NAILS POSITION PLUMBED PARTITION AT FLOOR
STEP 1
WITH SOLE PLATE LAGGED, DRIVE WEDGES BETWEEN TOP PLATE AND CEILING, DRIVE LAGS INTO CEILING JOIST
SOLE PLATE LAGGED TO FLOOR JOISTS OR SUBFLOOR (REMOVE NAILS)
STEP 2

Figure 11–15
The partition framework installed, ready for covering.

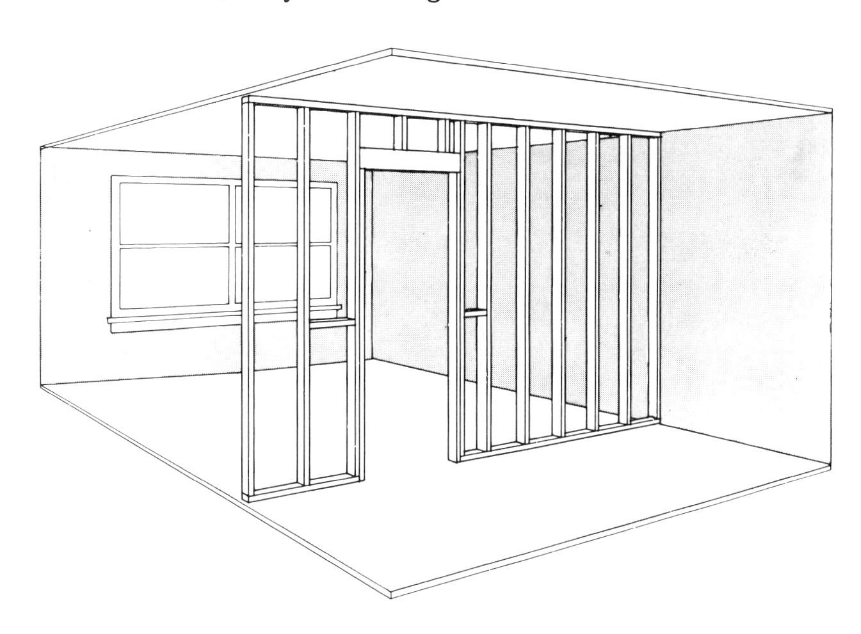

Figure 11–16
Framing details.

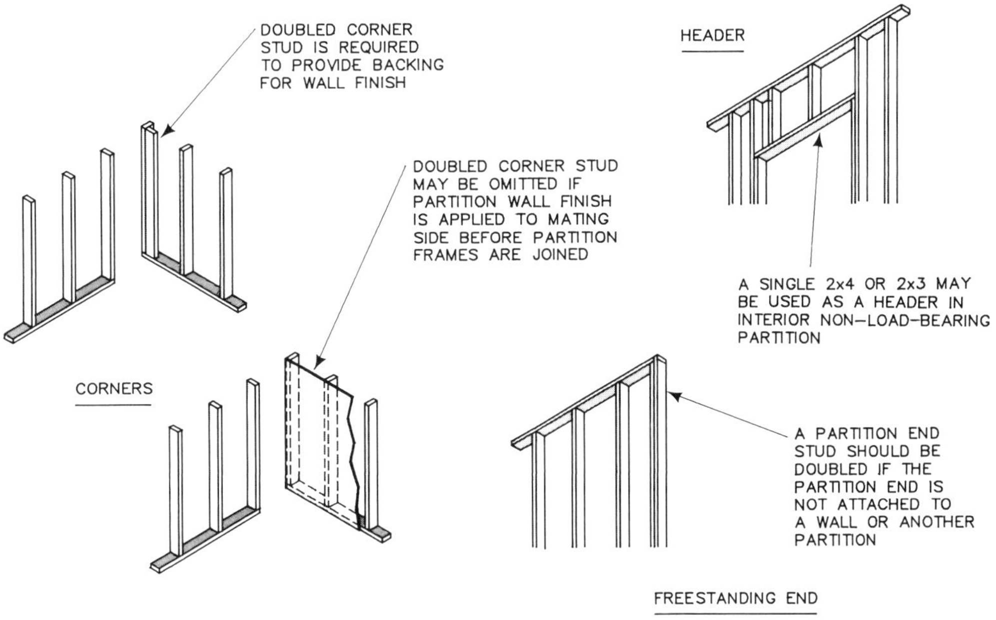

DOUBLED CORNER STUD IS REQUIRED TO PROVIDE BACKING FOR WALL FINISH

HEADER

DOUBLED CORNER STUD MAY BE OMITTED IF PARTITION WALL FINISH IS APPLIED TO MATING SIDE BEFORE PARTITION FRAMES ARE JOINED

A SINGLE 2x4 OR 2x3 MAY BE USED AS A HEADER IN INTERIOR NON–LOAD–BEARING PARTITION

CORNERS

A PARTITION END STUD SHOULD BE DOUBLED IF THE PARTITION END IS NOT ATTACHED TO A WALL OR ANOTHER PARTITION

FREESTANDING END

12 Gypsum Board

Gypsum board, wallboard, drywall, plaster board, and Sheetrock are all the same thing—the most widely used material for finishing interior walls. No other material is as practical, inexpensive, and easy to use. Properly selected and installed, gypsum board meets residential building codes for fire retardation. Gypsum board is easy to cut and install. Its only drawback is that it is heavy and sometimes hard to handle in full panels.

STANDARD PANELS

Gypsum board panels consist of a core of noncombustible gypsum plaster sandwiched between layers of heavy paper. The back side is covered with rough paper; the face is covered with a smooth, hard, gray, white, or cream-colored paper that can be painted. Standard thicknesses are ⅜″, ½″, and ⅝″. Standard panels are 4′ wide and 8′, 10′, or 12′ long. The face paper is wrapped around the long edges to make joint finishing easier. The long edges are tapered to make taping joints easier.

Gypsum board can be applied in one or two layers, but the usual practice is one layer. Two-layer construction produces a physically stronger and more soundproof wall.

In most locations, ½″ gypsum board must be used for ceilings, and should be used on walls, although some building codes permit using ⅜″-thick panels on walls. The cost saving obtained by using ⅜″ instead of ½″ is slight. Half-inch gypsum board makes a more solid and substantial wall. Some metropolitan areas require ⅝″ fire-resistant gypsum board (see page 117) for all interior walls. Check your local building code carefully on this point.

Quarter-inch-thick gypsum board can be used for a new finished surface over an existing wall, or as the surface layer in double-layer application. It is not strong enough to be used alone as a single layer, and that application is not permitted by virtually all building codes.

ADDITIONAL VARIETIES OF GYPSUM BOARD

Several other kinds of gypsum board are used in residential construction in addition to the basic board described above.

Backer board. Used for the base layer in two-layer walls. It is made with square or tongue-and-groove long edges. (*Note:* In small jobs it is usually more practical to use ordinary gypsum board for the base layer as well as the finish layer.)

Sound-deadening board. When installed under regular gypsum board in a wall, sound-deadening board can significantly reduce noise transmitted through the wall. It is one of the easiest sound control products to install (see Chapter 15).

Moisture-resistant board (tile backer board). This gypsum board is manufactured for use as a base for adhesive-mounted ceramic and other tiles in kitchens and bathrooms. The core consists of noncombustible gypsum combined with an asphalt/wax emulsion for water resistance. This board may also be used simply with the surface painted except where it would be in direct contact with water, as in a shower or tub area.

Firestop gypsum board. While all gypsum board is noncombustible, firestop gypsum board—also called fire-resistant or "Fire Code" board—has textile glass filaments reinforcing the gypsum to keep the wall core intact even after the chemically combined water in the gypsum has been released. Fire-resistance ratings are based on how long a particular wall, ceiling, or floor construction system can contain a fire and meet certain other requirements. Firestop gypsum board is also made in foil-backed and moisture-resistant versions. The foil backing acts as a vapor barrier when the board is used for the interior finished surface of an exterior wall (see Chapter 16).

WORKING WITH GYPSUM BOARD

The basic tools needed in working with gypsum board are shown in Figure 12–1. The blade of the Stanley Surform tool shown will be ruined for woodworking use by the gypsum, so keep a separate tool for that purpose. With the front knob removed, the Surform easily gets into cutouts for electrical boxes, in addition to smoothing cut

Figure 12–1
Basic tools. (From left) Utility knife and steel tape, hammer, Surform rasp and 100-grit garnet paper, tin snips, 3″, 6″, and 10″ joint or taping knives, hawk, gypsum board T-square.

Figure 12–2

Cutting gypsum board. (A) Work from the face side. Locate the cut line, taking three measurements. Draw the cut line using a straightedge. (B) Score the line with a utility knife. Cut through the paper and into the gypsum. (C) Snap the board backward. Cut through back paper. (D) Clean up the edge.

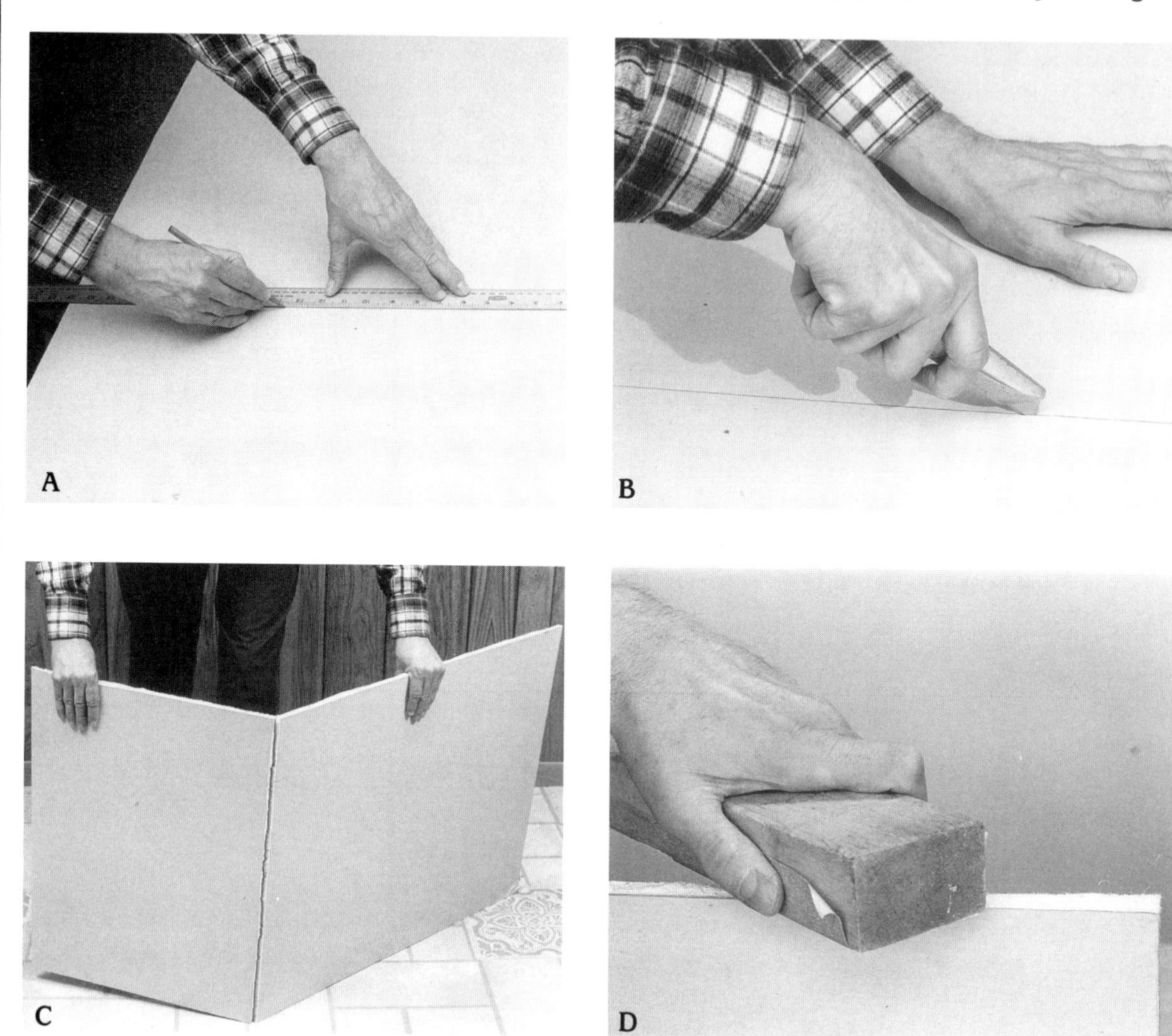

edges. As an alternative, use 100-grit paper wrapped around a block for sanding. The straightedge shown is called a drywall T-square. The tin snips are needed for cutting metal corner bead and expanded plaster lath. An abrasive-edge saber saw blade can be used to cut gypsum board, but the teeth of a keyhole saw or other saws or blades will be destroyed too quickly to be useful. The hawk is used to hold a supply of joint compound while you're up on the ladder, so you don't need to climb down to scrape each new knife-load out of the bucket.

Cutting

Gypsum board is cut by scoring the paper on the face side of the panel with a sharp utility knife. Then snap the board backward and slice through the paper on the back, as shown in Figure 12–2. The face scoring cut must be deep enough to go completely through the paper and slightly into the gypsum core. Use the Surform tool or coarse-grit sandpaper on a wood block to smooth ragged edges in order to achieve better fits between panels.

Figure 12–3
Cutting an opening. (A) Outline the opening on both sides. Lines are drawn extra-heavy for clarity. Cut through the paper on both sides; cut diagonal lines on back (or on front and back). (B) Tap the gypsum board on front and waste will drop through. (C) Clean up edge.

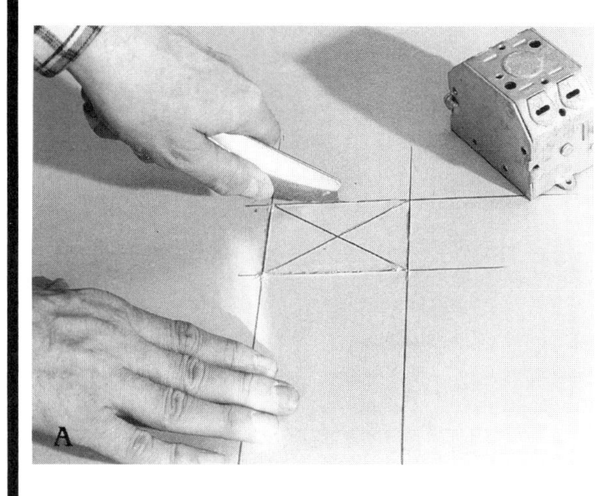

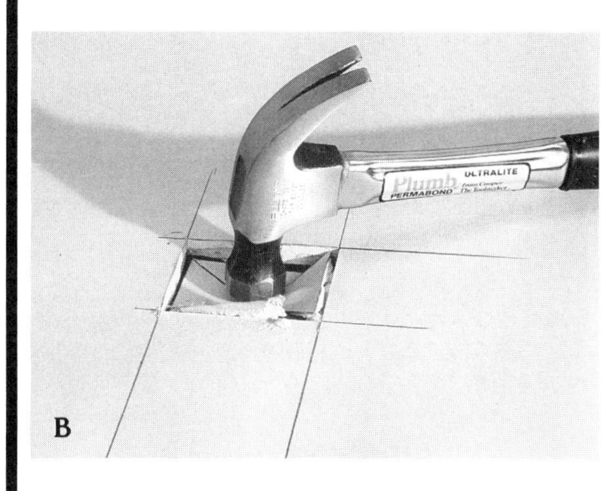

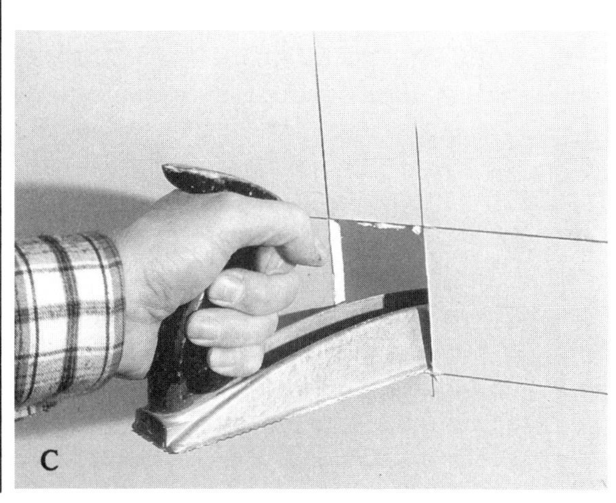

Making Cutouts

Small openings for outlets, switches, and the like can be made several ways. Sawing is one way, although as noted above the seemingly soft gypsum core can dull a blade quickly. Another way is to score the opening with a knife and break out the center, as shown in Figure 12–3.

To make a cutout, mark its outline on the face side of the gypsum board and drive finish nails through at the corners to indicate the corner locations on the other side. Score the cutout outline with a knife on both sides, and score corner-to-corner diagonals on the back side. A quick hammer blow on the front will snap out the gypsum in four pieces. The same technique can be used for cutting a corner.

Purchased metal templates for outlets and switch boxes provide another way to locate openings without making measurements. Be sure the power is turned off, then snap the template in position in the box mounted on the wall framing. Position the gypsum board panel and press its back side against the template. Prongs on the front of the template mark the box shape and location on the back of the panel. Proceed to drive nails at the corners, score both sides, and break out the center as before.

Cutouts for doorways and windows should be made after the full panel is fastened to the wall and across the opening. Score the opening outline plus diagonals, and break out the center. The raw edges will be covered by jambs and casing moldings.

Using Fasteners

Nails or screws (Figure 12–4) or a combination of nails (or screws) and adhesive can be used to apply gypsum board to wood framing. If nails are used they should be annular-ring gypsum board nails, and a technique called double nailing should be employed. Double nailing, explained below, prevents nail popping—a nail working its head out, above the surface—the bane of gypsum board walls and ceilings. Gypsum board screws have self-drilling points, special threads, and Phillips recessed bugle heads. They should be driven with a power screwdriver or a screwdriver attachment for an electric drill. Gypsum

Figure 12–4
Gypsum board fasteners.

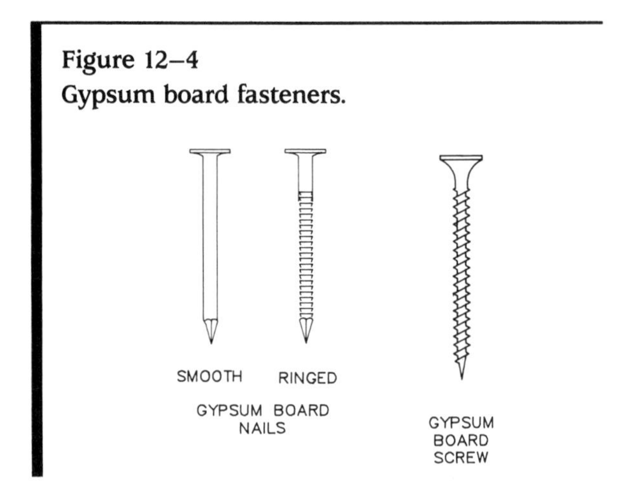

Table 12–1
GYPSUM BOARD NAIL AND SCREW SELECTION

Nail and Screw Description	Applications
1¼″ annular-ring nail	½″, ⅜″, and ¼″ gypsum board ½″ and ⅜″ gypsum backer board, to wood frame
1⅜″ annular-ring nail	⅝″ gypsum board to wood frame
2½″ 7d nail	⅝″ fire-resistant gypsum board layers over ¼″ gypsum sound-deadening board two layers ⅝″ fire-resistant gypsum board to wood studs
1⅞″ 6d nail, 13 gauge	⅜″ and ¼″ gypsum board over existing surface, wood frame
1⅞″ 6d nail, 13 gauge	⅝″ fire-resistant gypsum board to wood frame
1⅝″ 5d nail, 13½ gauge	½″ fire-resistant gypsum board to wood frame
1¼″ gypsum board screw	Single layer ⅜″ or ½″ gypsum board to wood frame
1⅝″ and longer gypsum board screw	Single layer ⅝″ gypsum board to wood frame All thicknesses gypsum board to wood frame under paneling, old gypsum board, or plaster finish wall

board mastic adhesive comes in cartridges. It is applied with the same sort of hand-operated gun used for caulking materials.

Table 12–1 gives nail and screw schedules; Table 12–2 gives recommended fastener spacing. Follow the directions on the cartridge for adhesive application.

Effects of Temperature

In cold weather, don't apply gypsum board with nails or screws below 40° F; it will be brittle and will crack with the pressure of fastening it. If you are using adhesive or taping joints, raise the room temperature to between 50° and 70° F for two days before you start, then keep the temperature in that range until the work is done and everything is dry. Provide adequate ventilation for drying.

Providing Support

For ⅜″ gypsum board, furring strips, studs, or joists must not be spaced more than 16″ on center. That is also best for ½″ or ⅝″ gypsum board, and essential for ceilings. However, wall supports for these thicknesses can be spaced up to 24″ on center if really necessary. If you nail a gypsum board ceiling to furring strips, the strips should be at least 2 × 2s. (The 2 × 2s available from lumberyards are usually warped. You can make your own by ripping 2 × 4s, but unless you are getting reasonably unwarped pieces, buy 2 × 3s.) If you are using gypsum board screws, you can use 1 × 2 furring strips if they are screwed (not nailed) to the ceiling joists above.

Taping

The flat joints between panels of gypsum board are filled to a smooth surface with special tape and joint compound. The name of the game in planning a gypsum board job is to keep this time-consuming joint taping to a minimum.

Table 12–2
GYPSUM BOARD NAIL AND SCREW SPACING

Gypsum Board Thickness and Application	Single Nailing	Double Nailing (Nails 2" Apart)	Screw Spacing (Framing 16" on Center)	Screw Spacing (Framing 24" on Center)
⅜" side walls	8" max.	12" max.	16" max.	Not recommended
½" side walls	8" max.	12" max.	16" max.	12" max.
½" ceilings	7" max.	12" max.	12" max.	12" max.
⅝" side walls	8" max.	12" max.	16" max.	12" max.
⅝" ceilings	7" max.	12" max.	12" max.	12" max.

APPLYING A SINGLE LAYER OF GYPSUM BOARD

Use ½" or ⅝" gypsum board for both walls and ceilings. When installed, all ends and edges of panels must be supported by framing or blocking. Some codes do not require blocking between studs to support the long side joint between two wall panels applied horizontally, provided you tape the joint and the studs are not more than 16" on center.

Apply gypsum board directly to wood framing members. Start with the ceiling, then do the walls. The ceiling is the difficult part of working with gypsum board. The length of the panels should run across joists, not with them. The ends will fall on joists that are 16" on center; if that is not the case, cut the panel to length so as to end on a joist. All the side edges must also be supported by framing, so you must put nailers between the joists. Use 2 × 3s. It will take two people to get a 4′ × 8′ panel up against the joists, and two T-braces to hold it there while you nail, screw, or glue it. The height of the T-brace should be about 1" more than the floor-to-ceiling height so as to wedge the panel tightly against the joists (see Figure 12–5). Use 2 × 3s for the braces.

Fasten all panels with nails or screws as specified in Tables 12–1 and 12–2. If you are using adhesive and nails, put a continuous bead of adhesive on joists and nailers, and use three nails per joist—one at the center and one at each edge to keep the panel against the joists until the adhesive sets. Cut panels accurately and butt them snugly, but don't force them together.

On walls, gypsum board can be applied vertically or horizontally on walls, as shown in Figure 12–6. Horizontal application, with the long edges across the studs, is preferred because there will be fewer joints to tape and the wall will be stronger. You can mix methods. Avoid panel end-to-end seams, called butt seams or joints, if possible, as they are more difficult to finish than tapered side edges.

Place nails or screws carefully. They should be at least ⅜" in from the edges and ends to avoid chipping the gypsum board. However, if the panel butts against another panel on framing (be sure the butt is centered), the nails or screws must be not more than ½" away from the edge to avoid missing the framing. In the field or central area of the panel, away from the edges, nails or screws should be centered on the framing. Don't guess frame positions; snap chalk lines to locate mid-panel framing.

Set nails slightly below the surface in a dimple formed by the hammer, as shown in Figure 12–7. Be careful not to break the face paper with the edges of the hammer head. If you mess up a screw or nail, remove it; don't plan to patch over it.

Single Nailing

Although suitable for closet interiors and other concealed locations, single nailing is generally not recommended; double nailing, described on page 122, is much better. For single nailing use only annular-ring gypsum board nails. Start nail-

Figure 12–5
Ceiling application. Panels must be applied across ceiling joists. Use T-braces to wedge each panel in place for fastening; also push the panel firmly upward in the immediate area as you drive nails.

ing in the center of the panel and work toward both ends, then nail the long sides and the ends, starting at the centers and working toward the corners.

Double Nailing

This is a superior method of putting up gypsum board, as it greatly reduces nail popping. Nail heads are covered with joint compound to hide them: wood frame shrinking or warping, or heavy vibration, may ultimately cause nails to loosen and "pop" above the finished gypsum board surface, or cause the compound over the nail head to come loose, like a manhole cover.

In double nailing, each panel is nailed twice. Use the same procedure as for single nailing, but space the nails no more than 12" apart. When the panel is completely nailed, drive a second set of nails 1½" to 2" away from each of the first set of nails. After the second set of nails is driven, strike each of the nails in the first nailing again to be sure the panel is fastened down tightly.

Using Screws

This method of installing gypsum board is better than nailing. The screws hold the board tightly, and you needn't worry about nail popping or about nailing a ceiling, which takes practice to do skillfully.

Use special gypsum board (drywall) screws. The screws have bugle (similar to flat-head) Phillips heads and sharp points for driving without pilot holes. A special power screwdriver called a screwgun is used to drive the screws. The screwgun has an adjustable depth stop control that gets the screw heads sunk just below the surface for compounding, but prevents you from breaking the paper or driving the screw right on

Figure 12–6
Wall application. Use either vertical or horizontal application on walls, whichever method results in fewer seams to be finished. You can mix methods.

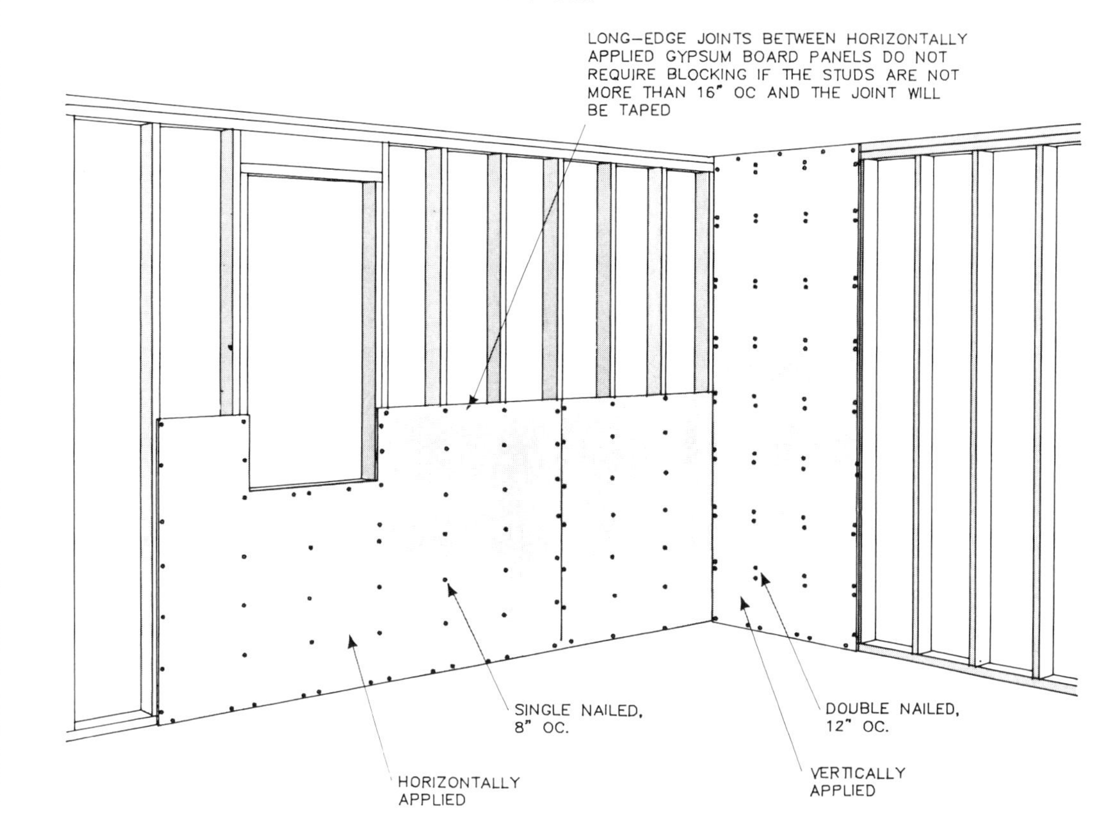
LONG–EDGE JOINTS BETWEEN HORIZONTALLY APPLIED GYPSUM BOARD PANELS DO NOT REQUIRE BLOCKING IF THE STUDS ARE NOT MORE THAN 16" OC AND THE JOINT WILL BE TAPED
SINGLE NAILED, 8" OC.
DOUBLE NAILED, 12" OC.
HORIZONTALLY APPLIED
VERTICALLY APPLIED

Figure 12–7
Correct nail setting.

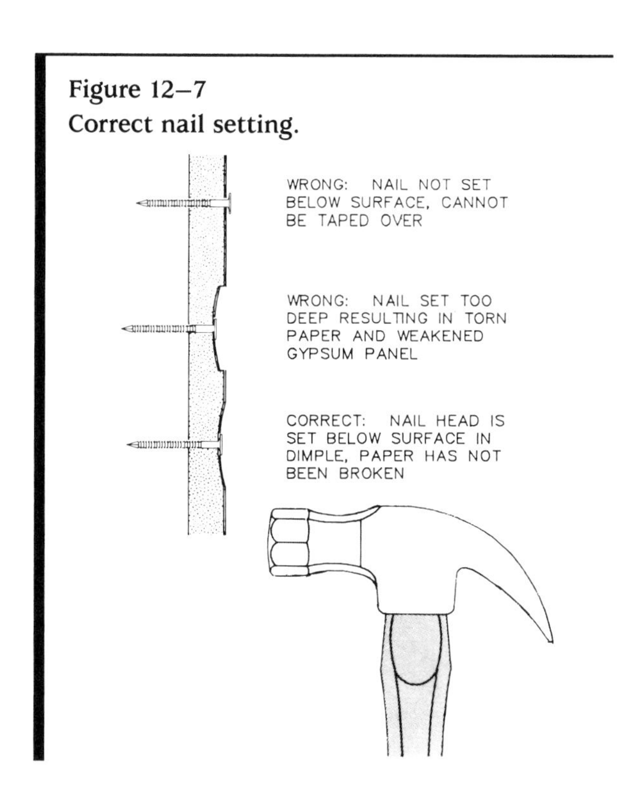
WRONG: NAIL NOT SET BELOW SURFACE, CANNOT BE TAPED OVER
WRONG: NAIL SET TOO DEEP RESULTING IN TORN PAPER AND WEAKENED GYPSUM PANEL
CORRECT: NAIL HEAD IS SET BELOW SURFACE IN DIMPLE, PAPER HAS NOT BEEN BROKEN

through the gypsum board. You can also buy an adapter (about $8) for use with any power drill (having a low speed) or high-torque power screwdriver. The adapter (Figure 12-8) drives screws to a preset depth and disengages.

With studs and joists 16″ on center, the screws should be spaced no more than 12″ apart on ceilings and 16″ on walls. With framing on 24″ centers, screws should not be more than 12″ apart on both walls and ceilings.

Applying Adhesive

Gypsum board can be applied to wood framing with special gypsum board (drywall) mastic adhesive in cartridges. Apply the adhesive in an intermittent or continuous bead; follow the directions on the cartridge. Secure the gypsum board to the framing temporarily with nails or screws, which can be widely spaced.

Figure 12–8
To drive gypsum board screws use a drywall power screwdriver or a special adapter for use with a cordless screwdriver or drill.

APPLYING A DOUBLE LAYER OF GYPSUM BOARD

Two-layer gypsum board walls provide a sturdier wall, greater fire-spread resistance and better soundproofing. The base layer should be ½″ thick for most applications; the top layer can be ¼″, ⅜″, or ½″ thick. The base layer is nailed, screwed, or glued to the framing; the face layer is laminated to the base layer with adhesive. Backer board or sound-deadening board may be used for the base layer in place of regular gypsum board. The joints of the two layers must be staggered and kept at least 10″ apart.

While the adhesive cures, hold the face layer in place with nails through scraps of gypsum board that can be easily broken away for nail removal.

Figure 12–9
Finishing a flat joint. Step 1: Lay self-adhesive fiberglass mesh joint tape directly on the gypsum board joint. With paper tape, first spread a bed layer of joint compound into the seam, then smooth the tape into it. Step 2: Apply a coat of joint compound to cover the tape and feather out 4″ on each side. Step 3: When dry (24 hours), wet-sand, then apply second coat of compound, feathered out to 6″ each side. Step 4: Repeat step 3, feathering out to 10″ on each side.

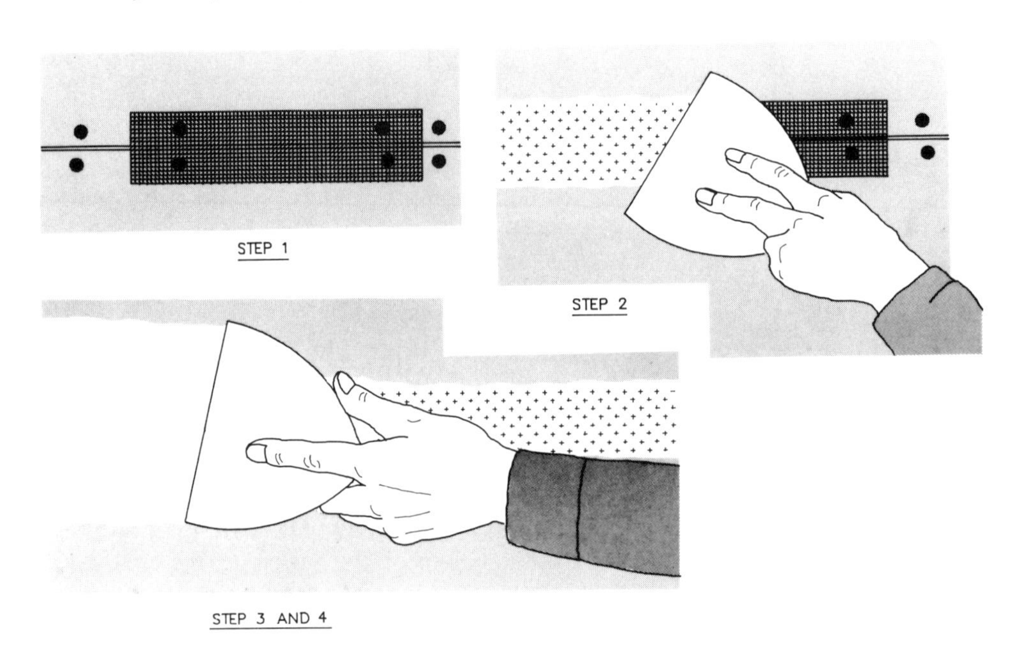

TAPING JOINTS

Flat joints between panels and inside corners are taped with joint tape and compound. Metal corner bead is used in place of tape when finishing outside corners. (Inside and outside corners may also be finished with wood molding.) Joint compound can be purchased in gallon pails. Finishing consists of bedding the tape in the compound, and troweling and sanding the covering compound into a smooth cover over the joint. Nail- and screw-head dimples in the field are also filled with joint compound to form a smooth, flat surface.

Joint tape comes in two forms—a 2″-wide paper tape or an adhesive-impregnated 2″-wide fiberglass mesh. The tape bridges the seam between the panels to form a monolithic wall and helps prevent cracks from forming. The fiberglass tape costs more, but does a better job of preventing later cracks. Seams between tapered or beveled edges are easy to tape and finish to an invisible joint when painted or covered over.

If you use paper tape, begin by laying down a bedding coat of joint compound with a 4″ taping knife (see Figure 12–9). Work the compound into the joint, spreading a continuous 4″-wide layer along the joint. Cut a strip of tape as long as the seam (use scissors) and press it firmly into the compound with your knife. Some but not all of the compound should be squeezed out. Smooth the surface and apply a thin layer of compound over the tape. Feather it out onto the panel surface 4″ on each side of the joint. Do this first step on all the joints to be taped and then quit for the day.

If you use fiberglass mesh tape, unroll and press the tape along the joint with your fingers. Cut the tape after it is in place. Now apply the first coat of compound over the tape, working it through the tape into the joint, and feather out 4″ on each side. Quit for the day.

After 24 hours the compound should be dry and ready for the next coat. Wet-sand the surface to knock off high spots with a damp, rough cloth wrapped around a block of wood. The advantage of wet sanding is no dust. Rinse the cloth often. When sanding be careful not to expose the tape.

Now apply another coat of compound and feather it out on each side of the joint 2″ beyond the first coat—that is, 6″ on each side of the joint. You are actually mounding the compound slightly over the joint and you want the mounding to be as invisible as possible. Don't try for a perfectly smooth coat yet.

After another 24 hours, sand and apply another coat feathered 10″ to each side. This time you can sand with a power sander to smooth the surface, but be very careful not to abrade the adjacent paper surface. These three coats should do it, although a partial fourth coat to fill the last of the irregularities may be required.

Finishing Inside Corners

Inside corners can be taped or hidden with wood molding as shown in Figure 12–10. Cut paper tape the length of the joint and crease it down the middle. Apply a layer of compound 1½″ wide on both sides of the corner. Bed the tape in the compound and feather. Be sure the crease in the tape is positioned at the center of the corner the full length of the joint. Continue with additional coats as for a flat joint. If you use fiberglass mesh tape, press it evenly into the corner and proceed as for a flat joint. The greater strength of the fiberglass tape makes a superior corner.

If you use wood molding, aim the nails so they pass through the gypsum board into the wood framing.

Finishing Outside Corners

Outside corners can be finished with metal corner bead or with wood molding (see Figure 12–11). Nail the bead to the corner in one piece top to bottom, with nails in both flanges opposite each other, not alternating. Be sure the bead is centered on the edge of the corner.

Spread joint compound with a 4″ knife, covering the bead flange, nails, and gypsum board 3″ out from the bead nose on both sides of the corner. Continue with additional coats as for a flat joint.

Figure 12–10
Inside corner joint finishing methods. (Left) Tape and compound is the preferred method for wall-to-wall and wall-to-ceiling joints. (Center) Use inside-corner wood molding on wall-to-wall inside corner joints. (Right) Use crown, bed, or cove molding at wall-to-ceiling joints.

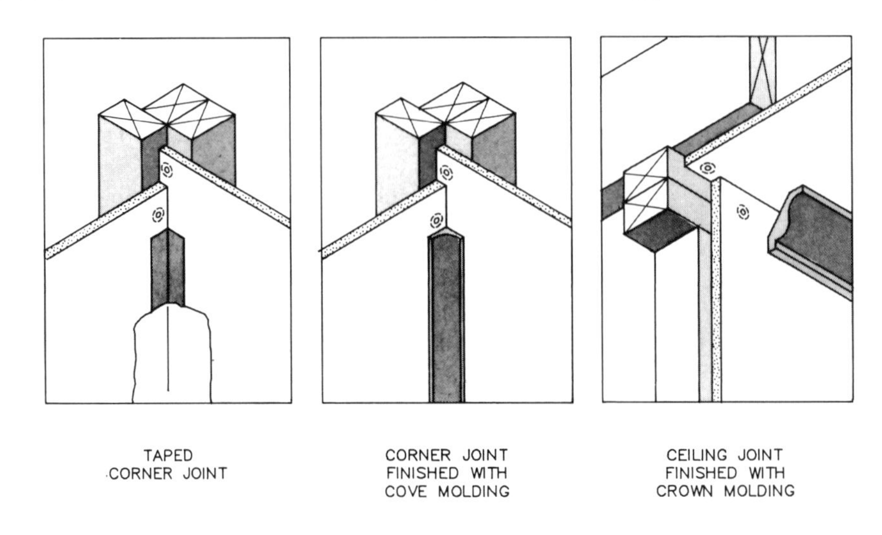

Figure 12–11
Finishing an outside corner. (A) Use outside corner molding. (B) Use outside corner metal bead and joint compound. (C) Nose of bead should remain clear of compound.

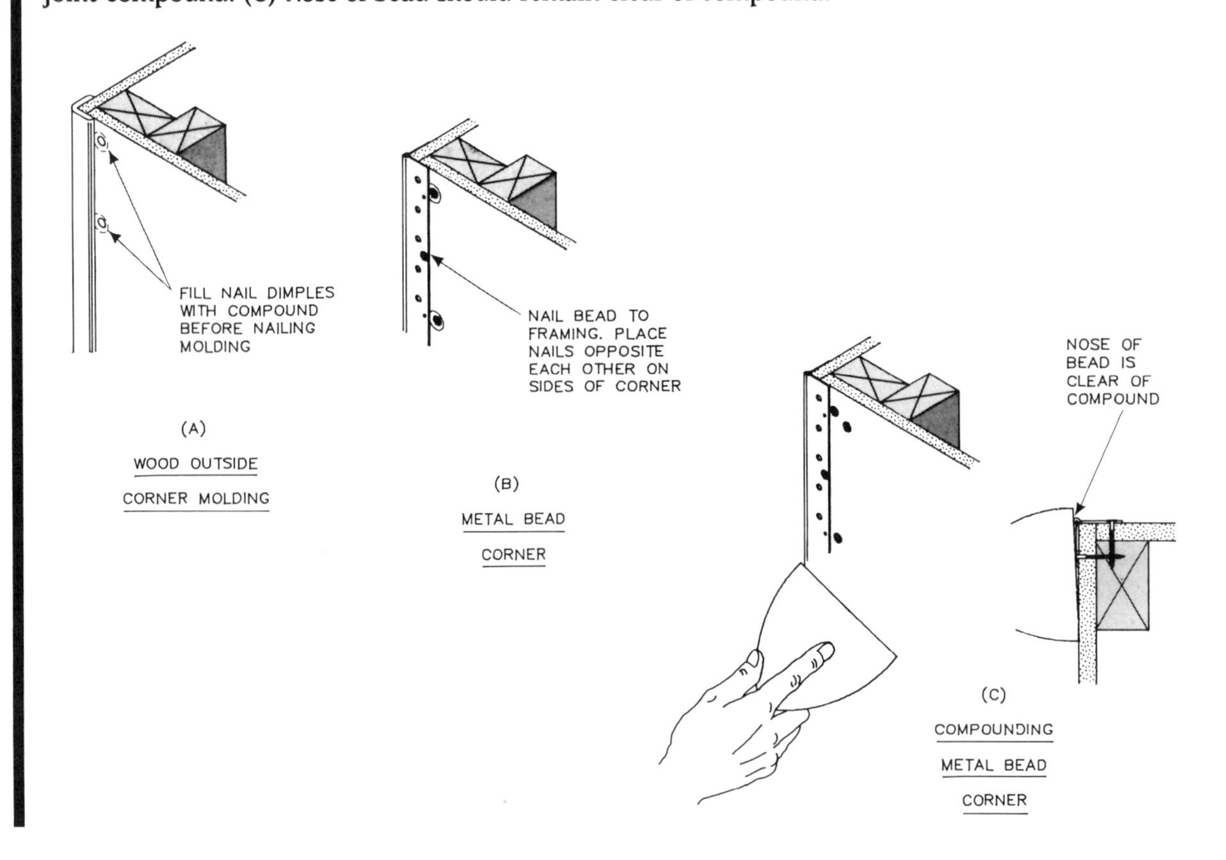

GENERAL TIPS

Do not locate gypsum board joints at the edges of openings such as doorways and windows; locate them midway across the opening even if it seems to be wasting gypsum board. The best way to handle an opening is to wall right over it, then cut out the opening. A gypsum board joint at the edge of an opening will be forever cracking.

Stagger end joints of horizontally applied gypsum board. Joints on opposite sides of a partition must not be located on the same stud, whether the gypsum board is applied horizontally or vertically. Otherwise, the result will be a weak wall.

Press firmly to keep the gypsum board tight against the framing when driving nails or screws.

Use your head to hold gypsum board against the ceiling for fastening. Wear a cloth cap to pad your head and to avoid transferring oil from your hair or scalp to the paper surface.

When applying gypsum board horizontally, install the upper piece first so you get a tight joint with the ceiling, which you should have already installed.

Leave a ½″ gap at the bottom, even if you have to cut off material to do it. This way you can get a foot-powered lifter under the panel to push it tight against the ceiling or other panel. You can improvise a lifter by placing a prybar or thin piece of wood across a dowel with one end under the edge of the panel. Step on the free end to lever the panel upward.

Bending and stooping to load your knife out of the can each time when taping is hard on the back, and on the legs if you are working on a ladder. Make or buy a hawk to hold a supply of compound.

APPLYING GYPSUM BOARD OVER MASONRY

There are several ways to apply gypsum board to masonry walls. For all methods, the masonry wall must be flat, with all mortar or concrete sticking out from the surface chiseled off. Don't use gypsum board in a basement where the dampness cannot be controlled or eliminated entirely.

Furred to Uninsulated Masonry

Attach furring with masonry nails, anchors in holes, or adhesive-backed anchors. Run wood furring horizontally at the floor line as shown in Figure 12–12. Then locate and fasten furring on 24″ centers horizontally or vertically, depending on how you plan to apply the gypsum board. The long dimension of the gypsum boards should be run at right angles across the furring. Center end joints over furring or add blocking. Support the gypsum board with blocking or furring around all openings and cutouts as close to the edge as possible.

Over Foam Insulation on Masonry Walls

The insulation may be urethane or extruded polystyrene foamboard; expanded-bead polystyrene is not recommended. Figure 12–13 shows how to make this kind of installation. If you are going to install the gypsum board horizontally, install wood furring to support the panel at all edges. Provide furring strips horizontally at the top and the bottom of the wall, midway on the wall to back up the horizontal joint between gypsum board panels, and vertically everywhere there will be a butt joint between panel ends. Also fur around all door and window openings. The furring should be nominally 2″ wide and *exactly* ⅓₂″ thicker than the foamboard. Use a nail gun and power-driven nails to secure the furring only to a poured concrete wall. With any kind of concrete or cinder block wall, use hammer-driven nails or screws driven into plastic masonry anchors. If the wall is old, do a test nailing to see what style of fastener will hold without damaging the wall.

If the gypsum board is to be applied vertically, furring is required at both the top and the bottom of the wall, at edge joints, and around any openings.

Cut the foamboard as necessary to fit without

Figure 12–12
Gypsum board applied to masonry with furring.

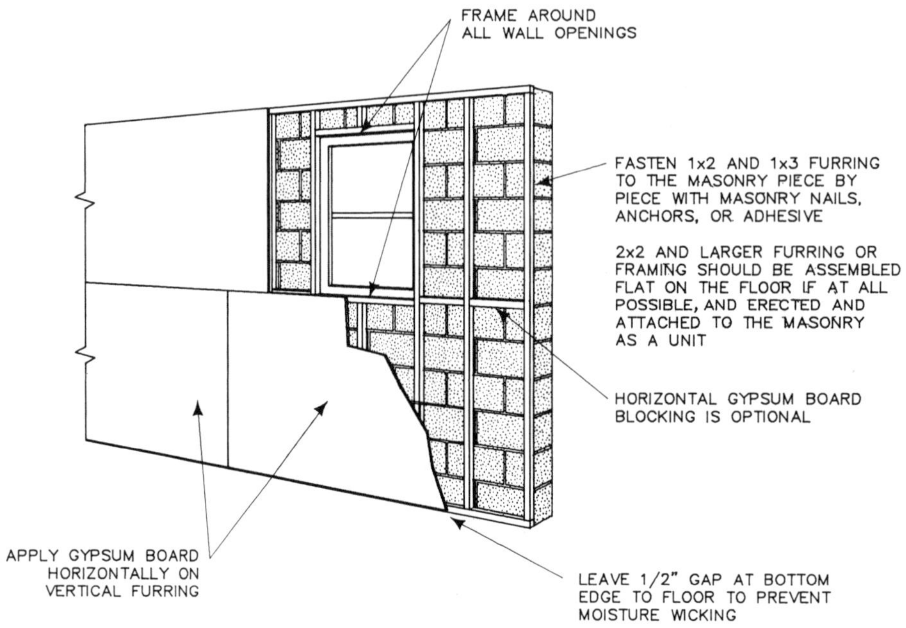

FRAME AROUND ALL WALL OPENINGS

FASTEN 1x2 AND 1x3 FURRING TO THE MASONRY PIECE BY PIECE WITH MASONRY NAILS, ANCHORS, OR ADHESIVE

2x2 AND LARGER FURRING OR FRAMING SHOULD BE ASSEMBLED FLAT ON THE FLOOR IF AT ALL POSSIBLE, AND ERECTED AND ATTACHED TO THE MASONRY AS A UNIT

HORIZONTAL GYPSUM BOARD BLOCKING IS OPTIONAL

APPLY GYPSUM BOARD HORIZONTALLY ON VERTICAL FURRING

LEAVE 1/2" GAP AT BOTTOM EDGE TO FLOOR TO PREVENT MOISTURE WICKING

Figure 12–13
Gypsum board applied to masonry over foamboard insulation.

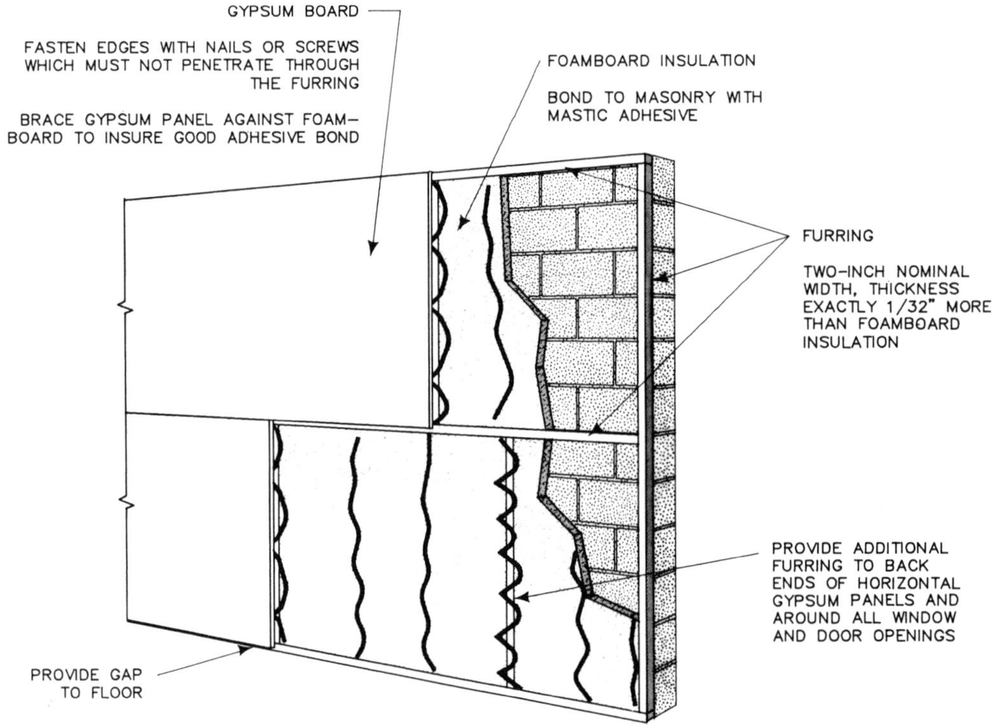

GYPSUM BOARD

FASTEN EDGES WITH NAILS OR SCREWS WHICH MUST NOT PENETRATE THROUGH THE FURRING

BRACE GYPSUM PANEL AGAINST FOAM-BOARD TO INSURE GOOD ADHESIVE BOND

FOAMBOARD INSULATION

BOND TO MASONRY WITH MASTIC ADHESIVE

FURRING

TWO-INCH NOMINAL WIDTH, THICKNESS EXACTLY 1/32" MORE THAN FOAMBOARD INSULATION

PROVIDE ADDITIONAL FURRING TO BACK ENDS OF HORIZONTAL GYPSUM PANELS AND AROUND ALL WINDOW AND DOOR OPENINGS

PROVIDE GAP TO FLOOR

forcing between the furring strips. Apply the foamboard to the clean masonry wall with adhesive. If the masonry has been painted, test-bond a sample with the adhesive to see if it will hold. Apply adhesive around the perimeter of the foamboard in a continuous ⅜" bead, and lengthwise every 16". Position the foam panel on the masonry wall with a sliding motion and hand-press the entire panel to ensure full contact of the adhesive and wall surface. With some adhesives (see cartridge directions), the foam panel has to be pulled off the wall momentarily to allow solvent flash-off and then be pressed into position again.

Apply the gypsum board to the foamboard with adhesive in the field, and with nails or screws into the furring along all the edges. Be sure you use an adhesive compatible with both the foamboard and the gypsum board. The nails or screws will be permanent. They should be spaced 12" apart, and just long enough to not penetrate through the furring. Leave a ⅛" minimum gap at the floor to prevent moisture wicking into the gypsum board. Temporary bracing is usually required to hold the gypsum board firmly against the foamboard insulation until the adhesive has set.

Figure 12–14
Gypsum board applied in a tub enclosure.

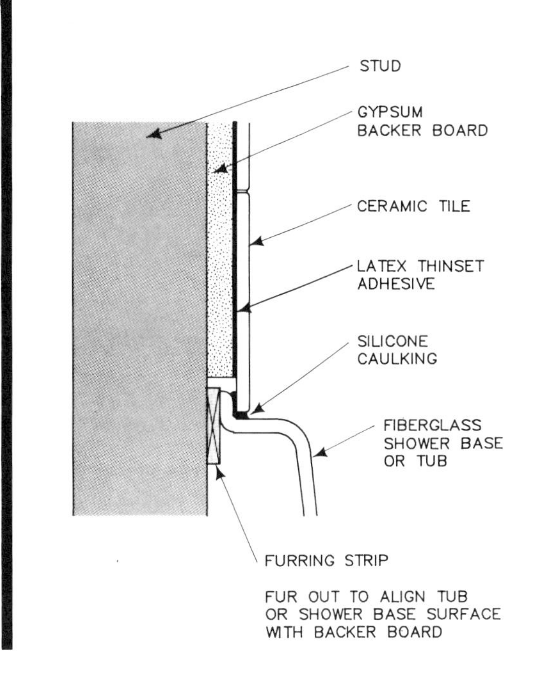

BATHROOM INSTALLATIONS

Moisture-resistant gypsum board is used in tub and shower enclosures as a base for adhesive-mounted ceramic tiles. Install shower pans and enclosures and tubs before putting up moisture-resistant board. Position tubs and shower enclosures so their lips will be flush with the face of the moisture-resistant board (see Figure 12–14). The tile can then be brought down over the board to form a seal over the lip joint.

Do not install a vapor barrier behind moisture-resistant gypsum board. The board itself provides the barrier. Apply moisture-resistant board horizontally to the framing, using nails spaced 8" or screws spaced 12" apart. Leave a ¼" gap between the board edge and the shower enclosure or tub top.

Tape joints and fill nail spots of the moisture-resistant gypsum board above the area to be tiled in the normal manner with joint compound. Raw edges of the board and cutouts, as well as joints and fasteners, should be caulked with the waterproof adhesive that will be used to set the tile.

13 Plaster and Gypsum Board Repair

On the surface, wet-plastered walls and gypsum board "dry" walls look about the same. The similarities go deeper. Both develop cracks (the crack patterns tend to be different) in the normal course of a house's aging, sometimes as a result of a house's structural problems. Both may be surface-gouged or have holes punched in them on occasion. And both can be damaged by water.

The repair techniques for plaster walls and gypsum board walls are not the same because the walls are constructed differently. However, the repair materials used—patching plaster, spackling paste, and joint compound—are the same. Chapter 12 explains how gypsum board walls are constructed; Figure 13–1 shows three wet-plaster wall construction methods.

TOOLS AND MATERIALS

Tools. Plaster and gypsum repair tools are shown in Figure 13–2.

Patching plaster. Patching plaster is used for large repairs. It is packaged as a dry powder, principally gypsum, which you mix with water, making up only what you can use in a few minutes. Patching plaster starts setting fast and its working time cannot be extended by adding more water. The advantage of using patching plaster for repairs is that it dries fast, so you can apply multiple coats a few hours apart. When using patching plaster, the repair surface must be wet.

Spackling paste. Spackling paste also comes ready-mixed. It is used fairly stiff, dries moderately fast (a few hours), and is used for filling cracks, surface gouges, and small holes. Ready-mixed spackling paste is also good for patching woodwork before painting. The repair surface must be dry.

Joint compound. Joint compound or joint cement is sold ready-mixed in one- and five-gallon containers. It is intended primarily for use in taping gypsum board joints, but it also makes an

Figure 13–1
Plastered wall construction. (A) Plaster on wood lath construction found in older houses. (B) Plaster on expanded metal lath. (C) Plaster on gypsum board lath.

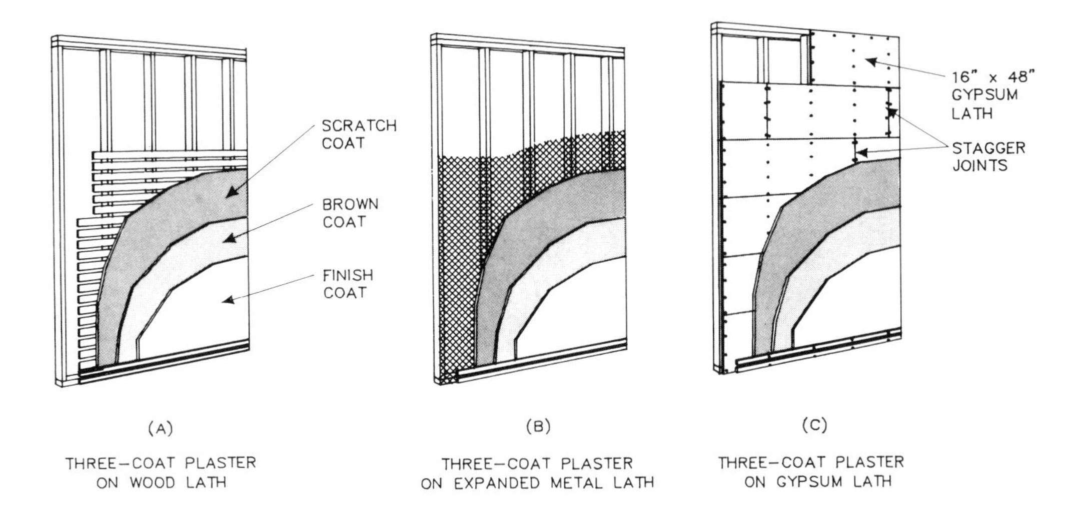

Figure 13–2
Tools for plaster and gypsum board repair.

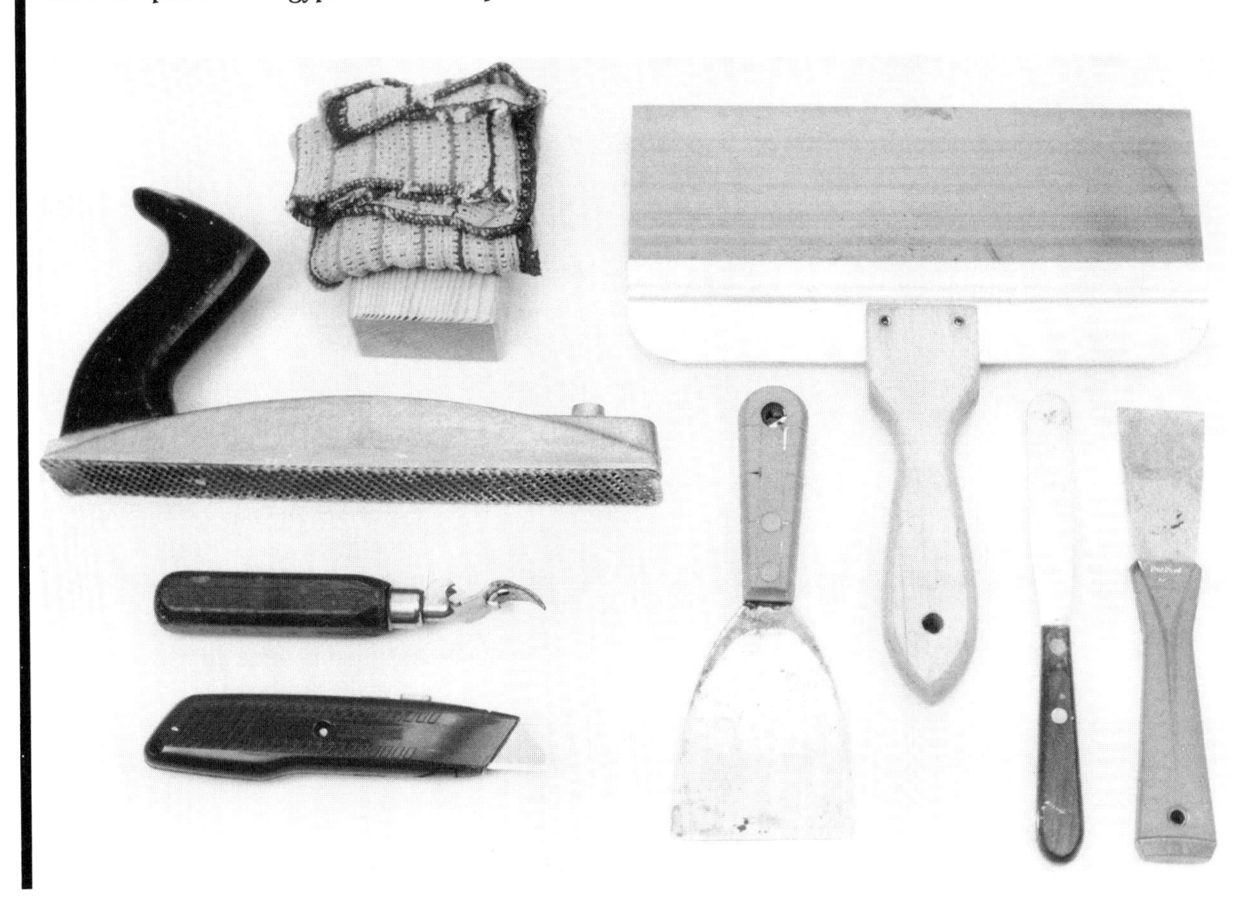

excellent crack filler and finish coat on repairs involving patching plaster because it is easy to work with, especially in feathering out the repair onto the surrounding smooth surface. It is much thinner than spackling paste, but it takes 24 hours to dry.

FILLING CRACKS

Cracks in gypsum board and plastered walls and ceilings are the result of movement of the supporting framing. This movement can be caused by normal settling, vibration, the wood framing drying out, seasonal expansion and contraction, or joist or beam deflection under a load. Cracks can also be caused by structural problems. In particular, a crack that won't stay repaired and keeps getting wider or changes seasonally may indicate a structural or foundation problem. If you have this kind of crack to deal with, investigate carefully to identify the cause.

Hairline cracks in plaster are hard to repair because it is difficult to get the patching material deep into the crack or, for that matter, into the crack at all. Drywall joint compound is the preferred material because it has a thinner working consistency and so can be forced into the crack more successfully.

With any of the repair materials, use a stiff-blade putty knife and heavy blade pressure, with the blade almost flat against the surface. Use crisscrossing strokes to fill the crack, then remove the excess from the surface and sand when dry.

If you can get the thickness of your putty knife blade into the crack, it's not a hairline crack. To make a successful repair, you must get the patching material packed into the crack in a way that will lock it in place. To begin, get all loose material out of the crack by scraping with a sharp hooked instrument such as an old beer can opener (not so common in this era of tab-opening can tops) or a pointed hook scraper. Don't use the corner of your putty knife or a screwdriver, because the plaster will damage the blade. If the crack is ¼" or more wide, undercut the edge on both sides so the crack is wider at the bottom than at the surface. Pack the material hard against the sides and bottom of the crack so that it will spread out and lock, or key, into the undercut. Because all patching materials shrink when they dry, you will have to apply a second layer to large cracks after the first layer has dried thoroughly.

Cracks in gypsum board walls most often occur at the edges of the panels where the joints have been taped. If there is a clean split in the tape and its covering joint compound, blow any debris from the crack and try filling it with joint compound, as in filling a hairline crack. This will work only if the tape is still tightly in place. If the tape is loose, it must be removed along with most of the joint compound, and the joint retaped from scratch. Taping over an old taped joint usually results in a wavy, uneven wall.

PATCHING SMALL HOLES

The first step is to prepare the edge of the hole for the patch. Break out all loose and crumbling plaster or gypsum board to get to firm supporting edges.

You must have something behind the hole to keep the patching material from falling through. For small holes, you can pack in wads of newspaper as shown in Figure 13–3. The paper only has to stay in position long enough for the first coat to set. You can also cut a scrap of window screen, wire cloth, or metal lath slightly larger than the hole and hold it in position behind the hole with a loop of string brought out and tied around a stick bridging the hole. After the first coat has set, cut the string. The procedure is illustrated with expanded metal lath in Figure 13–4.

Another way to back up patches is to pass a piece of wood or gypsum board through the hole and glue it to the back of the wall with panel adhesive. Leave a finger hole in this backup piece for handling.

A hole to be backed by newspaper or expanded lath should be flared—that is, the surface edges

Figure 13–3
To patch a hole, trim the gypsum board back to solid material. Make the hole edges alternately flared and undercut to lock in the patch. Use wadded newspaper for backing.

cut wider than the back edges—to keep the patch from falling in. With a flared hole, butter in the patching material so it also wraps around the back edge of the hole, keying it so it cannot fall out. The edge of a hole backed by wood should be alternately flared and undercut for keying because you cannot rely on adhesion to the wood to hold the patch in place.

With the hole backed, you can proceed with the patching two ways: you can trowel in patching plaster, spackling paste, or joint compound in thin layers, or you can cut a piece of gypsum board to fill the hole and tape the edges. You can also use a combination of both ways, laying in a gypsum board patch thinner than the wall thickness and troweling patching material over it.

For a troweled repair, mix enough patching plaster for the first coat to a stiff paste, wet the edges of the hole, and butter the plaster into the hole and around the back edge. Smear a thin layer on the newspaper or wood backing; force plaster through the mesh of a screen backing. When the plaster begins to harden, scratch the surface to provide a tooth for the next coat. Wait until the plaster is dry before applying the next

Figure 13–4
To back a patch with expanded metal lath, flare the edges of the hole. (A) Cut the lath larger than the hole. (B) Attach a string and pass lath through the hole. (C) Use the string to hold lath, and butter in the first patching plaster coat. Cut the string when the first coat is dry, then add the finish coat.

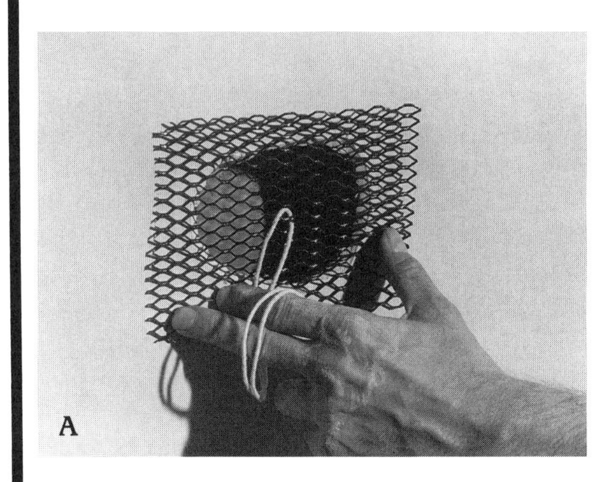

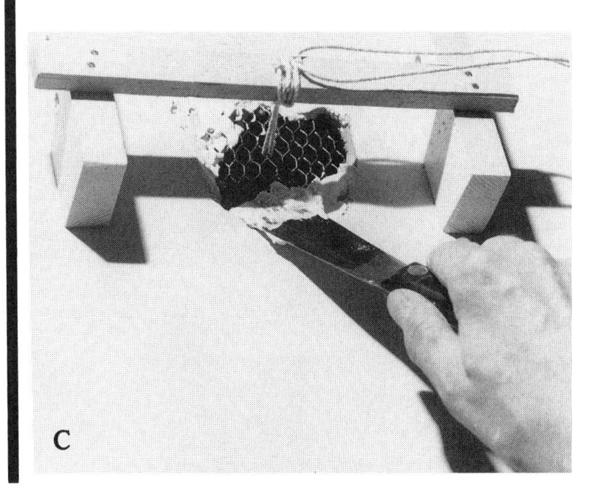

coat, but when you do, wet the old surface first with a sponge or mister. If the final coat over plaster is to be joint compound, do not wet the surface before applying it.

PATCHING A LARGE HOLE WITH GYPSUM BOARD

To fill a hole entirely with a gypsum board patch, first cut the hole to the shape of a rectangle or triangle and fasten something behind the hole to hold the patch in place. Cut the patch to fit the hole, butter its edges and back with plaster, spackling paste, or joint compound, and press it in place. Fill the edge spaces completely, then remove excess material from the surface. If the patch surface lies below the wall surface, when the edge filler is dry, build up the surface with joint compound.

Figures 13–5, 13–6, and 13–7 show how to patch holes in gypsum board more than a foot across and large holes in plastered walls where the lath is damaged or missing. First, cut back the hole to the edges of studs or joists. Attach 1 × 2 cleats to the side of the studs or joists to support the patch or new lath.

Don't attempt to trim the edges of the hole to the centerline of the stud or joist to save the trouble of putting in the cleats. You will not end up with a very firm attachment for the patch, and you will have weakened the rest of the wall. Use gypsum board screws to attach the cleats; don't

Figure 13–5
To patch a large hole, draw lines at the edges of studs or joists and at the top and bottom of the patch.

Figure 13–6
Cut out the wall and attach nailers to the studs or joists.

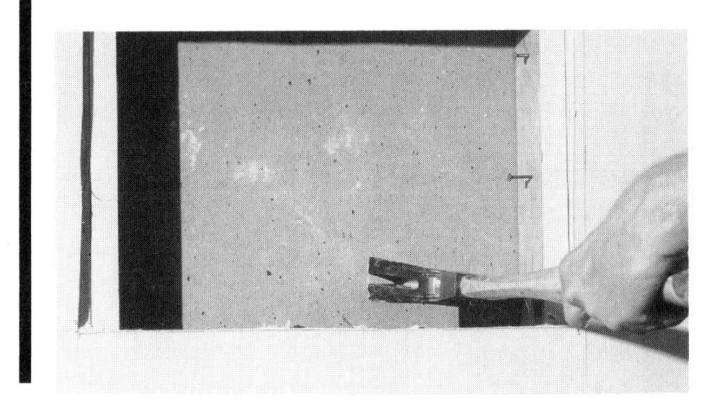

Figure 13–7
Attach a gypsum board patch to the nailers with gypsum board screws, then finish the joint with tape and compound.

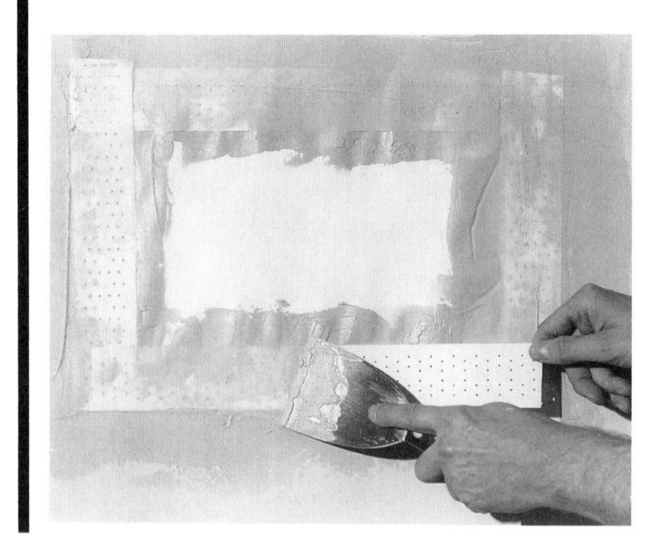

use nails. Hammering, particularly hammering on the side of framing members, will loosen nails holding the old wall in place. For additional patch edge support, you might want to attach a wood strip or 2× lumber across the hole edges between the framing. Use screws or mastic adhesive.

Now cut the patch to fit, butter the edges with spackling paste or joint compound, and fasten it to the cleats with gypsum board screws. To complete the repair, tape the edges.

If you can get at the hole from the back, such as a ceiling hole accessible from the attic, metal lath can be tacked in place or even held down against a ceiling with bricks.

PATCHING A PLASTERED CEILING

Figures 13–8, 13–9, 13–10, and 13–11 show the steps in repairing a large water-damaged area in a ceiling constructed of wet-plaster layers over gypsum board lath. The captions describe how the patch is installed and finished.

Figure 13–8
A water leak at the chimney turned a small area of the plaster of this ceiling to mush and caused the finish coat of plaster to separate over a wide area.

Figure 13–9
Trimming back to sound plaster revealed the water-damaged area to be much larger than was originally apparent. The plaster was trimmed to the joists and nailers were screwed to the sides of joists and the bridging. It is important to cut away all of the damaged plaster before starting the repair.

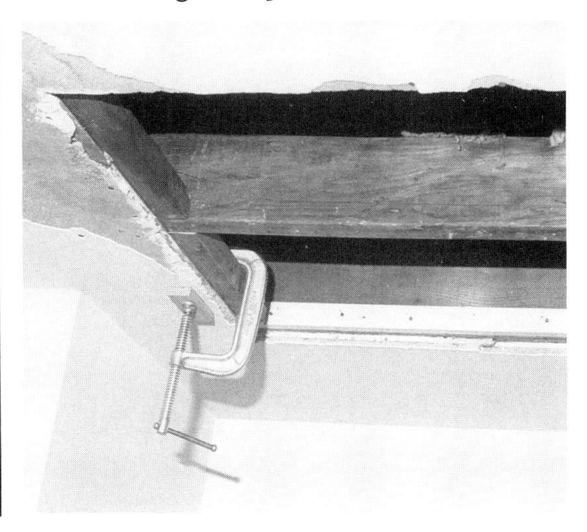

Figure 13–10
The repair patch consisted of two layers of gypsum board. The first was nailed to joists and attached to the bridging with plasterboard screws. The edge of the ceiling was also screwed to the bridging.

Figure 13–11
The second piece of gypsum board was attached with panel adhesive and screws, then the joint was taped.

If you find that the gypsum board panel has pulled away from the framing, reattach the whole area. Use drywall screws instead of nails this time, if possible, spaced as for new work (see Chapter 12). Press a wall panel tightly against the studs as you drive new screws or nails. Use a T-bar to wedge a loose ceiling firmly up against the joists before you start any refastening.

FIXING POPPED NAILS

The nails that are so carefully dimpled below the surface and covered with taping compound in gypsum board wall construction don't always stay put. Popped nails—the protrusion of the nail heads above the surface—result from some of the same causes as cracks: movement, vibration, and drying out of the framing. Also, many older drywall partitions were not nailed with annular-ring nails.

If more than one nail has popped in a small area, tap or press against the wall to see if the gypsum board has pulled away from the stud. If it hasn't, the cure is to drive gypsum board screws into the stud 1½″ above and below the popped nail, and then drive the popped nail back below the surface, using a nailset. Finish off with joint compound to cover the nail and screw heads. When driving gypsum board screws, use a cordless screwdriver or a power drill and a gypsum screw-driving attachment to set the head below the surface.

REPAIRING LOOSE PLASTER

When large areas of a plastered wall or ceiling are loose, it may be that the nails holding the wood, metal, or gypsum lath have either popped from the framing or rusted through. Or it may be that the lath is still firmly attached to the framing but the keys—the globs of plaster that reach through the lath openings to lock the plaster to the lath—have broken off. Often this problem occurs in conjunction with a hole in the plaster, and the loose areas should be taken care of before the hole is patched.

The separated plaster must be reattached, but don't nail. Nailing will only loosen more plaster. Instead, use gypsum board screws and, to spread the stress in the plaster, plaster washers.

Figure 13–12
Repairing loose plaster with plaster washers.

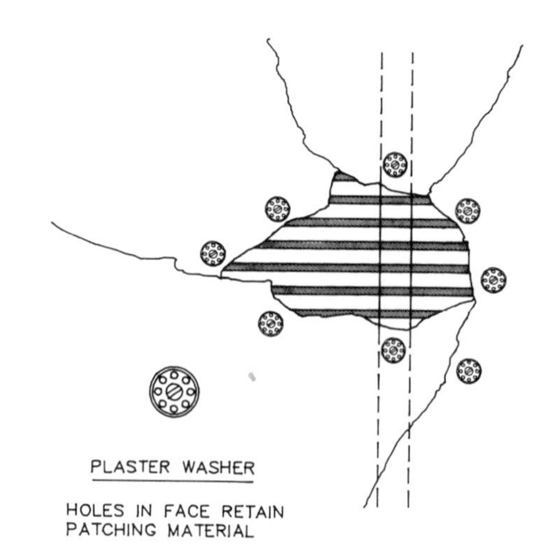

The screws must go into wood lath or framing to be effective. Drill pilot holes for the screws and place a plaster washer under each screw head. (See Figure 13–12.)

Plaster washers are not a home center item, but are available at building supply houses and in small quantities from Charles St. Supply Co., 54 Charles Street, Boston, MA 02114.

DEALING WITH RECURRING CRACKS

Some cracks in gypsum board and plaster walls and ceilings simply won't stay repaired—they open up again time after time. Before repairing such a crack one more time, it's a good idea to find out what is causing it. Cracks at the corners of interior doorways usually mean that the weight of the partition is causing the floor joists below it to sag. If you can get underneath, the cure is a supporting post (see Chapter 6). If the cracks are in an outside wall, you may have a foundation problem and you should seek competent help. Cracks across the middle of a ceiling are common and are usually caused by too much weight on the joists above. The cure may be as simple as moving some things out of the attic or overhead storage space.

Even if you can't completely correct the cause, a persistently recurring crack often can be repaired with a product such as Krack-Kote (Figure 13–13). With this material you don't attempt to fill the crack, just hide it. As shown in Figures 13–14 and 13–15, a polymerized oil mastic is brushed over the crack and an open-mesh glass fiber fabric is pressed into the mastic and the surface smoothed. After two hours, a second coat is brushed on and feathered out smooth.

The crack can open and close under the mesh, but the flexible material gives and stretches as the wall or ceiling works. When dry, the solid-appearing repair can be painted or papered over. The only problem with using Krack-Kote is that you have to do a good job smoothing it while it is wet because you can't sand it afterward.

Figure 13–13
Repairing a persistent crack. Brush Krack-Kote polymerized oil mastic over area of crack. No particular effort need be made to fill the crack.

Figure 13–14
Press open-mesh glass-fiber fabric into mastic and smooth the surface.

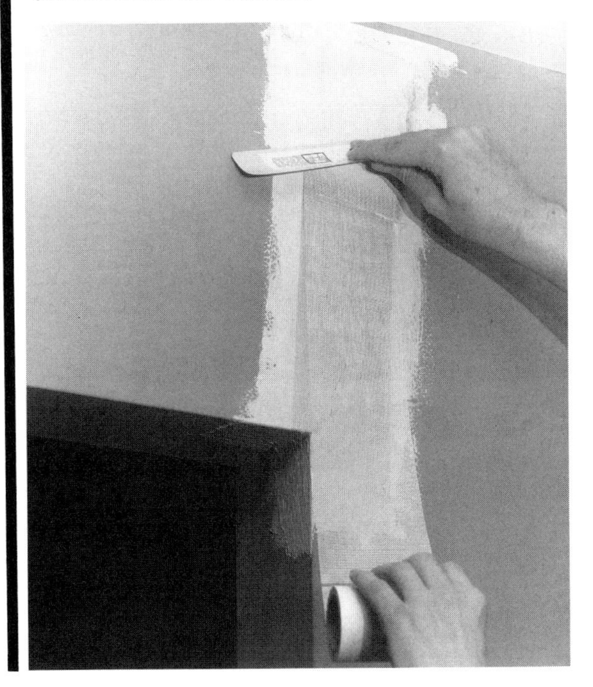

Figure 13–15
Let dry, then brush and feather out a second coat. The applicator tool comes with the kit.

OTHER METHODS

Skimming

Skimming is a plaster wall repair finishing technique in which a thin coat of joint compound is spread over the whole wall surface after repairs have been made to holes, cracks, and gouges. With several coats, it is possible to have a wall looking so good you won't want to paper it. Keep each coat thin. Apply with a wide trowel and wet-sand between coats.

Liner

If all else fails, or if you want a smooth surface on a concrete block wall, you can cover the wall with a heavy-duty lining material that is sold by wall-covering stores primarily to provide a flat surface for wallpaper.

The lining is sold like wallpaper in double or triple rolls under several brand names and in several weights. It is superior to the lightweight canvas once used for this purpose. Ready-mixed vinyl adhesives are used to hang the liners, which are butted together, not overlapped at edges. You cannot cover just part of a wall; you must cover the entire wall, corner to corner, to avoid both an edge that is difficult to feather with joint compound and a change in surface texture that will remain visible under paint.

The porous lining is not supposed to be the best surface for painting, but it is possible to paint some liners successfully. Use an oil-base paint unless the directions with the liner specify otherwise.

14 | Paneling

Wall paneling is available in a wide choice of styles, sizes, and materials. Traditionally, paneling was wood; then the term broadened to include materials that at least looked like wood. Today, paneling is considered to be any rigid material with a finished, decorative surface that is applied to a wall in standard-size sheets or relatively large pieces. There are, of course, many other wall-surfacing materials, such as ceramic, marble, and slate tile, acoustic tiles and panels, and resilient and wood parquet tiles, some of which are more often used as floor surfacings.

Four hundred years ago paneling meant carefully crafted constructions of wood planks held by vertical stiles and horizontal rails. Today this is still the most prestigious style of paneling a room, and the method of construction has changed little in four centuries. Planks of solid wood are today used vertically, horizontally, diagonally, or in mixed patterns.

The job of paneling now most commonly means applying 4′ × 8′ prefinished plywood or hardboard panels or sheets to walls. The variety of prefinished paneling available is extensive (see Figures 14–1 and 14–2 for two popular pan-

Figure 14–1
The handsome appearance of traditional paneling—random-width boards running vertically—is recreated here by 4′ × 8′ hardboard panels with a wood-grain surface and molded V grooves between the "planks." (Photo courtesy Masonite Corporation)

Figure 14–2
Deeply embossed "Oak and Cane" paneling can easily be used to create a formal wall. Its integral frame requires no ceiling or base molding for a finished appearance. (Photo courtesy Masonite Corporation)

eling patterns). Real wood veneer panels are made with domestic and exotic hardwood and softwood finish surfaces. Panels can also imitate marble, stucco, and other materials, or they can be embossed to simulate traditional English or French paneling, leather, cane, bamboo—or just about any material that could conceivably be used to cover a wall.

PANELING HEIGHT

In most contemporary installations, when a room is paneled the walls are covered from floor to ceiling; this is called room height paneling. Using prefinished paneling, it is an easy way to finish walls attractively and to cover a lot of defects quickly.

Walls in rooms with ceilings higher than 8′ can be paneled floor to ceiling, but the added panel height can be overwhelming. On the other hand, open two-story spaces, stairwells, and similar areas can be very attractively paneled to soaring heights.

If your ceiling is more than 8′ high and you

Figure 14–3
Standard 8′ panels can be stretched to cover walls almost 9′ high with tall baseboard and ceiling moldings. See Chapter 10 for a variety of baseboard and crown molding treatments.

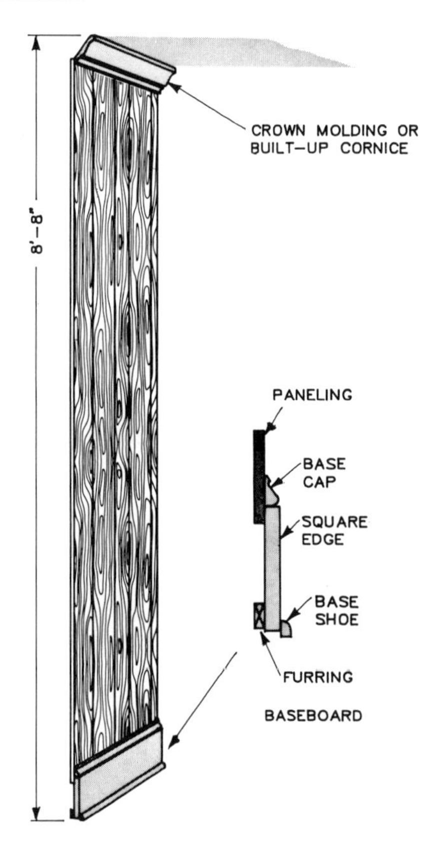

Figure 14–4
Stretching panel height by the use of a chair rail to hide the joint between full and part panel.

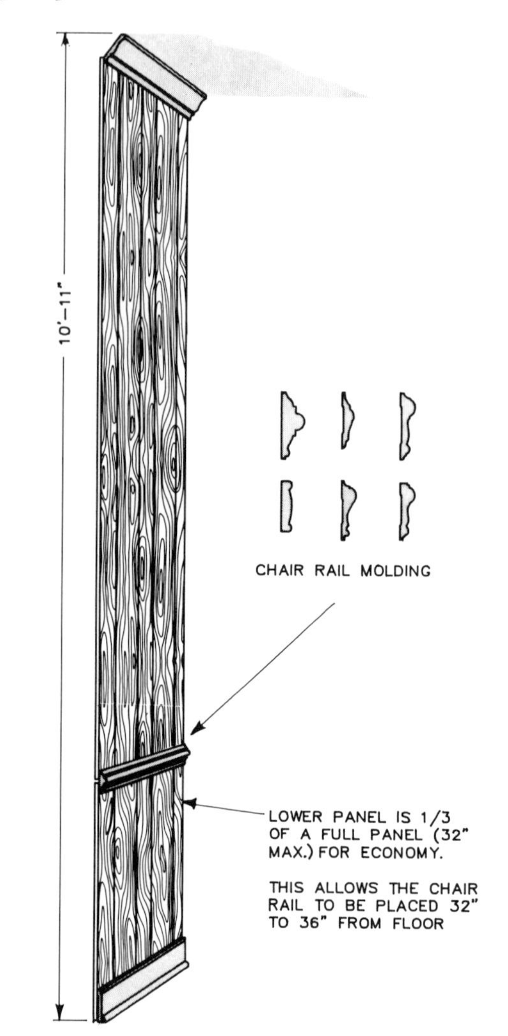

Figure 14–5
Paneling a high wall, with the horizontal seam between panels hidden by a high shelf.

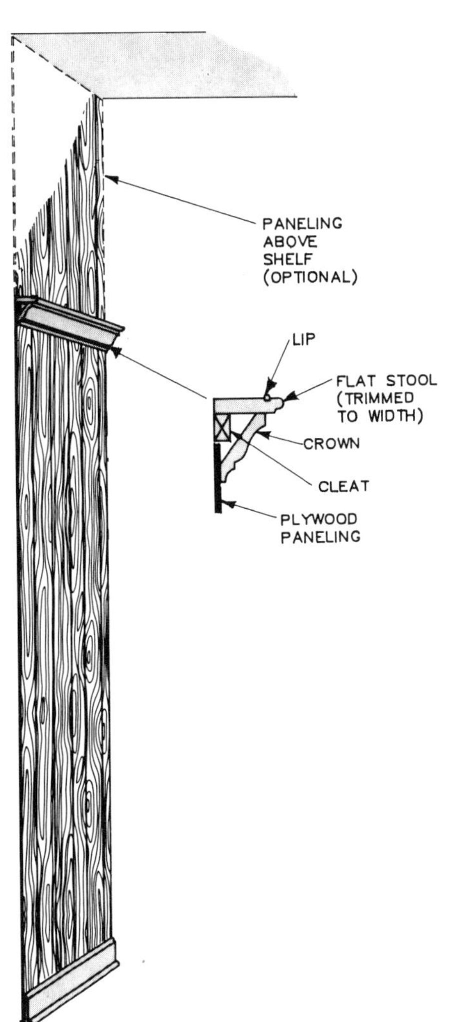

plan to panel full height with 4′ × 8′ panels, you can add some decorative feature to the wall to avoid a horizontal seam. For a ceiling up to 9′ high, a combination of tall baseboard and crown molding could stretch a single panel to cover, as shown in Figure 14–3. For higher ceilings, part panels can be used below a chair rail and full panels above, or full panels topped by part panels, with the seam hidden by molding or a shelf, as shown in Figures 14–4 and 14–5. When combining panels vertically, be careful to line up grooves, and try to match grain and figure above and below the rail. A few patterns are available in panels longer than 8′.

Paneling applied to a height of 6′ or 7′ on a wall is called partition-height paneling or simply paneling (see Figure 14–6). This treatment is very Early American in appearance with an 8′ ceiling and may be a good solution to paneling a wall when the room's ceiling is significantly higher than 8′. Partition paneling can be run out from the wall as a room divider; it is then called screen paneling. Partition paneling must be capped with some kind of molding; you will want to match your door and window casings, or at least choose a style that is compatible.

Wainscoting is a third style of paneling. This paneling is usually 32″ to 36″ high and is capped with a chair rail. Wainscoting offers good protection to a gypsum board or plaster wall in a hallway or playroom. Thirty-two inches is a good wainscot height because it is one-third of an 8′ panel. With baseboard and chair rail, a third-

Figure 14–6
Partition-height paneling.

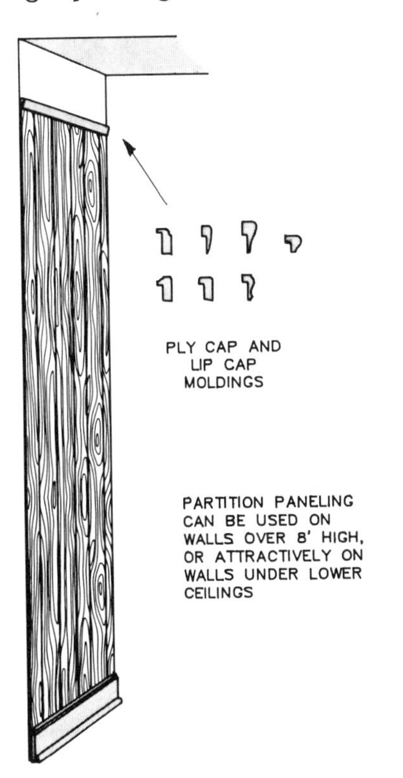

Figure 14–7
Wainscoting installation styles. (T&G: tongue-and-groove.)

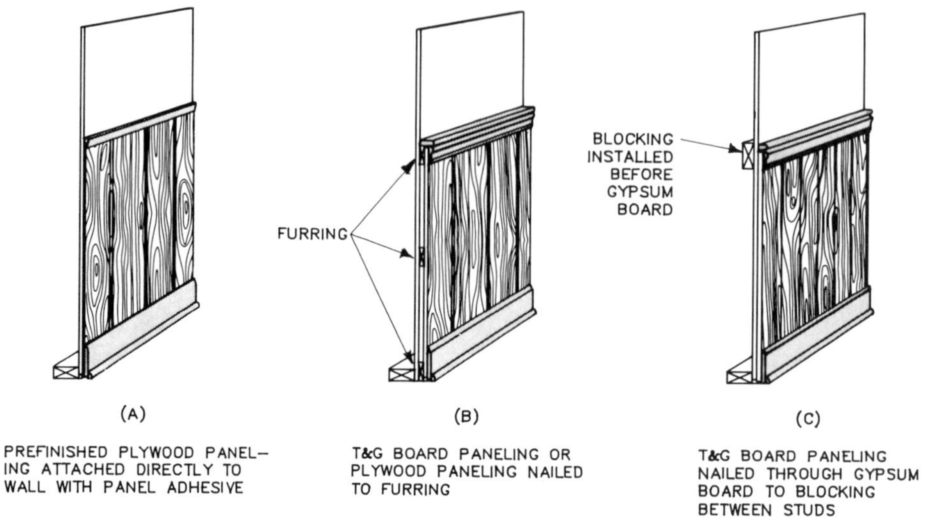

FURRING

BLOCKING INSTALLED BEFORE GYPSUM BOARD

(A) PREFINISHED PLYWOOD PANELING ATTACHED DIRECTLY TO WALL WITH PANEL ADHESIVE

(B) T&G BOARD PANELING OR PLYWOOD PANELING NAILED TO FURRING

(C) T&G BOARD PANELING NAILED THROUGH GYPSUM BOARD TO BLOCKING BETWEEN STUDS

panel can be stretched easily to a 36″ wainscot height. Three wainscot designs are shown in Figure 14–7.

INSTALLING PLYWOOD AND HARDBOARD PANELING

Whether plywood and hardboard paneling is used over an existing wall in a remodeling project or as sheathing over bare studs in new construction, installation is not difficult. But paneling is not a slap-it-up job. It takes planning, patience, and some know-how to do a job you will be happy with when it is finished.

Most paneling is 1/4″ thick, but there are also panels 1/8″, 5/32″, and 5/16″ thick. Paneling thinner than 1/4″ should not be applied directly to studs or furring, but only over a flat supporting surface such as gypsum board, plaster, or plywood sheathing. If at all possible, install even 1/4″ and thicker panels over gypsum board. It makes a sounder wall, and your building code may require gypsum board for sanitary, safety, or fire-resistance reasons.

ESTIMATING MATERIAL REQUIREMENTS

To determine how many panels you need for a room, add up the widths in feet of all the walls and divide by 4. For example, a 10′ × 15′ room will require a minimum of 13 panels:

$$\frac{10 + 10 + 15 + 15}{4} = 12.5, \text{ or 13 whole panels}$$

This is the most paneling you will need.

Draw a plan of the walls of the room and lay out panel widths on your plan as you intend to place them on the walls (see Figure 14–8). If at all possible, plan the layout to have all panels butt factory-edge to factory-edge. Panels with grooved faces have half-grooves at edges to form a full-width groove at butted edges. If you must rip a panel to a narrower width, your cut edge won't have a prefinished half-groove. Plan to put

Figure 14–8
Planning and estimating material requirements for wall paneling. With careful planning and panel ripping, you can save money.

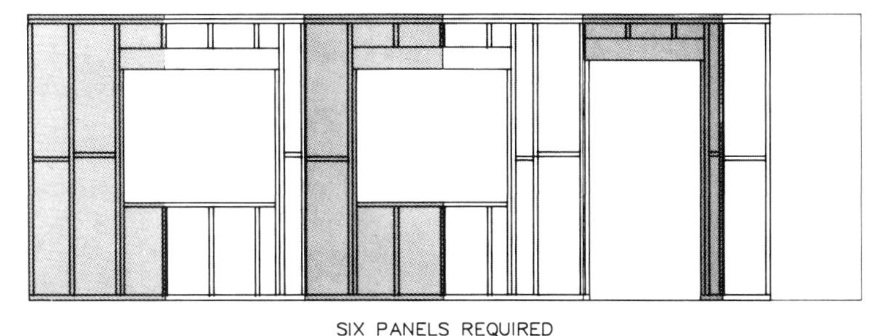

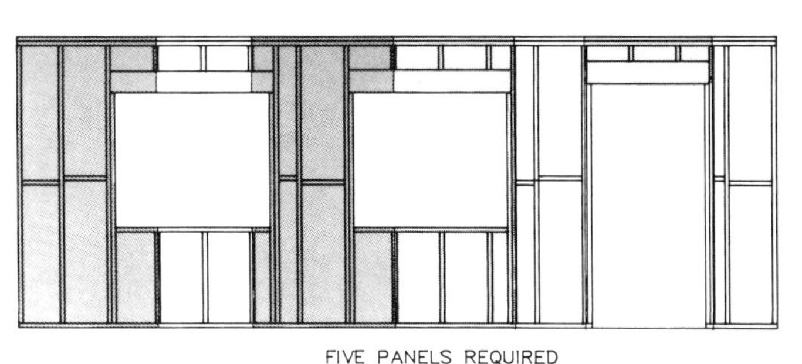

your edge in a corner or above a doorway. As doors and windows are not paneled over, you may be able to save some paneling, but be careful not to try to figure it too tight—you want to avoid having horizontal seams as you fit the last small pieces to finish the job.

CUT–PANEL METHOD
PANELS MUST BE RIPPED ACCURATELY FOR FIT, SOME GROOVES WILL BE HALF–WIDTH. THIS METHOD MUST BE USED IF STUDS ARE NOT ACCURATELY SPACED 16" OC.
THE NUMBER OF PANELS NEEDED CAN BE REDUCED, BUT CAREFUL PLANNING IS REQUIRED FOR MAXIMUM SAVINGS.
FIVE PANELS REQUIRED

PANELING A ROOM

If you are paneling a large room, store the panels in the room for 48 hours so they can adjust to the temperature and humidity of the room. You might want to stand them against the walls to work out the most pleasing arrangement of grain and coloring.

Sawing Plywood and Hardboard Paneling

If you rip panels with a portable circular saw, use a fence—the factory edge of a 6″-wide strip of plywood works well—clamped to the paneling to guide the saw. Since the saw blade rotates so the teeth bite upward, turn the panel good side down to rip it. On a table saw or with a handsaw, the teeth bite with the downward movement, so cut with the good side of the panel facing up. When using a saber saw, the good side should be down, except that you can saw from either side if you use a blade specifically recommended for sawing paneling, such as a Sears #28761 Super Fine Finish blade.

Attaching Paneling

Paneling can be attached to gypsum board, plaster, or studs with either nails or adhesive. Paneling can be applied directly to masonry only with adhesive. It is more usual to attach wood furring to a masonry wall so that panels can be nailed up.

Nails. Colored paneling nails are available to match all kinds of panels. They should be used rather than finishing nails, even if you plan to

Table 14–1
NAILING SCHEDULES FOR PANELING

Paneling Material	Panel Thickness	Backing	Frame Spacing	Nail Size	Nail Type	Nail Spacing Edge	Nail Spacing Center	Notes
Prefinished plywood	All	Studs	16"	1"	Ringed panel nail	6"	12"	Nails must penetrate ¾" into framing
		½" Gypsum board	16"	1⅝"		6"	12"	
Siding and other plywood	¼", 5/16"		16" (max)	4d				
	⅜" to ½"	Studs	24" (max)	6d	Finish	6"	12"	6d nail if gypsum board
	⅝" to ¾"		24" (max)	8d				8d nail if gypsum board
	Texture T1–11		24" (max)	8d				
Prefinished hardboard	Under 3/16"	—	—	—	—	—	—	DO NOT NAIL
	3/16" (min)	Studs	16"	1"	Ringed panel nail	4"	8"	
		½" Gypsum board	16"	1⅝"		4"	8"	
Hardboard panel siding	7/16"	Studs	16" (max)	6d	Box	6"	12"	
		½" Gypsum board	16" (max)	6d	Box	6"	12"	

sink and putty over the nail heads. These hardened annular-ring paneling nails can be driven flush or countersunk (wear protective goggles or safety glasses when using them), and they hold better than finishing nails. Nails 1" long are used for attaching paneling directly to wood; 1⅝" nails are used for panels placed over gypsum board and for trim—they are long enough to be driven into studs or furring. Table 14–1 gives nailing schedules for paneling.

Adhesive. Panel adhesive may be used instead of nails if the support surface is clean and sound. A few nails will still be needed to hold the panel to the wall while the adhesive cures. Panel adhesive comes in cartridges that are used in a gun-type applicator. Some panel adhesives adhere only to wood and cannot be used to apply paneling to gypsum board or plaster. Read the instructions on the cartridge before buying.

Figure 14–9 shows the proper pattern of adhesive application. With most adhesives, the panel is pressed to the wall, then separated for a spe-

Figure 14–9
Adhesive application for panel on studs.

APPLY CONTINUOUS BEAD AT TOP AND BOTTOM, AND AT PANEL EDGES, INCLUDING SUPPORTED EDGES OF LARGE OPENINGS IN PANEL. LOCATE BEAD 1/2" FROM EDGE OF PANEL.
APPLY TWO PARALLEL BEADS (ONE AT A TIME) AT MEETING PANEL EDGES, NOT A SINGLE WAVY BEAD AS SHOWN AT CENTER.
APPLY 3" BEADS SPACED 3" ON ALL INTERMEDIATE SUPPORTS

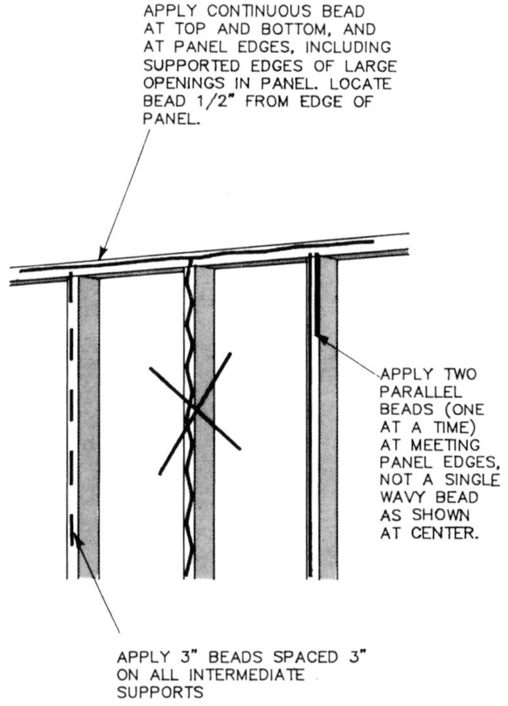

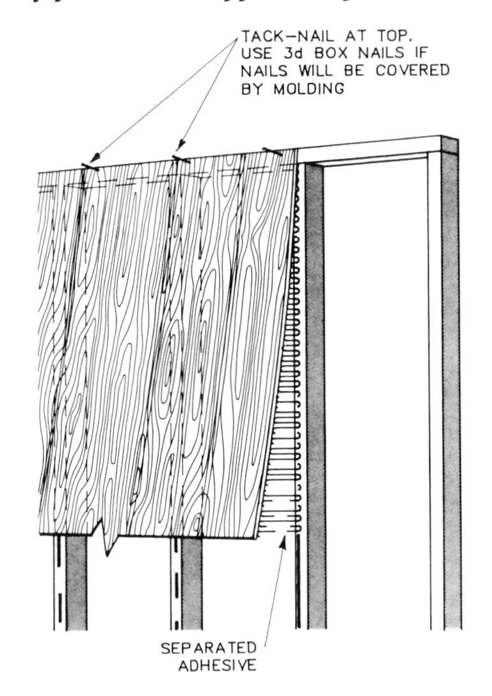
TACK–NAIL AT TOP.
USE 3d BOX NAILS IF
NAILS WILL BE COVERED
BY MOLDING
SEPARATED
ADHESIVE

Figure 14–10
Installing panel over adhesive. Panel tack-nailed at the top is pressed against adhesive on the studs, then separated before being finally pressed and tapped into place.

cified time before being permanently pressed down. This is usually done by tack-nailing the panel at the top to maintain accurate positioning, then pulling the bottom out 6″ to get the required separation. Figure 14–10 shows this technique. To be sure of good adhesion after pressing the panel to the support, hammer against a smooth softwood block laid over the face of the panel. If any edge warps out at this point it should be nailed down. Table 14–2 gives adhesive schedules for paneling.

Adding Insulation

If you are planning to add thermal insulation at the same time you install paneling, you can do it in several ways, as shown in Figure 14–11. If the studs are bare, staple foil-faced R-11 fiberglass insulation between the studs, with the foil facing inward (R-values are explained in Chapter 16). If the wall has been finished with gypsum board and there is no insulation inside the wall, you can drill a row of holes around the top of the wall, one hole for each stud cavity, and pour in vermiculite

Table 14–2
ADHESIVE SCHEDULES FOR PANELING

Paneling Material	Panel Thickness	Backing	Frame Spacing	Adhesive	Application Procedure	Notes
Prefinished plywood	All	Studs	16″ (max)	Panel	Procedures differ. Follow adhesive manufacturer's instructions.	
		Gypsum board	N/A	Panel	Procedures differ. Follow adhesive manufacturer's instructions.	Be sure adhesive is rated for use with gypsum board.
Hardboard						
Normal humidity	All	Gypsum board	N/A	Panel	Apply continuous perimeter bead ½″ from edge, 3″ intermittent beads offset—spaced 6″ over rest of panel.	Be sure adhesive is rated for use with gypsum board.
Moist locations	All	Gypsum board	N/A	Tile-bond	Apply adhesive to whole back of panel with 3/16″ notched trowel. Leave ridges only.	Avoid applying excessive adhesive.

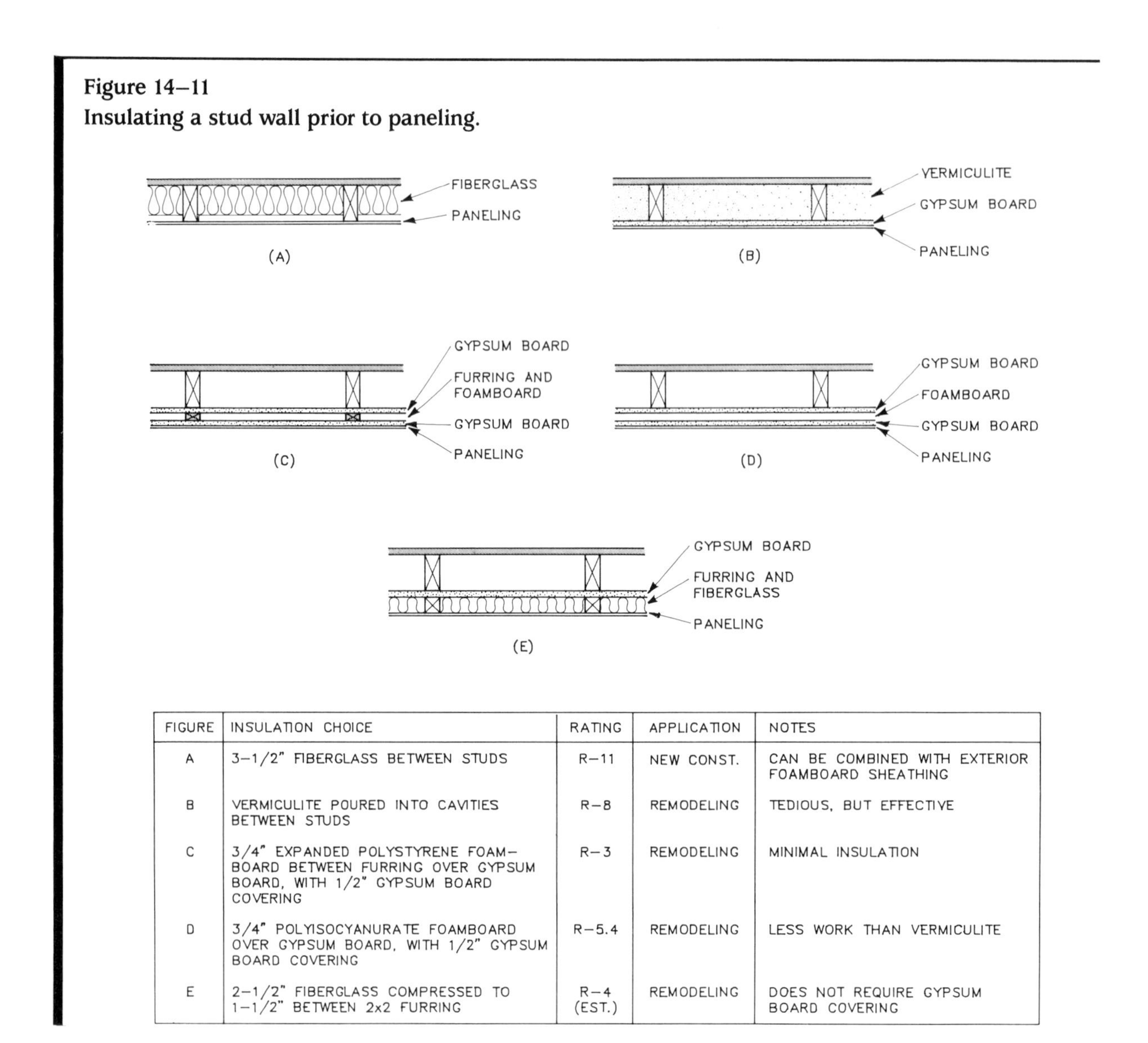

FIGURE	INSULATION CHOICE	RATING	APPLICATION	NOTES
A	3-1/2" FIBERGLASS BETWEEN STUDS	R-11	NEW CONST.	CAN BE COMBINED WITH EXTERIOR FOAMBOARD SHEATHING
B	VERMICULITE POURED INTO CAVITIES BETWEEN STUDS	R-8	REMODELING	TEDIOUS, BUT EFFECTIVE
C	3/4" EXPANDED POLYSTYRENE FOAM-BOARD BETWEEN FURRING OVER GYPSUM BOARD, WITH 1/2" GYPSUM BOARD COVERING	R-3	REMODELING	MINIMAL INSULATION
D	3/4" POLYISOCYANURATE FOAMBOARD OVER GYPSUM BOARD, WITH 1/2" GYPSUM BOARD COVERING	R-5.4	REMODELING	LESS WORK THAN VERMICULITE
E	2-1/2" FIBERGLASS COMPRESSED TO 1-1/2" BETWEEN 2x2 FURRING	R-4 (EST.)	REMODELING	DOES NOT REQUIRE GYPSUM BOARD COVERING

Figure 14–11
Insulating a stud wall prior to paneling.

(see Chapter 16). You can nail foamboard insulation on the surface of the wall, but you must then cover the foamboard with ½″ gypsum board for fire safety before you apply the paneling. If you panel on furring, you can put foamboard between the furring strips and apply ½″ gypsum board before you panel. You could place fiberboard sheathing or fiberglass between the furring without the need for gypsum board under the paneling, but this wouldn't add much to the wall's R-value, and it will not be as good a wall construction as one with gypsum board. The R-values for these constructions are compared in Table 14–3.

APPLYING PANELING TO VARIOUS SURFACES

Paneling Directly on Studs

Figure 14–12 shows this installation. The edges of the studs must provide a flat surface for the paneling. Check them with a straightedge (a 6″-wide strip ripped from the long edge of a plywood panel makes a good straightedge), or a string stretched the length of the wall top, center,

Table 14–3
THERMAL R-VALUE ADDED TO A WALL WITH ¼″ PLYWOOD OR HARDBOARD PANELING

Method of Application	Added Wall Thickness	Added R-Value	Notes
Paneling nailed or glued directly to wall	¼″	0.31	R=0.25 for hardboard.
Paneling on 1×3 furring, ¾″ extruded polystyrene foam-board between furring	1″	3.3	This construction is not allowed in most building codes and is not recommended.
Paneling on 1×3 furring, ¾″ fiberglass acoustical ceiling panels between furring	1″	3.3	Expensive. Look for a deal on damaged panels.
Paneling over ½″ gypsum board on 1×3 furring, ¾″ polystyrene between furring	1½″	3.8	Superior paneling method compared with those above.
Paneling on 2×3 studs, 2½″ fiberglass blanket between studs	2¾″	7.8	Next method produces a better wall.
Paneling over ½″ gypsum board on 2×2 studs, 1¼″ polyisocyanurate foamboard between studs	2¼″	9.8	Major improvement in wall insulation.

Figure 14–12
Paneling directly on studs.

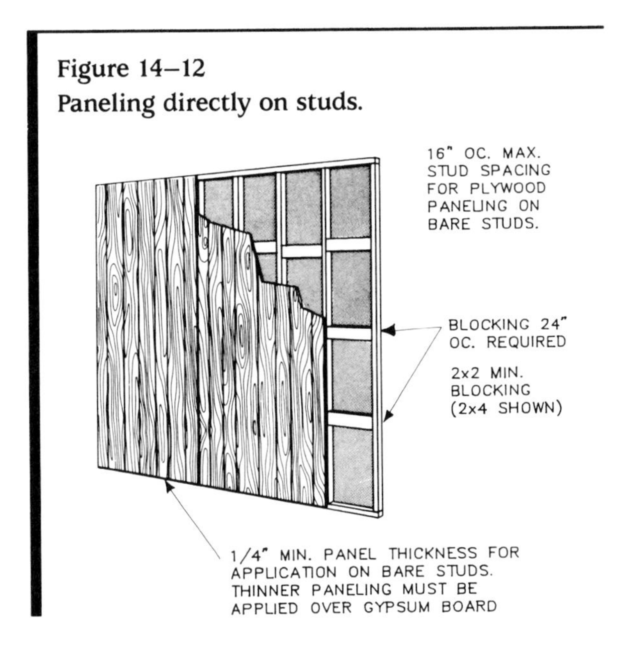

and bottom as shown in Figure 14–13. Shim or trim any studs that are out of alignment.

For additional wall strength, fire resistance, and sound deadening, ½″ gypsum board is recommended as a backing behind any prefinished paneling, as shown in Figure 14–14. This installation of paneling backed by gypsum board may be required by your local building code.

Paneling on Plaster and Gypsum Board Walls

Apply paneling directly on these surfaces by nailing through the plaster or gypsum board into studs and plates if the wall is flat. Locate the studs in the wall and draw lines on the wall marking their locations. Next locate both ends of the studs;

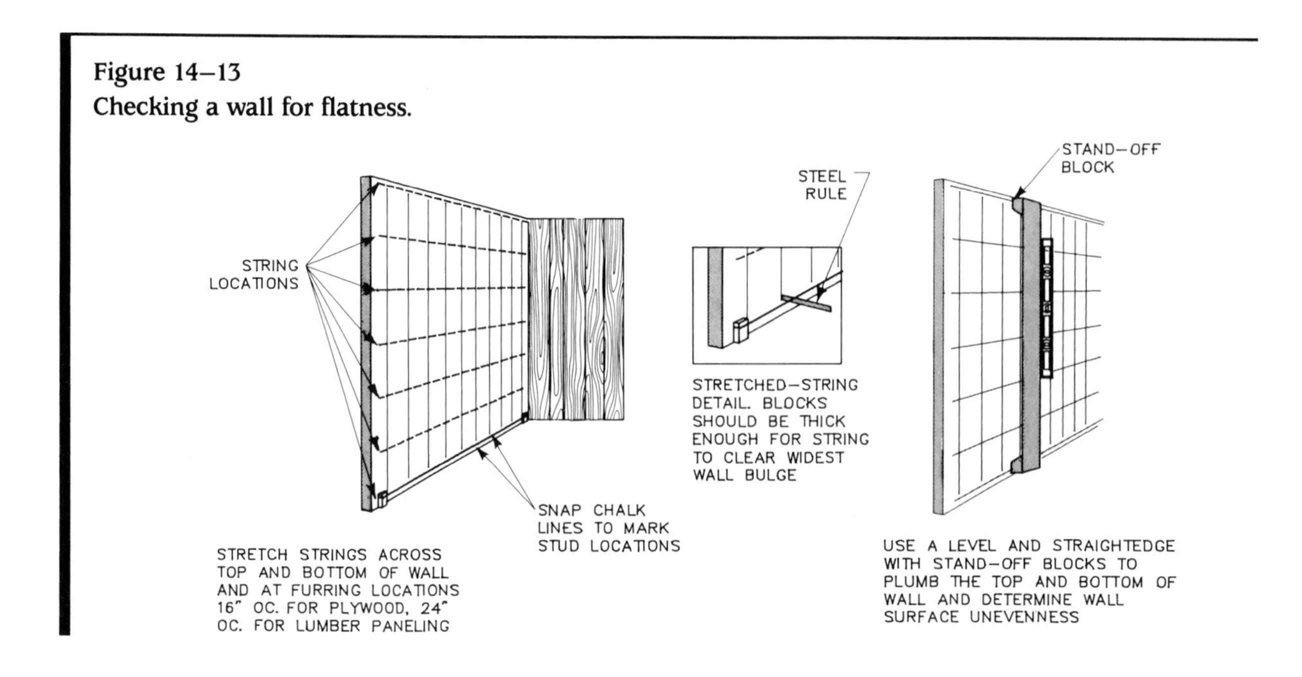
STRING
LOCATIONS
SNAP CHALK
LINES TO MARK
STUD LOCATIONS
STRETCH STRINGS ACROSS
TOP AND BOTTOM OF WALL
AND AT FURRING LOCATIONS
16" OC. FOR PLYWOOD, 24"
OC. FOR LUMBER PANELING
STEEL
RULE
STRETCHED-STRING
DETAIL. BLOCKS
SHOULD BE THICK
ENOUGH FOR STRING
TO CLEAR WIDEST
WALL BULGE
STAND-OFF
BLOCK
USE A LEVEL AND STRAIGHTEDGE
WITH STAND-OFF BLOCKS TO
PLUMB THE TOP AND BOTTOM OF
WALL AND DETERMINE WALL
SURFACE UNEVENNESS

Figure 14–13
Checking a wall for flatness.

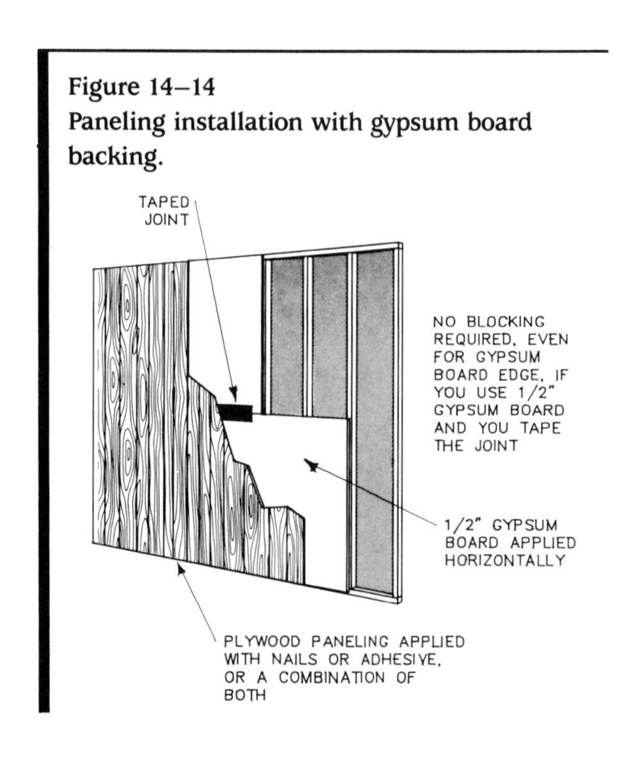
TAPED
JOINT
NO BLOCKING
REQUIRED, EVEN
FOR GYPSUM
BOARD EDGE, IF
YOU USE 1/2"
GYPSUM BOARD
AND YOU TAPE
THE JOINT
1/2" GYPSUM
BOARD APPLIED
HORIZONTALLY
PLYWOOD PANELING APPLIED
WITH NAILS OR ADHESIVE,
OR A COMBINATION OF
BOTH

Figure 14–14
Paneling installation with gypsum board backing.

they are not always straight, and do not assume they are all on 16″ centers. If the surface is painted and sound but not papered, you can use panel adhesive and do not have to worry about where the studs are located.

If a plaster or gypsum board wall has depressions, you can shim the low areas of the wall to a flat surface with hardboard or plywood, as shown in Figure 14–15. To do this, you have to know where your panel edges will fall so they will be solidly backed.

Furring Irregular or Damaged Walls

Furring can be used to provide a flat surface on which to attach the paneling when the wall itself is not sufficiently flat to be a usable base. Furring can also provide support for gypsum board and the paneling when foamboard insulation is to be applied under the paneling. Use 1 × 2 or 1 × 3 furring strips, or rip ⅜″, ½″, or ¾″ sheathing plywood into 2″-wide furring strips.

Nail strips horizontally on the wall for panels to be installed vertically, as shown in Figure 14–16. Locate the bottom strip ¼″ clear of the floor and the top strip ¼″ lower than the top of the paneling. Space the rest of the furring on 16″ centers.

Attach furring to plaster or gypsum board walls with box nails driven into the studs and framing. Nail into the high areas first, then drive shingles behind the furring in the low spots. Insert the shims from each side to prevent tilting the furring, then nail through both the furring and the shingles.

Add short vertical pieces to support the edges of the paneling every 48″ or wherever panel edges fall. Attach the vertical strips with panel adhesive if they don't lie on studs. (Use nails in plaster or gypsum board to hold the strips while the adhesive sets.)

For panels to be installed horizontally, locate long furring strips to support the long panel edges, with vertical strips over stud locations and at panel ends, as shown in Figure 14–17.

Figure 14–15
Leveling a wall without furring.

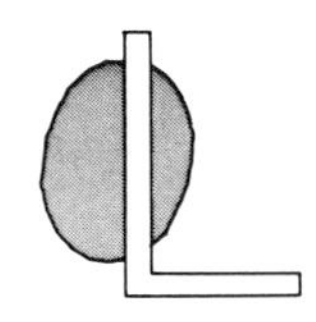

MEASURE DEPTH OF DEPRESSION WITH STRAIGHTEDGE

BRING CENTER OF DEPRESSION LEVEL OF WALL WITH PLY-WOOD PATCH BEDDED IN MASTIC ADHESIVE

FILL IN THE REST OF THE DEPRESSION WITH SPACKLING COMPOUND

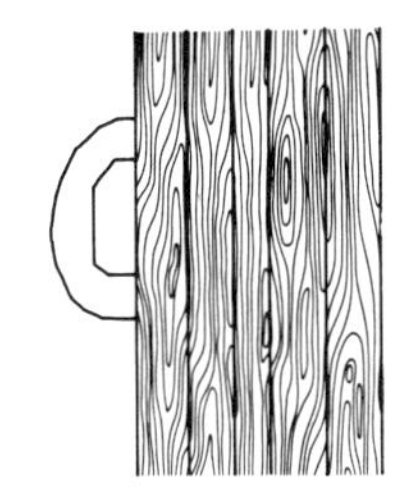

Figure 14–16
Horizontal furring applied to a wall to support vertical paneling.

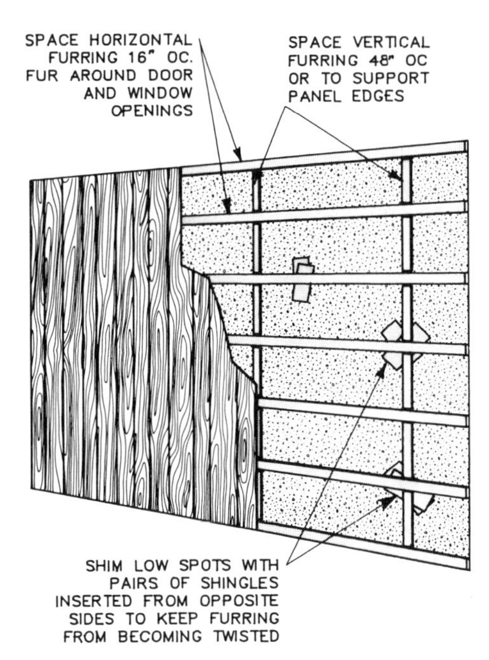
SPACE HORIZONTAL FURRING 16″ OC. FUR AROUND DOOR AND WINDOW OPENINGS
SPACE VERTICAL FURRING 48″ OC OR TO SUPPORT PANEL EDGES
SHIM LOW SPOTS WITH PAIRS OF SHINGLES INSERTED FROM OPPOSITE SIDES TO KEEP FURRING FROM BECOMING TWISTED
FASTENERS: USE BOX NAILS TO NAIL INTO STUDS, SCREWS AND EXPANSION ANCHORS TO FASTEN FURRING TO GYPSUM BOARD OR PLASTER

Figure 14–17
Vertical furring applied to a wall to support horizontal paneling. Vertical butt joints between panel ends may be inconspicuous enough to leave alone, or they may be covered with batten molding.

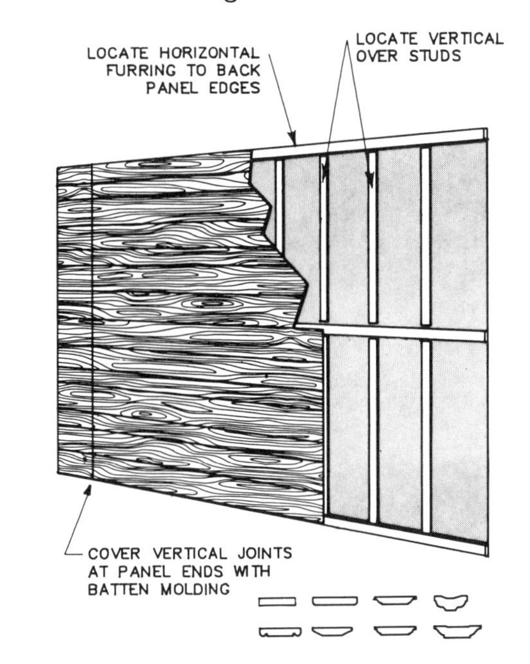
LOCATE HORIZONTAL FURRING TO BACK PANEL EDGES
LOCATE VERTICAL OVER STUDS
COVER VERTICAL JOINTS AT PANEL ENDS WITH BATTEN MOLDING

Paneling on Masonry Walls

You can apply paneling directly to a masonry wall with panel adhesive if the wall is exceedingly flat and smooth, and very dry. As these conditions are seldom met, and as outside masonry walls almost always need thermal insulation to help make the paneled room livable, it is usual to apply furring to the walls before paneling.

Masonry walls are furred in the same way as gypsum board and plaster walls. If the base of the wall is below grade, the furring, gypsum board, and paneling must all be spaced ¼″ clear of the slab floor to prevent moisture wicking. Exterior masonry walls above or below grade must be waterproofed before furring is applied. Do not cover up a dampness problem with paneling. Humidity can cause condensation on the inside of exterior masonry walls. Apply a 6-mil polyethylene film vapor barrier to protect the paneling. Attach furring to masonry with adhesive anchors or screws in plastic anchors rather than with masonry nails. Shimming while using masonry nails is difficult.

PUTTING UP THE FIRST PANEL

The procedure for positioning panels is essentially the same for all walls. Plan your panel layout to minimize cutting and waste, and to keep non-factory-edge joints short (locate them over a doorway, for example) and inconspicuous. If possible, start paneling where you can butt a whole panel into a corner without having to cut the other edge of the panel to fit around an opening.

Never assume that the room corner is straight or plumb in relation to either wall, and don't assume that door jambs and window frames are plumb. Put the first panel into place and butt its edge to the adjacent wall in the corner. Check the exposed edge with a carpenter's level to see if it is plumb (Figure 14–18). If you are nailing, the edge must also be located at a stud, but if you are using adhesive, that is not necessary. If the panel edge is not plumb, it means the corner is not plumb. Trim the corner edge of the panel to fit in the corner with the other edge plumb.

If the corner panel butts brickwork or stonework, or an uneven or out-of-plumb wall, the panel must be accurately fitted against it. Use a compass to mark the panel for cutting, as shown in Figure 14–19. When marking the panel, be sure that you have it wedged up to its installed

Figure 14–18
Installing the first panel.

STEP 1. WEDGE PANEL INTO PLUMB POSITION AGAINST CEILING, SCRIBE TOP OF PANEL TO CEILING IF MOLDING WILL NOT BE USED.
GENERALLY, THE TOP OF A PANEL SHOULD BE SCRIBED AND TRIMMED BEFORE SCRIBING TO A WALL.
CHALK LINES SNAPPED ON GYPSUM BOARD TO MARK STUD LOCATIONS.
STEP 2. WITH TRIMMED PANEL WEDGED AGAINST CEILING AND PLUMB, SCRIBE THE PROFILE OF EACH BRICK ON THE PANEL. IF NAILING, LOCATE THE SCRIBED LINE SO THE OTHER EDGE OF THE PANEL WILL FALL ON A STUD.
BLOCK AND SHINGLE WEDGES

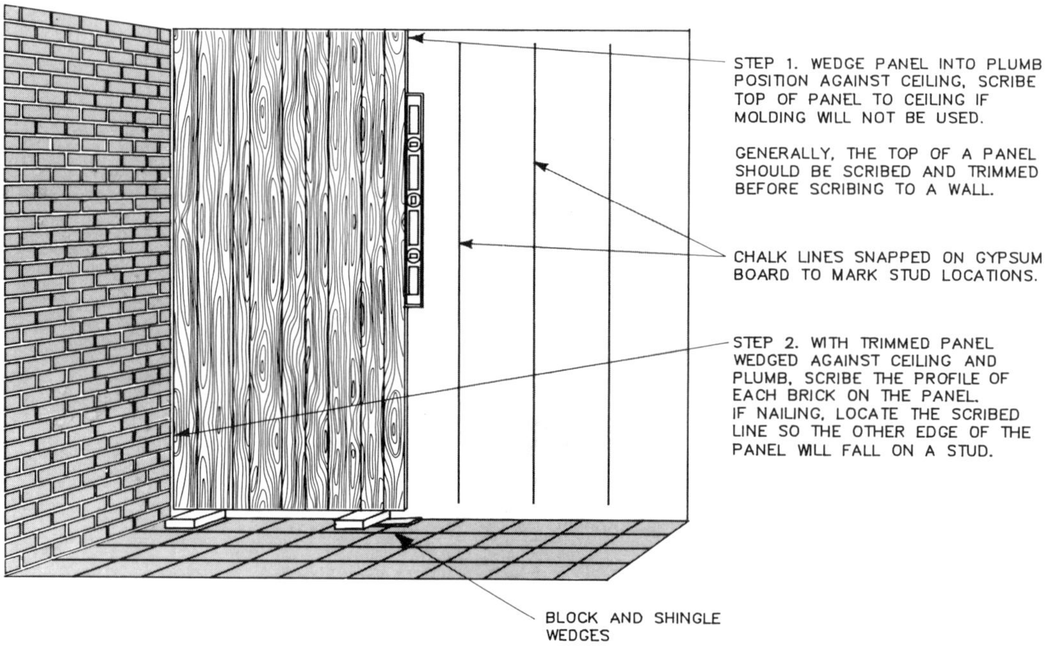

Figure 14–19
Scribing the panel edge to fit neatly against an irregular wall.

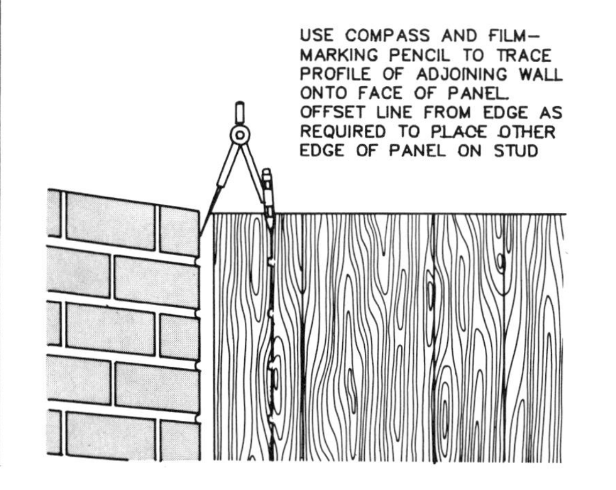
USE COMPASS AND FILM-MARKING PENCIL TO TRACE PROFILE OF ADJOINING WALL ONTO FACE OF PANEL.
OFFSET LINE FROM EDGE AS REQUIRED TO PLACE OTHER EDGE OF PANEL ON STUD

Figure 14–20
Jacking a panel into position for scribing, nailing, or applying adhesive.

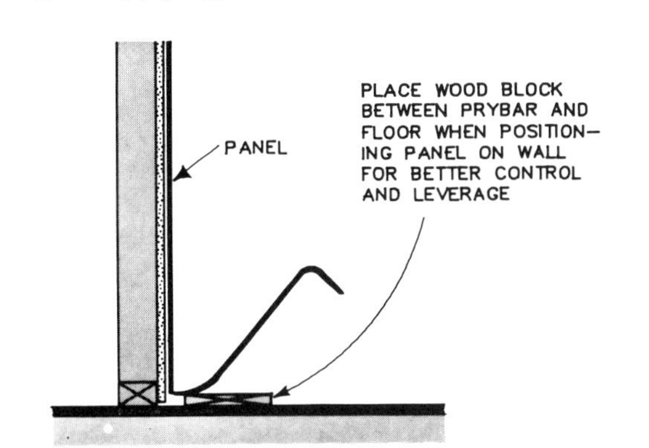
PANEL
PLACE WOOD BLOCK BETWEEN PRYBAR AND FLOOR WHEN POSITIONING PANEL ON WALL FOR BETTER CONTROL AND LEVERAGE

Figure 14–21
Wedging a panel into position.

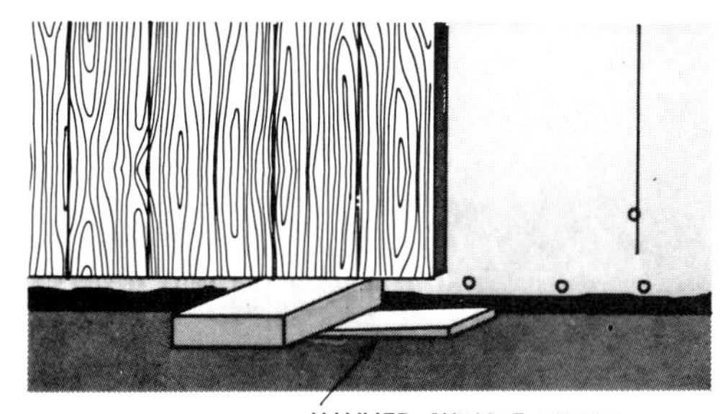
HAMMER SHINGLE WEDGE UNDER BLOCK TO RAISE AND PLUMB PANEL.
YOU MAY HAVE TO TRIM THE LENGTH OF AN 8' PANEL TO FIT AN 8' WALL.

height and that it is plumb. Panels can be jacked and wedged as shown in Figures 14–20 and 14–21.

Cutouts

Cutouts for switch and outlet boxes can be located by measurement or by using pronged templates that fit into the box after the cover plate is removed (and the power shut off). The templates have prongs on the front that mark the panel when it is pressed against it. The box cutout outline then can be traced on the panel back and sawed to outline. Another method is to rub the edges of the box with carpenter's chalk and press the back of the panel against the box to transfer the outline.

To comply with the National Electric Code, outlet and switch boxes must be repositioned when you install paneling so that, to quote from the code, "In walls and ceilings of wood or other combustible material, outlet boxes and fittings shall be flush with the finished surface or project therefrom" (Article 370–10). That means that when you put in paneling, even when you apply the paneling with adhesive to a gypsum board wall, you must move the outlet and switch boxes forward so that the front edge of the boxes is flush with the front surface of the paneling. *Caution:* Shut off the power while you do this.

INSTALLING ADDITIONAL PANELS

Fit and install successive panels one at a time. Do not trim a panel until the one it fits against has been glued or nailed permanently in place. As you start each new wall, check that the first panel is plumb before attaching it to the supporting furring or wall surface.

Tip: *Both plywood and hardboard panels expand and contract slightly with changes in humidity. If you install paneling in the winter, leave a 1/32" gap between panels.*

Figure 14–22
Trimming a door jamb or window side jamb to match paneling thickness.

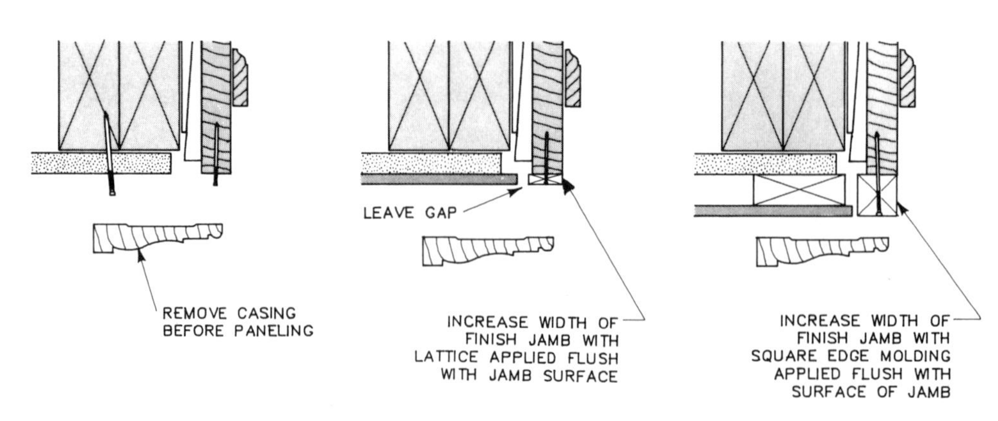
LEAVE GAP
REMOVE CASING BEFORE PANELING
INCREASE WIDTH OF FINISH JAMB WITH LATTICE APPLIED FLUSH WITH JAMB SURFACE
INCREASE WIDTH OF FINISH JAMB WITH SQUARE EDGE MOLDING APPLIED FLUSH WITH SURFACE OF JAMB

Finishing the paneling around doorways and windows requires increasing the thickness of the finish jamb to cover the exposed paneling edge before you replace the casing. Various techniques are shown in Figures 14–22 through 14–25. If the paneling is applied directly to the wall, you can usually use pine lattice wood (ripped for economy) to finish the edge. Nail the lattice flush with the edge of the jamb. For paneling applied over furring, thicker trim is required. This can be ripped from 1″ select pine; you might find square-edge molding with the correct width.

Figure 14–23
Trimming a traditional window sill for paneling applied directly to the wall.

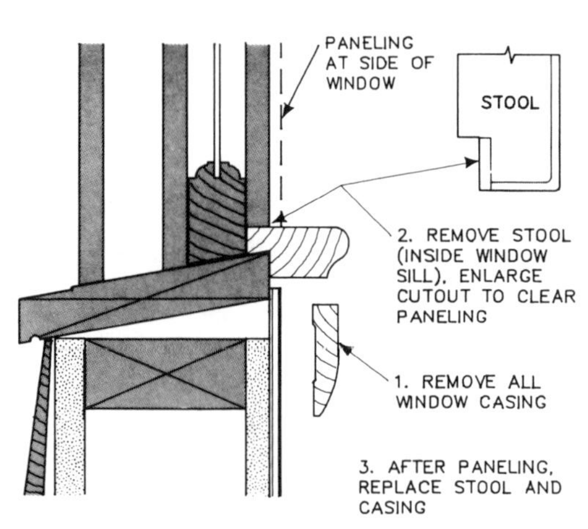
PANELING AT SIDE OF WINDOW
STOOL
2. REMOVE STOOL (INSIDE WINDOW SILL), ENLARGE CUTOUT TO CLEAR PANELING
1. REMOVE ALL WINDOW CASING
3. AFTER PANELING, REPLACE STOOL AND CASING

Figure 14–24
Trimming a traditional window sill for paneling applied over furring.

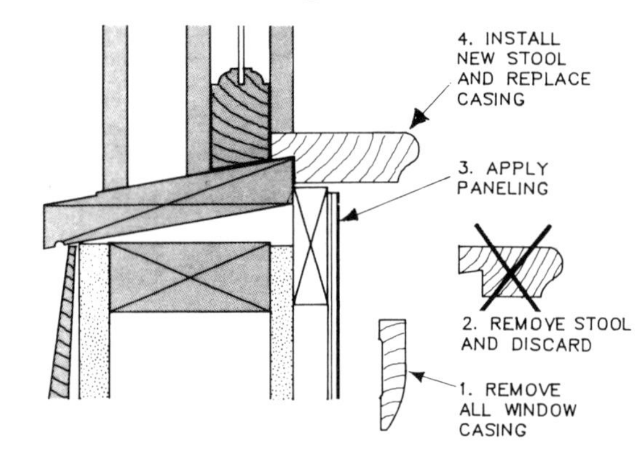
4. INSTALL NEW STOOL AND REPLACE CASING
3. APPLY PANELING
2. REMOVE STOOL AND DISCARD
1. REMOVE ALL WINDOW CASING

Figure 14–25
Trimming a contemporary window frame for paneling applied directly to the wall.

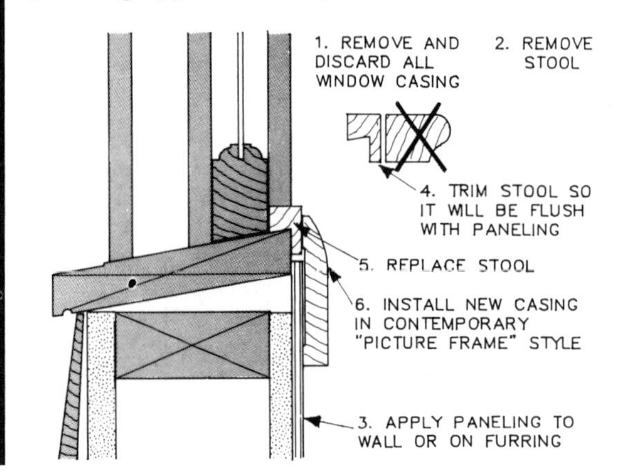
1. REMOVE AND DISCARD ALL WINDOW CASING
2. REMOVE STOOL
4. TRIM STOOL SO IT WILL BE FLUSH WITH PANELING
5. REPLACE STOOL
6. INSTALL NEW CASING IN CONTEMPORARY "PICTURE FRAME" STYLE
3. APPLY PANELING TO WALL OR ON FURRING

Figure 14–26
Stock moldings for use with paneling.

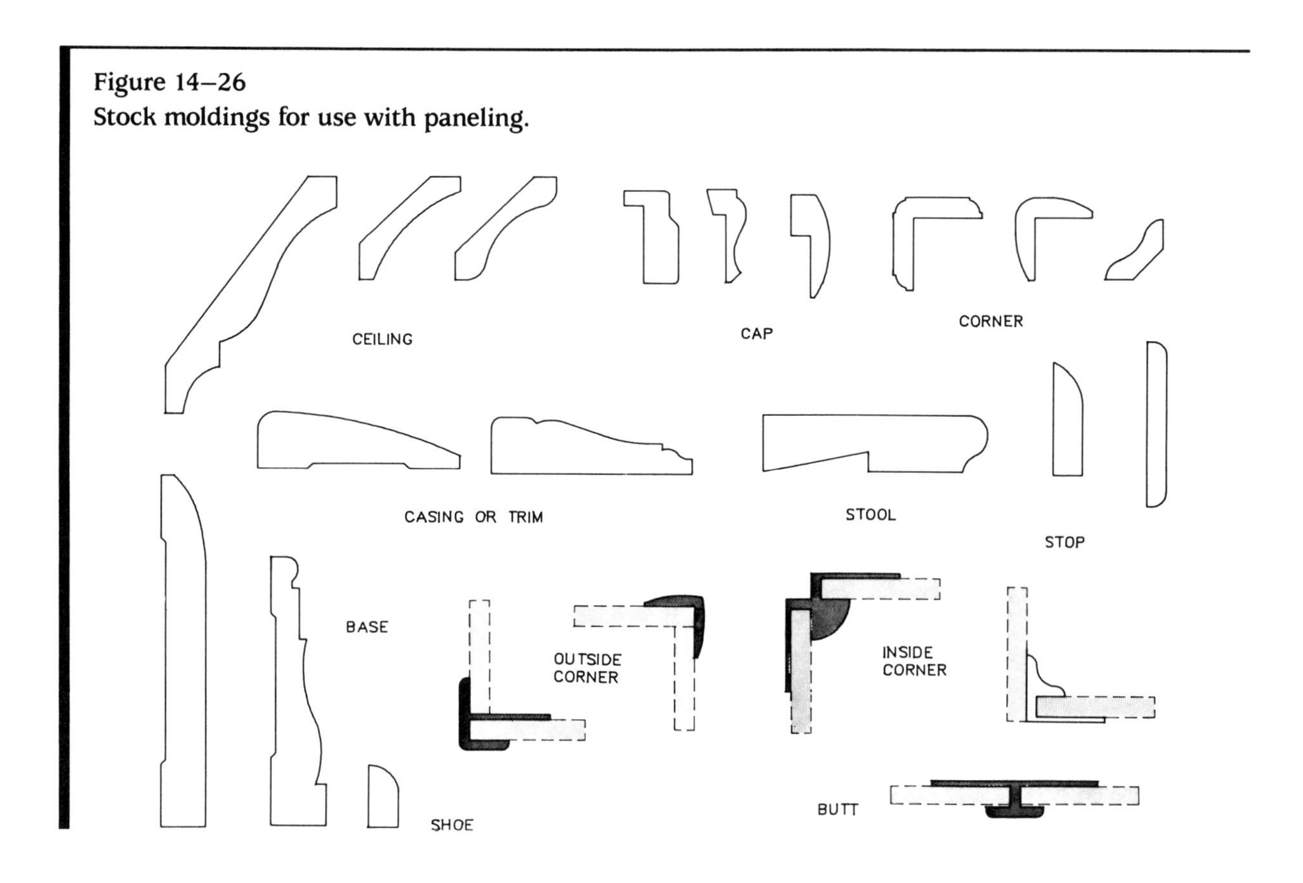

TRIM

Stock prefinished wood or plastic moldings are available to match most prefinished paneling. Figure 14–26 shows many of the stock moldings. In addition, you can use stock pine molding finished to match the paneling. The molding should be finished before installation.

Outside Corners

Outside corners can be finished in several ways. If outside corner molding is used, panel edges should be lapped as shown in Figure 14–27. Another way to make a neatly finished corner is shown in Figures 14–28 and 14–29. The panel edges are brought together but not lapped. Either ¼″ quarter-round molding or ¼″ × ¼″ square molding can be glued into the corner. If the paneling is on bare studs, you can tack-nail the molding until the glue dries, then pull the nails or set them below the surface and fill the holes. If the panels are over gypsum, do not tack-nail molding; instead, use masking tape to hold it in place until the glue dries. If the two panel edges do not meet accurately, use larger stock and trim it to fit after installation. A spokeshave with a slanted blade (slanted so the blade does not protrude on one side) is a good tool for this trimming.

Inside Corners

The neatest way to finish inside corners is to butt the two panels. If the corner is square in only one direction, lay in the other panel first, trimming the edge to fit roughly. Trim the second panel edge if necessary, and stain the edge before installing the panel. Various methods are shown in Figure 14–30. You can also use quarter-round or cove molding in place of the moldings shown.

Top Edge

Paneling can be butted against the ceiling without any molding, but this usually requires that the top edges of all panels be scribed for an

Figure 14–27
Using lapped panel edges covered by outside corner molding is the simplest way to finish off outside corners in new construction. (A) Paneling directly on studs. (B) Paneling over gypsum board.

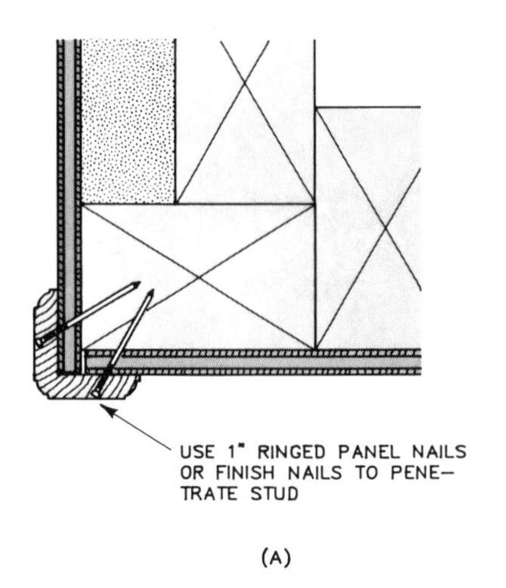
USE 1" RINGED PANEL NAILS OR FINISH NAILS TO PENETRATE STUD
(A)

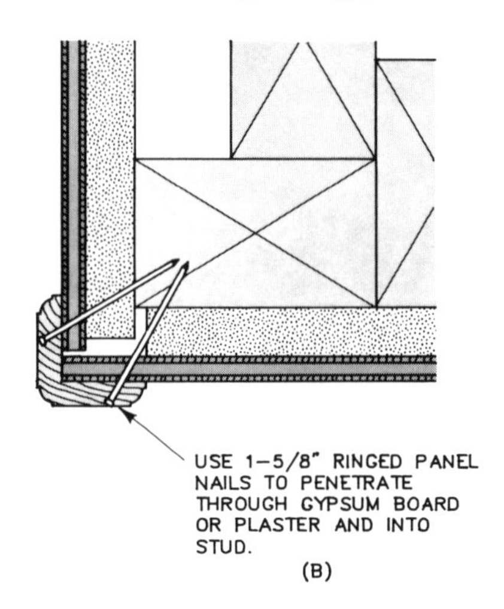
USE 1-5/8" RINGED PANEL NAILS TO PENETRATE THROUGH GYPSUM BOARD OR PLASTER AND INTO STUD.
(B)

Figure 14–28
Trimming outside corners with square and quarter-round strips for paneling on studs.

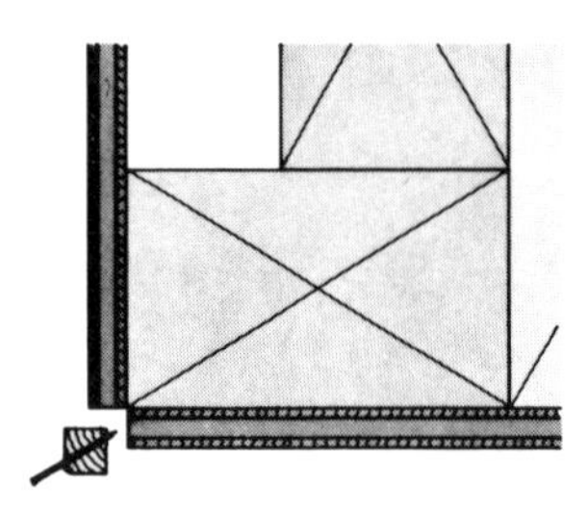

SQUARE CORNER STRIP

WHEN PANELING, ALIGN EDGES CAREFULLY WITH CORNER (START PANELING AT THE CORNER). GLUE AND NAIL SQUARE STRIP INTO CORNER, TRIM 1/4" STRIP FLUSH WITH ACTUAL 7/32" THICKNESS OF PANEL WITH A SPOKE SHAVE.

TIP: ANGLE THE BLADE IN THE SPOKE SHAVE TO PREVENT IT FROM TRIMMING THE PANEL.

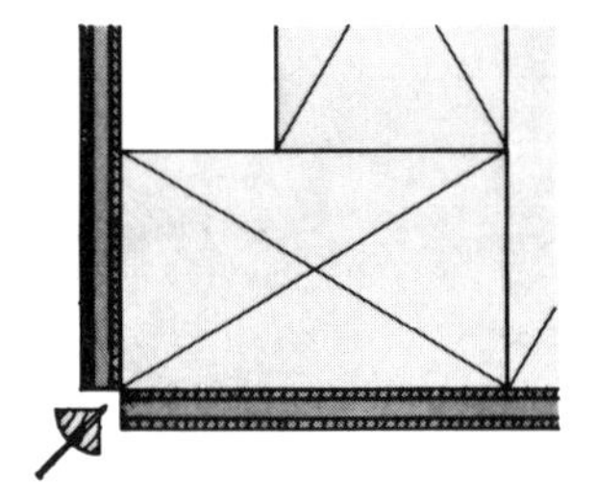

QUARTER-ROUND CORNER STRIP

IT IS DIFFICULT TO TRIM A SQUARE STRIP FLUSH WITH THE PANEL SURFACE WITHOUT GOUGING THE PREFINISHED PANEL SURFACE. THE TASK IS EASIER WITH QUARTER-ROUND AS ONLY THE EDGE OF THE TRIM HAS TO BE SHAVED.

accurate fit to the ceiling (Figure 14–31). The most widely used top-edge covering is a crown, bed, or cove molding. Partition paneling can be topped by a simple cap or crown molding, or a more elaborate combination of crown molding and shelf (see Figure 14–32). Wainscoting is usually trimmed with a chair rail along the top edge.

Baseboards

Baseboard and shoe moldings finish off the bottom edge of the paneling. As shown in Figure 14–33, the original baseboard can be removed and replaced over the paneling or left in place to act as furring to support the paneling and a new

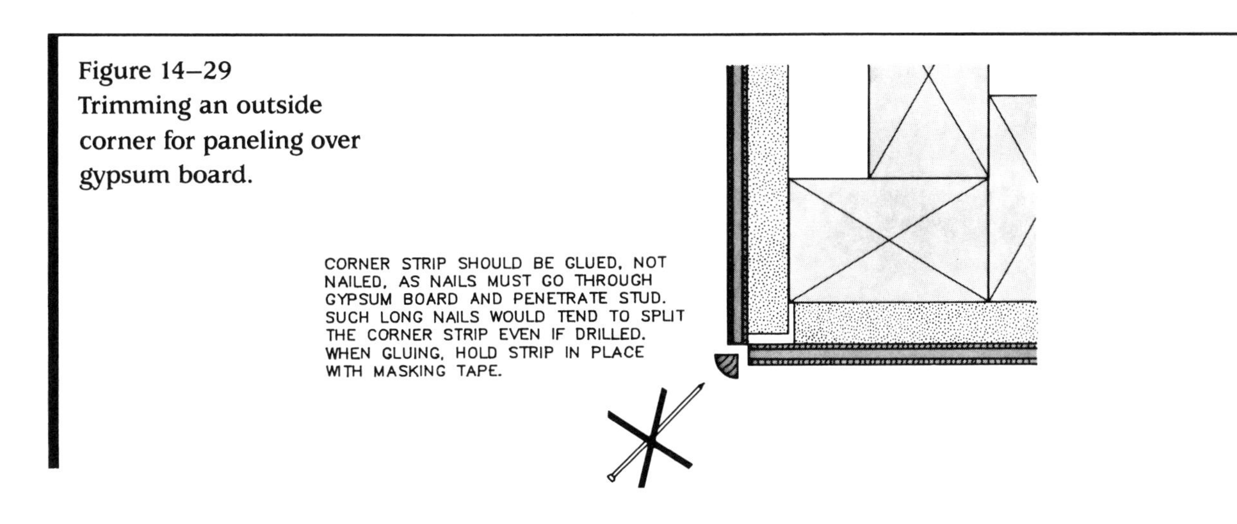

CORNER STRIP SHOULD BE GLUED, NOT NAILED, AS NAILS MUST GO THROUGH GYPSUM BOARD AND PENETRATE STUD. SUCH LONG NAILS WOULD TEND TO SPLIT THE CORNER STRIP EVEN IF DRILLED. WHEN GLUING, HOLD STRIP IN PLACE WITH MASKING TAPE.

Figure 14–29
Trimming an outside corner for paneling over gypsum board.

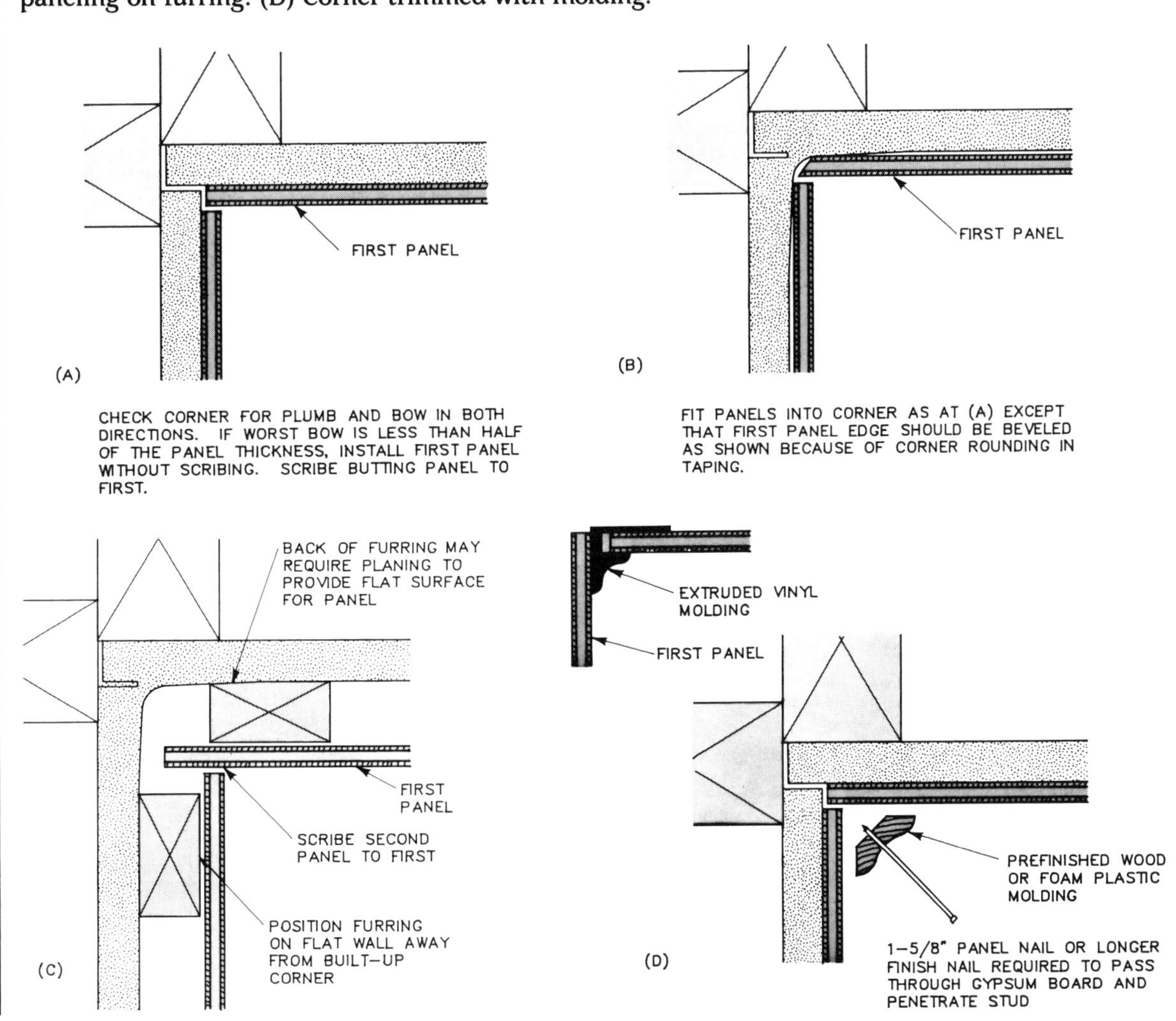

Figure 14–30
Trimming inside corners. (A) Butted paneling on bare studs or over gypsum board with untaped corner. (B) Butted paneling applied directly to gypsum board or plaster with finished corner. (C) Butted paneling on furring. (D) Corner trimmed with molding.

Figure 14–31
Scribing panel to ceiling. This is required if ceiling molding is not used.

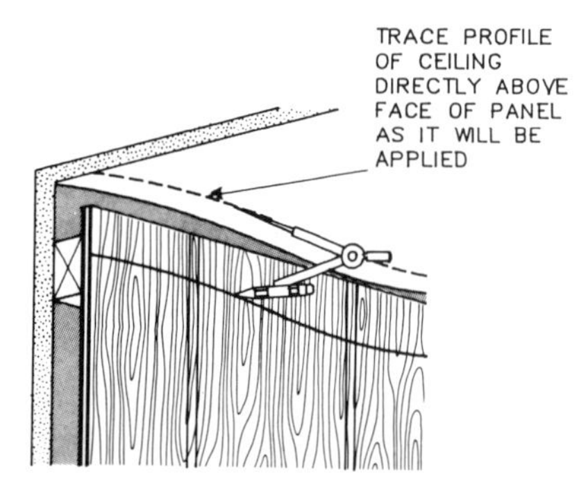

baseboard. Moldings and trim are discussed in greater detail in Chapter 10.

SOLID BOARD PANELING

Solid board paneling consists of boards ⅜″ to ¾″ thick and 3″ to 12″ wide, with either tongue-and-groove or lap edges. Typical interior paneling patterns are shown in Figure 14–34. Exterior sidings can also be used, both unfinished and prefinished, and even shingles and shakes. Strip and plank flooring offer further choices. Decorative woods may also be used—cherry, mahogany, walnut, pecky cypress, knotty pine, exotic imports—anything you can afford. Purchase

Figure 14–32
Trimming top edge of paneling with molding. (Left) Crown molding. (Right) Use molding with two finished edges for an edge that stops short of the ceiling.

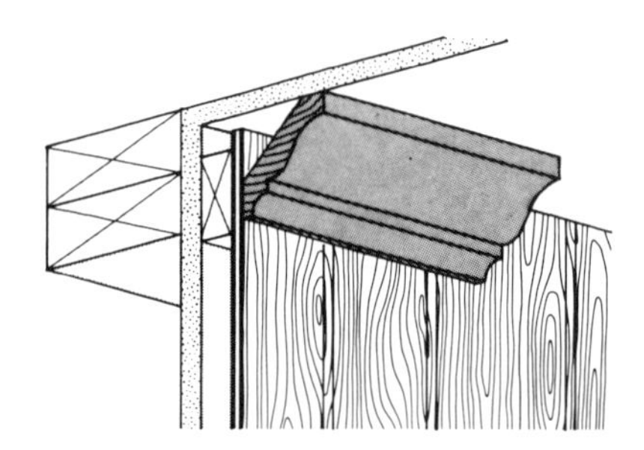

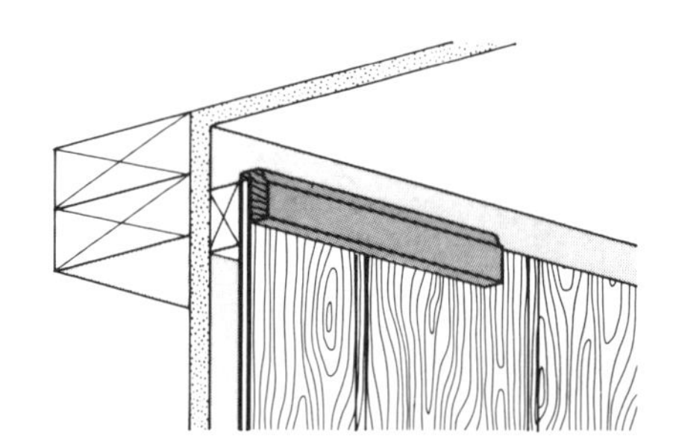

Figure 14–33
Baseboards over paneling.

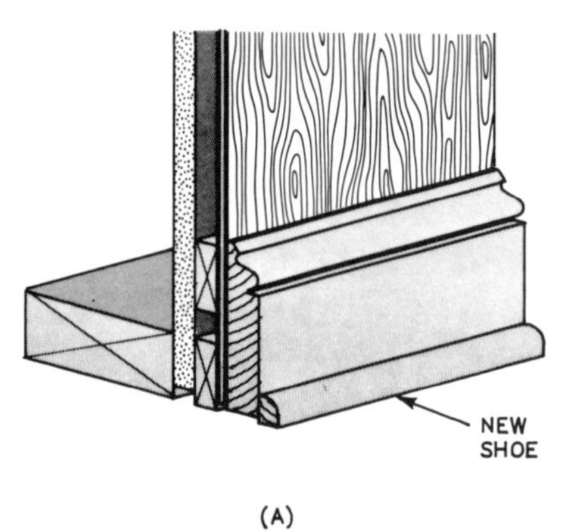

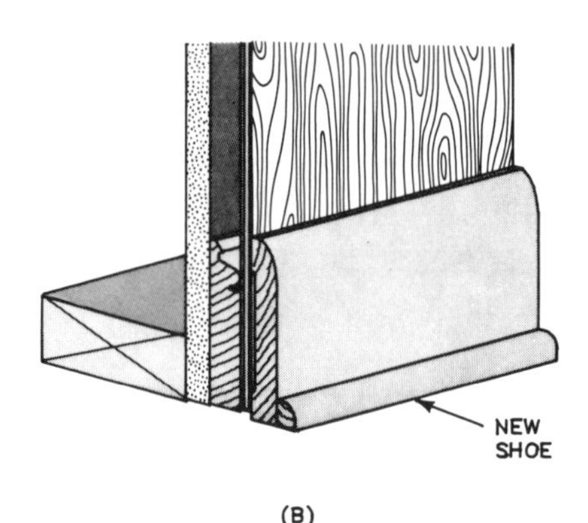

Figure 14–34
A selection of board paneling patterns. (SPP: Southern Pine Paneling)

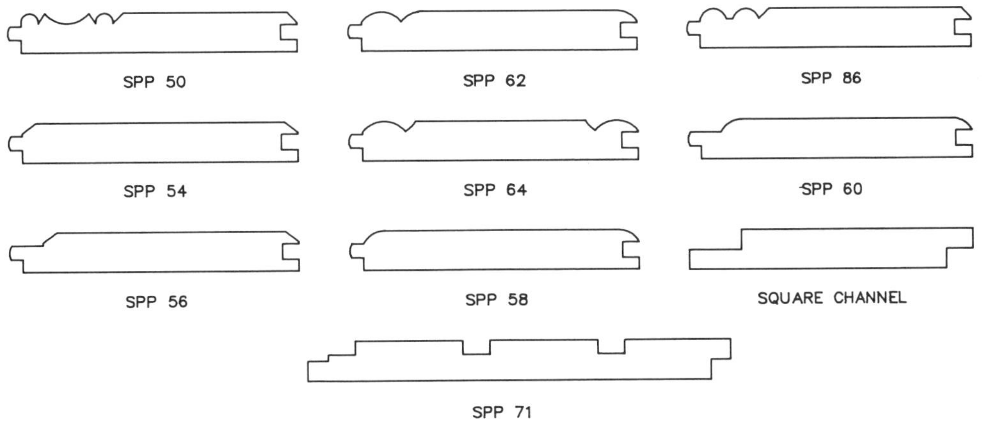

SPP 50
SPP 62
SPP 86
SPP 54
SPP 64
SPP 60
SPP 56
SPP 58
SQUARE CHANNEL
SPP 71

Figure 14–35
Blocking for installing vertical board planking on existing vertical studs.

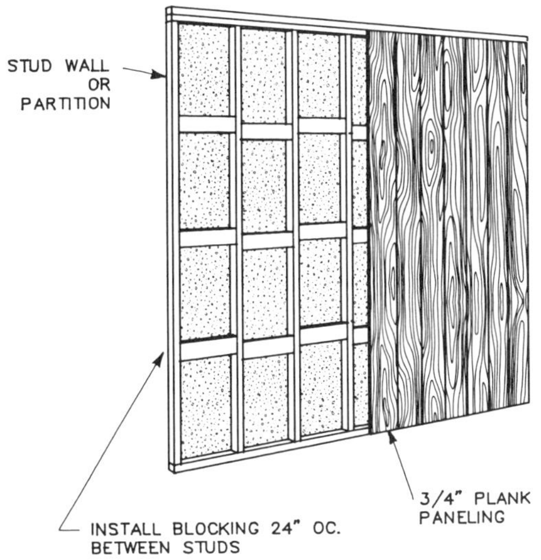

STUD WALL OR PARTITION
3/4" PLANK PANELING
INSTALL BLOCKING 24" OC. BETWEEN STUDS

Figure 14–36
Horizontal furring on masonry for vertical board paneling.

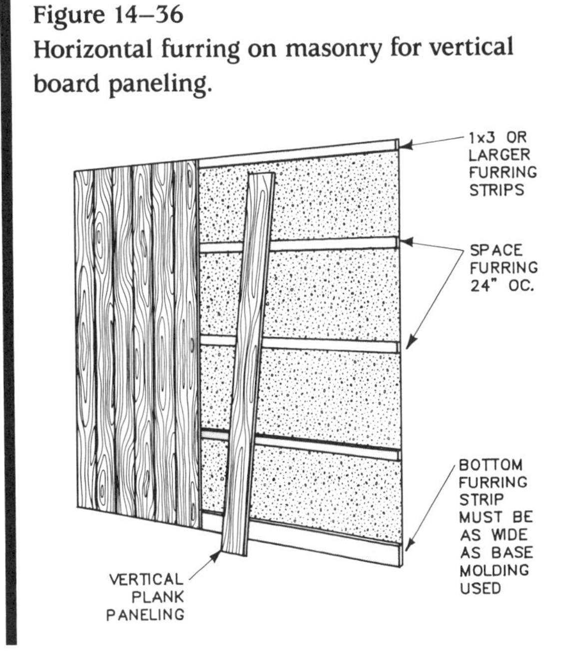

these woods as S2S (surfaced on two sides) board lumber and rip, shape, or rout them into plank paneling in your own shop.

Vertical Application

Blocking must be installed between studs in an unsheathed wall before paneling, as shown in Figure 14–35, and should not be more than 24" on center. Staggering 2 × 4 blocking as shown permits nailing through the sides of the studs into the ends of the blocking. Board paneling installed over gypsum board or plaster must be fastened with adhesive, plus nails into the top and sole plates of the wall. For vertical paneling on a masonry wall, furring strips are applied horizontally, as shown in Figure 14–36.

Installation

Start paneling at an inside corner. Plumb, scribe, and trim the first board to fit into the corner (see

Figure 14–37). If you are using tongue-and-groove boards, be sure the tongue side is facing outward. Fasten the first board before fitting the next. Do not cut a hole in a board for an outlet or switch until you are ready to fit and install that particular board.

Nailing technique is shown in Figure 14–38. Before nailing, tap each board into position with a block of wood cut from the same paneling with an edge that matches the board edge (Figure 14–39). This will protect the tongue from damage. However, do not wedge boards tightly together. Leave the joints a little loose to allow for seasonal expansion. Check often with a level to ensure vertical alignment (Figure 14–40). If you buy unfinished board paneling, or make your own, finish both sides before installation to avoid seeing unfinished edges and to minimize warping.

Horizontal Application

Applying horizontal board paneling is essentially the same as applying vertical paneling. On a framed wall, nails should be driven into studs. Furring on a masonry wall should run vertically, spaced 16″ to 18″ on center. Be careful to keep planks level. End-butted joints should fall on studs or furring and be staggered.

Figure 14–38
Toenailing plank paneling.

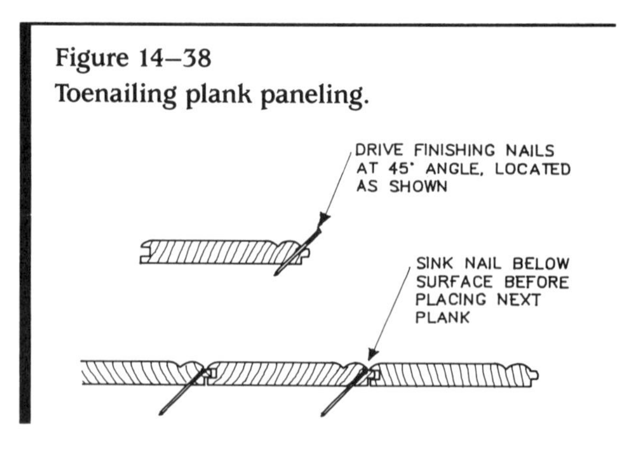

Figure 14–39
Hammering plank paneling without damaging edge.

USE SHORT SECTIONS OF RIPPED PLANKING AS PROTECTIVE PADS WHEN HAMMERING PLANK INTO PLACE

Figure 14–37
Scribing and sawing first board to fit against uneven corner wall.

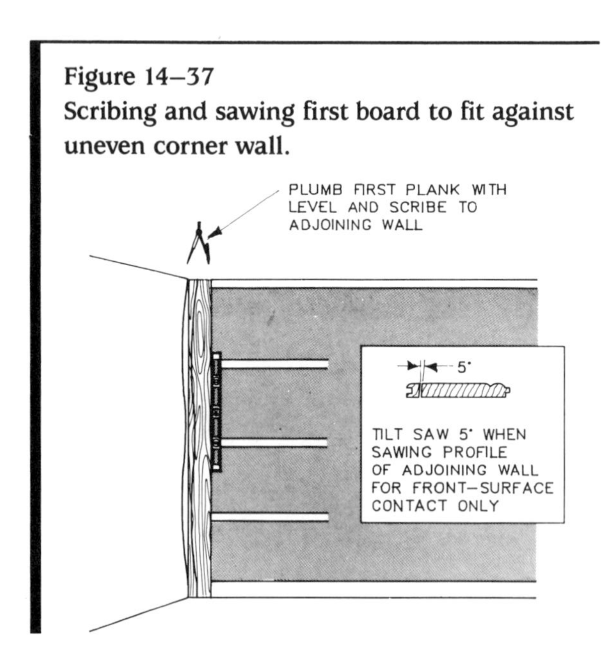

Figure 14–40
Check planks often with a level to keep them plumb.

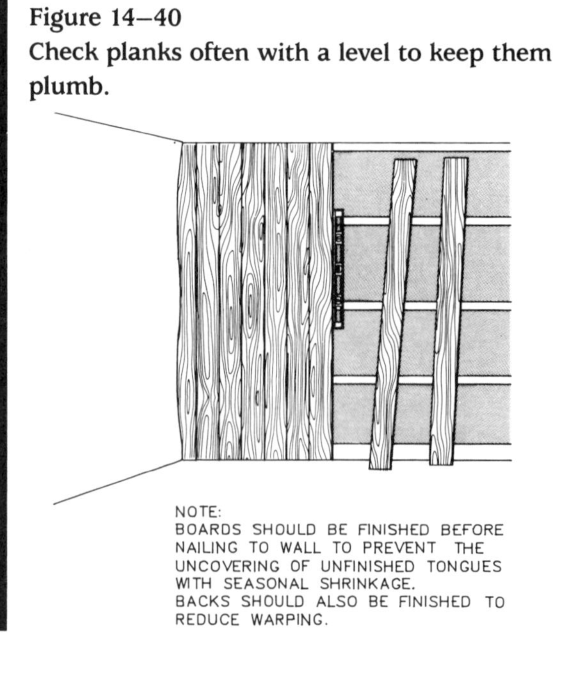

NOTE:
BOARDS SHOULD BE FINISHED BEFORE NAILING TO WALL TO PREVENT THE UNCOVERING OF UNFINISHED TONGUES WITH SEASONAL SHRINKAGE.
BACKS SHOULD ALSO BE FINISHED TO REDUCE WARPING.

15 Sound Control

There are three sources of unwanted noise in a house: (1) Noise from outside (traffic, aircraft, neighbors, children at play, barking dog, and the like); (2) noise originating in another part of the house and transmitted through the walls, floor, or ceiling; and (3) noise generated inside the same room. You can do a lot to control this unwanted noise, but before you start you need to know something about the nature of sound and how it is transmitted, and about how sound control works. Sound control is not soundproofing. If something is waterproof, no water can get into it. Similarly, if a room or a structure is soundproof, absolutely no unwanted sound can penetrate into it or escape from it. That degree of sound isolation is neither practical nor affordable for everyday living spaces. In fact, it is not even desirable, for it cuts off all sense of contact with the outside world.

What is useful, achievable, and affordable is sound control, which can reduce extraneous, unwanted noise to a tolerable background level. The most widespread method of residential sound control—and one of the easiest to accomplish—is to install acoustical tiles or panels on the walls and ceiling, as shown in Figures 15–1, 15–2, and 15–3.

DEFINING AND MEASURING NOISE

Noise is unwanted sound. The music from one person's stereo set or the sounds of his shop tools running and cutting can easily be very disturbing noise to another person.

Sound is produced by a vibrating object. Sound travels in all directions from the source as a pressure wave in the air—the same way ripples travel from a pebble in a pond. The greater the intensity of the vibration—also called the amplitude, pressure, or magnitude—the louder the sound. Sound/noise also has frequency, or pitch, as in a high-pitched squeal or a low-pitched rumble.

Sound control is the art of putting something between your ears and the source of the sound to reduce its intensity, or doing something to reduce the sound at its source. Or, if the ears and the sound source are in the same room, treating the room to reduce the amount of sound bouncing off the walls, floor, and ceiling.

Installing noise control material in a house is less expensive when it is part of new construction than if it is added on. A variety of materials and

Figure 15–1
Cement block walls reflected every bit of noise produced in this home woodworking shop. The open ceiling transmitted most of the noise to the floor above.

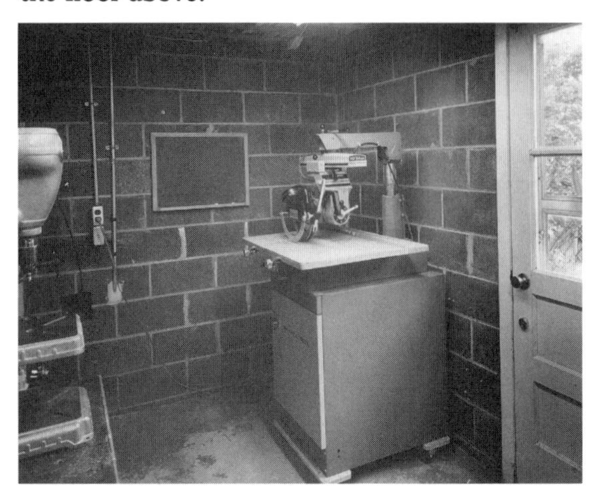

Figure 15–2
The walls were furred out to install paneling, then a grid for a suspended ceiling was hung in place.

Figure 15–3
Acoustical panels on key areas of the walls and over the entire ceiling brought the noise problem well under control.

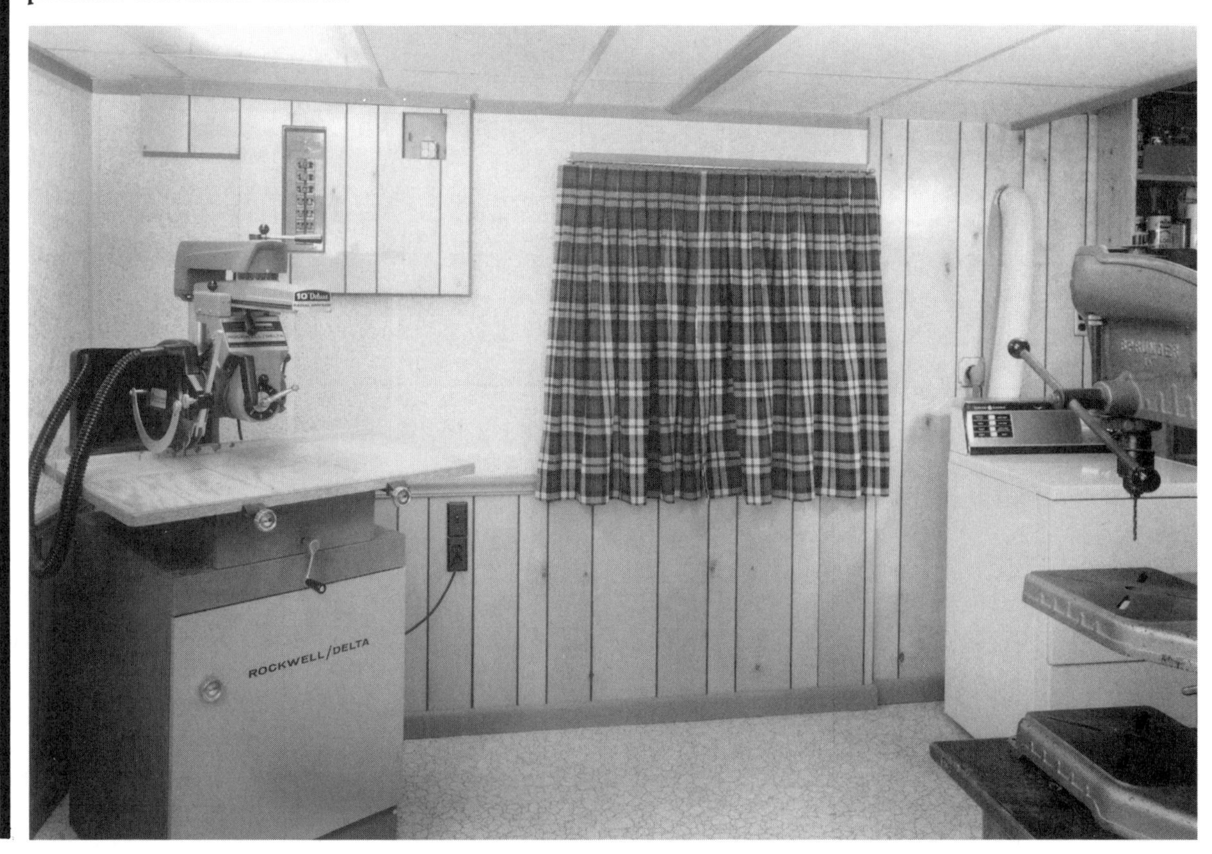

techniques are used, some common, some special.

Sound Intensity Measurement

A decibel (dB) is a measurement of sound intensity, or sound level. Zero (0) dB is the level of sound at the threshold of normal hearing. A whisper is around 10 dB, normal conversation 25 dB, a quiet radio 40 dB, a loud radio 80 dB, a router 100 dB, a radial-arm saw or a jet taking off 108 dB. A scale of sound intensities is given in Table 15–1.

Your ears cannot detect a difference of 1 dB, but a 5 dB increase or decrease in sound level is noticeable. Any change of 10 dB doubles or halves the sound level. Measuring noise levels and improvements by ear is difficult. A sound level meter, available at electronics stores, is used to measure sound intensity. Use one to compare noise sources, as in Figure 15–4, and to measure before and after levels as you work on a noise problem.

Figure 15–4
An inexpensive sound-level meter can be a great help in identifying sources of noise and evaluating efforts at noise reduction. For example, although they sounded different, the meter showed that this shop vac was making as much noise as the radial-arm saw it was connected to.

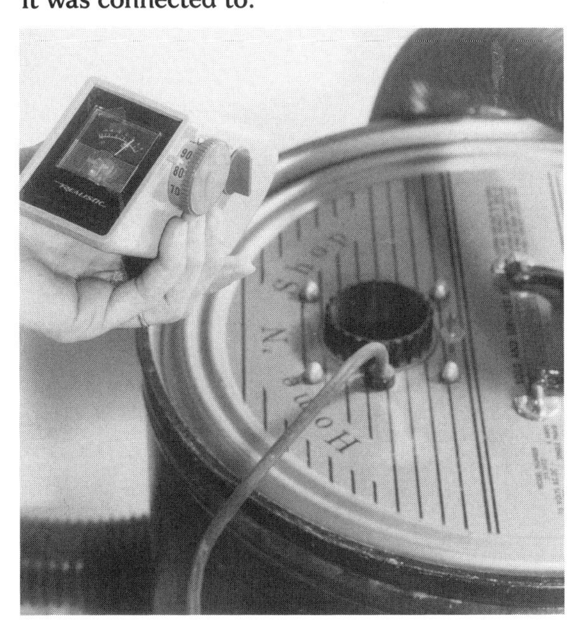

Most sound travels both through the air and through solid objects. For example, the sound of your voice is airborne sound until it strikes a wall and becomes structure-borne. The wall vibrates and reradiates the sound as airborne sound in an adjacent room or area, but at a lower level because some of the sound intensity has been dissipated inside the wall. At each physical barrier the cycle repeats until the sound dissipates completely.

Direct and Reflected Sound

Direct sound travels through the air in a straight path from source to receiver at a speed of about 1,130 feet per second at normal room temperature. Direct sound diminishes in intensity as the distance between source and receiver is increased.

Reflected sound is sound that strikes and bounces off surfaces on its way to the receiver. When a sound wave strikes a surface, its direction changes in the same way a bounced ball would. Some of the sound intensity is lost in the process of bouncing, and having a longer path to travel, the sound will not be as loud as direct sound when it reaches a hearer. It will also take slightly longer getting there, sometimes producing a ringing effect.

ACOUSTIC MATERIALS

There are two kinds of acoustic material. One kind blocks or attenuates sound transmission, the other kind absorbs sound. No single material acts in both ways effectively, but two materials can be combined in a sandwich to accomplish both. Sound-attenuating materials have a high sound transmission loss; they are used to reduce noise as it passes from one room to another room. Sound-absorbing materials are used to reduce sound levels in the same room as the source of the sound. Sound-attenuating materials include almost anything that is dense, has no air passages through it, and doesn't vibrate at audi-

Table 15–1
SCALE OF SOUND INTENSITY

Description	Level dB	Typical Sources	Physical Effects
Deafening	150		Short exposure can cause hearing loss
	140	Jet plane takeoff	
	130	Artillery fire Riveting	
	120	Siren at 100′ Jet plane (passenger ramp) Thunder, sonic boom	Threshold of pain
	110	Woodworking shop Accelerating motorcycle Hard rock band	Threshold of discomfort
	100	Subway (steel wheels) Power lawnmower Outboard motor	
Very Loud	90	Truck (unmuffled) Train whistle Kitchen blender Pneumatic jackhammer	
	80	Daisy wheel printer Subway (rubber wheels) Noisy office Average factory	Too loud for telephone use in immediate vicinity
Loud	70	Average street noise Dot matrix printer Freight train at 100′ Average radio	
	60	Noisy home Average office Normal conversation	
Moderate	50	Quiet radio Average home Quiet street	
	40	Private office Quiet home	
Faint	30	Quiet conversation Broadcast studio	
	20	Empty auditorium Whisper	
Very Faint	10	Rustling leaves Soundproof room Human breathing	
	0		Threshold of audibility

ble frequencies—materials like concrete, brick, gypsum board, and sheet lead.

Sound-absorbing materials are lightweight and fibrous—fiberglass insulation batts, acoustical ceiling tiles and panels, upholstery foam, horsehair padding, sawdust—any material of some thickness that includes thousands of tiny air pockets and passages that can trap the sound waves.

The ability of a material or a system of materials to block the transmission of sound from one area to another is called transmission loss (TL) and is measured in decibels. The transmission loss of wall or floor/ceiling construction can be improved by increasing mass, by filling cavities with sound-absorbing materials, and by breaking the sound vibration path. The basic idea is shown in Figure 15–5.

Sound transmission class (STC) is a single number rating of the airborne sound transmission performance of particular types of wall or floor/ceiling construction. STC rating numbers are obtained by testing sample constructions under standard laboratory conditions. Typical wall constructions and ratings are shown in Figure 15–6.

In real life these carefully determined numbers are at best loose estimates of what you will get with a particular construction. Building a sound-deadening barrier is an art, not a science. Table 15–2 gives some real-life interpretations of the effect of various STC ratings. From a practical standpoint in residence noise control, aim for an STC of 45 or 50.

Figure 15–7 shows floor/ceiling noise-blocking constructions that can be built with readily available materials, along with the STC rating of each construction.

If you are going to successfully block sound coming out of a room, the walls, ceilings, and

Figure 15–5
Sound transmission. Sound passing through a wall from one room to another can be reduced by modifying the construction of the wall.

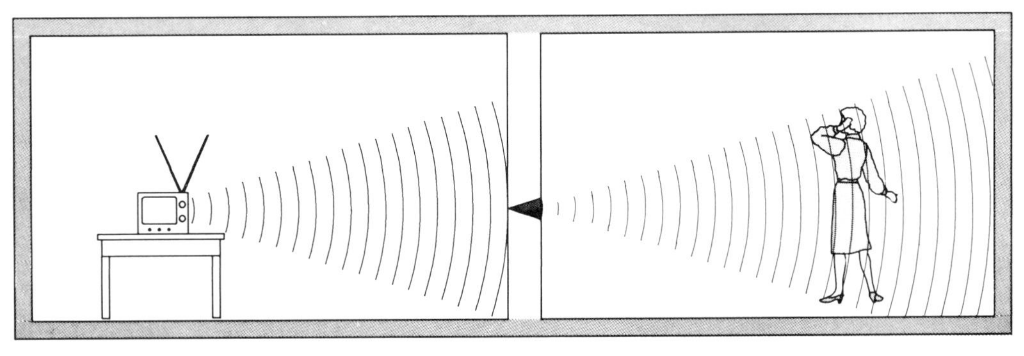

SOUND TRANSMISSION THROUGH A STANDARD GYPSUM BOARD AND STUD PARTITION CAN INTERFERE WITH NORMAL CONVERSATION

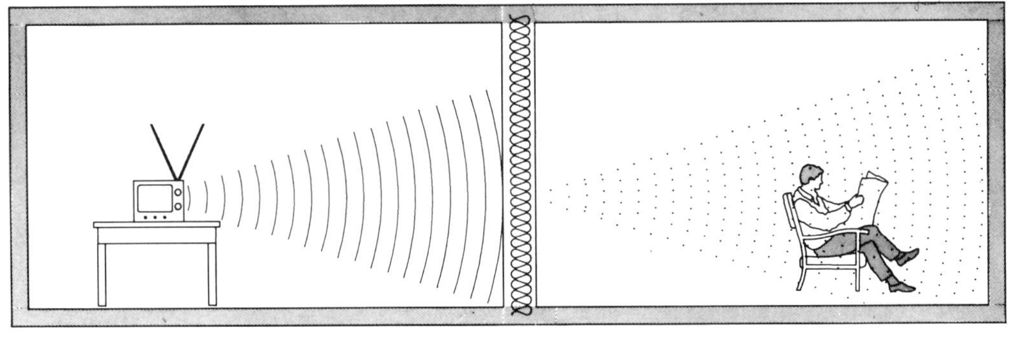

SOUND TRANSMISSION THROUGH THE PARTITION CAN BE REDUCED SIGNIFICANTLY WITH SOUND-BLOCKING CONSTRUCTION

floors must all have the same STC rating. Walls rated STC 52 and a ceiling rated STC 40 make no sense. The sound will come through the ceiling. If both walls and ceiling have an STC 52 rating, but there's a door in the wall rated STC 27, the sound will come through the door.

BLOCKING NOISE

Increasing Mass

To improve the noise-blocking performance of a wall, increase the mass. Heavy materials block sound better than light materials. For example, placing sound-deadening board under the gypsum board, applying a second layer of gypsum board, or even adding wood or hardboard paneling to a wall provides increased sound transmission loss.

Sound-deadening board comes in 4′ × 8′ panels that can be nailed to studs or joists. There are two kinds—a board made of organic fibers and a form of gypsum board. Ordinary gypsum board is applied over the sound-deadening board to provide a finished surface. It must be attached to the sound-deadening underlayer only with adhesive so there is no hard connection

Figure 15–6
STC ratings of typical wall constructions.

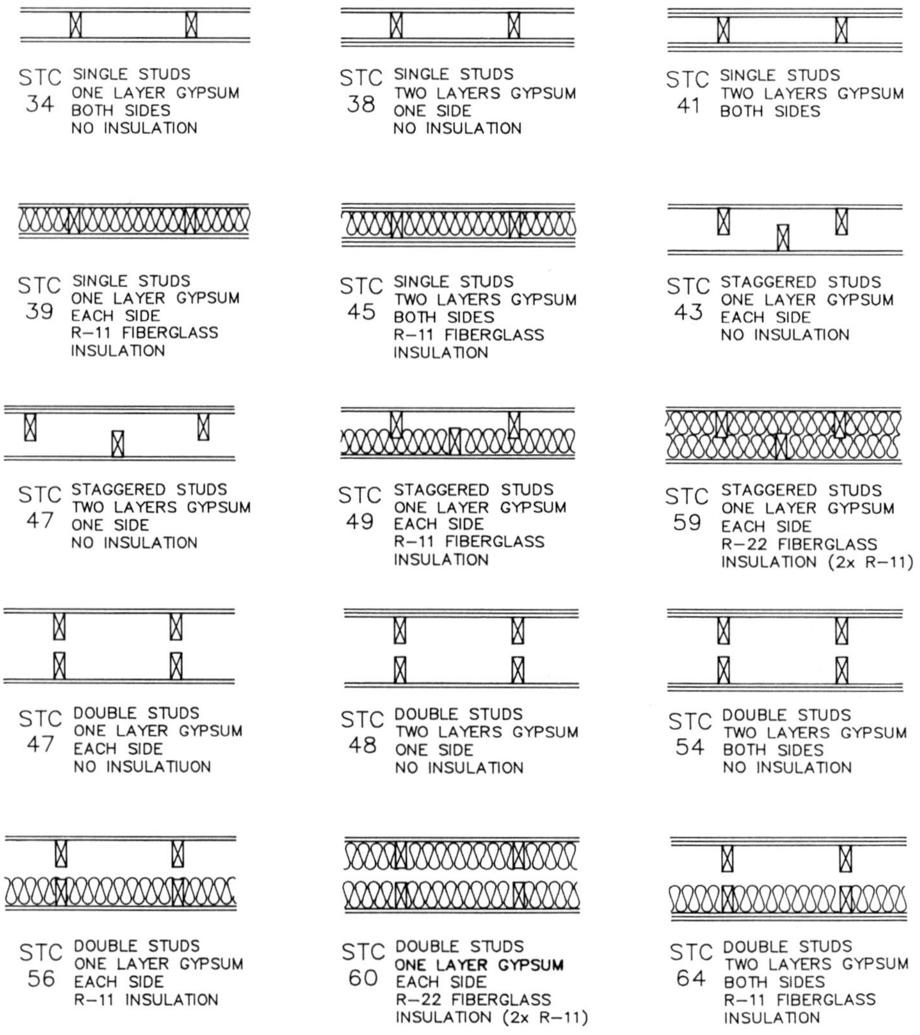

STC 34 SINGLE STUDS ONE LAYER GYPSUM BOTH SIDES NO INSULATION

STC 38 SINGLE STUDS TWO LAYERS GYPSUM ONE SIDE NO INSULATION

STC 41 SINGLE STUDS TWO LAYERS GYPSUM BOTH SIDES

STC 39 SINGLE STUDS ONE LAYER GYPSUM EACH SIDE R–11 FIBERGLASS INSULATION

STC 45 SINGLE STUDS TWO LAYERS GYPSUM BOTH SIDES R–11 FIBERGLASS INSULATION

STC 43 STAGGERED STUDS ONE LAYER GYPSUM EACH SIDE NO INSULATION

STC 47 STAGGERED STUDS TWO LAYERS GYPSUM ONE SIDE NO INSULATION

STC 49 STAGGERED STUDS ONE LAYER GYPSUM EACH SIDE R–11 FIBERGLASS INSULATION

STC 59 STAGGERED STUDS ONE LAYER GYPSUM EACH SIDE R–22 FIBERGLASS INSULATION (2x R–11)

STC 47 DOUBLE STUDS ONE LAYER GYPSUM EACH SIDE NO INSULATIUON

STC 48 DOUBLE STUDS TWO LAYERS GYPSUM ONE SIDE NO INSULATION

STC 54 DOUBLE STUDS TWO LAYERS GYPSUM BOTH SIDES NO INSULATION

STC 56 DOUBLE STUDS ONE LAYER GYPSUM EACH SIDE R–11 INSULATION

STC 60 DOUBLE STUDS ONE LAYER GYPSUM EACH SIDE R–22 FIBERGLASS INSULATION (2x R–11)

STC 64 DOUBLE STUDS TWO LAYERS GYPSUM BOTH SIDES R–11 FIBERGLASS INSULATION

Figure 15–7
STC ratings of typical floor-and-ceiling constructions.

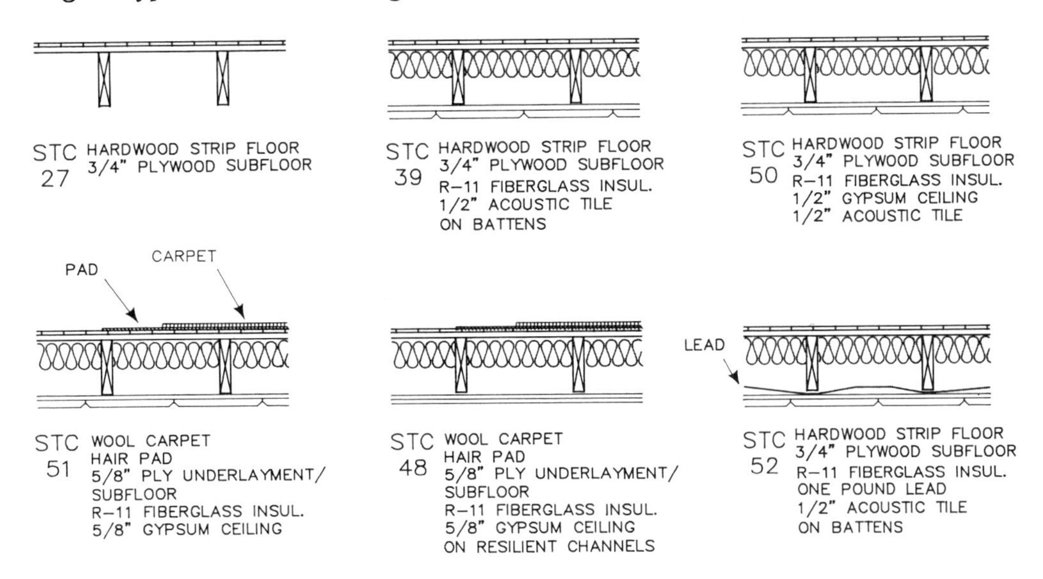

to the studs or joists. You may have to search a bit for a lumberyard that stocks sound-deadening board or will special-order it for you.

Fill hollow wall cavities with sound-absorbing material to reduce sound transmission. Fiberglass thermal insulation is very good for this. It does not matter how densely you pack it in, as long as the cavity is completely filled. Filling the cavities this way produces about 8 dB of absorption.

Do not depend on foamboard to absorb or block sound. Foamboard has closed bubble cells that make it watertight, but sound cannot enter them and become trapped. On the other hand, any foam product you can blow through is a good sound absorber. Upholstery foam is good because it has burst bubbles—only the intersections between bubbles remain when it is cured. The tiny air passages that are left between trap the sound.

Table 15–2
WHAT DO STC NUMBERS MEAN?

Transmission Loss (STC Rating)	What You Will Hear
25	Normal speech can be understood quite easily through a wall.
30	Loud speech can be understood fairly well. Normal speech can be heard but not easily understood.
35	Loud speech is audible but not intelligible. Normal speech may be heard faintly.
42	Loud speech is audible only as a murmur.
45	Must strain to hear loud speech. Loud singing can be heard faintly.
48	Some loud speech barely audible. Most loud musical instruments heard only faintly.
50	Loud speech not audible. Radio at full volume may be heard but only faintly. Rock band heard faintly, if at all.

Breaking the Vibration Path

Walls transmit sound vibrations from one face to another through the studs; ceilings and floors transmit sound to one another by vibrations through the joists. Eliminate framing members common to both sides and you have broken the path. In walls stagger the studs, or build two walls with a gap in the center, as illustrated in Figure 15–8.

If you build a second wall, construct it on one side, close to but not touching the existing wall. It

Figure 15–8
Split-wall construction. Insulation has been omitted from the non-load-bearing partition for clarity.

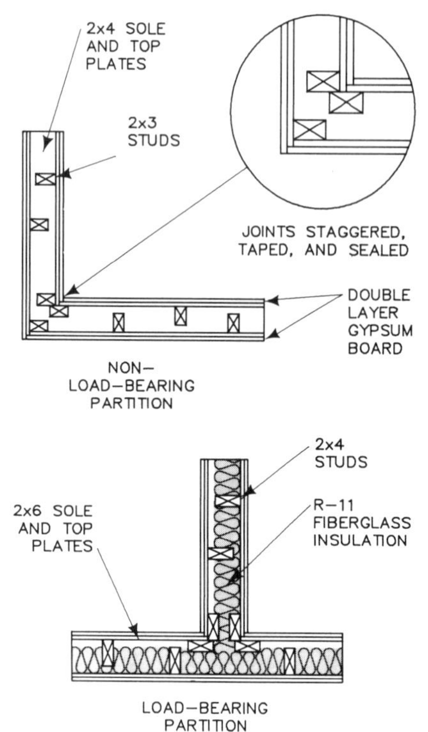

helps to add insulation batts between the new studs (these batts could touch the existing wall). Also, install a layer of sound-deadening board beneath the new wall's gypsum board.

Sealing the Wall

A wall must be essentially watertight; the perimeter of the wall should be sealed as if you were planning to prevent water from flowing from one room to the other. If water could get through the wall, noise will also get through.

Use a nonhardening, permanently resilient butyl or silicone caulking compound, applied in two beads (for new construction) or a continuous bead, to seal the sole and top plates at each side of the wall. The end studs should also be sealed to adjacent walls before you apply the finish wall. At corners where multiple layers of gypsum board and sound-deadening boards butt, use joint compound and tape to seal the staggered layers. Soundproofing can be designed into exterior walls in new construction; it is difficult to retrofit it into existing walls.

Adjusting Doors and Windows

Use solid-core wood or metal doors instead of hollow flush doors or wood frame and panel doors. Gasket the doors at top and sides with soft weatherstripping, and use a drop-type threshold closure at the bottom. Replace sliding and folding doors with hinged doors. The doors to rooms on opposite sides of a hall should not be directly opposite one another. A significant decrease in noise transmission through a door opening can be obtained by mounting doors back to back in the jamb and sealing both as described above. This installation is shown in Figure 15–9.

Figure 15–9
Double-door installation creates an airlock in between that produces a sound transmission loss greater than twice the loss through a single door.

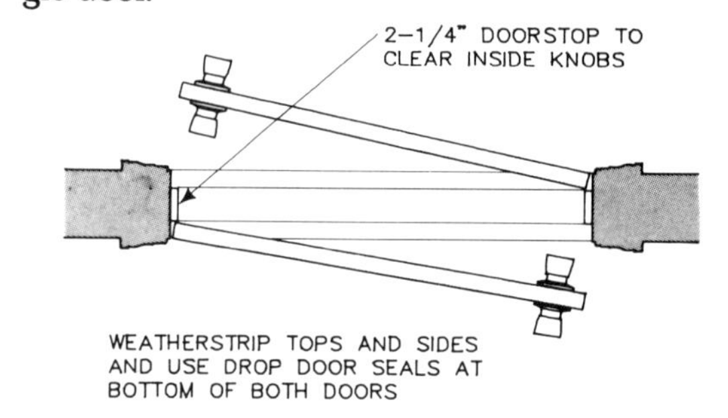

It is difficult to reduce sound transmission through a window beyond adding a second sash, such as a storm window, and having all the sashes tightly sealed in the frames.

Repositioning Electrical and Plumbing Connections

Light switch and outlet boxes should not be located back to back, as that will provide a noise path through the wall. Instead, they should be

Figure 15–10
Stagger switch and outlet boxes locations in opposite walls to minimize sound transmission through the wall.

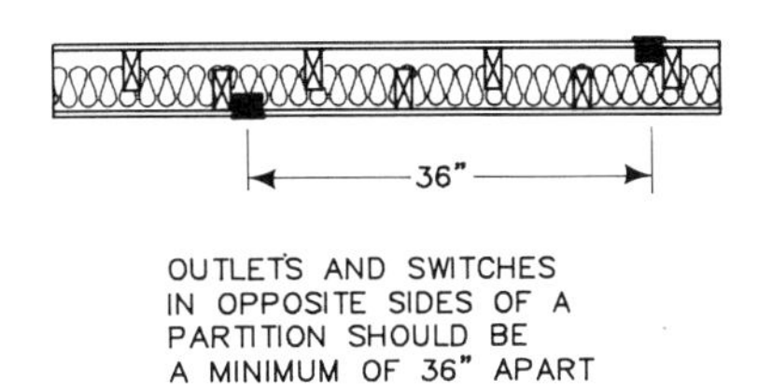
36"
OUTLETS AND SWITCHES
IN OPPOSITE SIDES OF A
PARTITION SHOULD BE
A MINIMUM OF 36" APART

offset at least 36", as shown in Figure 15–10. Each box opening in a wall must be sealed. From a noise control standpoint, surface wiring is better than hidden wiring. Surface ceiling fixtures are also preferred over recessed electrical fixtures. Also caulk all openings around pipes and other connections to plumbing fixtures in the framing if possible, and in the wall surface (see Figure 15–11).

Electrical equipment, particularly a telephone, should not be mounted on an interior wall unless you want to hear the telephone ring in both rooms. The whole wall can be set into vibration, amplifying the sound.

Figure 15–11
Caulk all openings in walls for electrical boxes and plumbing connections to block sound transmission.

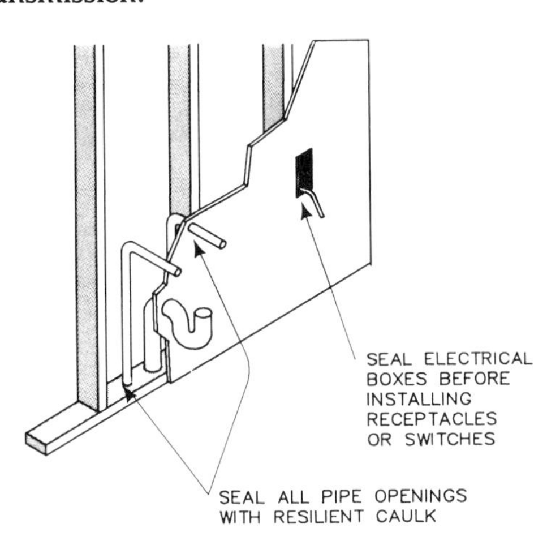
SEAL ELECTRICAL
BOXES BEFORE
INSTALLING
RECEPTACLES
OR SWITCHES
SEAL ALL PIPE OPENINGS
WITH RESILIENT CAULK

Lining and Wrapping Ducts

Heating and air-conditioning ducts readily transmit sound. Ducts can be lined with sound-attenuating duct liner insulation; they can also be wrapped. Both steps will reduce fan and air movement noise, as well as external sound picked up and transmitted through the duct system.

USING ACOUSTIC CEILING TILES AND PANELS

Acoustic ceilings not only absorb sound to reduce reflection, they also reduce the transmission of noise through them. The STC rating can be confusing, however. Acoustic ceiling products are rated for sound transmission loss between two rooms with a ceiling-high partition separating them, so that there is no common space above the ceiling between the two rooms.

The sound transmission loss of a ceiling can be improved by placing fiberglass insulation batts on the backs of the ceiling panels or between the joists. On the other hand, recessed ceiling lighting fixtures, particularly fluorescents, seriously compromise the STC of a ceiling.

DEALING WITH A NOISY ROOM

Controlling sound and noise inside a room with sound-absorbing materials must be approached cautiously. You can easily go too far. On the one hand, a room with all hard walls and no carpet will be a "live" room. Sound will bounce off all the hard surfaces and echo and re-echo in your ears in a manner ranging from annoying to deafening. At the other extreme, a heavily carpeted room with all the ceiling and wall surfaces covered with sound-absorbing materials will be a "dead" room, with practically no echoing sound.

Most of the time, a ceiling of acoustic tiles or panels will provide a reasonable amount of

Figure 15–12
Noise absorption. Treating the walls and ceiling of a noisy room with sound-absorbing material reduces both the noise inside the room and noise transmitted out of the room.

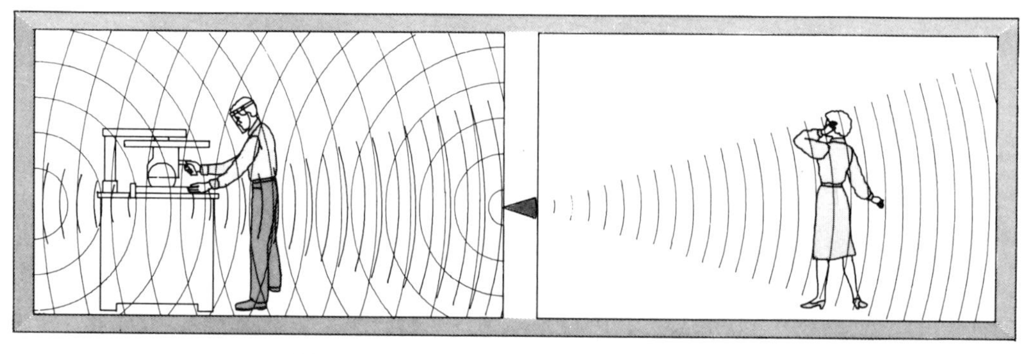

SOUND REFLECTING OFF HARD-SURFACED WALLS AND CEILING INCREASES THE NOISE LEVEL IN THE SOURCE ROOM AND IN THE RECEIVING ROOM.

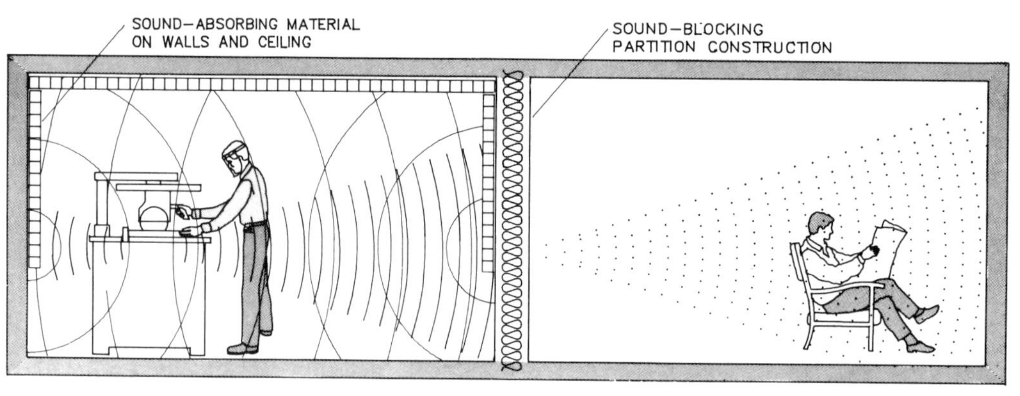

SOUND-ABSORBING MATERIAL ON THE WALLS AND CEILING OF THE SOURCE DECREASES THE NOISE LEVELS IN BOTH ROOMS.

sound absorption for any room of the house, except possibly a woodworking shop. A good approach is to do the ceiling first and try the room for a while. Then, if the room still seems too live, add sound-absorbing areas to the walls. The concept is illustrated in Figure 15–12.

If you are "soundproofing" your woodworking shop, there are sources of noise you should muzzle as much as you can. You need to get as much absorption as possible, but the tools, wood standing around, and shelves all either increase the liveness of the room or limit the amount of sound-absorbing material you can install.

For any noisy room, such as a home shop, a rule of thumb is that you need to apply sound-absorbing material on at least half of the total ceiling, wall, and floor area of the room. Start with an acoustical ceiling (installation is described in Chapter 17). The next step is to add sound-absorbing areas to the walls, or to cover the floor with a carpet and pad, or carpet, pad, and acoustic underlayment. For walls, choose acoustic tiles or panels that can withstand being poked or bumped into.

SOUND ABSORPTION

The STC ratings of materials and various methods of construction are for transmitted sound. The sound reflection characteristics of various materials are expressed as noise reduction coefficient (NRC) values. The NRC values of the materials used in constructing and furnishing most homes are given in Table 15–3.

Figure 15–13
High-efficiency noise reduction panels.

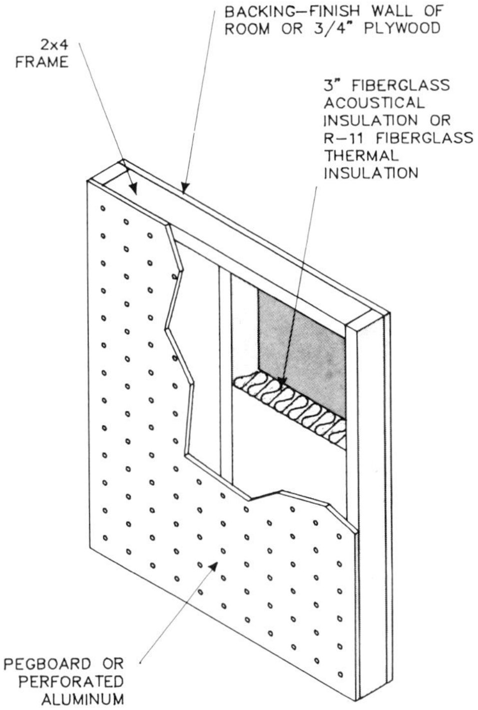

Sound absorption can be added to a room (in addition to the ceiling treatment) with one or two high-absorption panels mounted on adjacent walls. If two panels are constructed as shown in Figure 15–13, the NRC value will be 0.95, compared with 0.45 for the same area of typical consumer tiles or panels. If pegboard is substituted for the perforated aluminum, the NRC value drops to 0.75. The facing should be painted before assembly; when repainting is required, the panel must be disassembled to keep paint off the insulation.

Table 15–3
NRC VALUES: SOUND-ABSORPTION COEFFICIENTS OF GENERAL BUILDING MATERIALS AND ROOM CONTENTS

Materials	NRC Value
Brick	
Unglazed	0.05
Unglazed, painted	0.00
Carpet	
⅛" pile height	0.15
¼" pile height	0.25
3/16" combined pile & foam	0.25
5/16" combined pile & foam	0.35
Ceiling	
⅝" mineral board ceiling	0.55
⅝" film-faced fiberglass ceiling	0.75
1½" glass-cloth-faced fiberglass ceiling	1.00
Concrete block	
Unpainted	0.35
Painted	0.05
Brick wall	0.30
Fabrics	
Lightweight, flat	0.15
Medium weight, opened drape	0.55
Heavyweight, opened drape	0.60
Matting	0.15
Floors	
Concrete or terrazzo	0.00
Resilient tile or sheet vinyl on concrete	0.05
Wood	0.10
Wood parquet on concrete	0.05
Glass	
Windows	0.05
Gypsum board, plaster	0.05
Marble or glazed tile	0.00
Hardwood or plywood paneling	0.10
Wall panels	
Fiberglass wall panels	0.80
Wood roof decking (from inside)	0.15
Room contents	
Person (each)	4.70
Upholstered chair	1.00 to 2.50
Other seating	0.10 to 0.15

IMPACT SOUND

Impact sound is caused by a floor or wall being set into vibration by direct mechanical contact—vibrating machinery, walking, running, tap dancing, a stereo speaker, and similar sources. The sound is then radiated by the floor or wall surface.

Although transmission of floor impact noise can be reduced by fiberglass between the joists, an acoustic ceiling, and sound-absorbing wall treatment in the room below, the best results can be obtained by modifying the floor itself. A carpet and pad help, but more can be done. A ½"-thick sound-deadening underlayment can be put down under the carpet pad. In new construction the best approach is an isolated floor—a floor supported on a resilient pad with no hard connection through the pad or to the walls.

16 Thermal Insulation

Thermal insulation is placed in the ceilings, walls, and floors of buildings to reduce heat flow out of a building in the winter and, to a lesser extent, to reduce heat flow into the building in summer.

Almost all materials used in house construction have some insulating value. Even the air in the spaces between studs resists the passage of heat to some degree. Filling these cavities with an insulating material increases the resistance to heat flow through the wall considerably. A typical installation is shown in Figure 16–1.

The rapid flow of heat out through the walls and roof of a house in winter and the flow in during the summer make the house uncomfortable in both seasons because of the resultant variations in temperature, even within individual rooms, and the possible inability of a heating or air-conditioning system to maintain or even achieve acceptably comfortable temperatures. A well-insulated house costs less to heat in winter and to air-condition in summer.

How much insulation is enough? Anytime fuel prices rise, there is a tendency to stampede in the direction of more insulation regardless of how much insulation already is in the building. More is not always better. Once you have some insulation, each additional inch, wherever applied, produces less and less energy savings per dollar invested in insulation.

Figure 16–2 shows minimum recommendations by Owens-Corning Fiberglas for ceilings, walls, and floors in six different insulating zones in 48 of the 50 states. The recommendations are based on the cost of insulation, climate conditions, energy costs, and the assumption that

Figure 16–1
Traditional foil-faced fiberglass thermal insulation applied between studs during remodeling of a family room.

Figure 16–2
R-value map. Find your zone on the map and the corresponding recommended R-values in the top table. Use the bottom table to select insulating materials with the required values.

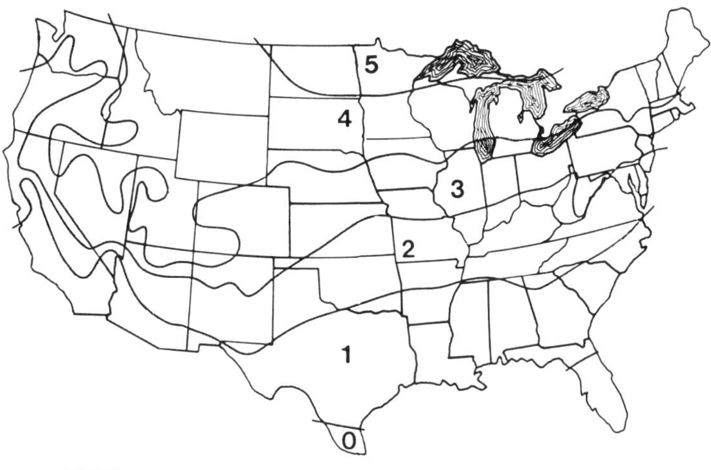

RECOMMENDED R–VALUES

HEATING ZONE	ATTIC FLOORS	EXTERIOR (MINIMUM)	WALLS (DESIRABLE)	FLOORS OVER UNHEATED SPACE
1	R–26	3–1/2"	R–11	R–11
2	R–26	SEE TABLE	R–19	R–13
3	R–30	FOR	R–19	R–19
4	R–33	MATERIAL	R–19	R–22
5	R–38	R VALUES	R–19	R–22

INSULATING MATERIAL THICKNESS REQUIRED FOR R–VALUE (INCHES)

R–VALUES	BLANKETS & BATTS		LOOSE FILL (POURED OR SPREAD)			
	FIBERGLASS	ROCK WOOL	FIBERGLASS	ROCK WOOL	CELLULOSE	VERMICULITE
R–11	3–1/2 – 4	3	5	4	3	5
R–13	4	4–1/2	6	4–1/2	3–1/2	–
R–19	6 – 6–1/2	5–1/4	8 – 9	6 – 7	5	–
R–22	6–1/2	6	10	7 – 8	6	–
R–26	8	8–1/2	12	9	7 – 7–1/2	–
R–30	9–1/2 – 10–1/2	9	13 – 14	10 – 11	8	–
R–33	11	10	15	11 – 12	9	–

NOTE: FOAMBOARD INSULATION MAY BE USED (WITH RESTRICTIONS) TO AUGMENT BLANKET, BATT, AND LOOSE FILL INSULATION.

heating costs will continue to rise. Use these recommendations whenever you have an opportunity to add insulation—for instance, during remodeling or when building an addition to your home.

VAPOR BARRIER

Whenever thermal insulation is installed in the path of heat flow, a vapor barrier must be used to prevent moisture condensation inside the wall, ceiling, or floor structure.

Warm air can contain more moisture than cold air. As air cools or is cooled, moisture condenses out. When warm air passing out through the walls or ceiling of a house is chilled by getting close to the outside, the moisture it carries condenses. The purpose of a vapor barrier at or close to the inside surface of a wall is to block the flow of most of the moisture in the air, keeping it inside the house, where it will remain in the air.

Moisture condensing inside a wall can in time do a lot of damage.

A vapor barrier such as polyethylene or foil facings on insulation does not totally block moisture vapor; it only reduces the flow. For a vapor barrier to be effective, there should not be any gaps in the coverage. During installation holes and tears should be patched. The flanges on fiberglass blankets should be fitted and stapled snug against the framing. The coverage should be as complete as possible. Figure 16–3 shows installation of a polyethylene vapor barrier to a wall. Ceiling installation is the same. The vapor barrier is always on the side facing the source of warm air—that is, facing the interior of the room.

INSULATING MATERIALS

Insulating materials are available in three forms: blankets or batts of glass or mineral fiber material, rigid or semirigid boards, and loose-fill fibers or pellets of mineral material.

Fiberglass

This blanket and batt insulation is inherently moisture-proof, vermin-proof, fungus-proof, and fire-resistant. It is a lightweight, versatile, low-cost, resilient, and easily installed insulation for use in wall, ceiling, and floor cavities that are open at the time of installation.

Both blankets (long rolls) and batts (4′ lengths) are made in several thicknesses and in widths designed to fit snugly between studs, joists, and rafters 16″ or 24″ on center. Both blankets and batts come in three forms: with a foil facing, which forms a vapor barrier, with a non-vapor-barrier kraft paper facing, or unfaced. The facings have edge flanges for stapling to framing. Installation is shown in Figure 16–4.

Fiberglass is also packed as loose-fill insulation to be blown or spread in attics, or blown into wall cavities through small access holes cut near the ceiling. The types of fiberglass insulation are summarized in Table 16–1.

Warning: A respirator should be worn when working with fiberglass insulation, and long sleeves and gloves should be worn for protection from the tiny fibers.

Figure 16–3
Applying a polyethylene sheet vapor barrier.

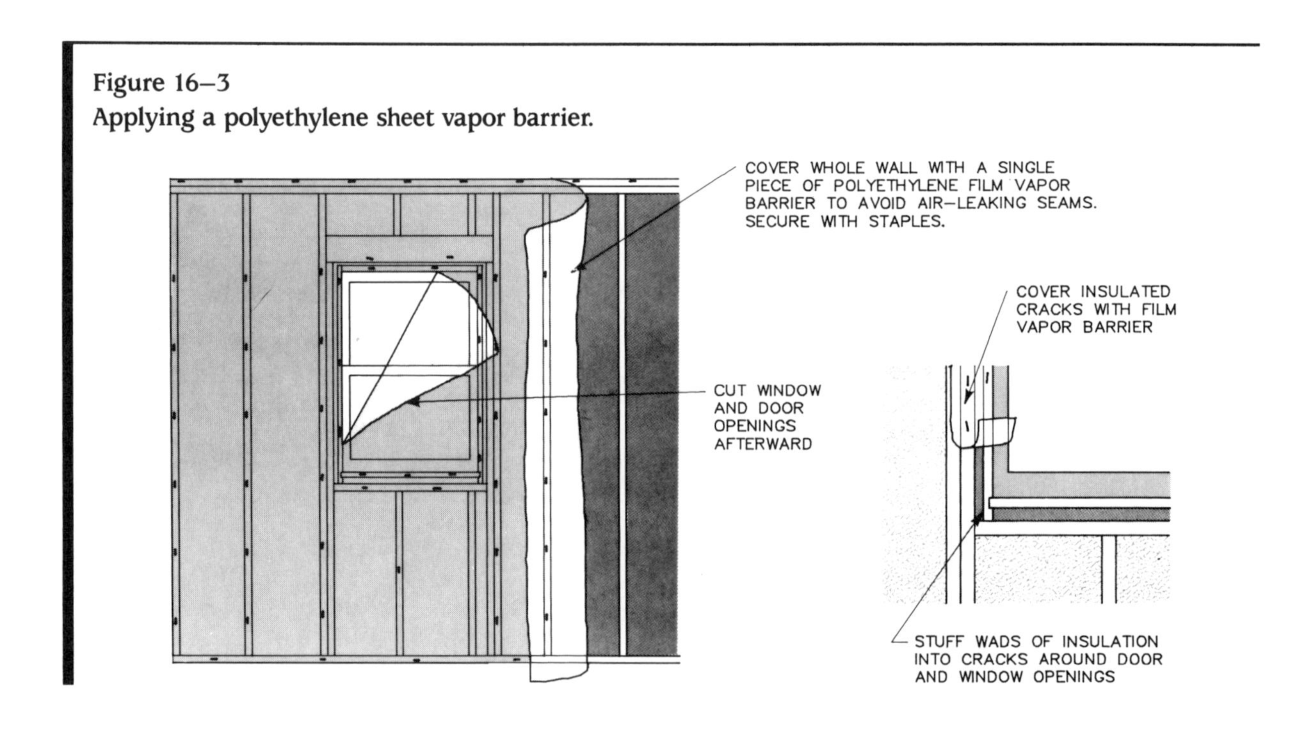
COVER WHOLE WALL WITH A SINGLE PIECE OF POLYETHYLENE FILM VAPOR BARRIER TO AVOID AIR-LEAKING SEAMS. SECURE WITH STAPLES.
COVER INSULATED CRACKS WITH FILM VAPOR BARRIER
CUT WINDOW AND DOOR OPENINGS AFTERWARD
STUFF WADS OF INSULATION INTO CRACKS AROUND DOOR AND WINDOW OPENINGS

Rigid and Semirigid Foamboard Panels

Inch for inch, polyisocyanurate, extruded polystyrene, and block-molded expanded polystyrene foamboards have higher R-values than any other residential thermal insulation material. Because of their high thermal efficiency, these panels are now widely used as wall, ceiling, roof, and foundation insulation in both new and renovation work.

Foamboard has a drawback, however. All foamboard panels are highly flammable. When installed on an interior surface, they must be covered with ½" or thicker gypsum board as a fire barrier.

Polyisocyanurate (Thermax) panels are normally made with the foam sandwiched between two sheets of paper or aluminum foil. Polyisocyanurate provides the greatest R-value for a given thickness.

Block-molded polystyrene board is made from tiny polystyrene beads. The beads are first expanded in a hot air and steam chamber, then fused under pressure and steam into a large block. The block is cut into pieces of the desired size by hot wires. Molded polystyrene board typically has a lighter density and greater water vapor permeability than the other boards. Where a vapor barrier is needed, polystyrene sheathing is available with foil on one side.

Extruded polystyrene (Styrofoam) is made by Dow Chemical Company. The extrusion process produces a board with a tough, dense skin on both sides. The high R-value and compression

Figure 16–4
Insulating a framed wall with blankets or batts.

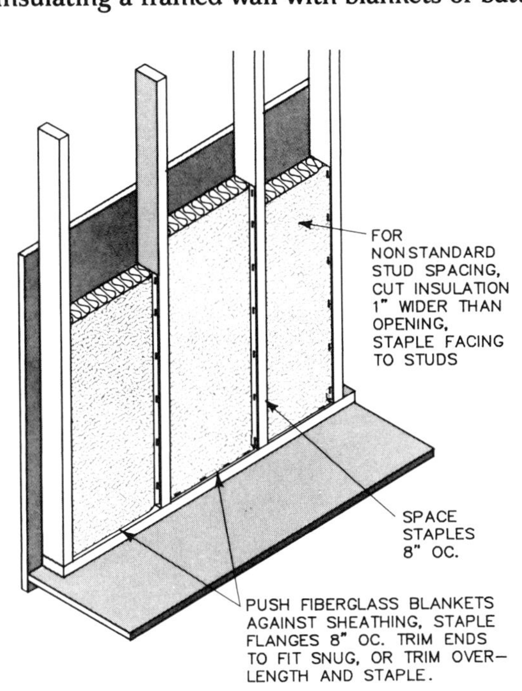
FOR NONSTANDARD STUD SPACING, CUT INSULATION 1" WIDER THAN OPENING, STAPLE FACING TO STUDS
SPACE STAPLES 8" OC.
PUSH FIBERGLASS BLANKETS AGAINST SHEATHING, STAPLE FLANGES 8" OC. TRIM ENDS TO FIT SNUG, OR TRIM OVER-LENGTH AND STAPLE.

Table 16–1
FIBERGLASS INSULATION PERFORMANCE AND PACKAGING

Form of Insulation	Use	Thickness	R-Value	Packaging	Notes
Fiberglass batts and blankets, foil-faced or kraft-faced	Between exposed studs, joists, and rafters; against crawl space walls	2½" 3½" 6½" 9½"	7 11 19 30	Blankets and batts in widths to fit between framing 16" or 24" on center; blankets in rolls, batts compressed in bales	Foil- and kraft-faced batts and blankets have flanges for stapling to framing. Separate polyethylene film provides better vapor barrier because it is continuous
Fiberglass batts, unfaced	Over existing attic insulation	3½" 6½" 9½"	11 19 30	Same as above	Requires separate vapor barrier
Loose-fill fibers (pellets)	Poured over existing attic insulation or blown into wall cavities	3½"	11	Compressed in bag	Requires separate vapor barrier

strength and low water absorption of this board make it suitable for use below ground level as foundation thermal insulation. Extruded polystyrene board is available with square edges or with tongue-and-groove edges to form joints that block air infiltration.

The physical properties and R-values of the three foamboard materials are compared in Table 16–2.

Loose-fill Pellets: Vermiculite

Vermiculite is a mineral, a form of mica. When crushed vermiculite is heated to 2200°F entrapped water expands to steam, forcing the mica layers apart, increasing the volume about twelve times. Inorganic vermiculite is lightweight, vermin-proof, rot-proof, and fire-resistant. In cavities it supports its own weight and does not settle. Vermiculite loose-fill insulation is an excellent general-purpose pouring-type insulation. It is easily poured over other types of insulation to add R-value.

Vermiculite pellets can be used to fill the cavities of a stud wall that was completed without insulation by pouring them in, as shown in Figure 16–5. Although pouring is a tedious operation, and slower than blowing in insulation, patience and a large-diameter funnel will get the job done quite effectively.

Each stud space must be filled individually. Use a 2″ carbide-blade hole saw to drill a hole near the top of each stud cavity. Make a paper or cardboard funnel to fit the hole, and ladle in the

Figure 16–5
Filling wall cavities with vermiculite.

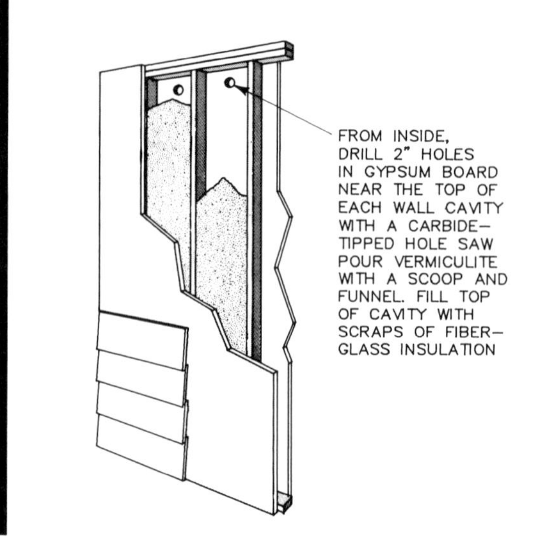
FROM INSIDE,
DRILL 2" HOLES
IN GYPSUM BOARD
NEAR THE TOP OF
EACH WALL CAVITY
WITH A CARBIDE–
TIPPED HOLE SAW
POUR VERMICULITE
WITH A SCOOP AND
FUNNEL. FILL TOP
OF CAVITY WITH
SCRAPS OF FIBER–
GLASS INSULATION

Table 16–2
FOAMBOARD INSULATION STANDARD SIZES AND PERFORMANCE

Material	Thickness	R-Value	AVAILABLE SIZES					T & G
			2 × 8	4 × 8	4 × 9	4 × 10	4 × 12	Edges Available
Extruded	¾″	3.8	X	X	X	—	X	YES
polystyrene[1]	1″	5.0	X	X	X	—	X	YES
Molded	¾″	3.0	X	X	X	X	X	YES
polystyrene[2]	1″	3.9	X	X	X	X	X	YES
Polyisocyanurate	½″	3.6	—	X	X	X	X	NO
	⅝″	4.5	—	X	X	X	X	NO
	¾″	5.4	—	X	X	X	X	NO
	1″	7.2	—	X	X	X	X	NO
	1¼″	9.0	—	X	X	X	X	NO

[1] Must be protected when used on foundation exterior.
[2] Also widely available 13½″ × 48″ for use between 1 × 2 furring.

vermiculite. When the cavity is filled up to the hole, stuff wads of fiberglass in to fill up the top of the cavity. Patch the holes; a good time for this insulation effort is just before repainting or wallpapering.

If your wall framing has firestops or other blocking running horizontally between the studs, insulation poured from the top will go only partway down the cavity. To check this, pour in a fair amount of insulation and then drill a small hole near the floor to see if the insulation is there. If not, you will have to tap the wall or use a stud finder to locate the horizontal blocking. Drill a 2″ hole below the blocking in each cavity to fill the lower half. Finish off by stuffing in fiberglass and patching the holes as described above.

Figure 16–6
Applying fiberglass batt insulation to the inside of a masonry wall.

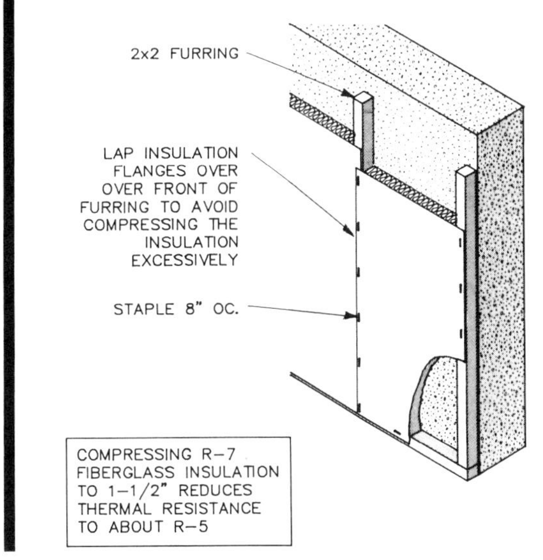
2x2 FURRING
LAP INSULATION FLANGES OVER OVER FRONT OF FURRING TO AVOID COMPRESSING THE INSULATION EXCESSIVELY
STAPLE 8" OC.
COMPRESSING R-7 FIBERGLASS INSULATION TO 1-1/2" REDUCES THERMAL RESISTANCE TO ABOUT R-5

MASONRY WALLS

Fiberglass blanket insulation can be applied to the inside of masonry walls, placed between 2 × 2 or 1 × 2 furring. Do not compress insulation; R-7 insulation will be reduced to an effective R-5 rating when compressed to the 1½″ depth of 2 × 2 furring, and economy insulation will be R-3 at best between 1 × 2 furring. If you do use blanket insulation, lap the flanges over the furring and staple, as shown in Figure 16–6. Although plywood and hardboard paneling can be applied directly to the furring over fiberglass insulation, applying gypsum board first results in a superior wall—a wall that doesn't sound hollow. (See Chapter 12 for gypsum board installation and Chapter 14 for paneling installation.)

Foamboard insulation can be applied to basement interior walls in either of two ways: over or under wood furring. In the first method, attach 1 × 2 or 1 × 3 furring to the masonry wall with power-actuated fasteners. Attach the foamboard to the furring with the manufacturer's recommended adhesive, then cover with ½″ or ⅝″ gypsum board. Make sure no gypsum board joints fall directly over foamboard joints. Use drywall screws long enough to penetrate into the furring.

In the second method, place the foamboard vertically against the inside of the wall and install 1 × 2 or 1 × 3 furring not more than 24″ on center. Secure the furring to the masonry through the foamboard with masonry nails or power-actuated fasteners 12″ on center, as shown in Figure 16–7. The nails or fasteners should penetrate the masonry 1″. Cover the foamboard sheathing with ½″ (minimum) gypsum board, staggering the gypsum board and sheathing joints. Fasten the gypsum board with nails, screws, or adhesive.

FIRST FLOORS AND CRAWL SPACE PERIMETER WALLS

Floors over crawl spaces can be insulated indirectly by insulating the foundation wall if the crawl space is unvented (not good construction practice). The floor itself can be insulated if the crawl space is vented. The header joist atop the perimeter walls of the crawl space should be insulated, using either method.

Fiberglass blankets with the vapor barrier facing inside can be stapled to the sill plate and

Figure 16–7
Applying foamboard insulation directly to the inside of a masonry wall.

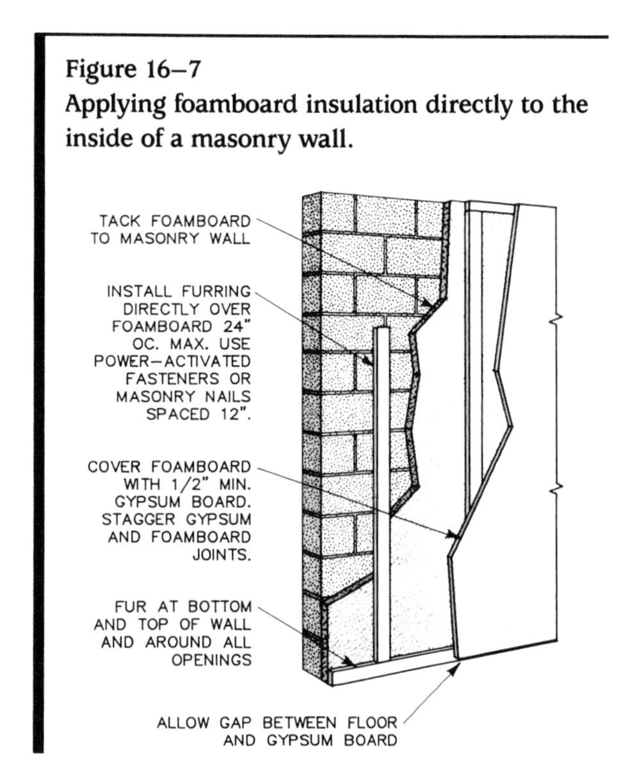

INSTALL FURRING DIRECTLY OVER FOAMBOARD 24" OC. MAX. USE POWER–ACTIVATED FASTENERS OR MASONRY NAILS SPACED 12".

COVER FOAMBOARD WITH 1/2" MIN. GYPSUM BOARD. STAGGER GYPSUM AND FOAMBOARD JOINTS.

FUR AT BOTTOM AND TOP OF WALL AND AROUND ALL OPENINGS

ALLOW GAP BETWEEN FLOOR AND GYPSUM BOARD

Figure 16–8
Insulating crawl space wall with fiberglass batts.

WEDGE BATTS OF INSULATION AGAINST HEADER

STAPLE FIBERGLASS TO SILL AND DRAPE DOWN CRAWL SPACE WALL

EXTEND INSULATION 2' ACROSS FLOOR OF CRAWL SPACE

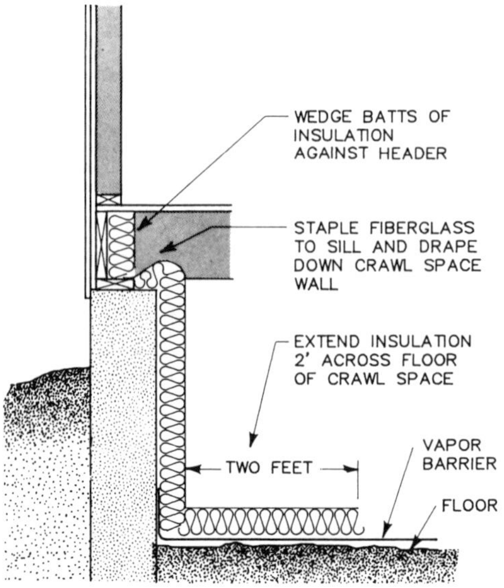

draped down the wall and 2′ across the floor of the crawl space, over a vapor barrier, as shown in Figure 16–8. Fiberglass insulation can be placed between floor joists and held in place with wedged wires sold for the purpose or with chicken wire stapled to the joists. The vapor barrier on the insulation should face up, toward the floor above.

Use foamboard on the inside surface of a crawl space wall, as shown in Figure 16–9, if the wall and the crawl space are dry. If the crawl space is subject to occasional water on the floor, use fiberglass batts instead. Do not apply foamboard to the underside of the floor joists because of the fire hazard.

Before applying the foamboard, install a 4-mil polyethylene vapor barrier on the floor and up the wall, attaching the top edge of the film to the sill plate. Attach a 3″ nailing strip the same thickness as the foamboard to the bottom of the wall with masonry nails or power-actuated fasteners. Install the foamboard with its lower edge resting on the nailer and attach it to the sill plate with galvanized roofing nails. Cover the sheathing and all but the lower ½″ of the nailer with ½″ gypsum board. Attach it with nails or screws.

POURED CRAWL SPACE FLOOR

FOAMBOARD INSULATION EXTENDING 2' FROM WALL

SPACER BOARD

NAILER FOR GYPSUM BOARD

Figure 16–9
Insulating crawl space wall with foamboard.

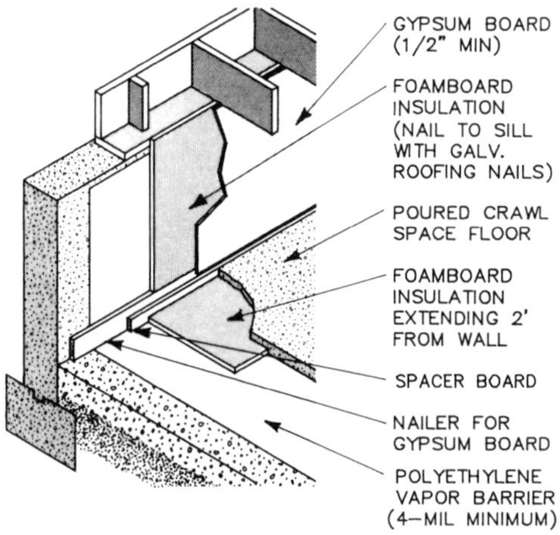

17 Acoustic, Metal, and Paneled Ceilings

The simplest finished room ceiling is painted gypsum board. Gypsum board has limitations, however. It does not absorb sound within a room; it does little by itself to block sound coming or going through the ceiling; it has no thermal insulating value to speak of; and it is not in itself particularly decorative. Three alternatives to a plain ceiling are acoustic tiles and panels, metal ceilings, and wood paneling. In addition to decorating a plain ceiling, they can cover up a lot of problems.

THE ALTERNATIVE MATERIALS

Acoustic tiles. These tiles are manufactured from cellulose fiber (wood) or mineral fiber (mineral wool spun from slag at high temperature). They can be attached directly to a gypsum board or plaster ceiling that is level, relatively smooth, clean, and structurally sound. If not, or if the joists are bare, ceiling tiles can be stapled to furring strips attached to the ceiling or joists. Ceiling tiles can also be mounted on stamped metal tracks with clips (Armstrong Easy UP Track system). An acoustical tile ceiling is shown in Figure 17–1.

Acoustic panels. Panels are manufactured from cellulose fiber, mineral fiber, or fiberglass. They can be laid into a metal grid suspended below the ceiling structure. In most grid-and-panel installations, the grid remains visible, as in Figure 17–2; in other installations, the grid is recessed and invisible. There are several advantages to a suspended ceiling: Access to wiring and plumb-

Figure 17–1
Acoustic tile ceiling. (Photo courtesy Armstrong World Industries, Inc.)

Figure 17–2
Acoustic panel ceiling.

ing under the floor above is maintained. Fluorescent lighting fixtures can be incorporated in the ceiling. Noise reduction and sound transmission ratings of the ceiling/floor or ceiling/roof deck structure can be improved beyond what could be accomplished by using equivalent tile. Thermal insulation can be incorporated in the ceiling structure.

Acoustic panels and tiles have perforated, patterned, textured, or fissured surfaces that permit sound to penetrate into the interior voids of the fibrous material. Panels and tiles are not all equal in acoustic performance and, except for some fiberglass panels, have insignificant thermal R-values.

Most of the lightweight, low-density panels and tiles are fragile, easily abraded or torn, and many are easily soiled. Some commercial panels and tiles do have hard, impact-resistant surfaces. Surfaces can be washed, carefully, but painting usually destroys acoustic properties.

Acoustic ceiling tiles have interlocking edges. The same tiles are used for both adhesive and staple application. Square-edge tiles have nearly invisible joint lines; beveled edges produce prominent tile patterns, as shown in Figure 17–3.

Mineral fiber and fiberglass panels and tiles are fire-resistant; cellulose fiber tiles are not.

Metal ceilings. Metal ceilings, popularly called tin ceilings, but actually made of pressed or embossed sheet-metal plates or panels, began as a cheap and quick fix for covering damaged plaster ceilings in the nineteenth century. The stamped sheet-steel ceilings rapidly became popular in new construction, and by the 1890s they were the preferred ceiling for most commercial and residential construction. Their popularity was sustained over this long period because of their design flexibility and low cost, as long as the customer wanted an ornate or at least pattern-decorated ceiling. When plain plaster and gypsum board ceilings became popular in the 1930s, metal ceilings faded out.

They are now once more popular—a result of a combination of Victorian nostalgia and practicality. Some of the many designs available today are shown in Figure 17–4. A stamped metal ceiling can hide messed-up plaster or bare joists, as well as almost any other kind of ceiling. The one disadvantage of the material is that it contributes no sound reduction inside a room and in fact may increase the reflected sound level (see Chapter 15).

Figure 17–3
Acoustic tile edge details.

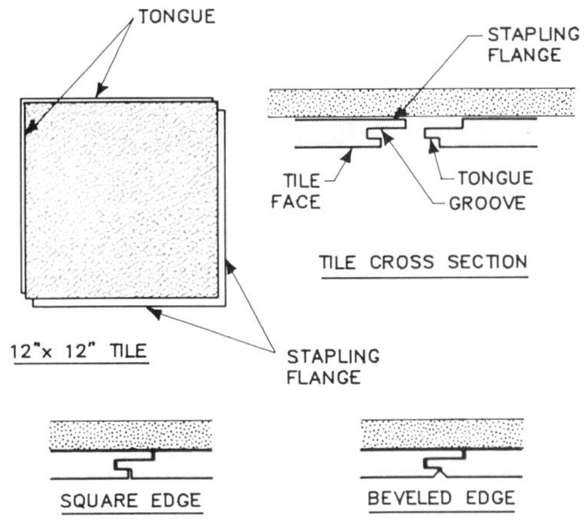

"Tin" ceiling panels, cornices, and trim made from original dies used close to a century ago are available from several companies. Special furring is required for their installation.

Wood paneled ceilings. Any of the wall paneling materials described in Chapter 14 can be applied to a ceiling, either nailed to joists or to furring installed over an old ceiling surface. Adhesive attachment is not recommended. Strip, plank, or parquet wood flooring can also be used on a ceiling.

PLANNING

The first step in installing any panel, tile, or stamped metal ceiling is to draw an accurate room plan as shown in step 1 of Figure 17–5. Measure the room at the ceiling level, or at the level a suspended ceiling will hang, depending on the kind of installation you intend to make. Use these measurements to draw a scaled plan of the ceiling on grid paper. Select the grid size and

Figure 17–4
Tin ceiling field plates. (Photo courtesy Chelsea Decorative Metal Co.)

Figure 17–5
Room ceiling plan and layout grid for tiles or panels.

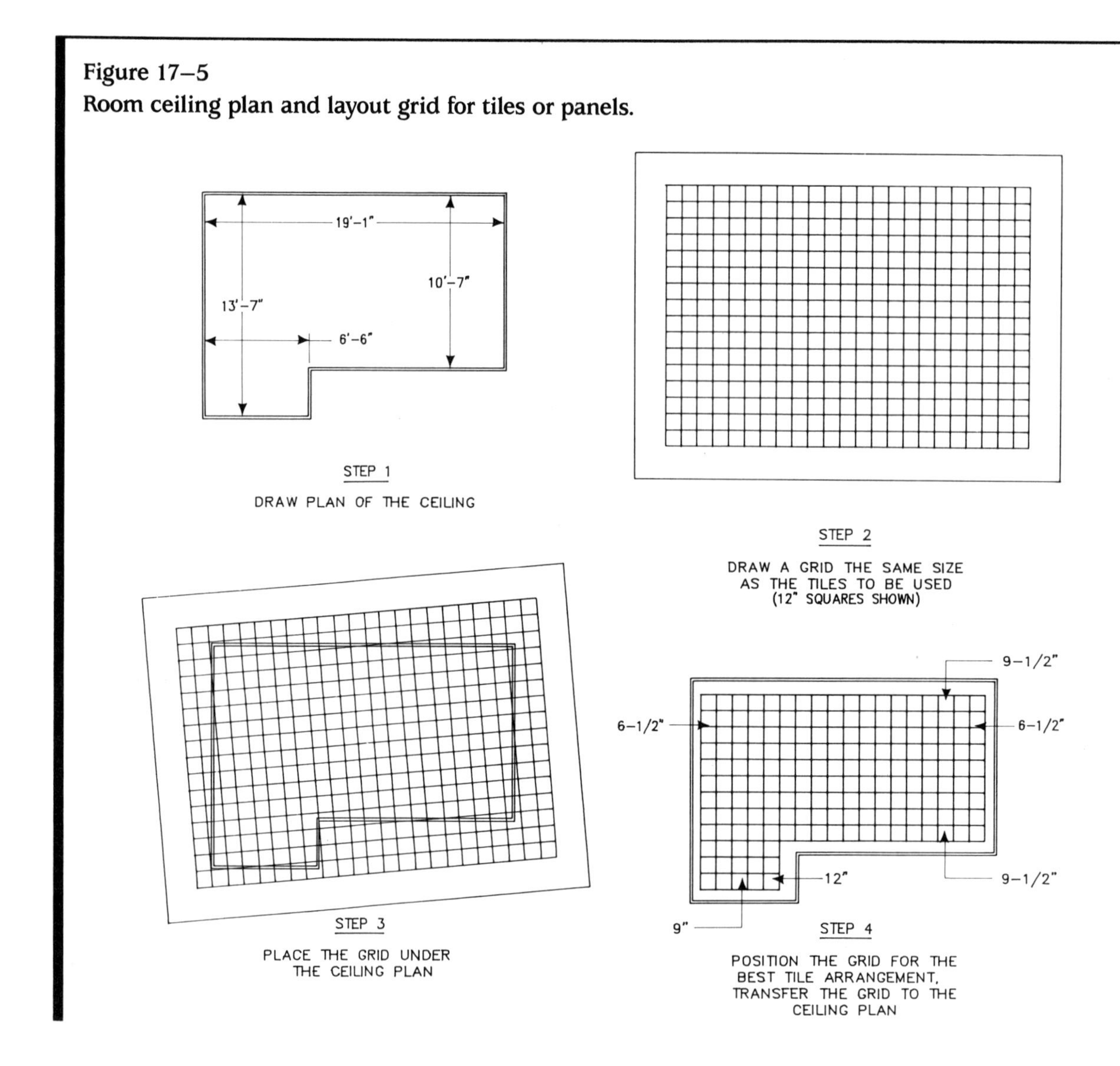
19'-1"
10'-7"
13'-7"
6'-6"
STEP 1
DRAW PLAN OF THE CEILING
STEP 2
DRAW A GRID THE SAME SIZE
AS THE TILES TO BE USED
(12" SQUARES SHOWN)
STEP 3
PLACE THE GRID UNDER
THE CEILING PLAN
9-1/2"
6-1/2"
6-1/2"
12"
9-1/2"
9"
STEP 4
POSITION THE GRID FOR THE
BEST TILE ARRANGEMENT.
TRANSFER THE GRID TO THE
CEILING PLAN

the scale for the drawing so that the grid squares comfortably relate to the size of the tiles or panels. Include all alcoves or corners and locate lighting fixtures, air ducts, columns, pipes, and other obstructions that the ceiling will incorporate or have to be worked around.

Do not assume that walls are straight or the same length on opposite sides of the room, or that corners are square. Measure and record the length of *each* wall. Use a stretched string to see if walls *at the new ceiling height* are straight. A good way to determine if corners are square is to hold a half sheet of ¼" plywood in the corner. Measure and record any out-of-square deviation.

Is the old ceiling flat and level? Use the string to check. The ceiling will probably be bowed downward in the center as a result of sagging of the joists above. You should generally plan to live with moderate overall bowing. Other high and low areas should be identified and marked (this step is not necessary for a suspended ceiling). If there is too much waviness in the ceiling, install furring, which you can shim level (or nearly so), and plan to install acoustic tiles with staples, not adhesive.

On another sheet of tracing paper, draw tiles or panels at the same scale as the ceiling drawing (Figure 17–5, step 2). Cover an area larger than the ceiling. Place this tracing under the ceiling plan and slide it around to get the best fit against the ceiling outline (steps 3 and 4).

If you don't get a good fit in an odd-shaped room (an L-shaped room, for example), try dividing the room into two rectangles to see if the arrangement works better.

Part-width border rows of acoustic tiles or panels should be of equal width on opposite sides and ends of the ceiling, unless the room is an odd shape. The width of border rows should be more than half a single tile or panel width. If on your first measurement the borders come out to less than half, reduce the number of full tiles or panels by one in that direction.

From your tile layout (Figure 17–5, step 4), you can determine the position of the center tile, which can be in any of four positions in relation to the center point of the room (see Figure 17–6).

APPLYING CEILING TILES WITH ADHESIVE

Lightweight acoustic tiles may be adhered directly to a level, structurally sound, clean gypsum board or plaster ceiling. New plaster must be thoroughly cured and dry. A finish coat of white plaster is unnecessary; the tiles can be cemented to the plaster brown coat if it is true and reasonably smooth. Porous and dusty old plaster should be primed with wall size. Wallpaper and painted surfaces should be tested for adhesion. To do this, install three or four pieces of a tile in different sections of the ceiling and pry them off after 48 hours. If the wallcovering or paint pulls away with the tile, it should be removed or the tiles installed on furring.

Figure 17–6
Center tile positions.

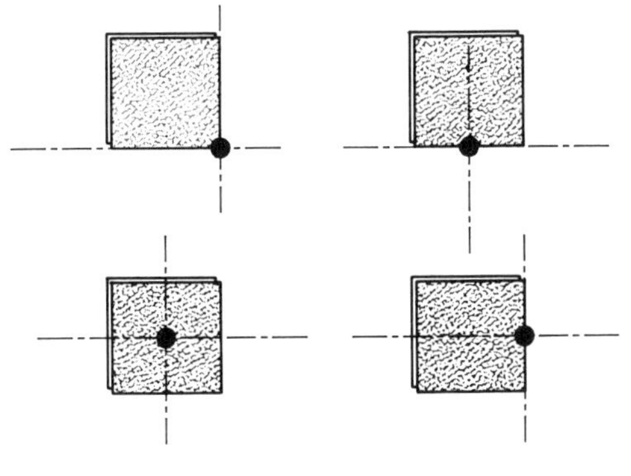

Types of Adhesive

There are two kinds of adhesive: brush-on and acoustic cement. The brush-on type allows the tile to be set tight against the ceiling, but staples are required to hold the interlocking tile to the surface until the cement dries. Acoustic cement comes in cartridges to be applied with a gun or in cans for putty-knife application. It can be used with either square-edge or interlocking tiles.

Layout

Snap a chalk line across the ceiling to locate the flange edges of the first row of tiles, whether full or part width (see Figure 17–7). If you are installing square tiles, snap a second chalk line perpendicular to the first line to locate the flange edge of the border row along an adjoining wall. Use 3-4-5 triangulation to get the second line at a right angle to the first; don't trust offset measurements from the adjacent wall.

Installing the Tiles

Brush the ceiling surface free of dust. If it is a hot day, powder your hands with baby powder to prevent smudging the tiles.

Apply a walnut-size glob of cement 2″ from each corner (Figure 17–8), and squiggle the tile into position against the ceiling. To do this, raise the tile to the ceiling close to the final position, avoiding unnecessary lateral movement once adhesive is in contact with ceiling (Figure 17–9). Press the tile to the ceiling and move the tile to mesh the tongue-and-groove joints.

Do not slide the tile in place. Sliding results in poor adhesion and adhesive getting on the tile edge. Tip the tile into place and give it a little movement as you position it. For better adhesion,

Figure 17–7
Determining position of the border tile rows.

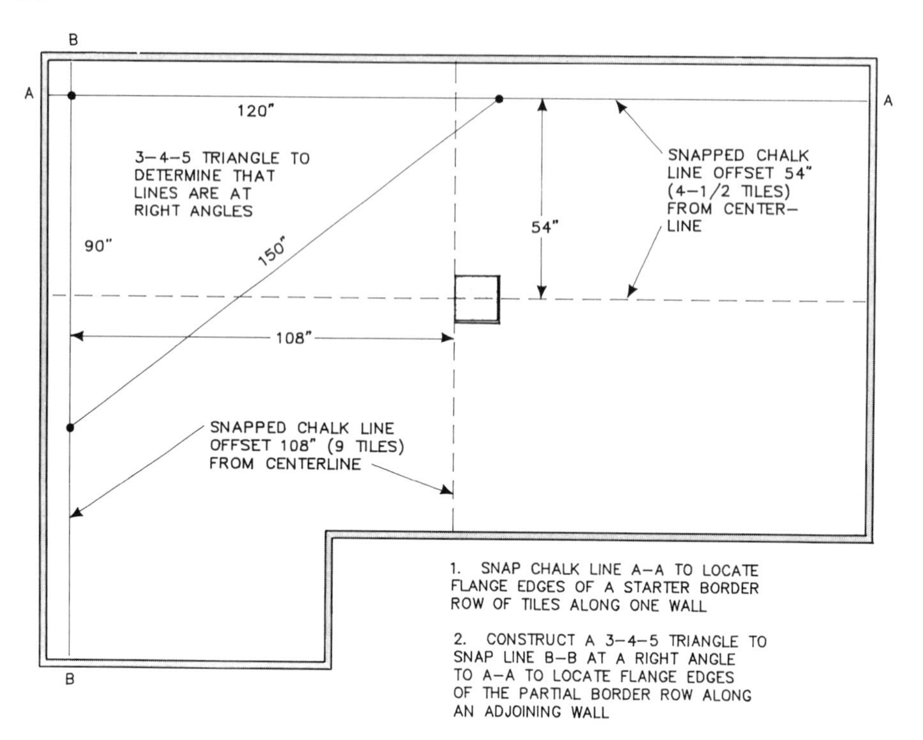
B
A
120"
3–4–5 TRIANGLE TO DETERMINE THAT LINES ARE AT RIGHT ANGLES
90"
150"
54"
SNAPPED CHALK LINE OFFSET 54" (4–1/2 TILES) FROM CENTER–LINE
A
108"
SNAPPED CHALK LINE OFFSET 108" (9 TILES) FROM CENTERLINE
B
1. SNAP CHALK LINE A–A TO LOCATE FLANGE EDGES OF A STARTER BORDER ROW OF TILES ALONG ONE WALL
2. CONSTRUCT A 3–4–5 TRIANGLE TO SNAP LINE B–B AT A RIGHT ANGLE TO A–A TO LOCATE FLANGE EDGES OF THE PARTIAL BORDER ROW ALONG AN ADJOINING WALL

Figure 17–8
Adhesive application patterns.

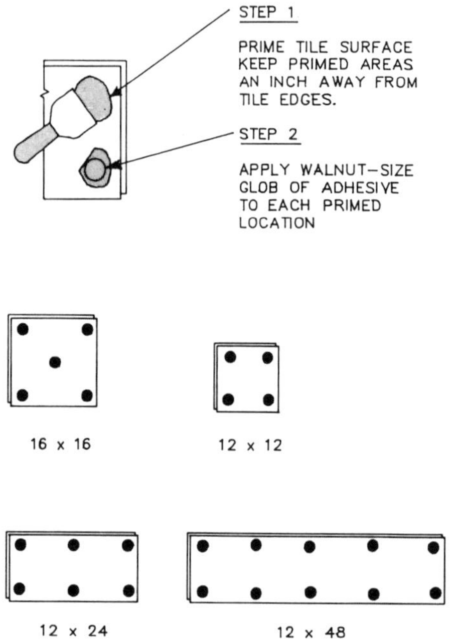

STEP 1

PRIME TILE SURFACE KEEP PRIMED AREAS AN INCH AWAY FROM TILE EDGES.

STEP 2

APPLY WALNUT–SIZE GLOB OF ADHESIVE TO EACH PRIMED LOCATION

16 x 16

12 x 12

12 x 24

12 x 48

Figure 17–9
Adhesive installation: placing a tile.

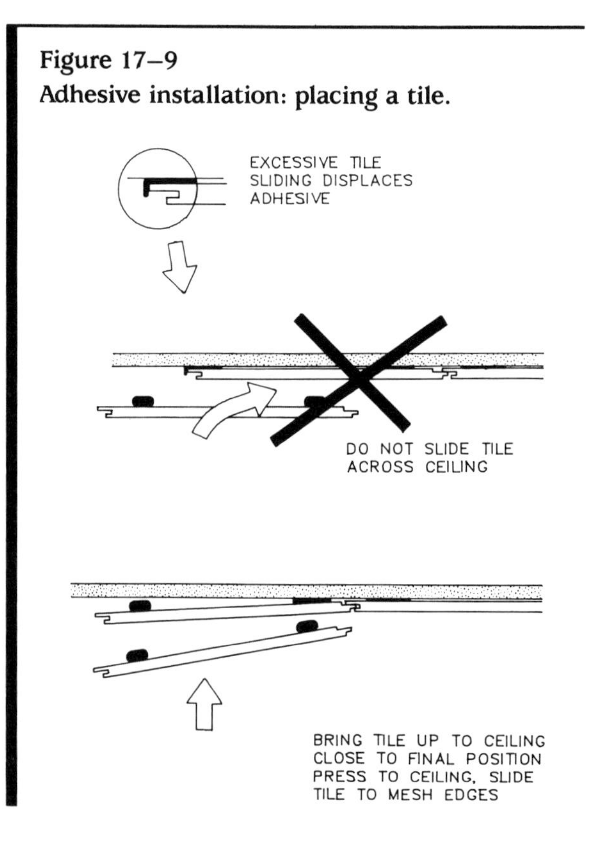
EXCESSIVE TILE SLIDING DISPLACES ADHESIVE
DO NOT SLIDE TILE ACROSS CEILING
BRING TILE UP TO CEILING CLOSE TO FINAL POSITION PRESS TO CEILING, SLIDE TILE TO MESH EDGES

take a small amount of adhesive on a putty-knife blade and prime spots on the back of the tile by forcing adhesive into the pores, before applying globs of adhesive over them. If you pull the tile off for any reason before the cement dries, apply new cement before replacing the tile.

The step-by-step procedure for tiling a ceiling is shown in Figures 17–10 through 17–14. The first tile goes into a corner. Both stapling flanges must be exposed. If it's a full tile, remove both tongues. If it's a part tile, in width or length, or both, trim it on the tongue sides. When cutting tile, cut with the face up. Use a sharp utility knife and guide the blade with a carpenter's square. Border row tiles are positioned against this first corner tile, then field tiles are successively placed against the borders, as shown in the illustrations.

APPLYING CEILING TILES ON FURRING STRIPS

Acoustic tile should be applied on furring strips: (1) when the ceiling is installed on bare joists

Figure 17–11
Trim and place the border tiles. Measure each tile individually. Be careful not to move previously placed tiles. Keep the flanges aligned with the chalk lines.

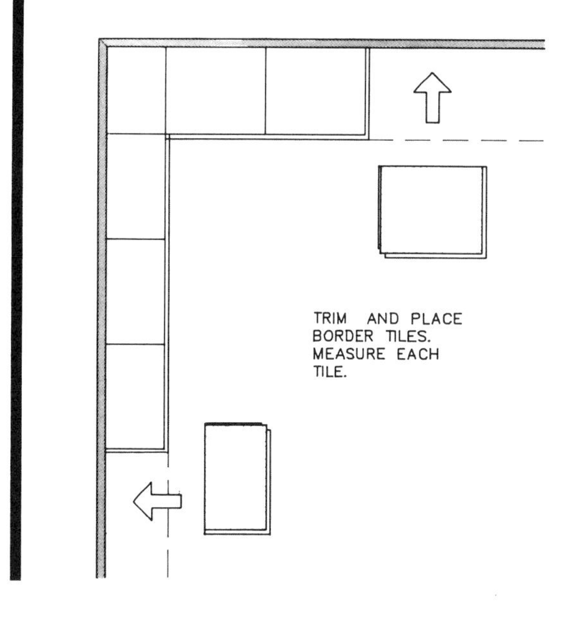
TRIM AND PLACE
BORDER TILES.
MEASURE EACH
TILE.

Figure 17–10
Place the corner tile. Trim a partial corner tile so its flanges line up with the snapped chalk lines. Apply adhesive and place the tile in the corner, carefully aligning the tile flanges to the snapped lines. If the border rows are full tiles, trim the tongues (not the flanges) facing the walls before fitting.

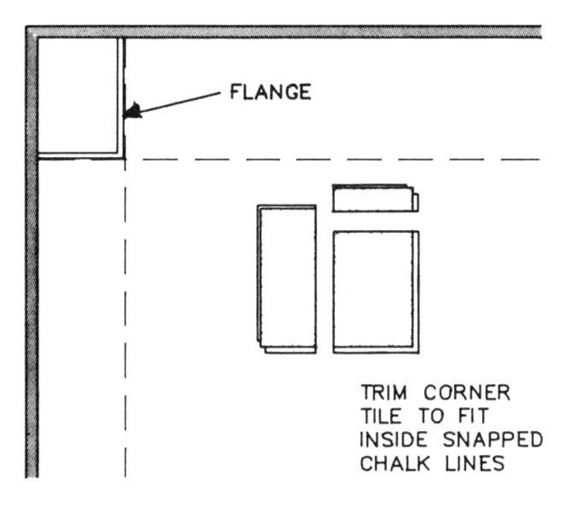
FLANGE
TRIM CORNER
TILE TO FIT
INSIDE SNAPPED
CHALK LINES

Figure 17–12
Install the field tiles. Use a sequence of tiles that maintains a more-or-less sawtooth work line diagonally across the ceiling.

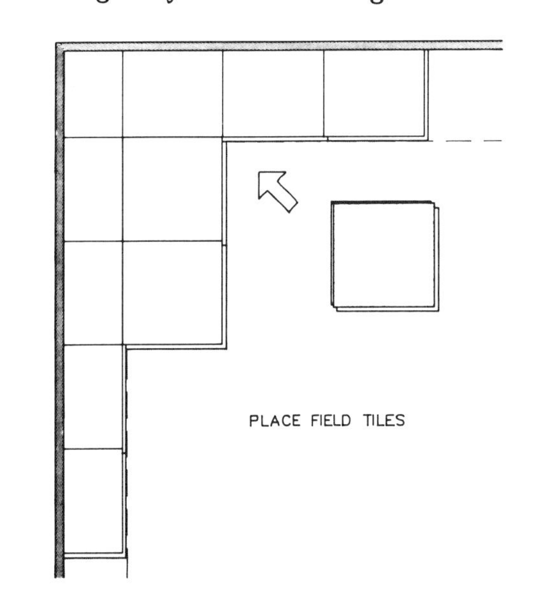
PLACE FIELD TILES

Figure 17–13
Install the last border rows. Check the fit dry before applying adhesive to the tiles.

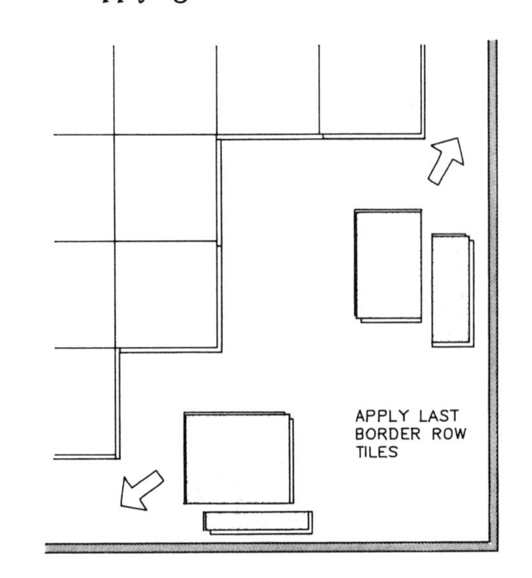

Figure 17–14
Install the last corner tile. Make a wire hook so you can remove the tile after checking the fit dry. To avoid damaging the surface, use a block of wood slightly smaller than the tile to press the tile into place.

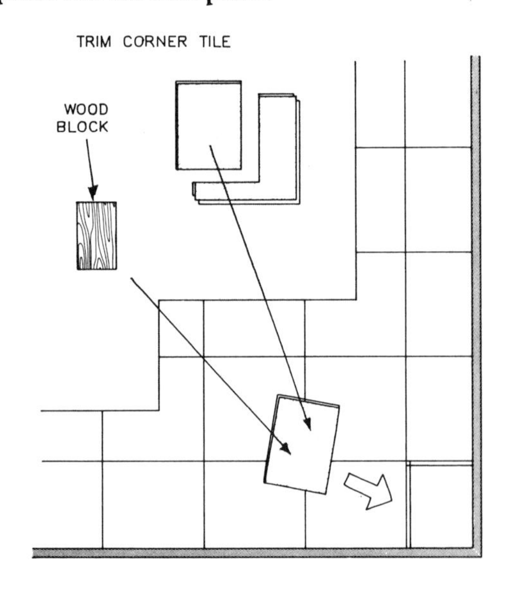

(Figure 17–15) or (2) when the ceiling is applied over a finished ceiling that is not flat or is in such poor condition that tiles cannot be applied with adhesive. If the old ceiling has been water-damaged, correct the problem before installing a tile ceiling.

The furring should be straight, kiln-dried 1×3s or 1×4s without bow or twist. Furring strips are installed across the joists. Attach furring with gypsum board screws long enough to penetrate the joists 1″.

Draw a plan of the ceiling and an overlay showing the tile pattern, the same as for a cemented tile ceiling. If the ceiling is finished, locate the ends of the joists (use nails or drill holes near the wall at each end of the ceiling, if necessary) and snap chalk lines between the ends.

Installing Furring

Secure a furring strip—*A* in Figure 17–16—to the ceiling or joists flush against the starting wall. Locate the centerline of the next furring strip a distance from the wall equal to the border tile width plus ½″ (for the flange). Mark off centerline positions (assuming 12″ tiles) 12″ on center for the remaining furring strips. Do this at opposite side walls and at points where you will be butting furring strips. This will avoid any creeping errors. Attach the furring strips across the field of the ceiling. Finish with a strip flush against the far wall. Butt the furring strips only at the joists, with two screws in each end, and stagger the butt joints.

If your ceiling or the underside of the joists isn't level, shim the furring as you put it up (see Figure 17–17). Drive shingle shims from each side of the furring so the strip won't be tilted. At low areas, use thinner furring strips ripped from ½″ or ⅝″ plywood.

Furring around Obstructions

Various obstructions are likely to be encountered that will have to be boxed in one way or another. Each obstruction can be handled individually, or all of the furring can be dropped to be positioned

Figure 17–15
Acoustic ceiling tile applied on furring attached to bare joists.

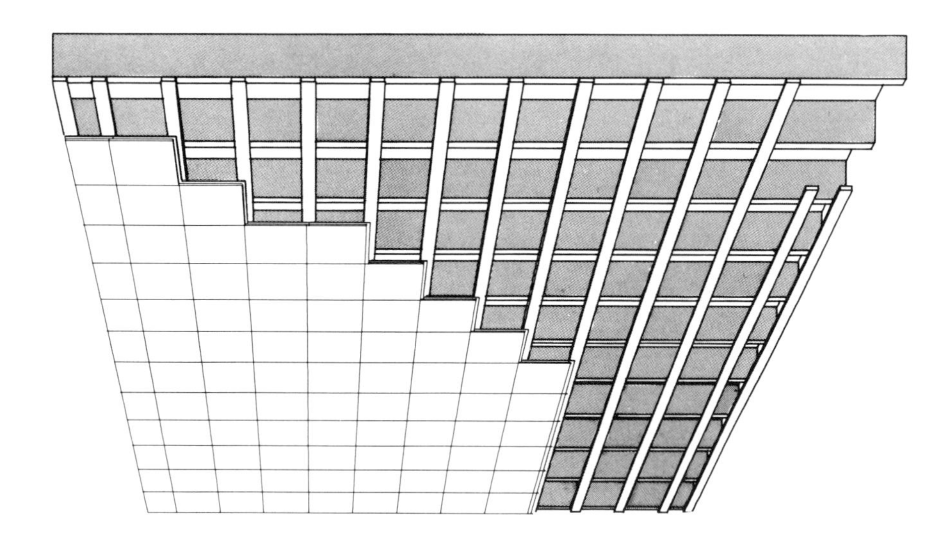

Figure 17–16
The first furring strip out from the wall (A) is spaced for the partial-width border row. Succeeding strips are spaced on center at the full tile or panel width.

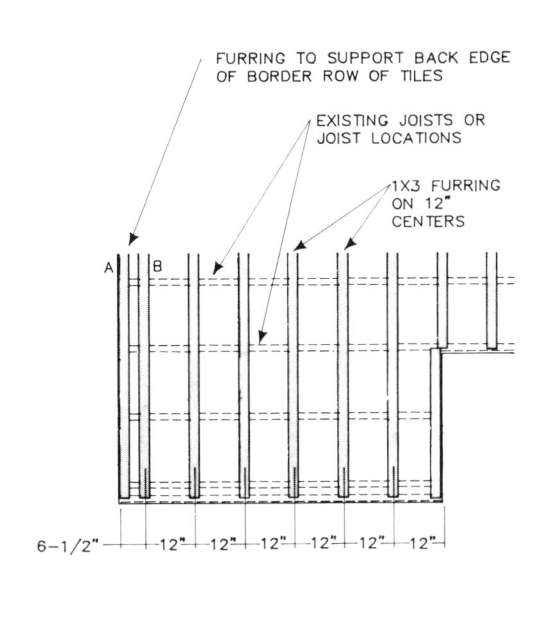

below most or all of the obstructions. Figures 17–18 through 17–21 show a number of ways to deal with obstructions.

Dropped furring. To clear obstructions you may prefer to drop the whole acoustic tile ceiling below the obstructions. This is done with furring suspended from hangers. An all-wood dropped furring system is shown in Figure 17–22. The steps in its construction are detailed in Figures 17–23 through 17–27. *Note:* This construction is not adequate to support a dropped gypsum board ceiling. A dropped, furred acoustic tile ceiling is advisable when there are many obstructions in the way of a ceiling directly below the joists; it is a more solid ceiling than a suspended panel ceiling. The disadvantage is lost access to wiring, plumbing, and ductwork that becomes hidden.

Applying Tile to Furring

Snap a chalk line down the centerline of the second furring strip. The flange edges of the border row of tiles must be aligned to this line, not to the wall. Snap a second chalk line perpendicular to the first line to locate the flange edge of

Figure 17–17
Shim furring as necessary to provide an even surface to attach tiles to. For furring placed over a finished ceiling, screws must be long enough to penetrate into joists.

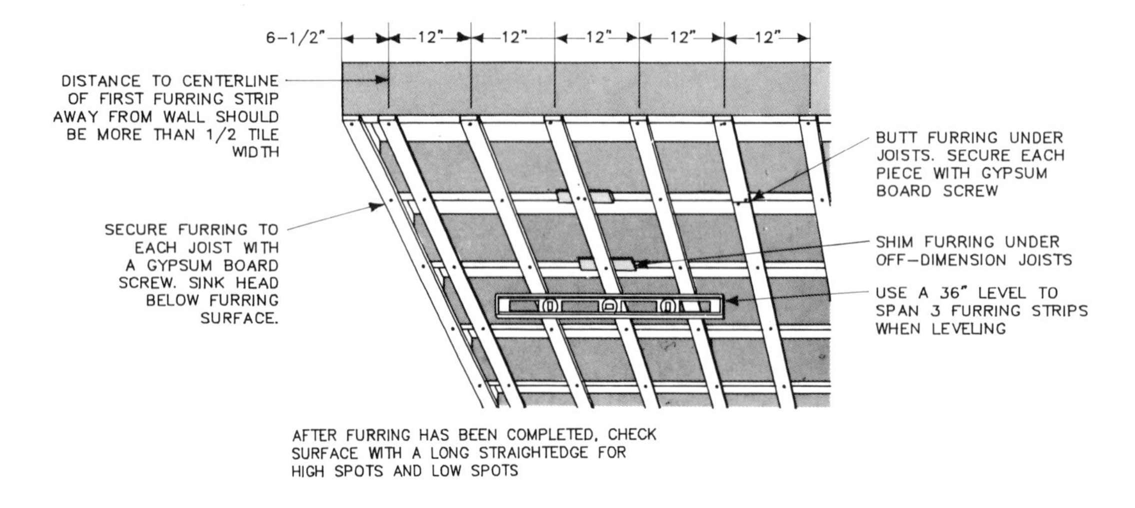

6-1/2"
12"
12"
12"
12"
12"
DISTANCE TO CENTERLINE OF FIRST FURRING STRIP AWAY FROM WALL SHOULD BE MORE THAN 1/2 TILE WIDTH
SECURE FURRING TO EACH JOIST WITH A GYPSUM BOARD SCREW. SINK HEAD BELOW FURRING SURFACE.
BUTT FURRING UNDER JOISTS. SECURE EACH PIECE WITH GYPSUM BOARD SCREW
SHIM FURRING UNDER OFF-DIMENSION JOISTS
USE A 36" LEVEL TO SPAN 3 FURRING STRIPS WHEN LEVELING
AFTER FURRING HAS BEEN COMPLETED, CHECK SURFACE WITH A LONG STRAIGHTEDGE FOR HIGH SPOTS AND LOW SPOTS

Figure 17–18
Basement windows set high in the foundation wall usually will be intersected by a dropped ceiling. The solution is to box the window. Be sure the box is big enough to accommodate curtains and a fully opened window.

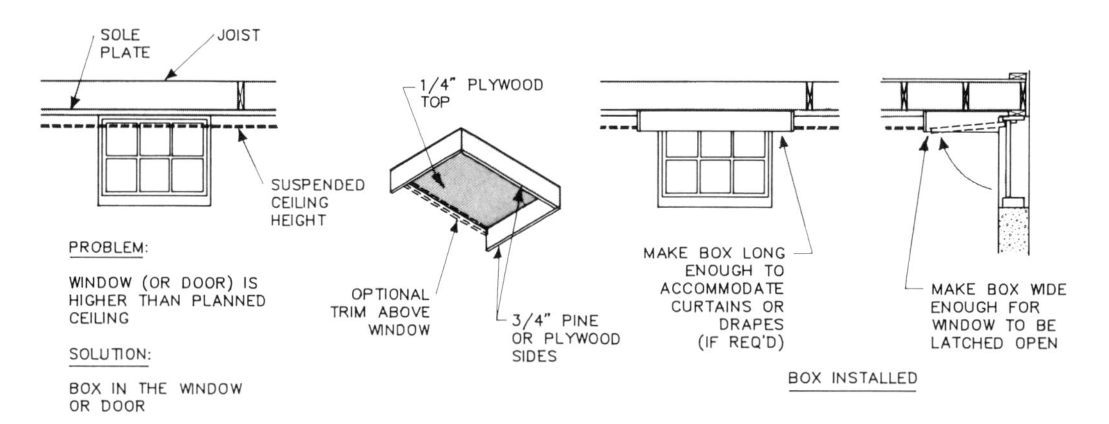

SOLE PLATE
JOIST
SUSPENDED CEILING HEIGHT
PROBLEM:
WINDOW (OR DOOR) IS HIGHER THAN PLANNED CEILING
SOLUTION:
BOX IN THE WINDOW OR DOOR
1/4" PLYWOOD TOP
OPTIONAL TRIM ABOVE WINDOW
3/4" PINE OR PLYWOOD SIDES
MAKE BOX LONG ENOUGH TO ACCOMMODATE CURTAINS OR DRAPES (IF REQ'D)
MAKE BOX WIDE ENOUGH FOR WINDOW TO BE LATCHED OPEN
BOX INSTALLED

the first tile of the adjoining border row on each of the other furring strips (see Figure 17–28).

Trim the tiles for the border row the same as cemented tiles. The first tile installed must have both stapling flanges exposed. Line it up with the chalk lines and attach it with three ½" or 9⁄16" staples into the furring, plus a box nail in the back corner. Install all the tiles in the first row, keeping them aligned with the chalk line, as in Figure 17–29.

Electric staplers tend to drive staples clear through the tile flange. You will never know until you try; if a ½" or 9⁄16" staple drives through, try a longer staple.

You usually won't need any more chalk lines; just fit the tiles and proceed across the ceiling as

Figure 17–19
Girders, soil pipes, and ductwork can be boxed with 1 × 2 cleats and ½″ C-D plywood as shown, whether the tile is to be applied with adhesive or staples. The boxing should touch and be attached to the joists only. Ductwork that sweats should be insulated before boxing.

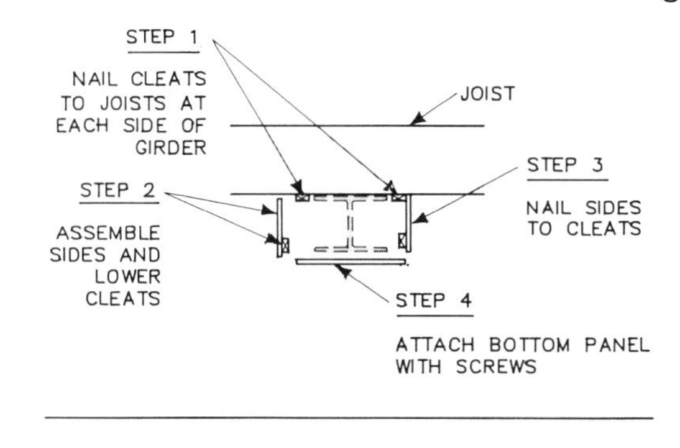

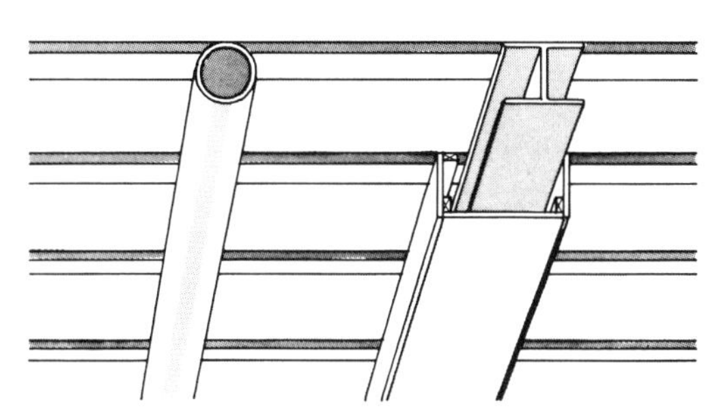

Figure 17–20
A pipe can be boxed in and the boxing painted to match the color of the tile. To prevent damage to the tiles, encase water-supply pipes that sweat in split-tube foam insulation.

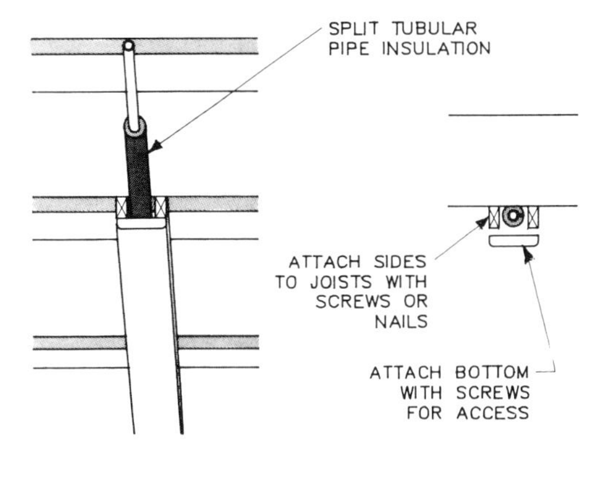

Figure 17–21
Wiring can be boxed in the same way as pipes, or it can be rerouted from junction box to junction box (splices outside boxes are not permitted by any electrical code) through holes drilled in the joists. If there is a great deal of wiring below the joists, all of the furring can be dropped ¾″ to 1″ by inserting blocks of wood or plywood between the joists and furring at every crossing.

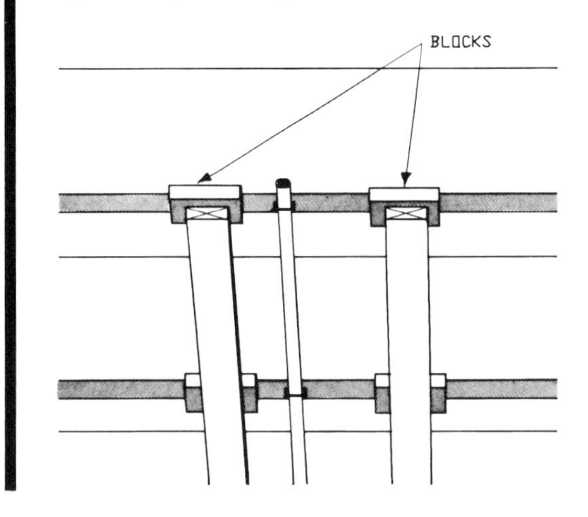

shown in Figure 17–30. Cut tiles for the last row to width, tip them into position, and secure them with one staple in the corner into each furring strip (see Figure 17–31). Face-nail each tile to the wall strip with one box nail placed close enough to the wall to be hidden by the wood molding that will be used to cover the ceiling edge. Secure the last tile in the corner with box nails placed to be hidden by molding.

Wood molding should be painted or stained before putting it up. Nail the molding to the furring strips or the wall studs as appropriate.

Square ceiling tiles are usually installed in a checkerboard pattern, with all joints running in continuous straight lines. However, an ashlar pattern, in which the joints in alternate rows are staggered as in a brick wall, can also be used. Tiles 12″ × 48″ should be installed as they run; the leftover tile end from one row becomes the partial first tile in the next row and so on, row to row. The asymmetrical staggered joint pattern produces a better-looking ceiling.

Figure 17–22
Furring for a dropped acoustic tile ceiling.

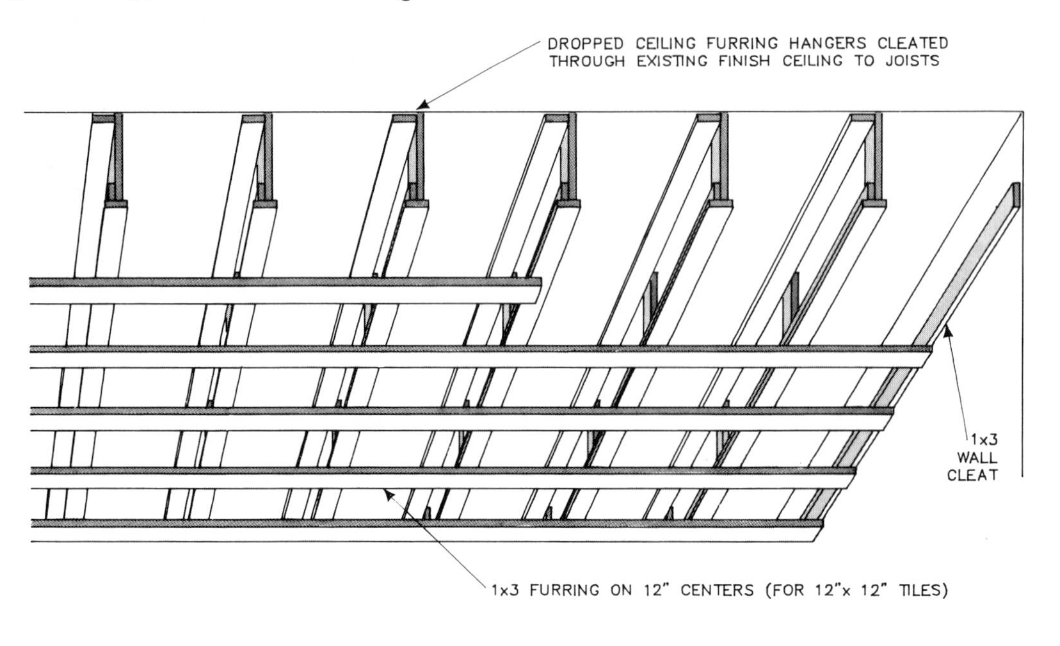

Figure 17–23
Stretch and snap string around the room 1¼″ above the desired ceiling height (¾″ for furring, plus ½″ for tiles) to locate the bottom edge of the wall cleats. Attach the wall cleats to the walls.

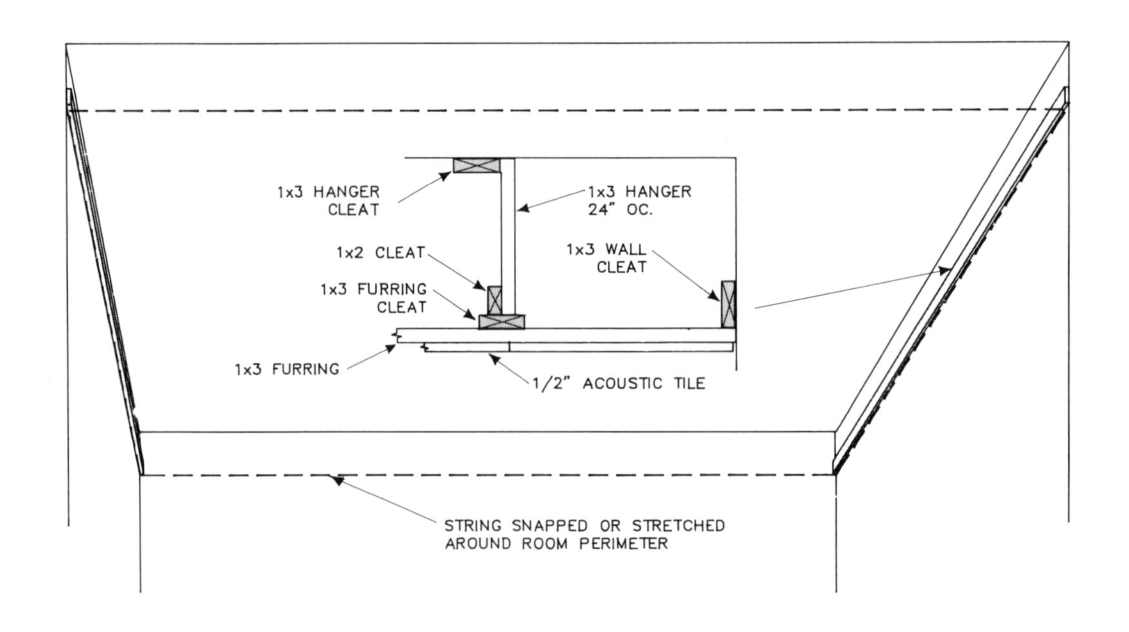

Figure 17–24
Attach hanger cleats to the ceiling with gypsum board screws into the joists. The cleats can run with the joists (one cleat under each joist) or across the joists (24″ on center with attachment to each joist). Use gypsum board screws.

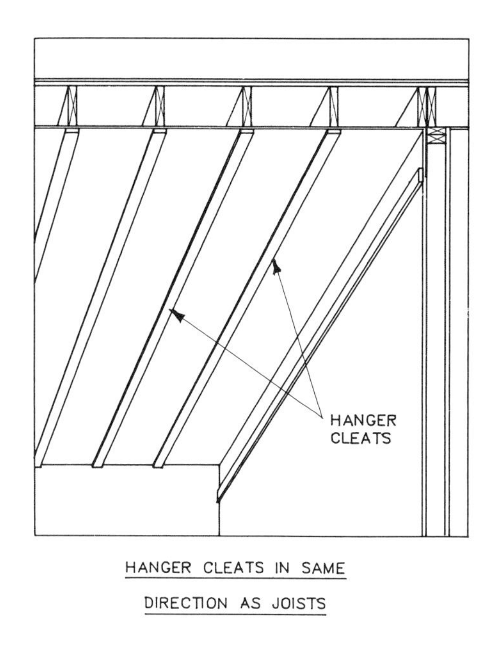

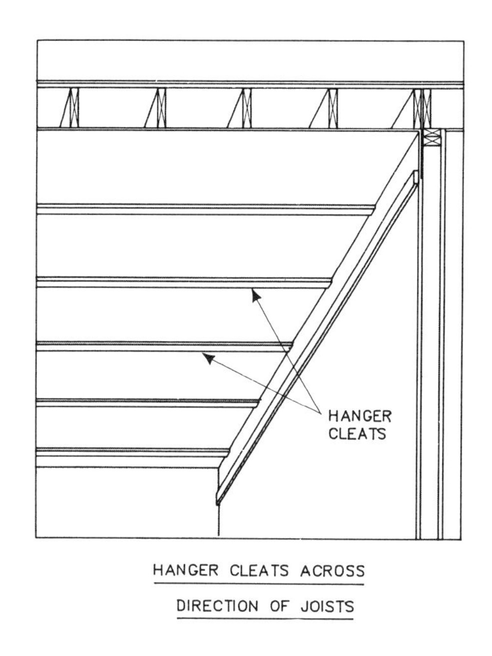

Figure 17–25
Cut hangers. Stretch strings from wall to wall under each hanger cleat at the level of the bottom of the wall cleats. At each hanger position measure the joist-to-string distance to find individual hanger lengths.

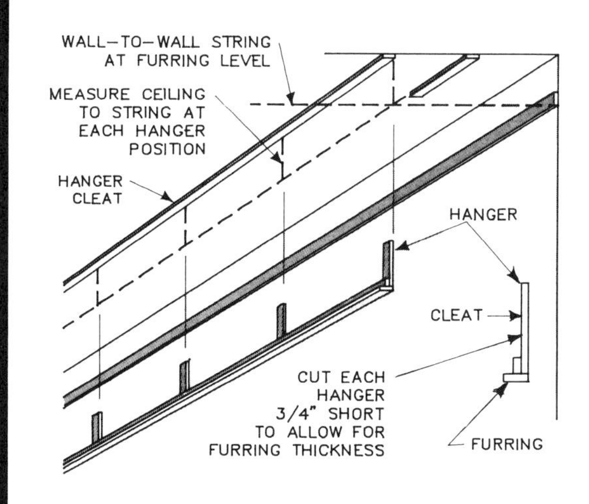

INSTALLING PANELS IN A SUSPENDED GRID

A suspended grid ceiling consists of lightweight acoustic panels (24″ × 24″ or 24″ × 48″) laid in an interlocking grid of metal supports. Fluorescent lighting fixtures can be incorporated in the ceiling, either supported by the grid or hung independently from the structural ceiling above. Figure 17–32 shows such an installation. Grid systems are assembled from three different lightweight stamped metal parts as shown in Figure 17–33. Basic assembly is shown in Figure 17–34.

Parts of a Suspended Ceiling Grid

Wall molding. This L-shaped trim is fastened to the room walls to form the decorative perimeter of the ceiling and to support the panel edges. Wall molding is available in 8′ and 10′ lengths.

Main runners. Main runners run from wall to wall. Although their ends rest on the wall moldings, their weight and the ceiling panel load are supported by a series of hanger wires attaching them to the structural ceiling above. These main runners have holes prepunched for the hanger wires, slots spaced at 6″ intervals for cross tees, and splice tabs for joining main runners end to end. Main runners are available in 8′ and 12′ lengths and can be spliced for longer runs.

Figure 17–26
Assemble furring and attach as shown to a finished ceiling or to bare joists.

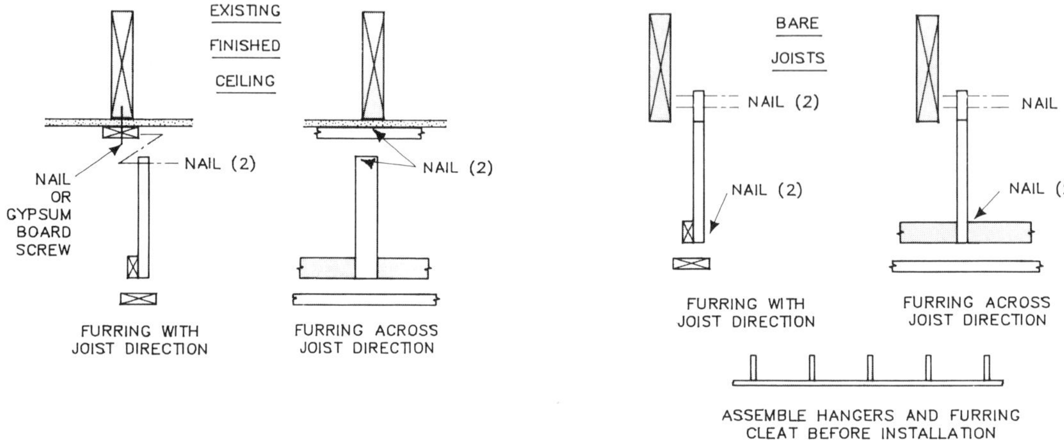

EXISTING FINISHED CEILING

NAIL OR GYPSUM BOARD SCREW

NAIL (2)

NAIL (2)

FURRING WITH JOIST DIRECTION

FURRING ACROSS JOIST DIRECTION

BARE JOISTS

NAIL (2)

NAIL (2)

NAIL (2)

NAIL (2)

FURRING WITH JOIST DIRECTION

FURRING ACROSS JOIST DIRECTION

ASSEMBLE HANGERS AND FURRING CLEAT BEFORE INSTALLATION

Figure 17–27
Furring for a recessed incandescent fixture.

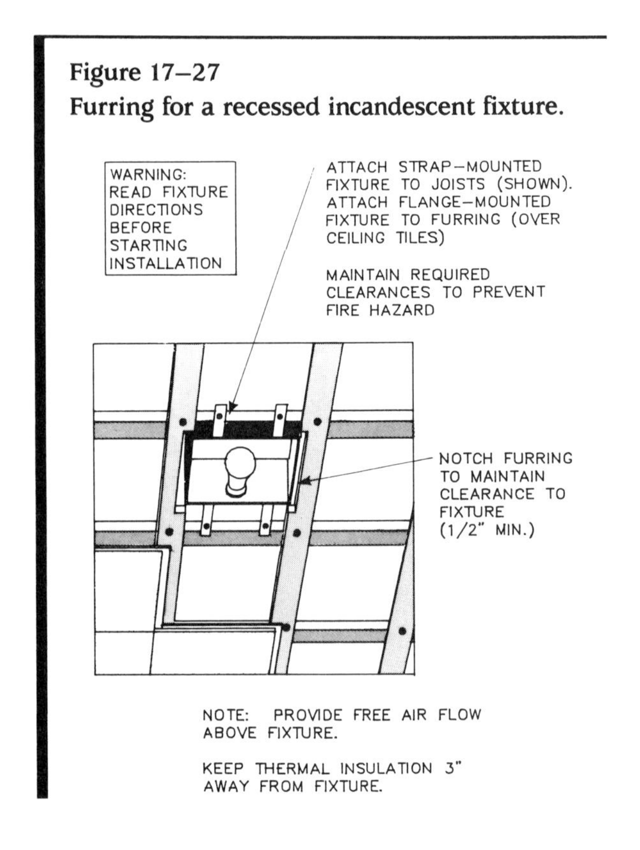
WARNING: READ FIXTURE DIRECTIONS BEFORE STARTING INSTALLATION
ATTACH STRAP-MOUNTED FIXTURE TO JOISTS (SHOWN). ATTACH FLANGE-MOUNTED FIXTURE TO FURRING (OVER CEILING TILES)
MAINTAIN REQUIRED CLEARANCES TO PREVENT FIRE HAZARD
NOTCH FURRING TO MAINTAIN CLEARANCE TO FIXTURE (1/2" MIN.)
NOTE: PROVIDE FREE AIR FLOW ABOVE FIXTURE.
KEEP THERMAL INSULATION 3" AWAY FROM FIXTURE.

Figure 17–28
Placing the corner tile: Trim a partial corner tile so its flanges line up with snapped chalk lines. Staple the tile to the furring, then drive a single box nail to secure the back corner. Position the nail so it will be covered with the perimeter trim.

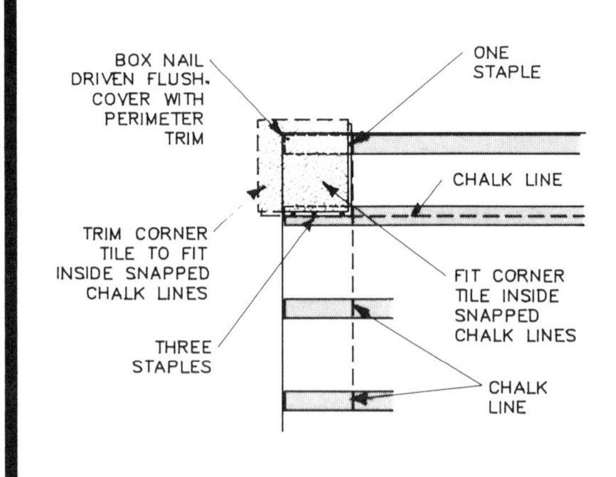
BOX NAIL DRIVEN FLUSH. COVER WITH PERIMETER TRIM
ONE STAPLE
CHALK LINE
TRIM CORNER TILE TO FIT INSIDE SNAPPED CHALK LINES
FIT CORNER TILE INSIDE SNAPPED CHALK LINES
THREE STAPLES
CHALK LINE

Cross tees. Cross tees snap into slots in the main runner to space the main runners and support the ceiling panels. The outer ends of cross tees are square-cut to length and rest on the wall molding. Cross tees are available in 2′ and 4′ lengths. The 4′ length has a slot at the center for attachment of another cross tee.

Planning

Draw a plan of the ceiling on tracing paper and place it over a scaled grid, as shown in Figure 17–35. The grid on the left of the figure is scaled for 24″ × 24″ suspended ceiling modules. Panels can run across the room or with the length, and they can run with or across the joists if exposed, as can the main runners. Position the drawing on the grid and locate the lighting fixtures. In this case four fixtures can be symmetrically located. Transfer the appropriate grid lines to the plan. The lines on the right half of Figure 17–35 are for 24″ × 48″ panels. Determine the location of the main runners. As shown, two full lengths and one partial length of main runner are required.

Either a main runner or a row of panels should run down the centerline of your room. For the

Figure 17–30
Installing field tiles: Use a sequence of tiles that maintains a more-or-less sawtooth work line diagonally across the ceiling.

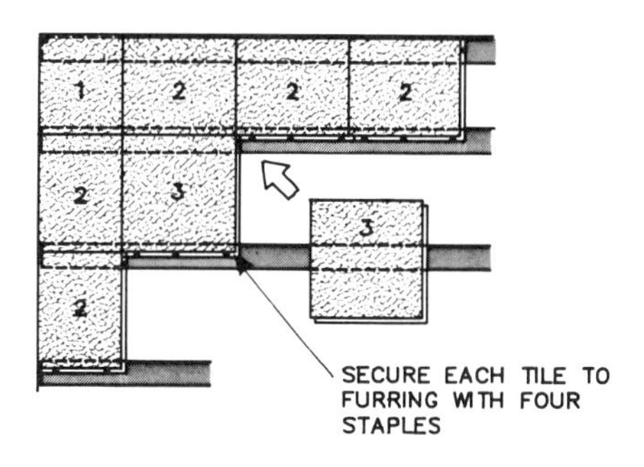
SECURE EACH TILE TO FURRING WITH FOUR STAPLES

1	2	2	2	2	2	2	2	2
2	3	4	5	6	7	8	9	
2	4	5	6	7	8	9		
2	5	6	7	8	9			
2	6	7	8	9				
2	7	8	9					
2	8	9						
2	9							
2								

INSTALL AND STAPLE FIELD TILES IN A SAWTOOTH OR STAIR–STEP SEQUENCE

Figure 17–29
Trim and place both partial border rows of partial perimeter tiles, keeping the flanges aligned with the snapped chalk lines. Drive two nails per tile.

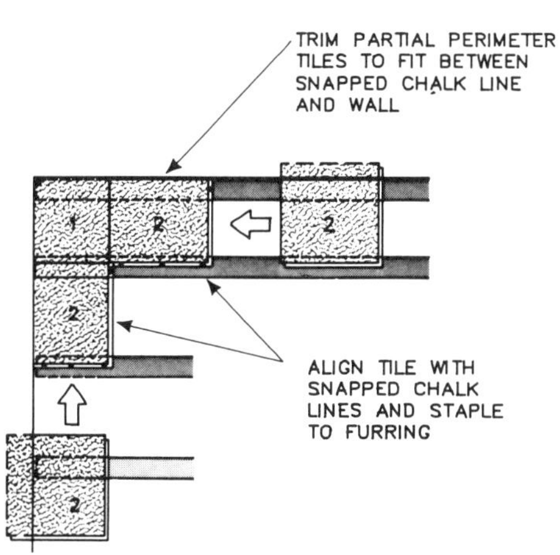
TRIM PARTIAL PERIMETER TILES TO FIT BETWEEN SNAPPED CHALK LINE AND WALL
ALIGN TILE WITH SNAPPED CHALK LINES AND STAPLE TO FURRING

Figure 17–31
The last border row: Nail corner of first tile in row as shown. Secure last corner tile after trimming with a single nail in the corner.

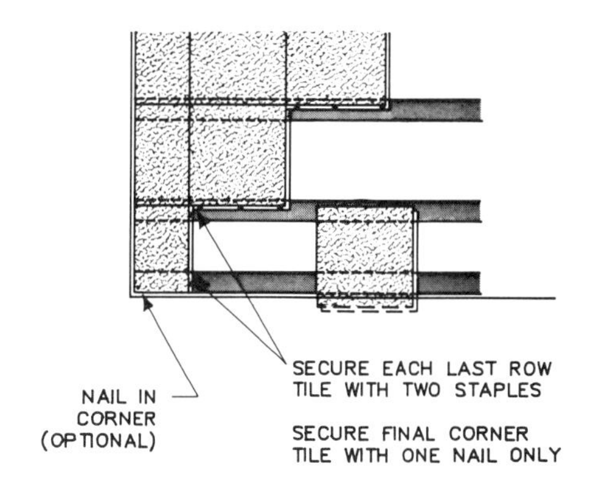
NAIL IN CORNER (OPTIONAL)
SECURE EACH LAST ROW TILE WITH TWO STAPLES
SECURE FINAL CORNER TILE WITH ONE NAIL ONLY

Figure 17–32
Suspended acoustic panel ceiling with fluorescent fixtures.

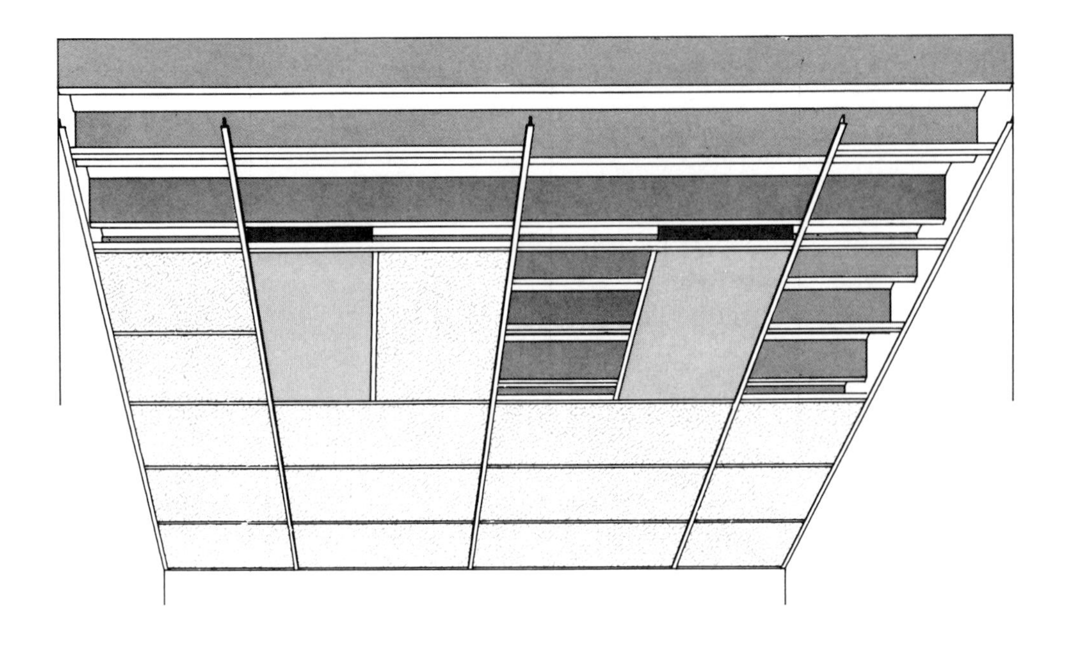

Figure 17–33
Suspended ceiling grid system parts.

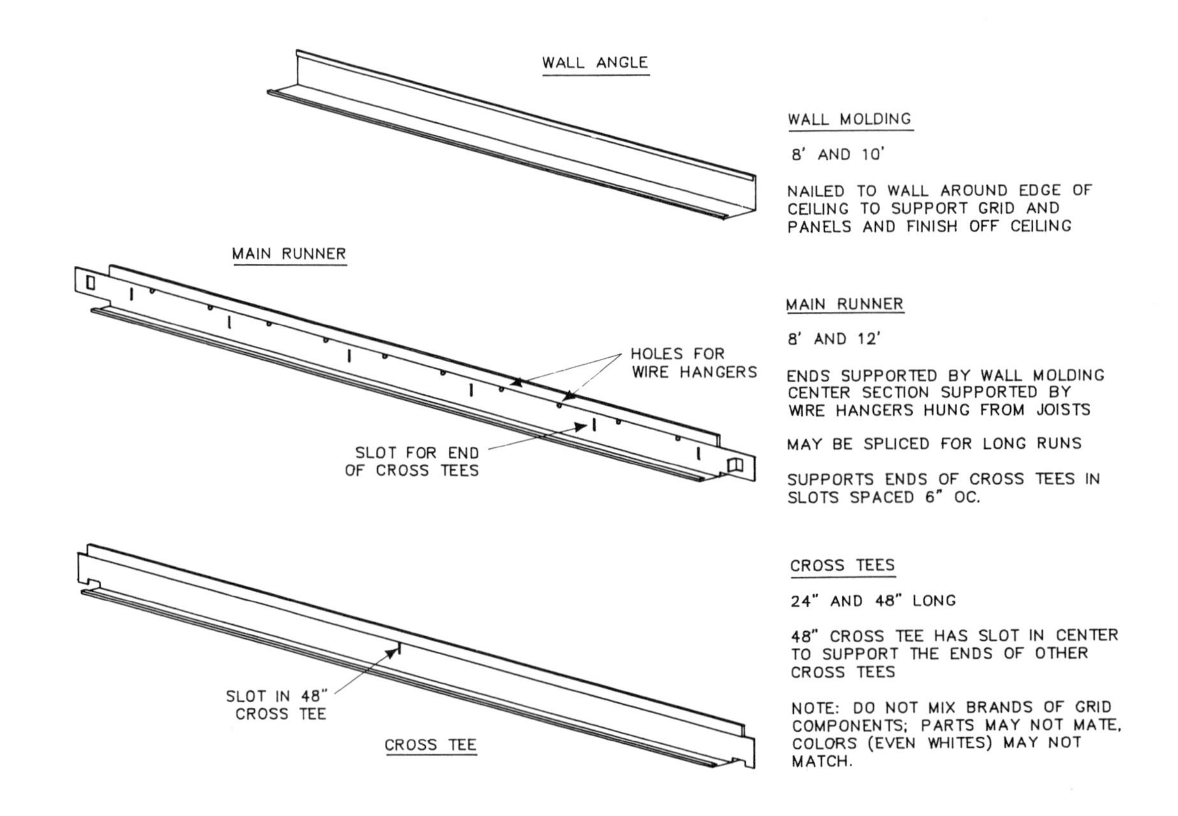

most pleasing appearance, border panels on opposite sides of the room should be equal in width (as measured from the wall to the closest grid member).

Border panels should be more than a half-panel wide. If you are installing fluorescent lighting fixtures, either attached to the ceiling above the new ceiling or supported by the grid, the required location of the fixtures becomes the number one priority in the ceiling design.

Figure 17–34
Suspended ceiling grid system assembly.

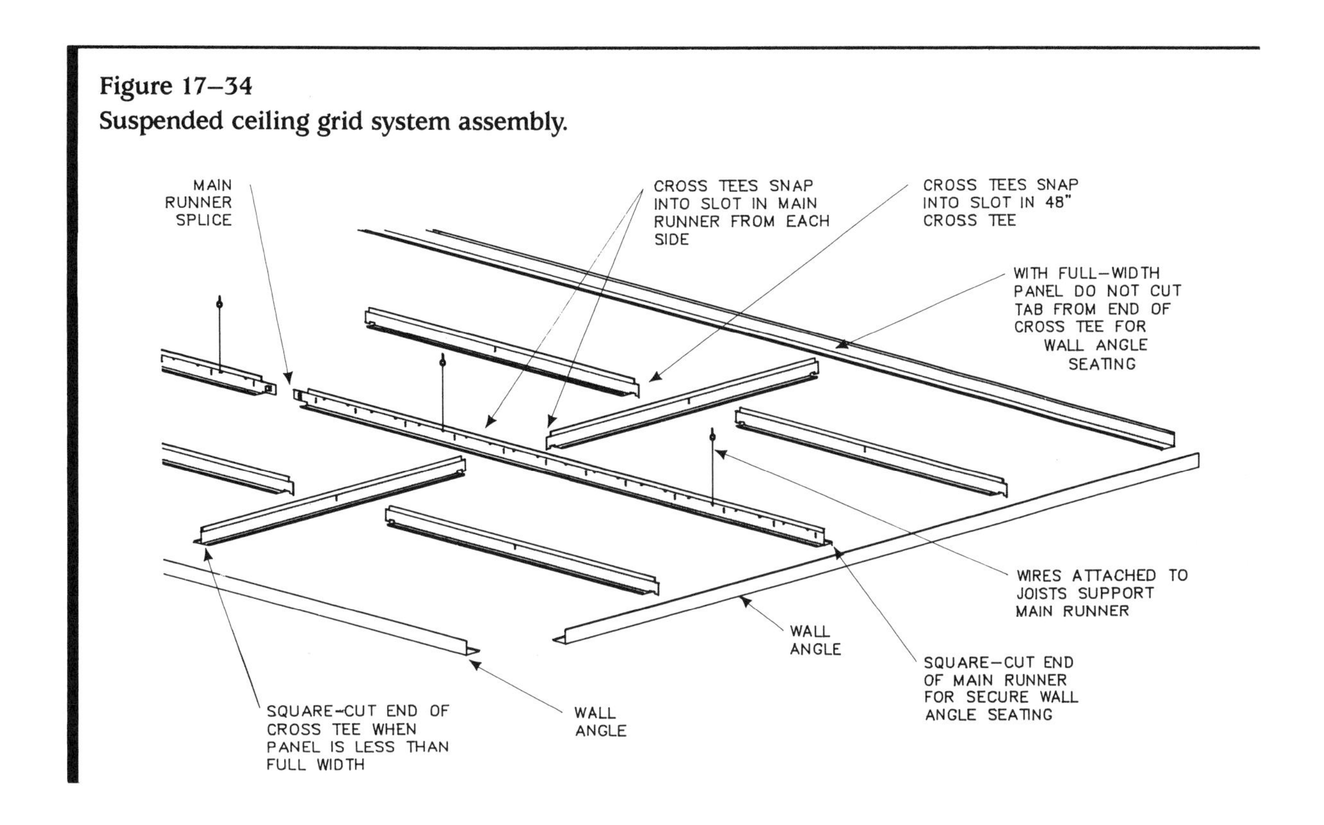

Figure 17–35
Planning a suspended ceiling.

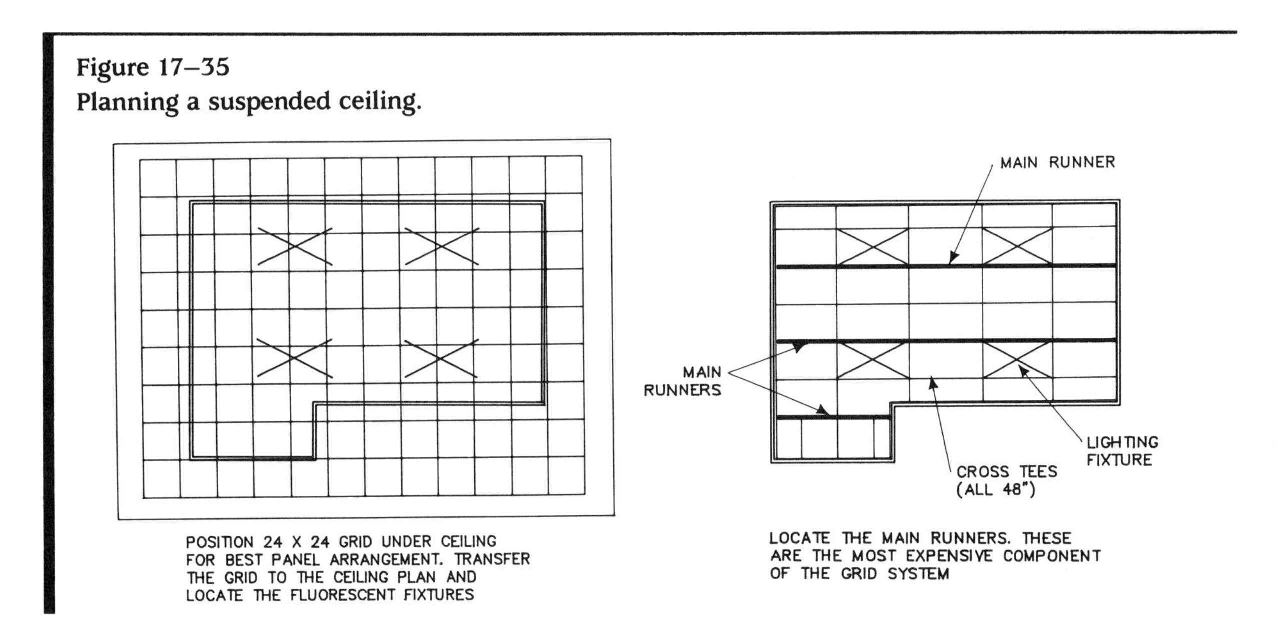

Main runners can be run across the room either way, and do not determine which way the panels are run. It is easier to hang the wires for the runners if you run them across the joist direction. They must be 4′ apart, and you will need one every 4′. Cross tees snap into main runners and can be arranged in various patterns to accommodate the panels.

Start by selecting the new ceiling height. It must be high enough to be above doorways. It should be high enough to clear the window frames, and it should be low enough to clear the lowest obstructions such as pipes, air ducts, or beams. These conflicting requirements may make it necessary to box some lower obstructions or around windows. A 3″ minimum ceiling drop is required to maneuver the panels into place in a standard drop-in grid system. Some systems require only 2″.

Installing Wall Molding

Start in a corner where you can start with long, unobstructed runs. Tack up the corner ends of the moldings; raise, level, and tack the far ends. Butt wall molding end to end. Use a level and another piece of molding extended under the two butted pieces to position the second piece.

To maintain a level around a boxed corner, run a piece of molding into the corner and fasten it. Then extend a short piece of molding diagonally across the corner with a level on it. Rest the molding piece on the second wall of the corner on this and match its end with the previous piece.

To fit wall molding at inside corners, trim both moldings square and let them lap. To fit outside corners, trim one end at 45° and square-cut the other piece, which will be placed over the mitered piece, or miter both pieces. These corner joints are shown in Figure 17–36.

To fasten molding to the walls, you can use nails or pan-head sheet-metal screws long enough to reach into the wall studs. For attachment to a masonry wall, use pan-head screws driven into plastic anchors inserted in holes drilled in the wall.

Figure 17–36
Wall molding corners for suspended ceiling.

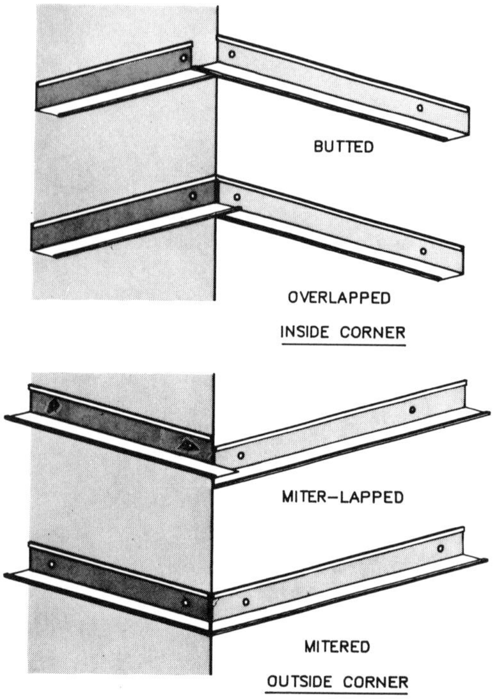

Installing Main Runners

From your plan, locate the main runner positions and stretch strings from wall to wall to mark the top edge of each main runner. Figure 17–37 shows how to establish the string end points. The ends of the runners will rest on the wall molding and be supported across the field by hanger wires attached to the joists or ceiling above. Locate the first hanger position for each runner by

Figure 17–37
Locating main runner string height for suspended ceiling grid.

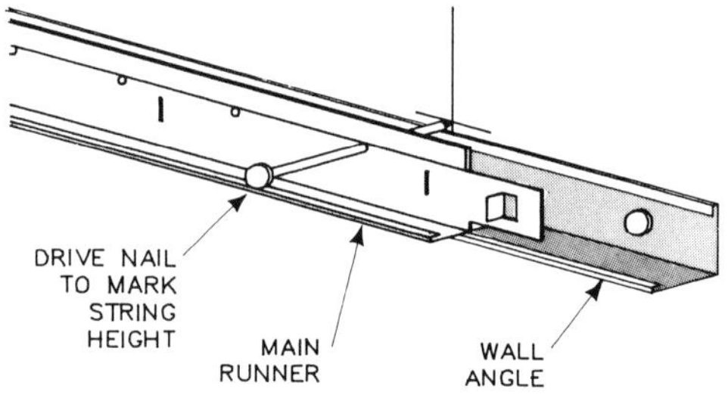

resting a length of runner on the wall molding, and use a level to make a mark above one of the holes in the runner but not more than 2′ from the wall. (Hanger holes are spaced every 4″ in a runner.) Succeeding hanger wires can be spaced 4′ apart and, at the other end, no more than 2′ from the wall. Attach wires at these points with screw eyes or appropriate anchors in the overhead ceiling or framing. The wires should extend about 6″ below the runner strings.

Stretch a string across the main runners at the position of the first set of cross tees. Main runners are punched every 6″ with slots for attaching cross tees. Cut one end from the first main runner so that the distance to one of its cross-tee slots is at the string. Make the cut square. The trimmed ends of the main runner butt the wall molding and rest on its flange. Splice or trim the main runner to length to fit against the other wall. Repeat these steps for other main runners.

Insert hanger wires in the nearest holes provided on the main runners and bend the wires up. Check with a string to be sure the main runner is level before securing the wires with three tight twists.

Installing Cross Tees, Lights, and Panels

Install cross tees by inserting their tabs in the appropriate main runner slots and lock them in place. Locking mechanisms differ: some snap down; others must be bent sideways with a screwdriver. To fit cross tees to the wall molding, cut them off square. The ends stay loose and rest on molding flange, like the main runners.

Install the lighting fixtures next. Make sure they are properly secured to the grid or the overhead framing as specified in their installation instructions. Make all switch and power connections to the fixtures at this time.

Finally, install the panels. Insert each one up through the grid by tipping it at an angle; then lower it so the edges rest on the main runner and cross-tee flanges with the finished side facing down.

If there are no lighting fixtures in the ceiling, you can improve acoustic isolation through the ceiling and floor above by fastening unfaced fiberglass thermal insulation between the joists if exposed, or by laying unfaced batts on top of the grid and panels. If there are lighting fixtures in the grid, this should not be done. Insulation must be kept away from the fixtures because of the fire hazard. Also, effective sound reduction will be impossible because the noise will pour through the fixtures.

INSTALLING A METAL CEILING

A metal ceiling consists of a field of plates 24″×24″ (or some multiple of 24″), filler trim plates, and a cornice. The decoration of the field plates is modular: there can be one design per 24″×24″ plate, or four, sixteen, or sixty-four per plate.

The design must appear to be centered in the room. Unlike acoustic ceiling tiles and suspended ceiling panels, field plates cannot be arbitrarily edge-trimmed to make the equivalent of border tiles or panels without seriously degrading the appearance of the ceiling. The field plates must be installed in whole modular design multiples. The centerlines of a ceiling may cross at any nailing button that is at the center, corner, or center of a side of a modular design unit. Typical ceiling pattern examples are shown in Figures 17–38 and 17–39.

If the modular field plate repeat pattern does not fill the ceiling neat to the cornice, filler trim plates are used to bridge the gap. These plates are embossed with an overall pebbly pattern, which can be trimmed to odd sizes.

Planning for a Metal Ceiling

Draw a ceiling plan as shown in Figure 17–40. Draw, place, and center a 24″×24″ scale grid representing field plates on the ceiling plan. If you don't like the fit, and your selected ceiling pattern is a 12″ or 6″ module, add those lines to your field plate overlay and see if you can get a better fit. If you do, you will have to cut the field plates, but their use will be in whole modules.

Draw the cornice projection; a 6″ cornice pro-

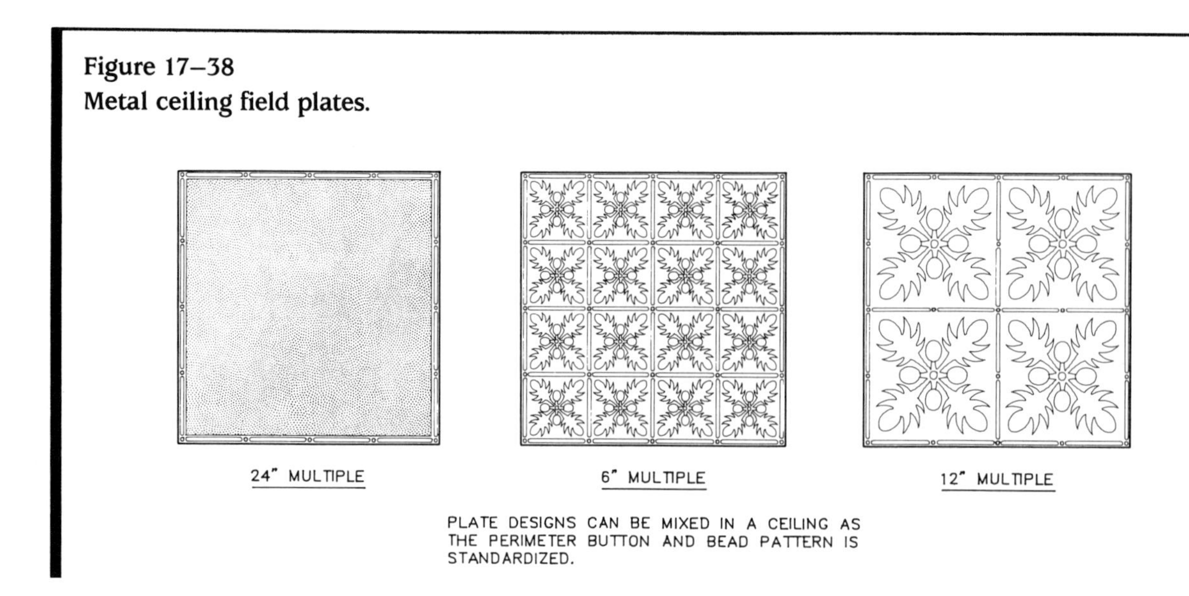
24" MULTIPLE
6" MULTIPLE
12" MULTIPLE
PLATE DESIGNS CAN BE MIXED IN A CEILING AS THE PERIMETER BUTTON AND BEAD PATTERN IS STANDARDIZED.

Figure 17–38
Metal ceiling field plates.

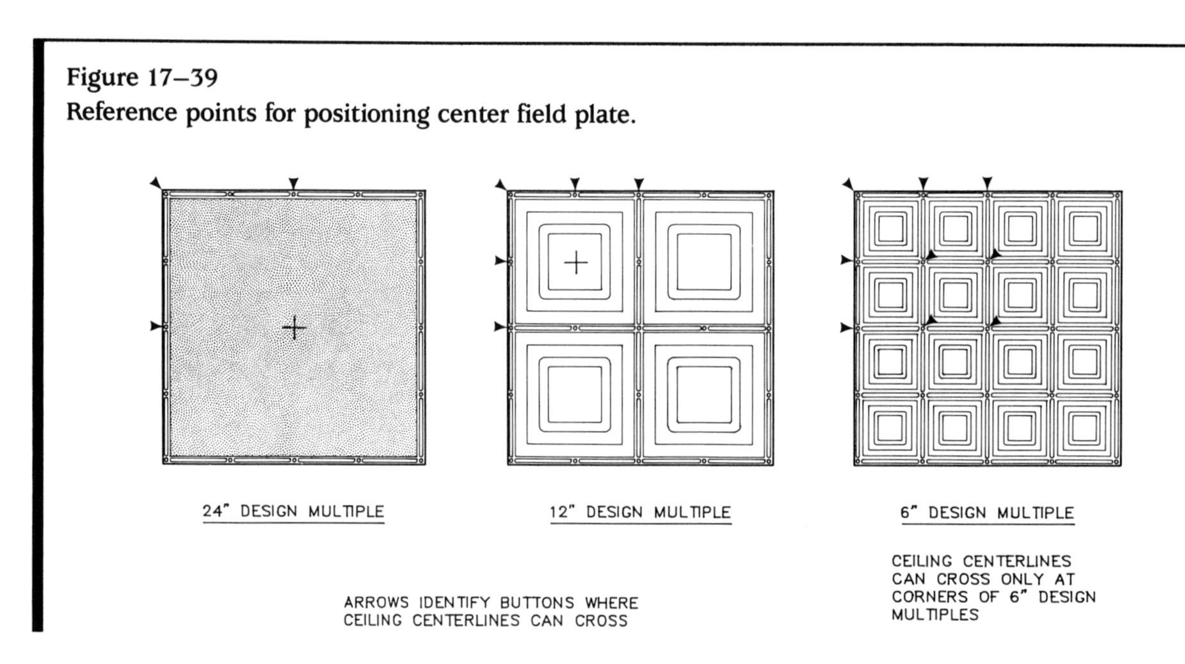
24" DESIGN MULTIPLE
12" DESIGN MULTIPLE
6" DESIGN MULTIPLE
ARROWS IDENTIFY BUTTONS WHERE CEILING CENTERLINES CAN CROSS
CEILING CENTERLINES CAN CROSS ONLY AT CORNERS OF 6" DESIGN MULTIPLES

Figure 17–39
Reference points for positioning center field plate.

jection is shown in steps 2, 3, and 4 of Figure 17–40. The fit of 24″ plates on the ceiling of the room (step 2) is not very good and creates awkward differences in border plate widths. The layout with a 12″ module (step 3) is better: only two plates have to be cut, but the borders are uneven and the fit in the alcove is poor. A layout with a 6″ module (step 4) has even border widths all around the room, but a lot of plates must be trimmed and the design seems too busy. However, the fit on this particular ceiling can be made nearly perfect with the use of 6″ multiple plates.

Partial plates can be trimmed with the rough-cut edge tucked under the border trim panels. (If you have a lot of plates to be cut, find an air-conditioning duct company with a sheet-metal shear and pay them to make the cuts.)

Installing Furring

The metal ceiling parts are fastened to furring with 1″ common nails or #6 × 3⁄4″ pan-head sheet-metal screws. The type of furring used and its arrangement depend on the particular size and

Figure 17–40
Planning a tin ceiling.

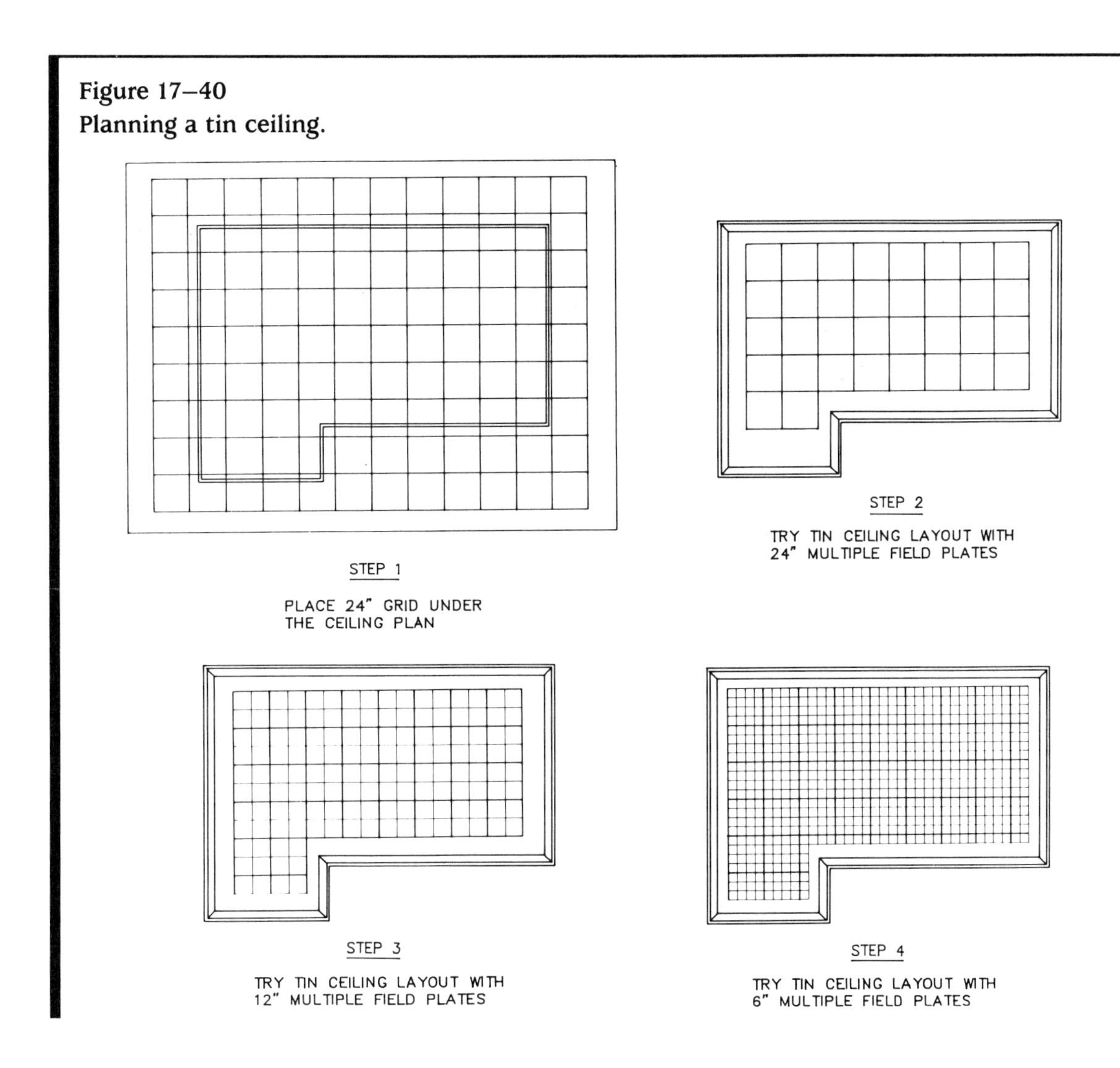
STEP 1
PLACE 24" GRID UNDER
THE CEILING PLAN
STEP 2
TRY TIN CEILING LAYOUT WITH
24" MULTIPLE FIELD PLATES
STEP 3
TRY TIN CEILING LAYOUT WITH
12" MULTIPLE FIELD PLATES
STEP 4
TRY TIN CEILING LAYOUT WITH
6" MULTIPLE FIELD PLATES

design of the ceiling panels being used. Furring must be spaced so that all plate edges are supported, and the furring must be shimmed level. Most metal ceiling panels are designed so they can be nailed at 6″ intervals. In fact, nails can be spaced 12″ in the center of the panels, but edges should be nailed every 6″.

Methods of furring. Furring can be accomplished several ways:

1. Locate 1 × 2 furring on the old ceiling surface, 6″ on center, running across the joists and with ends butted at joists. This method will provide backing for all possible nailing points. As panels are installed, smear glue on short blocks of 1 × 2 and insert them under panel edges as backing between furring.
2. Locate 1 × 3 furring on the old ceiling surface 12″ on center to coincide with the edges of the panels, beginning at the centerline, and add cross furring between to back the rest of the nails; this scheme is shown in Figure 17–41. The problem with this method is there is nothing in the ceiling structure to nail the cross furring to. A plaster ceiling in good condition might provide enough backing for nailing the panel, but a single-layer gypsum board ceiling probably would not.
3. Install double-layer furring as shown in Figure 17–42. Nail furring 24″ on center across the joists. Nail a second furring

Figure 17–41
Furring on 12″ centers. This method is marginal on a single-layer gypsum board ceiling.

LOCATE FIRST 1x3 FURRING STRIP
AT CENTER OF CEILING
SPACE THE BALANCE
OF THE FIELD PLATE
FURRING STRIPS 12"
ON CENTER
LOCATE LAST
FURRING STRIP
TO SUPPORT
CORNICE
24" BY 24"
OR LONGER
FIELD PLATES
TRIM PLATE
LAPPED UNDER
CORNICE
TIN
CORNICE

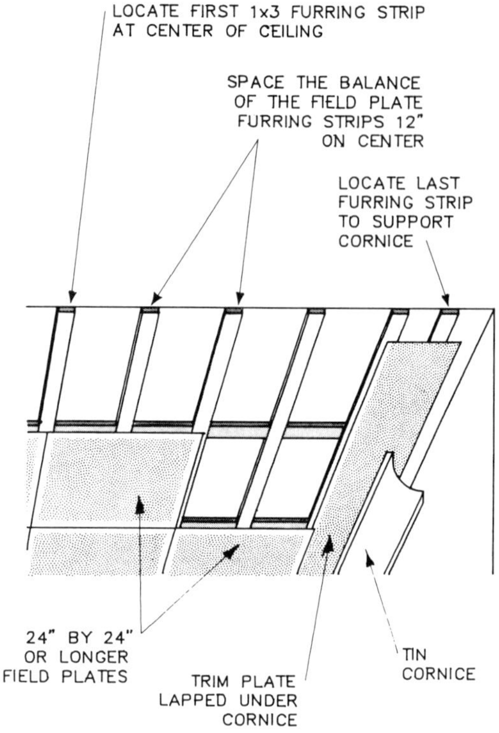

Figure 17–42
Double-layer furring, second layer on 12″ centers. This is the preferred method.

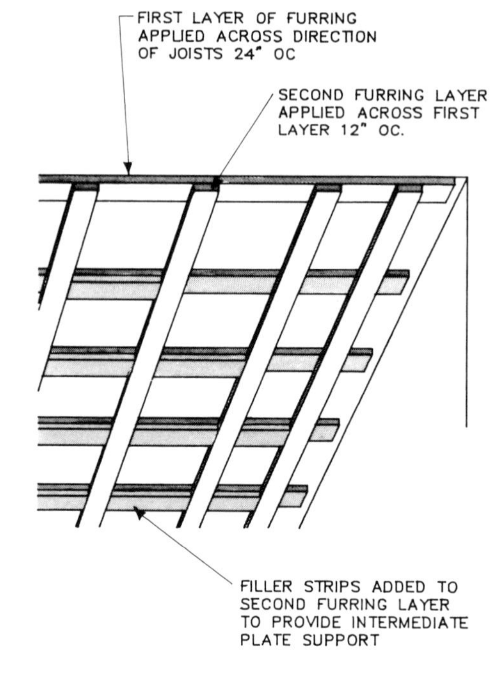
FIRST LAYER OF FURRING
APPLIED ACROSS DIRECTION
OF JOISTS 24" OC
SECOND FURRING LAYER
APPLIED ACROSS FIRST
LAYER 12" OC.
FILLER STRIPS ADDED TO
SECOND FURRING LAYER
TO PROVIDE INTERMEDIATE
PLATE SUPPORT

across the first furring, parallel to the joists but on 12″ centers. Fill in between the furring on both layers. Fasten the furring together with gypsum board screws.

4. Sheath the entire ceiling with ½″-thick sheathing-grade plywood rather than using close-spaced furring. This is probably the fastest way to get the job done and will result in a quieter ceiling. The cost is also reasonable: for a 12′ × 16′ plywood ceiling sheathing would cost about 25 percent more than furring strips 6″ on center, but you can put sheathing up without having to make center-to-center measurements.

Laying out the ceiling. Strike a chalk line lengthwise on the centerline of the ceiling, then mark off and strike chalk lines parallel to this line every 6″ or 12″, depending on the method of furring you are going to use. Strike a line crosswise on the centerline of the room, then mark off and strike chalk lines to mark seam locations. If you are sheathing the ceiling, do this after sheathing. The chalk lines must be accurate; the plates don't stretch and you can't fudge.

Putting Up the Metal Ceiling

Begin the ceiling by installing the plates forming the main part of the ceiling. *Note:* Wear heavy work gloves. The edges of metal ceiling plates are extremely sharp and can slash your hands badly.

Before putting a plate up, wash it with paint thinner to remove the oil coating. The edges of the plates overlap; the nailing beads or buttons form rows of ball and socket joints. Start tacking up plates at the far end of the room so seams will lap away from view. Nails are driven through both buttons. Drill pilot holes for screws. When you are sure the panels are straight and aligned properly, drive a nail in each 6″-spaced button around the perimeter of each plate, except at the edges of the main ceiling, where nails should be only started (they have to be removed temporarily). Drive a nail through the button at the center of each 24″ square.

Install the cornice next. Small cornices (up to 6″ projection) are nailed 6″ on center to furring on the ceiling and into every wall stud. Larger cornices should be supported on wood brackets shaped to the cornice profile. Miter the corners and nail wood brackets on each side of the miter, as shown in Figure 17–43. Also install wood brackets to back up intermediate joints where the cornice pieces overlap (see Figure 17–44).

Cutting the miters involves nibbling the metal until you get it right. Try to match patterns in the corners. Once you get left and right ends that fit together accurately, use them as templates for the rest of the miters before installing any.

For a cornice corner joint that is easier to close, add folded tabs to one of the pieces and lap the other piece over the tabs, as shown in Figure 17–45. Cut off one of the mating pieces ¼″ beyond the line to leave material for tabs. Now, make a series of short cuts ¼″ deep and ¼″ apart

Figure 17–43
Installing a metal cornice.

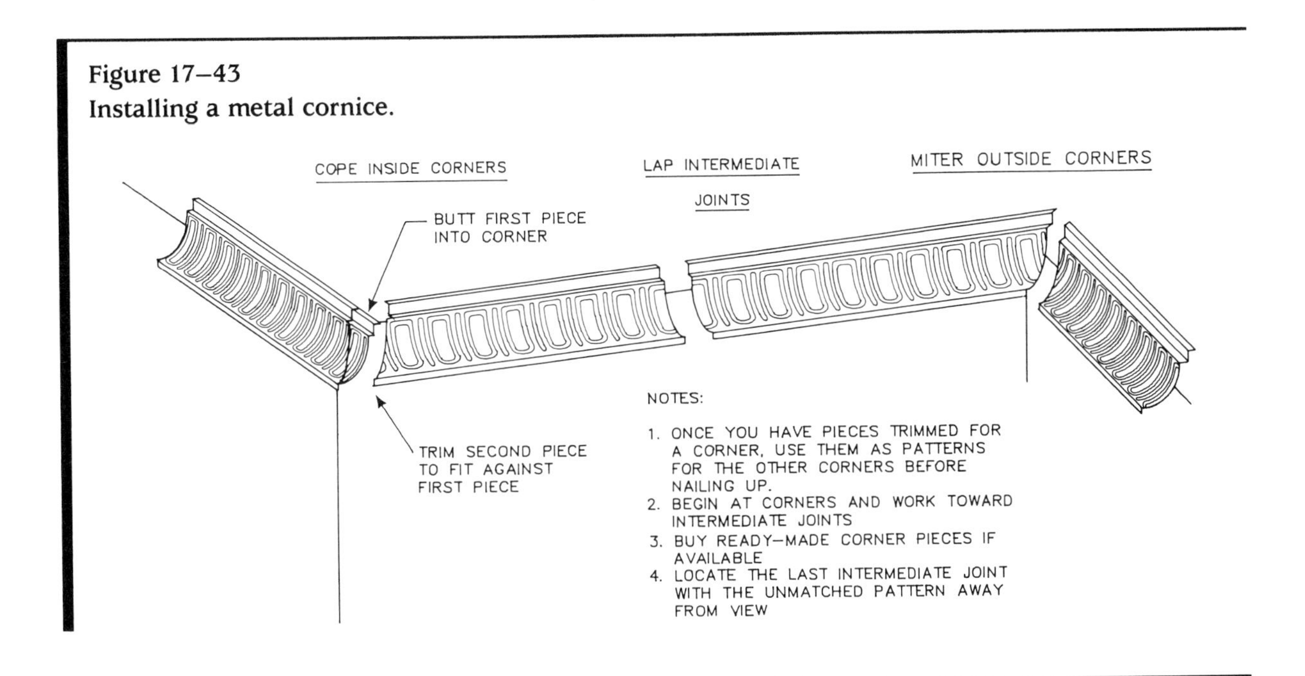

Figure 17–44
Back up large metal cornices with wood blocking at corners and all intermediate joints.

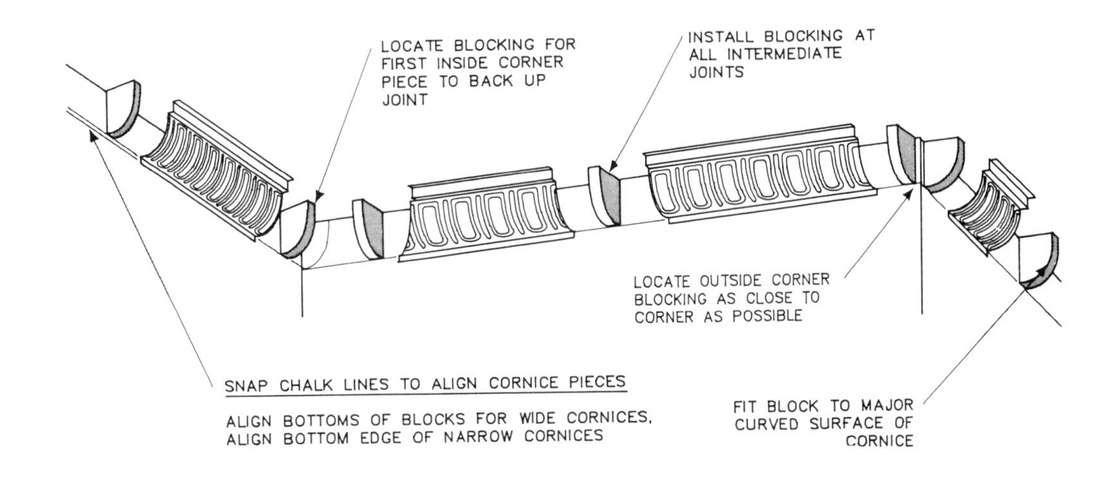

to create tabs. Bend the tabs 90° in or out, depending on whether it is an outside or inside corner. This will produce a tight miter without gaps.

When installing the cornice, begin at the corners and work toward a factory edge somewhere on the wall where a possible mismatch in the pattern continuity won't be too noticeable.

If you are using filler plates between the field plates and the wall, they must go up before the cornice. Cut the filler into strips of the required width. They should overlap the cornice nailing flange and extend 1" under the main ceiling plates as shown in Figure 17–46. Pull the temporary nails at the edges of the main plates to slip the filler under. Miter fillers at the corners or use a mitered lap joint. Finish the job by renailing or screwing the outer edges of the main ceiling plates up through the filler strips into the furring or sheathing.

In place of a cornice, a molding strip can be used to finish the edge of the ceiling. Some moldings are mitered at corners; with others, ells and tees are available. Ells and tees are installed first, and molding is installed over them.

Finish installation by closing the seams. Tap along the seam edges using a light hammer and a flat-head nail with the head held against the seam. This is called "caulking" the seam in the metal ceiling trade.

Figure 17–45
Mitered outside cornice corner.

TABS ON MITERED EDGE WHEN POSITIONED BEHIND OTHER MITERED CORNICE PRODUCE A NEATER OUTSIDE CORNER

CHALK LINE

Finishing

A metal ceiling can be left in its silvery tin color by applying a polyurethane varnish. If you want to paint the ceiling, use an oil-base metal primer as a first coat. (It is a good idea to wipe over the plates with paint thinner to remove any oil on the surface before applying the primer.) Then roll or brush on a finish coat of oil-base paint.

Figure 17–46
Installing a metal cornice using a filler strip.

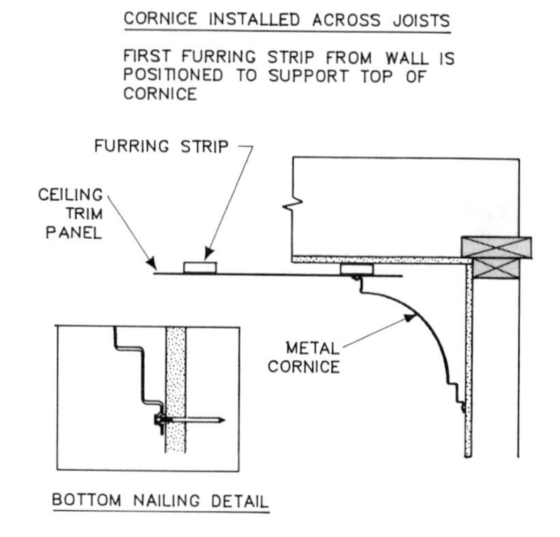

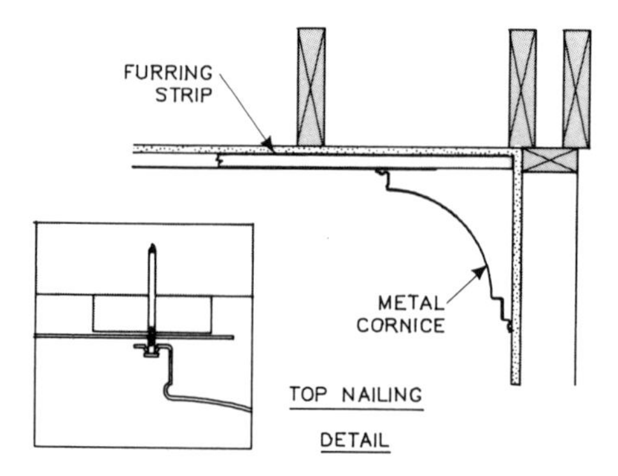

18 Installing Wood Flooring

In renovation, as in new construction, you should hold off laying the finish flooring until just about everything else has been completed, to keep it from getting messed up.

Many materials are used for finish flooring—hardwood, softwood, resilient tile and sheet flooring, ceramic and clay tile, natural stone and brick, and carpet; all have advantages and disadvantages. Wood is the traditional flooring material. It is durable and comfortably warm, even friendly in appearance. The attractive figure and grain of wood and its wide range of colors and possible application patterns make it the popular choice. Softwoods used include yellow pine, white pine, Douglas fir, and hemlock. Hardwoods include domestic woods such as oak, maple, beech, cherry, walnut, pecan, and birch, and exotic woods such as iroko, rosewood, and teak. Today, much wood flooring is not solid wood but plywood.

Wood finish flooring is available in three forms: strips, planks, and blocks, also called parquet squares or simply parquet (Figure 18–1). These materials can be laid on a variety of bases;

Figure 18–1
Prefinished parquet flooring is both rich and varied in appearance. (Photo courtesy Bruce Hardwood Floors)

the two most common are a joist and subfloor structure, and a concrete slab with or without a wood subflooring structure. Proper preparation of the base is essential to obtain a sound floor that will not develop squeaks or loose or warped boards. Wood flooring products, their fastenings, and appropriate substructures are listed in Table 18–1.

Some wood flooring can also be laid over an existing wood floor or over resilient tile or seamless flooring, but not a ceramic tile, slate floor, or a rubber tile floor—those must be taken up. An old floor that is left down must be sound and flat, with no loose pieces.

SUBSTRUCTURE FOR FLOORING

Finish wood flooring can be laid over:

1. Plywood panels fastened directly to joists or laid over an old floor.
2. Underlayment panels laid over plywood or over an old subfloor or finish floor.
3. Combined plywood and underlayment (APA Sturd-I-Floor grade) panels fastened directly to joists.

These materials are described in Chapter 2. Installation details are given here.

Plywood Subfloor

Install plywood with the long dimension of each panel running across the joists and with the panel continuous over two or more joist spans (see Figure 18–2). Locate the panel end joints over the framing. Allow ⅛″ spacing at the panel ends and ¼″ at the panel edges, unless the panels are marked otherwise. Support the panel side edges with 2 × blocking between the joists. Space nails 6″ apart along the panel edges and 10″ apart at intermediate supports with 6d deformed-shank nails for ½″ panels or 8d nails for greater thickness. See Chapter 3 for information about nails.

Plywood panels may also be glued to joists.

Figure 18–2
Installing a plywood subfloor.

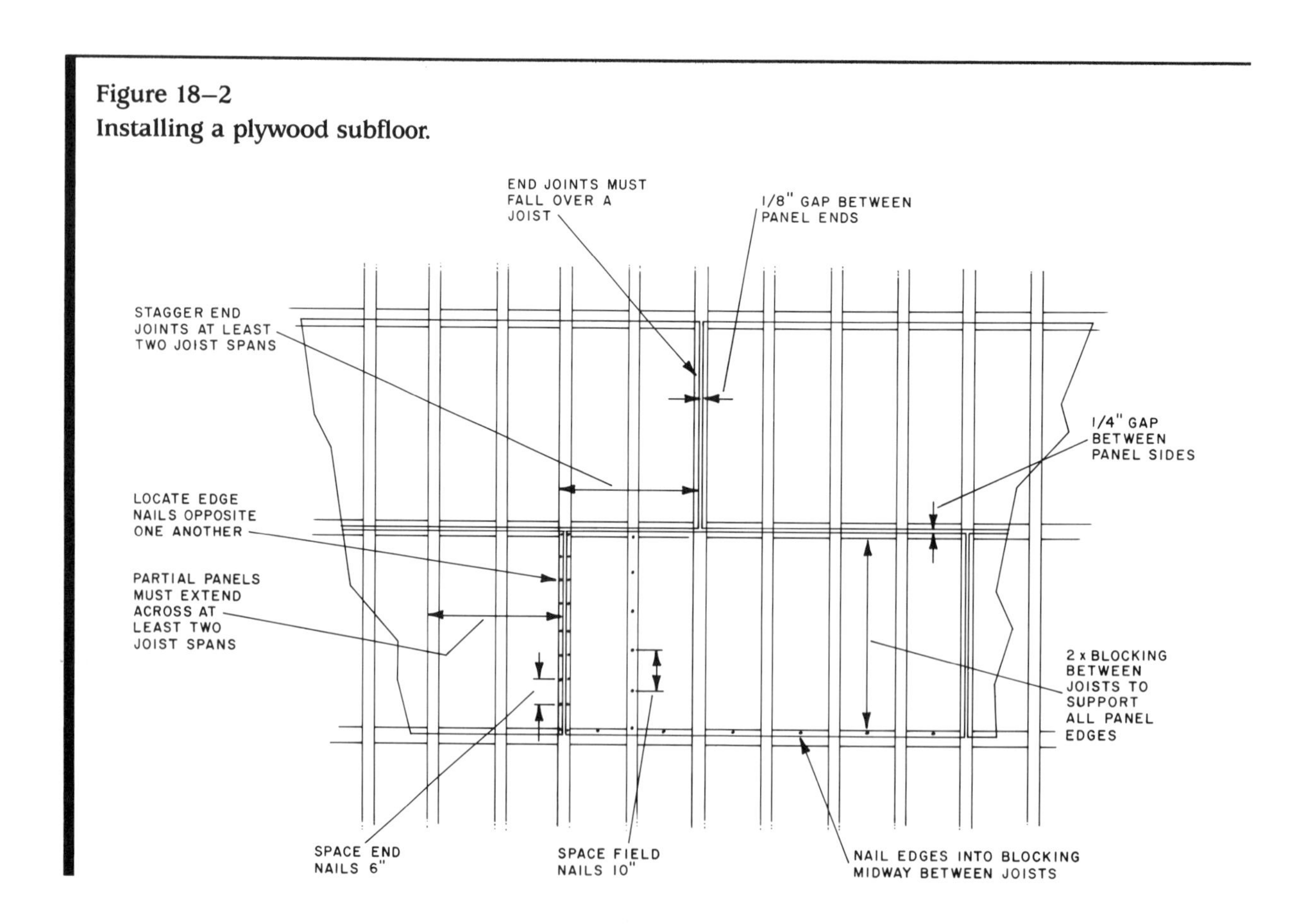

Table 18–1
WOOD FINISH FLOORING PRODUCTS

Finish Flooring Product	Unit Size	Thickness	HOW FASTENED				SUBSTRUCTURE				
			Nails	Screws	Adhesive	Pegs	Plywood	T&G Boards	Joists (Bare)	Sleepers	Concrete Slab (Direct)
Tongue-and-groove strip flooring	1½″, 2¼″, 3¼″	25/32″	X		X		X	X	X[1]	X	X
Square-edge strip flooring	1½″, 2″	5/16″	X				X	X[1]			
Solid wood plank	2¼″ to 8″ wide, random lengths		X	X		X	X	X	X[2]		
Plywood plank	3″ to 8″ wide, random widths, random lengths and 96″ length	3/8″, 5/8″			X		X	X[1]			
Hardwood block (parquet)											
Laminated block	12″ × 12″	½″			X						
Unit block	12″ × 12″	25/32″			X						X
Slat (tile) block (also sold loose)	6″ × 6″ to 36″ × 36″	5/16″ to 2″			X						X[3]

[1] Not the most satisfactory substructure.
[2] Not recommended, but many old floors are planked in this way.
[3] In 25/32″ thickness only.

Use an adhesive that meets APA specification AFG-01, and apply it in accordance with the manufacturer's recommendations. The product label should give the specification and use information. If particleboard or waferboard panels with sealed surfaces and edges are used, use only solvent-base glues. Apply a continuous line of adhesive on the joists and a continuous or intermittent line of glue in the groove of tongue-and-groove panels. Use 6d ring- or screw-shank nails spaced no more than 12″ apart at the panel ends and intermediate bearings. (Your local code may require closer nail spacing.) If plywood thicker than ¾″ is used, also nail it as described above for nail-only installation.

Underlayment

Underlayment panels require a subfloor; they cannot be installed directly on joists because they do not have the strength to span the space between joists without flexing, sagging, or even cracking. If laid over an existing floor, all nails in the old floor must be set below the surface (a tedious job with the face-nailed flooring that is found in many old houses), loose boards must be firmly nailed or screwed down, and any ridging in the old surface must be sanded or planed off.

The installation shown in Figure 18–3 assumes a subfloor of plywood panels, but the layout and nailing requirements are the same if a properly prepared old floor is used. Underlayment provides the proper surface for resilient tiles or sheet flooring, or for a pad and carpet.

When installing underlayment, stagger panel end joints with respect to each other and offset all joints with respect to the joints in the subfloor (see Figure 18–3). Butt panel ends and edges to a close but not tight fit (allow 1/32″ space). Space

Figure 18–3
Placing underlayment on a plywood subfloor. Underlayment panels can also be oriented at right angles to the subfloor panels as long as all joints are offset in the two layers.

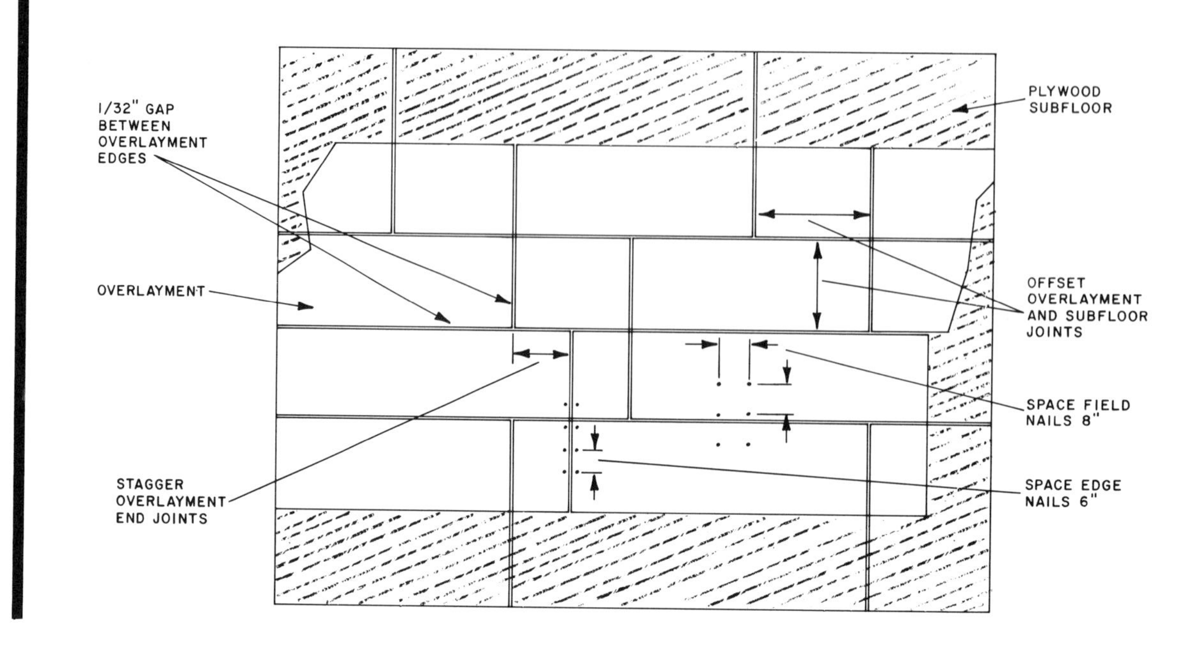

nails 6″ apart along panel edges and 8″ apart each way throughout the remainder of the panel with 3d ring-shank nails for thicknesses of ½″ or less, or 4d nails for panels ⅝″ and ¾″ thick. Nails must be sunk below the surface. After nailing, fill and thoroughly sand the edge joints. Lightly sand any surface roughness, particularly at joints and around nail heads or other fasteners.

Combined Subflooring and Underlayment

Sturd-I-Floor panels are intended to be covered directly with resilient flooring or with a pad and carpet. They can be used under finished wood flooring, but a plain plywood subfloor is just as good for that purpose and cheaper. The panels are installed as described above for plywood, with the long dimension running across supports and with the panel continuous over two or more spans. If the panel edges are not tongue-and-groove, they must be supported by 2× lumber blocking installed between joists. Protect the surface against damage until the pad and carpet or other finish flooring is installed. Stagger the panel end joints and locate them over the framing. Allow ⅛″ spacing at the panel ends and edges, unless the panels are marked otherwise. For nailed floors, space nails 6″ apart at the panel edges and 10″ apart at intermediate supports. Use 6d ring- or screw-shank nails for panels ¾″ thick or less and 8d nails for thicker panels. With 1⅛″ panels, 10d common nails may be used if supports are well seasoned. Nails must be driven flush or slightly below the surface.

WOOD STRIP FINISH FLOORING

Strip flooring is solid wood flooring. There are two types: tongue-and-groove and square edge. Both are sold in bundles containing random-length pieces, prefinished or unfinished. Prefinished is recommended.

Tongue-and-Groove Strips

Tongue-and-groove strip flooring is customarily end-matched, meaning that the ends are also tongued and grooved. The flooring is always blind toenailed, as shown in Figure 18–4. The dimensions of prefinished tongue-and-groove flooring strips are typically 25/32″ thick by 1½″, 2¼″, or 3¼″ wide. The strips are hollow-backed, so that only the outer edges rest on the subfloor. This allows them to lie flat and not rock on slightly uneven surfaces.

Strip flooring is usually applied over structural subflooring. It can also be applied directly to sleepers over a concrete floor without plywood or plank subflooring. (Sleepers are random-length short 2×4s laid on a concrete slab to provide nailing support for the strip flooring.)

Strip flooring looks best running the long way in a room. Don't change directions from room to room. If the subfloor consists of boards laid diagonally across the joists, the floor will be stronger if the finish strip flooring is laid to run at right angles to the joists. If the board subflooring runs straight across the joists, lay the finish flooring across the subflooring to run in the same direction as the joists. With plywood subflooring, you can lay the strip flooring in any direction you want, but perpendicular to the joist direction will give you the strongest floor system. The end joints of tongue-and-groove strip flooring do not have to be located over joists.

Figure 18–4
Tongue-and-groove strip flooring and concealed or blind nailing technique.

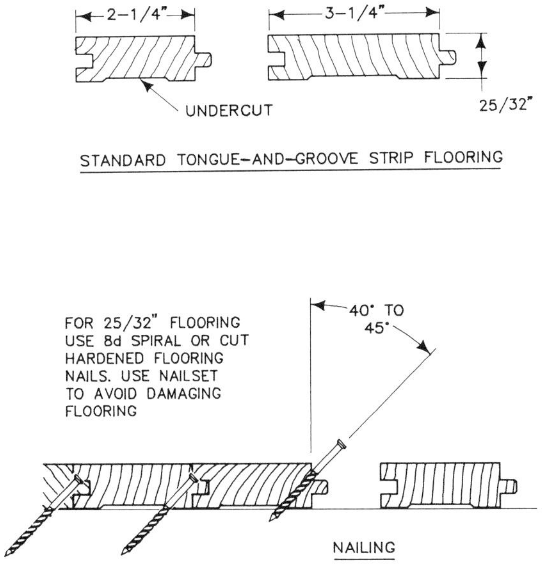

Rent a nailing machine. It takes a lot of skill to blind-nail hardwood flooring without messing up the edge. It does not take a lot of skill to use a floor-nailing machine, which can be rented from flooring suppliers, some home centers, and similar sources. Some nailing machines drive flooring staples instead of nails. See Table 18–2 for nail sizes and a nailing schedule.

How to lay tongue-and-groove strip flooring. The method of laying tongue-and-groove flooring is shown in Figure 18–5. If you are renovating, remove the baseboards. If the flooring will go through a doorway and continue in another

Table 18–2
STRIP FLOORING NAIL SIZES AND NAILING SCHEDULE

Flooring Type	Flooring Strip Size	Nail Size	Nailing Method	Nail Spacing
Tongue-and-groove	25/32″ × 1½″ 25/32″ × 2¼″ 25/32″ × 3¼″	7d or 8d	Toenail	One nail, 10″ to 12″ on center
Square edge	5/16″ × 1½″ 5/16″ × 2″	3d or 4d	Face-nail	Two nails, ⅜″ from opposite edges, 6″ to 8″ on center

Figure 18–5
Laying tongue-and-groove strip flooring.

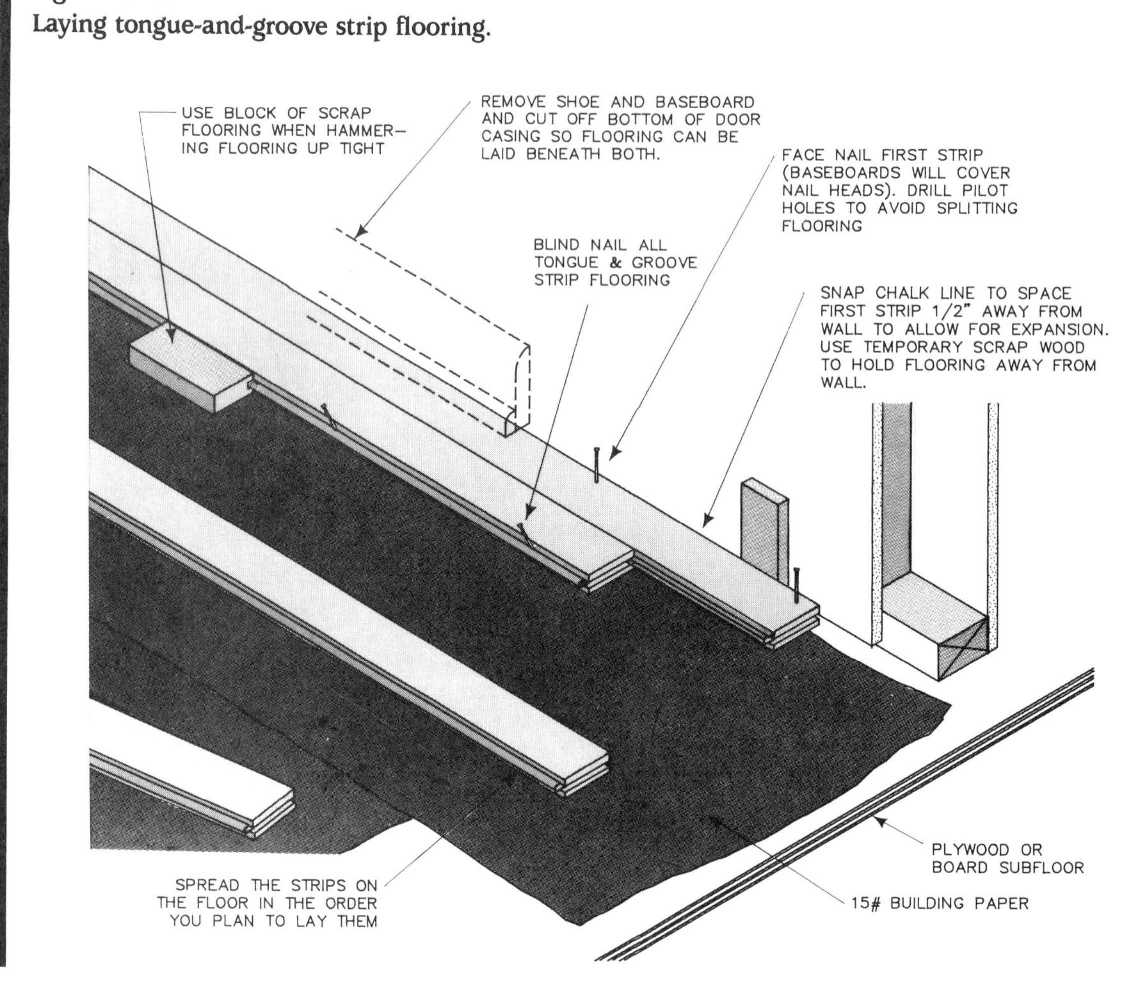
USE BLOCK OF SCRAP
FLOORING WHEN HAMMER-
ING FLOORING UP TIGHT
REMOVE SHOE AND BASEBOARD
AND CUT OFF BOTTOM OF DOOR
CASING SO FLOORING CAN BE
LAID BENEATH BOTH.
FACE NAIL FIRST STRIP
(BASEBOARDS WILL COVER
NAIL HEADS). DRILL PILOT
HOLES TO AVOID SPLITTING
FLOORING
BLIND NAIL ALL
TONGUE & GROOVE
STRIP FLOORING
SNAP CHALK LINE TO SPACE
FIRST STRIP 1/2" AWAY FROM
WALL TO ALLOW FOR EXPANSION.
USE TEMPORARY SCRAP WOOD
TO HOLD FLOORING AWAY FROM
WALL.
PLYWOOD OR
BOARD SUBFLOOR
15# BUILDING PAPER
SPREAD THE STRIPS ON
THE FLOOR IN THE ORDER
YOU PLAN TO LAY THEM

room, lay the flooring continuously, without a sill. Cut off the door casing so you can slip the flooring under it. Lay a handsaw blade horizontally on a scrap piece of flooring to make the cut at the right height. The finished result will look better than fitting the flooring around the casing. Sweep and vacuum the subfloor. Check for nails or anything else sticking up; remove them or drive them securely below the surface. Fasten down any loose boards. Sand or plane off any ridges.

Roll out 15-pound building paper with edges lapped 2″. The paper keeps dirt, dust, and even spilled water from filtering through. Tack the paper to keep it from moving about. Remove the tacks as you come to them when installing the flooring.

Spread the wood strips on the subfloor in the order you plan to nail them. Joints should be no closer than 6″ in adjacent rows. Use a mix of long pieces and short pieces throughout, except in doorways and other heavy-traffic locations; there use pieces long enough to avoid any joints in the traffic path.

Nail the first strip ½″ away from the wall with the tongue facing out. The gap at the wall leaves room for expansion and will be covered by the baseboard. Drill pilot holes and face-nail the wall edge of this first strip. The first strip and all other strips are then blind-toenailed through the tongue.

Tip: *Fill the gap between the wall and the first strip with a ½″-thick piece of wood to keep the flooring from creeping toward the wall as you blind-nail the first few strips; then remove it.*

You will have to drive the blind nails in the first few boards by hand. Drive them at a 45° angle. Use a nailset to drive them home so as to avoid damaging the top edge of the strip. Strip flooring is laid tight. Use a short piece of cut-off flooring as a driving block to fit the strips against each other. If you do a good job of nailing now, the floor won't squeak in later years. Stand on the strips when you are fitting, aligning, and nailing them.

Rip the last strip to fit with a ½″ gap at the far wall, the same as for the beginning strip. Drill pilot holes for face-nailing the wall edge of this strip. Wedge the strip with a prybar to get it tight against the previous strip, then nail.

If you find it necessary to reverse direction—for instance, to lay flooring in a closet—turn a strip around and connect it to the previous strip with a spline, a narrow strip that fits into the groove in both pieces of flooring. Drill pilot holes and toenail through the spline.

Square-Edge Strip Flooring

Square-edge strip flooring is face-nailed to subflooring, as shown in Figure 18–6. Square-edge flooring is thinner than tongue-and-groove—only 5⁄16″. Widths are 1½″ or 2″. It is best laid over board or plywood subflooring. When laid over old finish flooring, its direction should be perpendicular to the old flooring. End joints do not have to be located over joists. It should be face-nailed with spiral flooring nails only, as specified in Table 18–2.

Figure 18–6
Square-edge strip flooring and face-nailing technique.

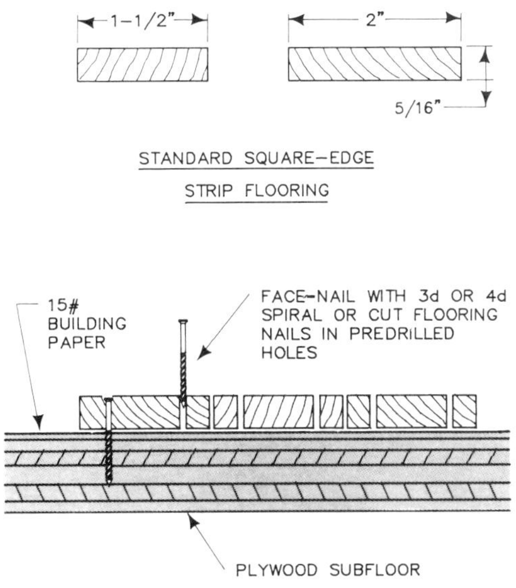

1-1/2″
2″
5/16″
STANDARD SQUARE-EDGE STRIP FLOORING
15# BUILDING PAPER
FACE-NAIL WITH 3d OR 4d SPIRAL OR CUT FLOORING NAILS IN PREDRILLED HOLES
PLYWOOD SUBFLOOR

WOOD PLANK FINISH FLOORING

Traditional plank flooring consisted of solid wood boards of random widths from 8″ to 14″ (sometimes even more in old floors) with shiplap, tongue-and-groove, or square edges. They were face-nailed or screwed to joists. Today, plank flooring is usually tongue-and-groove or square-edge solid boards, or tongue-and-groove lengths of plywood, as shown in Figure 18–7. They may be installed with nails, screws, or mastic.

Figure 18–7
Tongue-and-groove plank flooring and pattern for face-screwed installation of square-edge planks.

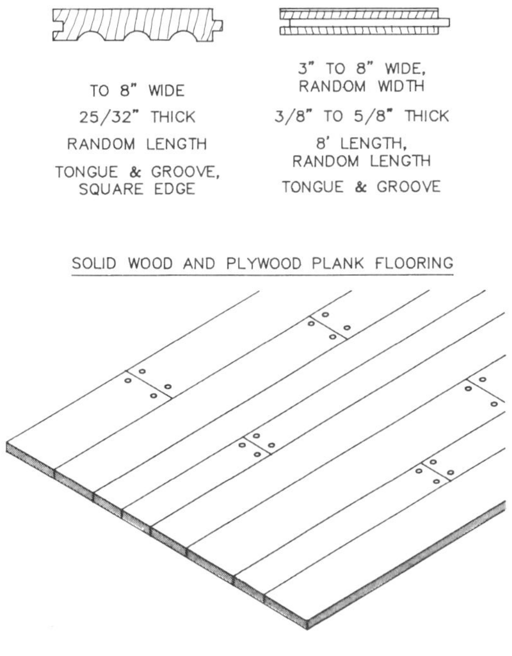

TO 8″ WIDE
25/32″ THICK
RANDOM LENGTH
TONGUE & GROOVE, SQUARE EDGE
3″ TO 8″ WIDE, RANDOM WIDTH
3/8″ TO 5/8″ THICK
8′ LENGTH, RANDOM LENGTH
TONGUE & GROOVE
SOLID WOOD AND PLYWOOD PLANK FLOORING

Solid Wood Planks

Modern random-width/random-length solid wood planking runs from 2¼" wide to 8" wide and is available prefinished. It is sold in bundles.

Tongue-and-groove plank flooring can be blind-nailed to wood and plywood subflooring in the same manner as tongue-and-groove strip flooring, explained earlier. Plank flooring can also be applied with mastic directly to concrete or resilient tile or sheet flooring that has a sound surface and is firmly attached to its underlayment. For installation on concrete, there must be a vapor barrier under or over the slab, as explained later in this chapter. Plank flooring should not be driven tightly together as strip flooring is, but laid with a 1/32" gap between planks at the face.

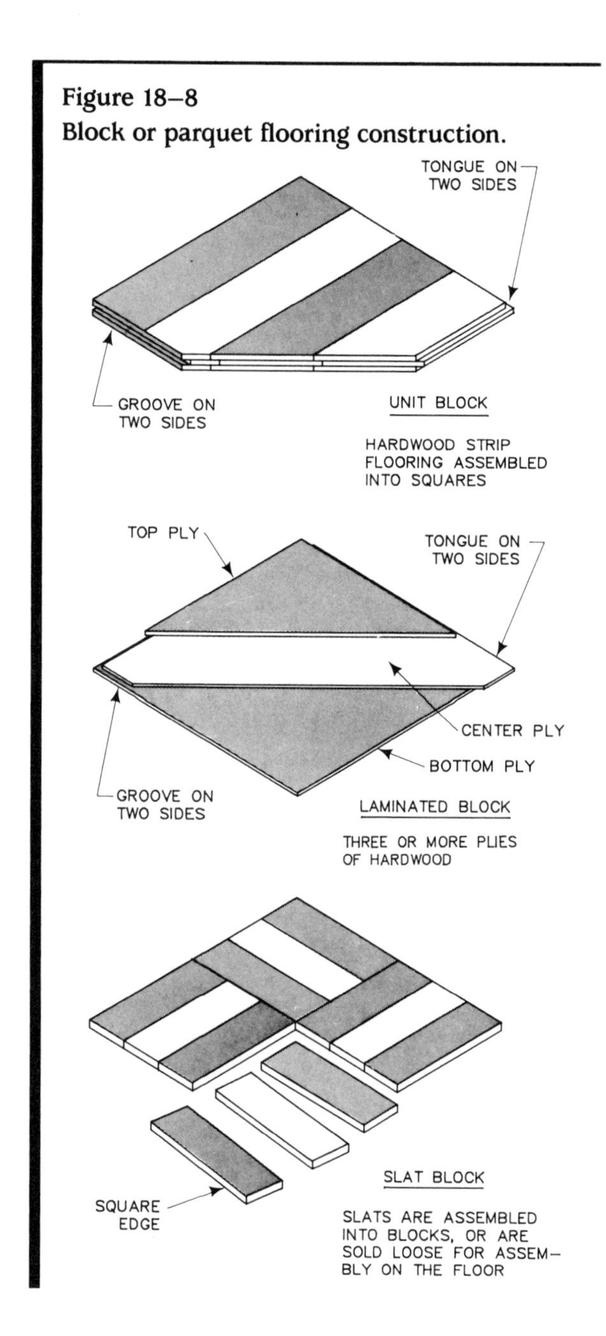

Figure 18–8
Block or parquet flooring construction.

Planks can also be screwed or nailed to the subfloor from the face. This is the appropriate way to install square-edge solid wood planking. If screws are used, sink them in counterbored holes. Locate two screws at each plank end, and space additional pairs of screws along the plank. Prefinished plank flooring comes with plugs that can be glued in screw holes flush with the surface.

You can use nails instead of screws for face-fastening. Cut steel nails look quite appropriate with wide-board flooring when the heads are driven flush with the surface. Drill pilot holes for nails to avoid splitting the plank. Use spiral nails if you want to drive the heads below the surface and fill the holes to match the plank.

Plywood Planks

Plywood planks are 3/8" or 5/8" thick, 8" wide, and 96" long. In appearance, they are strip flooring, with tightly butted or beveled face edges. The edges and ends of the planks are tongued and grooved. The thickness of the hardwood face veneer is generally 1/16" or less. Plywood plank flooring is applied only with mastic, so excellent subfloor preparation is very important.

HARDWOOD BLOCK FLOORING

There are three kinds of hardwood block flooring, or parquet: solid unit, laminated, and slat (see Figure 18–8). All are laid with mastic adhesive. Most varieties have tongue-and-groove edges for positive alignment on the floor.

Most wood block flooring is manufactured as prefinished assembled squares that you lay much as you would resilient floor tiles. It is also manufactured as prefinished pieces that you can

Figure 18–9
Block flooring patterns.

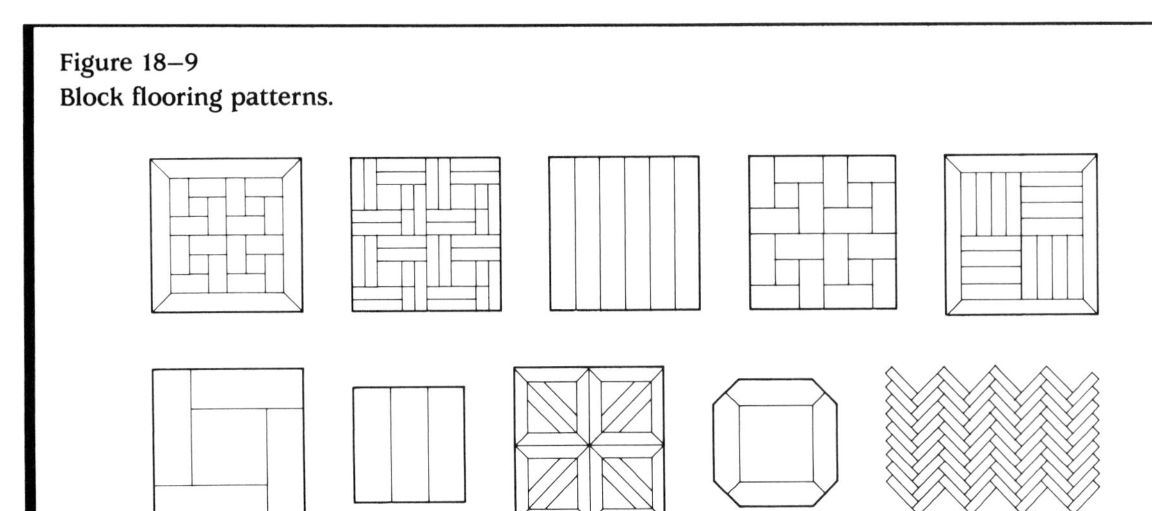

assemble into various patterns on your floor. Examples are shown in Figure 18–9.

Laying Block Flooring

To lay a pattern with edges square to the walls, start by locating the center of the room, then snap perpendicular chalk lines running through the center; the steps are shown in Figure 18–10. Position dry tiles along the chalk lines to determine how blocks will have to be cut at the walls of the room (see Figure 18–11). If the border pieces will be less than a half-block wide, shift the center point half the width of one block. This will increase the width of the border blocks; they will look better, and be easier to handle.

Use the mastic adhesive recommended by the flooring manufacturer or supplier. When you spread the mastic, you will put down the right amount if you use a trowel with the notch pattern specified for that adhesive. Lay the blocks in a pyramid pattern starting at the center of the room and working toward the sides and corners. Lay each block neatly in place; avoid excessive sliding. Tap the block edges lightly, using a piece of scrap wood, to assure a tight fit between blocks, and tap the face lightly, again with a scrap of wood to protect the block.

Blocks also can be laid in a diagonal pattern. Use the method shown in Figure 18–12 to lay out the first row of blocks, then work from the center, as in laying a square pattern.

LAYING A FINISH FLOOR OVER A CONCRETE SLAB

Finish flooring can be laid directly on the surface of a new concrete slab if the slab is cured, dry, and has had a vapor barrier installed under it. If insulation has been installed around the perimeter, heat loss to the outside air and the ground will be greatly reduced (see Figure 18–13). Although the slabs of basementless houses can be and are insulated to prevent this loss, the typical patio, carport, or garage slab is not. If you want to insulate the slab for a warm floor, you will have to put down a subflooring system of some kind. This will also produce a floor that is more comfortable to walk on than one that is laid directly on the slab.

Strip flooring can be laid on sleepers or sleepers plus plywood subflooring. Resilient tile or sheet flooring or carpet can be laid on a "universal" subfloor (explained later). All require a vapor barrier, and all but the directly laid flooring can be insulated.

Vapor Barriers

If the slab has been poured over a vapor barrier, a second vapor barrier is not needed above the slab. Vapor barriers once consisted of layers of

Figure 18–10
Laying out guidelines to lay block flooring in a pattern square to the walls.

STEP 1

LOCATE CENTER OF ROOM BY STRETCHING DIAGONAL STRINGS FROM THE CORNERS

STEP 2

LOCATE WALL CENTER POINTS AND SNAP CHALK LINES BETWEEN WALL CENTERS, PASSING THROUGH THE ROOM CENTER

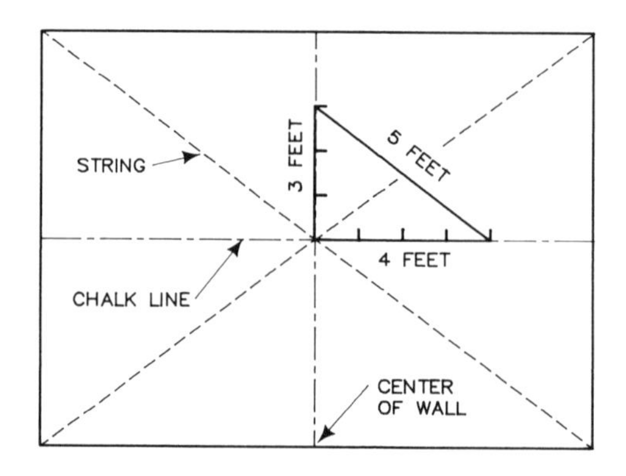

STEP 3

MEASURE OFF A 3–4–5 RIGHT TRIANGLE TO BE SURE CHALK LINES ARE PERPENDICULAR

IF CHALK LINES DO NOT PASS THROUGHT CENTER OF ROOM, OR IF THEY ARE NOT PERPENDICULAR, THE ROOM IS NOT SQUARE AND YOU WILL HAVE TO ALIGN THE BLOCK FLOOR FOR A BEST FIT IN THE ROOM.

Figure 18–11
The block alignment at left produces unattractively narrow borders. Shifting the alignment so the centerline runs through the middle block, as at right, produces wider borders.

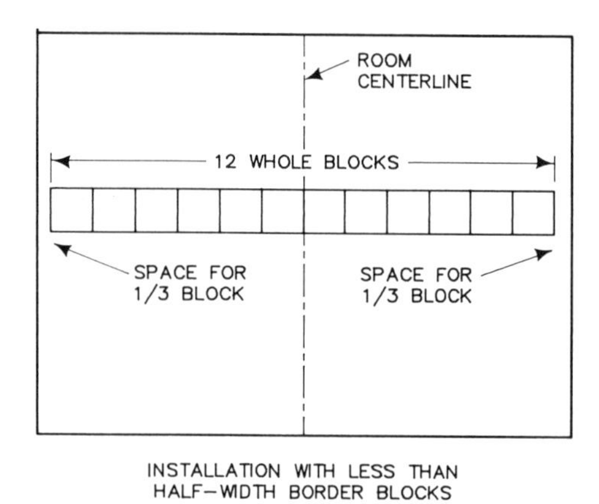

INSTALLATION WITH LESS THAN HALF–WIDTH BORDER BLOCKS

ROOM CENTERLINE

11 FULL BLOCKS

SPACE FOR 2/3 BLOCK

SPACE FOR 2/3 BLOCK

INSTALLATION WITH MORE THAN HALF–WIDTH BORDER BLOCKS

Figure 18–12
Laying out guidelines for diagonal block pattern.

STEP 1

DRIVE NAILS A AND B IN CORNERS EQUALLY SPACED FROM THE WALLS

STEP 2

MEASURE DISTANCE A TO B AND MARK OFF SAME DISTANCE A TO C AND DRIVE NAIL

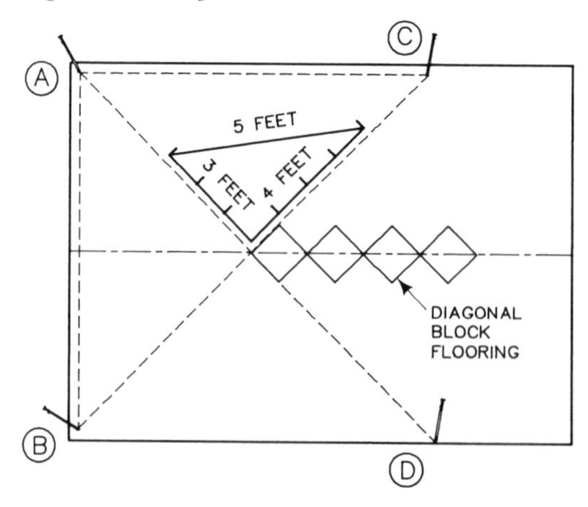

STEP 3

MEASURE DIAGONAL DISTANCE A TO C AND MARK FLOOR AT MID–POINT

STEP 4

SNAP CHALK LINE A–D , DRIVE NAIL AT D

STEP 5

MEASURE OFF A 3–4–5 RIGHT TRIANGLE TO BE SURE CHALK LINES ARE PERPENDICULAR

Figure 18–13
Slab and foundation configurations for vapor barrier.

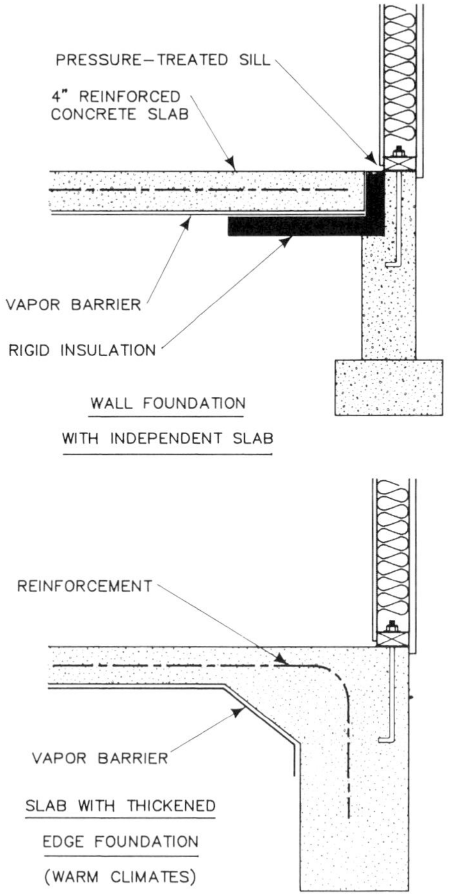

asphaltic mastic and asphalt-impregnated felt. Today, 6-mil polyethylene film is used instead, either laid on the concrete in cold mastic or without mastic but with wide overlapping joints between strips.

Almost all of the heat loss through a concrete slab at or near grade is through the 2′ to 3′ around the portion of the perimeter exposed to the weather. The loss through the center area is negligible. Glass-fiber insulation board should be placed under the floor in a 2′-wide strip around the edge.

The center area is better left uninsulated and clear to provide some air circulation and thus prevent dry rot. Do not depend on the insulation to be a substitute for the polyethylene film vapor barrier.

Slab Preparation

All the wood flooring systems described below start with preparing the slab surface. Chip off high spots, or sand them flush with a coarse silicon carbide sanding disk. Repair cracks and level low spots with patching cement. Apply two heavy coats of a masonry sealer such as UGL Dry-lock.

To get any adhesive to adhere to the concrete, the slab must be primed. An acrylic-rubber-base primer works best with a mastic adhesive. Check the mastic instructions for primer recommendations.

Composite Floor with Split Sleepers

This flooring system is good only for use under $25/32''$ tongue-and-groove hardwood strip flooring. It is based on a method devised by the National Oak Flooring Manufacturers Association (NOFMA) as an alternative to their hot-mastic system. The NOFMA method, shown in Figure 18–14, uses 2 × 4 sleepers set in a layer of mastic

Figure 18–14
Composite floor with sleepers (NOFMA).

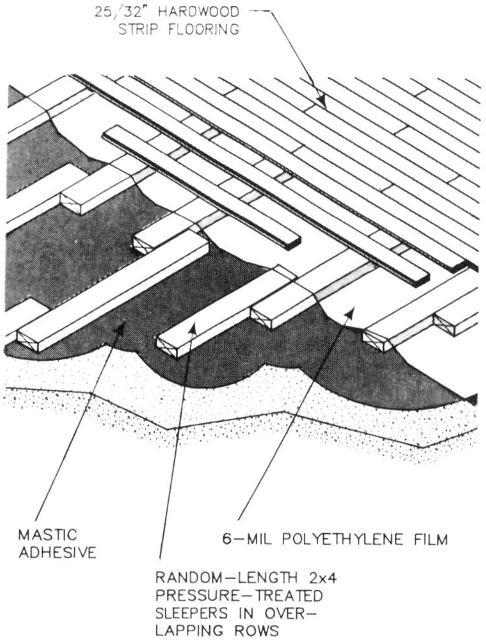

Figure 18–15
Improved composite floor with split sleepers.

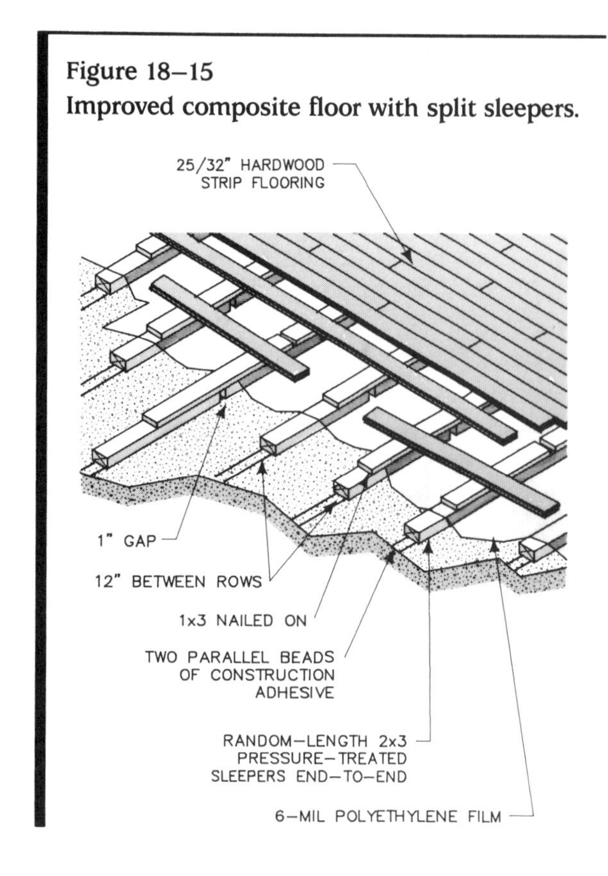

troweled over the entire surface of the slab. The strip flooring goes directly on top of a polyethylene vapor barrier laid over the sleepers.

The improved composite floor uses split or two-layer sleepers to provide an airspace between the vapor barrier and the finish flooring. The system is shown in Figure 18–15 and is constructed as follows.

Sweep the primed slab clean and snap chalk lines on 12″ centers to locate the sleepers, crosswise to the direction you want the finish flooring to run. Cut random-length (18″ to 36″) bottom sleepers from pressure-treated 2 × 3s; lay these sleepers in parallel beads of construction adhesive, leaving 1″ gaps between the ends. Lay additional sleepers around the perimeter of the room. Anchor the sleepers with short concrete nails sufficient to prevent any movement of the sleepers after the flooring is down.

Next, cover the sleepers with a 6-mil polyethylene film, lapping the joints over a sleeper. Now nail on random-length top 1 × 3 sleepers, which can be fir or pine or anything handy. Lap the top and bottom sleeper joints.

Hardwood tongue-and-groove strip flooring can now be blind-nailed to the top sleepers. Do not lay strip flooring with an adhesive; it will never remain in place. Prefinished flooring is a better choice than unfinished: sanding, staining, and varnishing a floor are not as easy as they may seem.

Wood Block Flooring Directly on Slab

Like strips and planks, parquet block flooring can be installed over a full wood subfloor. They can also be laid on concrete. The following is a method used to lay parquet flooring directly on a concrete slab that does not have a vapor barrier underneath. It is illustrated in Figure 18–16.

Coat the slab with primer. When the primer is dry, trowel on a coat of cold-stick mastic using a straight-edge trowel. Let this dry 30 minutes, then spread strips of 4-mil polyethylene film, lapping not more than 4″. Don't worry about occasional blisters or bubbles. Trowel cold-stick mastic on top of the polyethylene and lay the parquet in the mastic.

Figure 18–16
Laying block flooring directly on a concrete slab.

TROWELED MASTIC ADHESIVE
6-MIL POLY. FILM
SECOND LAYER OF TROWELED MASTIC ADHESIVE

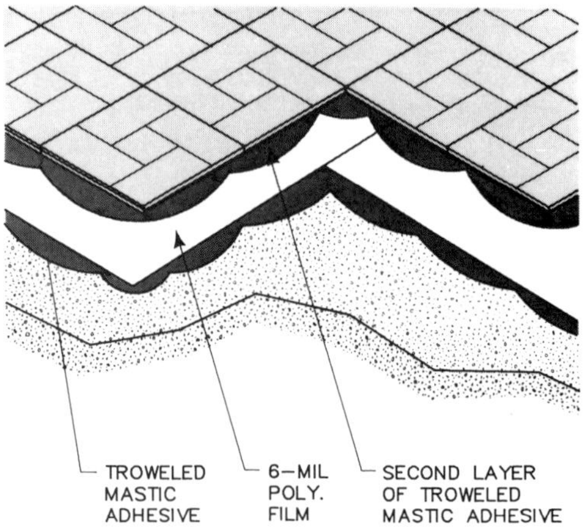

Universal Subfloor System

This subfloor system, shown in Figure 18–17, supports plywood panels on 2 × 4 sleepers; it can be used with any kind of finished floor. Cut pressure-treated 2 × 4s into random 2′ to 4′ lengths for sleepers. Discard any pieces that are warped. Short pieces are used to minimize any troubles that might develop from warping after installation of the floor. Glue sleepers to the slab with two parallel ⅜″ beads of construction adhesive. Locate the sleepers end to end on 16″ centers, with 1″ gaps between ends. Add a row of sleepers across the ends of the rows, as well as wherever the edges of the plywood subflooring will fall.

Fit 1½″-thick, 14″-wide glass-fiber board insulation between the sleepers all around the outside perimeter of the floor. Lay a 6-mil polyethylene vapor barrier over the 2 × 4s, stapling it in place. Lap joints 12″, always over a sleeper, in order to get a good seal.

Lay panels of ¾″ sheathing plywood rated for 16″ floor spans with the face grain running across the sleepers. Nail the plywood subflooring to the sleepers using 4d deformed-shank nails spaced 6″ along edges and 10″ elsewhere.

Strip, plank, or parquet wood flooring can be laid directly on this subfloor, as can resilient tiles and sheet flooring, and carpet. However, rated underlayment is required for nonwood flooring materials. (See Chapter 20 for ceramic tile floor substructure requirements.)

Figure 18–17
Universal subfloor/slab system.

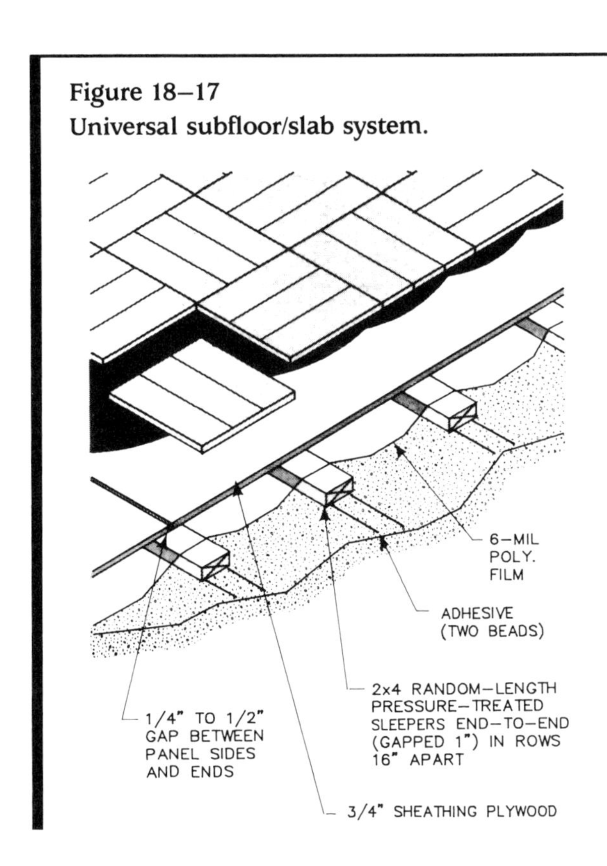

Figure 18–18
Plywood subflooring nailed to slab.

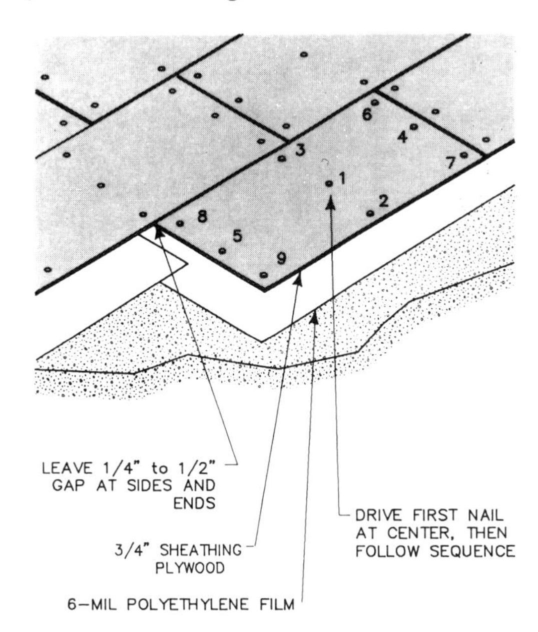

Plywood Subflooring on Concrete

As shown in Figure 18–18, a ¾″ span-rated sheathing plywood subfloor can be laid directly on a sound concrete slab and fastened with hardened masonry nails. Use nails with heads, not cut nails, driven with an engineer's hammer so the heads are flush or below the surface. Use a lapped polyethylene sheet vapor barrier under the plywood. Space the panels and use the nailing sequence as shown in the illustration.

Any flooring except carpet and resilient flooring can be laid on the plywood subflooring; for those two floorings, underlayment must be put down first.

19 Repairing Wood Floors

Floors laid on subflooring and joists are almost certain to require repair at some point in their lives. Floors creak and squeak, and floorboards come loose, warp, split. These, however, are minor problems, easily made right. Floors that sag or slope, or seem springy underfoot, can be indicators of more serious structural problems that require attention and often professional help.

GETTING RID OF SQUEAKS

Floors that squeak and stairs that creak when nobody is there are usually normal—all structures expand and contract with the changes in temperature and traffic from day to night. But the kitchen floor that squeaks in front of the refrigerator when you are doing a little raiding, or the step that announces your approach when you are trying not to wake anybody, can be annoying. These are problems you can fix.

A squeak is caused by one board rubbing against another board or against a nail. The squeak can be anywhere in the floor structure. The permanent cure is to stop the piece of wood from moving, which is sometimes easier said than done. A temporary cure can at times be made with a little lubrication so the wood will at least move quietly.

If the squeak is in a floor over an unfinished ceiling, go downstairs. Have someone walk on the spot where the squeak occurs. From below, pinpoint the location and chalk an *X* to mark it.

Sagging Joists

Worst problems first. Is it the joist? When your helper walks over the noisy spot, does the joist deflect, or is the noisy spot spread over a large area? The joist may be weak and sag under the live load. Unless your problem is rot or termites, the cure for a sagging or flexing joist is to install a support post, either a wood 4×4 or a concrete-filled steel post commonly called a Lally column. Use an adjustable jack post to raise the joist or joists to a firm position, as shown in Figure 19–1. Then install a permanent post alongside. Fasten the top of the post to the joist or floor structure and the bottom to a footing that will spread the weight before removing the jack post.

Some building codes allow a jack post to be used as a permanent repair post (but not a pri-

Figure 19–1
Using a jack post to raise joists to install a permanent support post.

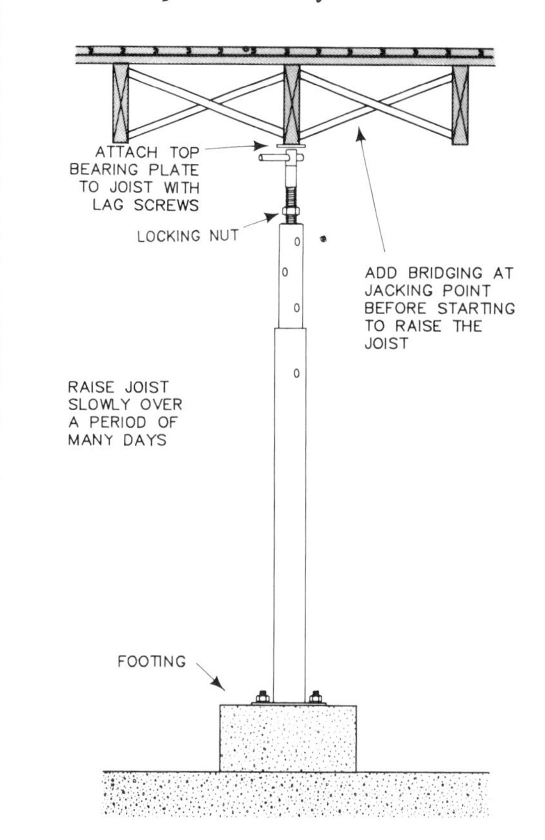
ATTACH TOP
BEARING PLATE
TO JOIST WITH
LAG SCREWS
LOCKING NUT
ADD BRIDGING AT
JACKING POINT
BEFORE STARTING
TO RAISE THE
JOIST
RAISE JOIST
SLOWLY OVER
A PERIOD OF
MANY DAYS
FOOTING

Figure 19–2
A jack post with a locking nut can sometimes be left in place. Screw the nut down against the post top to lock adjustment; fasten the top and bottom plates securely.

mary structural post) if a locking nut is included on the threaded shank. The nut must be run a short way up the shank to allow the jack to be adjusted (see Figure 19–2). Once the required height has been reached—usually after raising the screw a half-turn per day over a period of time—the nut is screwed firmly down against the top of the upper jack section to lock the adjustment in place. The top bearing plate must be lag-screwed to the overhead structure and the bottom plate fastened to an appropriate footing.

If the joist is rotten or termite-damaged, it must be replaced. The method is explained later in this chapter.

Subfloor Problems

If a floor squeak is directly over a solid joist, there could be a gap between the joist and subflooring, or between the subflooring and the underlayment or the finish flooring. If the floor upstairs does not bulge, try hammering a thin wood wedge between the subflooring and the joist. A piece of cedar shingle makes a good wedge. Put one in from each side if the gap is big enough (Figure 19–3, top).

If there is a long gap, as there might be if the joist is badly warped, clamp a piece of 2×4 or 2×6 alongside the joist tight up against the floor, then fasten it to the joist with lag screws (Figure 19–3, bottom). When this brace is in place and secure, nail the flooring to it from the top.

If the squeak is located between two joists, nail a piece of 2×6 between the adjacent joists, directly under the squeaking spot. Now insert wedges or nail down the flooring, as appropriate.

Figure 19–3
Correcting a floor squeak caused by a single warped joist.

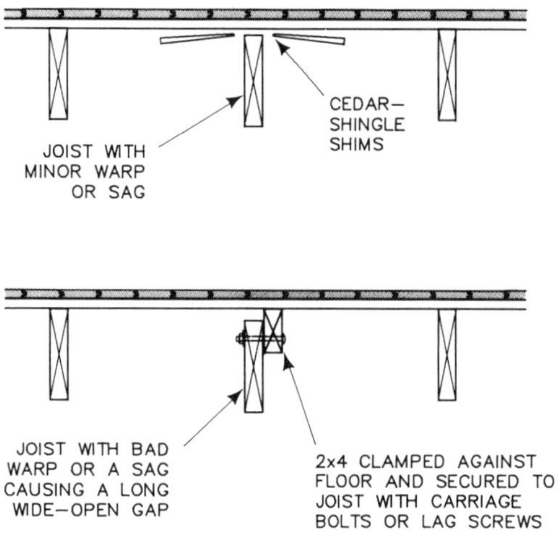

JOIST WITH MINOR WARP OR SAG

CEDAR-SHINGLE SHIMS

JOIST WITH BAD WARP OR A SAG CAUSING A LONG WIDE-OPEN GAP

2x4 CLAMPED AGAINST FLOOR AND SECURED TO JOIST WITH CARRIAGE BOLTS OR LAG SCREWS

Figure 19–4
Securing looser square-edge and tongue-and-groove floorboards with screws.

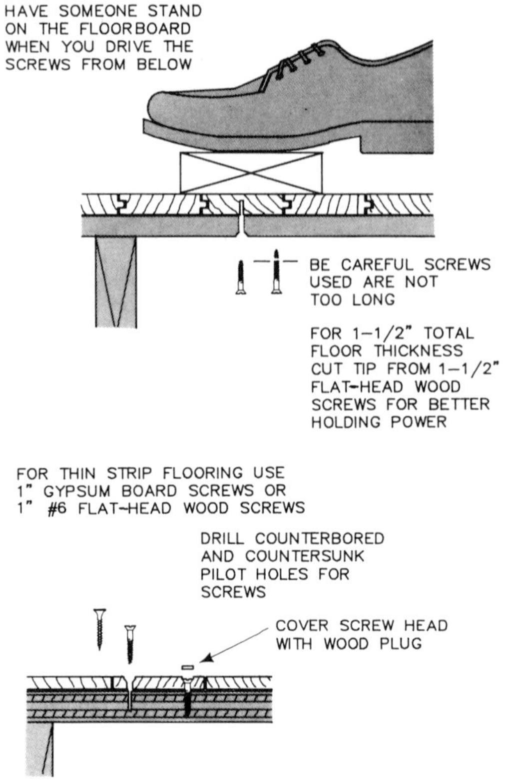

HAVE SOMEONE STAND ON THE FLOORBOARD WHEN YOU DRIVE THE SCREWS FROM BELOW

BE CAREFUL SCREWS USED ARE NOT TOO LONG

FOR 1-1/2" TOTAL FLOOR THICKNESS CUT TIP FROM 1-1/2" FLAT-HEAD WOOD SCREWS FOR BETTER HOLDING POWER

FOR THIN STRIP FLOORING USE 1" GYPSUM BOARD SCREWS OR 1" #6 FLAT-HEAD WOOD SCREWS

DRILL COUNTERBORED AND COUNTERSUNK PILOT HOLES FOR SCREWS

COVER SCREW HEAD WITH WOOD PLUG

Finish Floor Problems

If there is an upward bulge in strip or plank flooring, it usually means that the finish floor is loose from the subflooring. The best cure is to screw them together from below, but only if the finish flooring is at least 25⁄32" thick. Measure the thickness carefully—you don't want your wood screws going through and coming out the top. Drill pilot holes for the screws; enlarge the holes in the subflooring for a slip fit. Now, with a helper standing on the spot to press the finish flooring down, run in the screws from below (Figure 19–4). For thin finish flooring, screw from the face as shown in the illustration.

Working from the top provides only two options. You can face-screw the finish flooring, or you can drive new nails through the squeaky spot to lock the finish flooring and subflooring together.

For any nailing through flooring use spiral flooring nails if at all possible. Drill pilot holes, about half the nail shank diameter to avoid splitting. With spiral nails, angling the holes doesn't add much hold-down power. In order for the nails to pull the floor down tight, the holes must be straight, as shown in Figure 19–5. Drive the nails flush, then set them below the surface and fill the

Figure 19–5
Nailing loose tongue-and-groove floorboards.

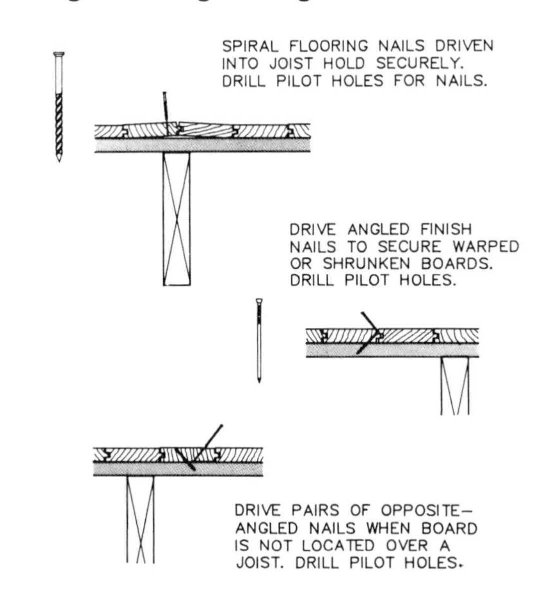
SPIRAL FLOORING NAILS DRIVEN INTO JOIST HOLD SECURELY. DRILL PILOT HOLES FOR NAILS.
DRIVE ANGLED FINISH NAILS TO SECURE WARPED OR SHRUNKEN BOARDS. DRILL PILOT HOLES.
DRIVE PAIRS OF OPPOSITE-ANGLED NAILS WHEN BOARD IS NOT LOCATED OVER A JOIST. DRILL PILOT HOLES.

holes. If you do use finish nails, they must be driven at an angle to have any holding power, as shown in the illustration.

Don't waste time trying to reset nails that have pulled loose in hopes that they will stay tight this time. If they pulled up once, they'll do it again. Remove them, drive new nails or screws nearby, and fill the old holes.

Is your squeak over a joist that you can face-nail into? Try tapping on the floor—it will sound solid over joists, hollow between them. Nails leave much smaller holes than screws, but they don't always work midway between joists, because the flooring may be too springy. Nails will be most secure when there's a joist directly underneath to back up the nailing. However, unless the subflooring has sprung loose from the joist, there is no particular need for the nails to go into the joist. Try small nails first. If nails don't work, use gypsum board screws in counterbored holes. After you drill the pilot hole, enlarge the hole in the finish floor for a loose fit. Plug holes.

Squeaks under Resilient Flooring

Resilient tile and sheet vinyl floors, whether over a conventional flooring system or over plywood, must be opened up if you can't get at and cure the squeak from below. Unless the flooring is cushioned, try to fix the squeak with small nails if you can localize it; screw heads are hard to conceal in tile or sheet materials.

If nails won't do the job, remove the tile. Use a hair dryer or heat gun to soften the tile and the adhesive, then gently pry the tile up with a wide-blade putty knife. After it's out, press the tile flat before you go to work on the squeak. Screw or nail the floor as described previously to correct the squeak.

Vinyl sheet or seamless floor covering presents more of a problem. To get at the subfloor, make an L-shaped cut in the floor covering and apply heat to make it flexible and to soften the adhesive. Gently pry up a triangular area, being careful not to crease the flooring by allowing it to bend sharply. When you have repaired the squeak, reheat and reglue the flooring.

REPAIRING AND REPLACING FLOORBOARDS

Plank Flooring Repairs

Square-edge planks are usually fastened to the joists with face nails or screws, or both. If a screw has pulled loose, leaving an enlarged hole in the joist with no visible support for another screw of the same size and length, you have several repair choices. Substitute a longer screw of the same diameter, a long gypsum board screw, or two gypsum board screws angled into the joist and covered by the same plug (see Figure 19–6). Or plug the hole with wood and drill a new pilot hole for the screw. The last method provides the greatest holding power. Using a larger-diameter screw in an old hole is not a good choice unless you are able to cover the head.

If you can get at the underside of the floor, it is better to pull loose floorboards tight to the subfloor from below. You can spread the load over more screws and none of them will show. Panhead tapping screws are a good choice for this job. The threads hold better than regular wood

Figure 19–6
Securing loose plank flooring with screws in counterbored holes.

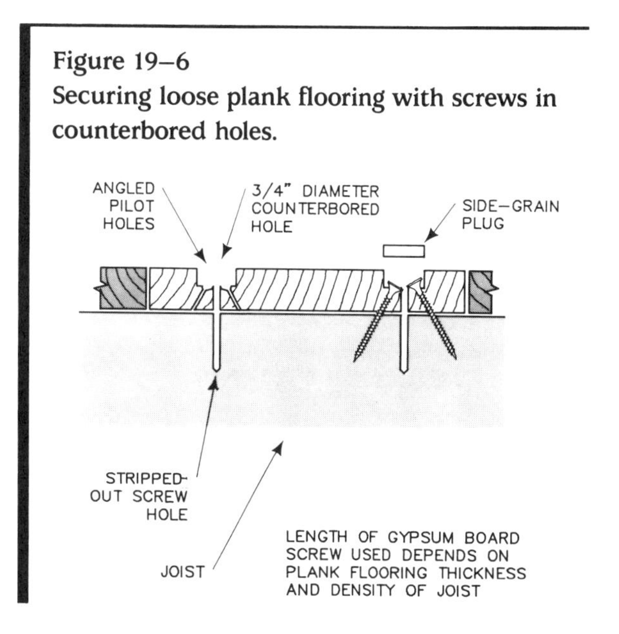

screws. Select screws the same length as the combined thickness of the finish and subflooring; use a washer or two under the head. After drilling pilot holes (the hole in the subflooring should be a clearance hole), start the screw, then remove it and snip off the tip. This way you get maximum thread engagement in the flooring.

Repair holes in the face of plank flooring with wood plugs cut to fit and glued in. Small holes and cracks can be filled with a latex patching compound.

Splinters

Glue splinters down with epoxy cement. Clean out dirt under and around old splinters. Plan to pin the repair with countersunk small finishing nails or, if it is at the edge of the board, use small wedges to hold the splinter while the cement cures.

Cracks between old flooring planks should be left alone if they come and go seasonally. If the planks are laid on joists without subflooring and you can get at the underside, tack wood strips to the undersurface to cover the cracks and block the draft. You could also fill the cracks with a felt weatherstripping material.

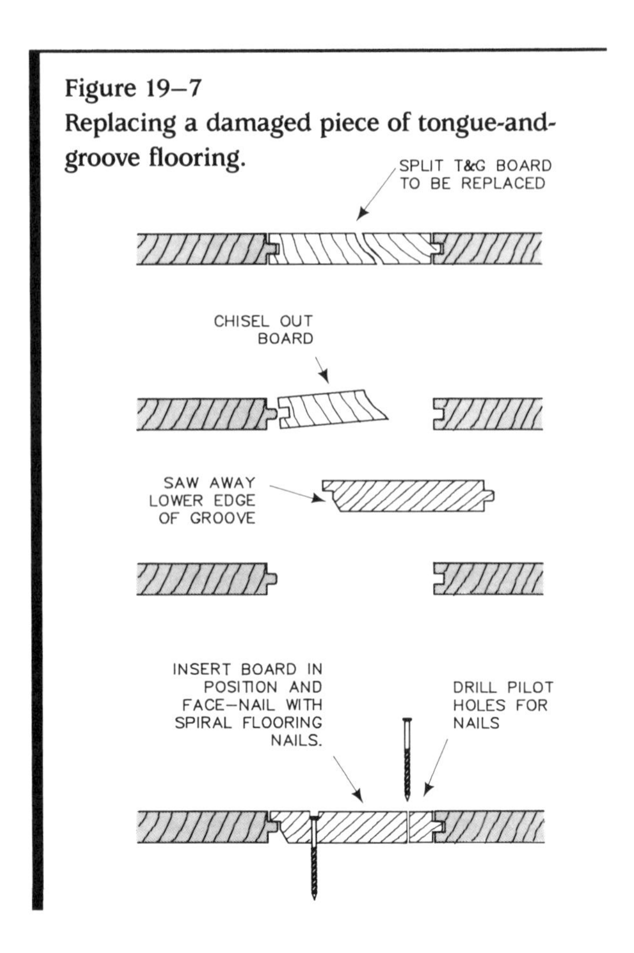
SPLIT T&G BOARD TO BE REPLACED
CHISEL OUT BOARD
SAW AWAY LOWER EDGE OF GROOVE
INSERT BOARD IN POSITION AND FACE-NAIL WITH SPIRAL FLOORING NAILS.
DRILL PILOT HOLES FOR NAILS

Figure 19–7
Replacing a damaged piece of tongue-and-groove flooring.

Replacing Tongue-and-Groove Flooring

You must destroy a damaged tongue-and-groove plank or strip to get it out. Removal and replacement are shown in Figure 19–7. Drill a row of holes across the board about an inch from the ends and chisel out the ends in pieces. Depending on how the grain runs, you may now be able to split the board. If not, drill another row of holes farther along the board and chisel out another section.

You can also take out a part of a damaged board. Drill rows of holes to define the ends and remove the section in between. Use a router to dress the remaining board ends. You can chisel the end square, but a router is easier. Set up a box fence for the router and tack it to the floor. Finish the corners of the cut with a chisel.

The replacement floorboard must be modified by cutting off the lower part of the groove edge so the board can be slipped into place, as shown in the figure. Drill pilot holes and face-nail the board. If you are replacing several adjacent boards, only the last one has to be face-nailed; blind-nail the others at the tongue edge as explained in Chapter 18.

Replacing Square-Edge Plank or Strip Flooring

Square-edge flooring is easier to replace than tongue-and-groove. If the flooring is nailed in place, you will still have to split out the old board, but you may be able to pry it out. Be careful about damaging the edges of adjacent boards when you are prying.

If you cannot find a good matching replacement for the board, of the required thickness, use thinner stock shimmed up with a piece of wood or plywood. The two can be glued together, if not too wide, or glued with a bead down the center only.

PROBLEMS WITH OLD FLOORS

Big cracks in foundation walls, foundation or other walls out of plumb, doors and windows that stick because the frames are out of square—these are significant problems. Along with sagging or sloping floors, they may indicate a major foundation problem or, if it is winter, frost heave. With these conditions, you might have more than a problem with the floor structure. Get help from an architect or structural engineer.

However, there are many lesser things that could be wrong, causing a problem that you can correct.

Joist Repair

Often a joist has been weakened by being notched or drilled through for wiring or plumbing at some time in the past, and it now creates a problem because of a change in load on the floor. The effects of joist notching are shown in Figure 19–8. A common load change results from the fixtures in an added bathroom—a home improvement often made without considering what load the joists were designed to bear. (A tub full of water may weigh half a ton.)

A joist that is sagging because it is notched, drilled, or splintered must be replaced, have the load shifted to another parallel joist, or be supported under the sag by a post. It cannot be repaired by strapping steel mending plates across the break or even nailing 2×4s to the sides of the joist. Figure 19–9 shows various ways to deal with damaged joists.

Figure 19–8
Notching reduces the depth of a joist, leaving less material to carry load stresses. Off-centerline holes will also weaken a joist significantly.

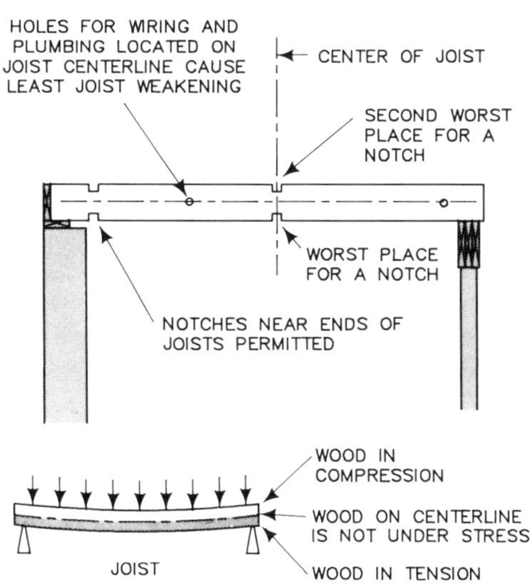

Bridging

It is also possible that the floor structure in an old house was never adequate for the original load. They didn't always do things better in the old days.

A floor that is too springy can sometimes be corrected by adding bridging between the joists, particularly if the joists are deep and long (2×10s or 2×12s spanning 16′ to 20′). The primary function of bridging is to keep joists from twisting or canting as they age and dry out over the years. Bridging may also help to stiffen a floor by transmitting loads on one joist to the adjacent joists for sharing. Figure 19–10 shows three kinds of bridging. Solid bridging can require a lot of work to cut and fit, because of the usual variations in joist spacing, twist, and cup. Putting in diagonal wood bridging is very difficult after the subfloor has been laid. Diagonal steel bridging is much faster and easier to install under an existing floor.

Joist Span and Girders

There might be nothing wrong with the joists except that they have too long a span. You may need a girder, a support beam that runs at right angles under the joists. You might want a competent contractor to put in a long girder. If you want to do it yourself, consult an architect to determine the proper size, location, and number and spacing of the support posts.

If a structural sag is localized, the repair can also be localized. If only a few joists are involved, support may be needed under only them; this

Figure 19–9
Repairing and replacing damaged joists.

CRACKED OR SAGGING
JOIST, ENDS SOUND:
JACK JOIST TO LEVEL
POSITION, LAG SISTER
JOIST TO ONE OR BOTH
SIDES, DEPENDING ON
JOIST LOADING
SISTER
JOIST
TEMPORARY
SUPPORT
STUB SISTER
JOIST
JOIST SOUND
EXCEPT FOR END:
SUPPORT JOIST END WITH
TEMPORARY POST: WORK
STUB SISTER JOIST ONTO
SILL OR GIRDER SUPPORT
LAG TO JOIST
JOIST ROTTED, CRACKED,
UNSOUND END TO END:
2X LEDGER LAGGED
OR NAILED TO GIRDER
ADJUSTABLE STEEL POST.
WOOD POST COULD ALSO
BE USED. A JOIST UNDER
A PARTITION SHOULD HAVE
A FOOTING UNDER THE POST.
INSTALL SISTER JOIST.
RAISE FLOOR TO LEVEL
BY JACKING UP ADJACENT
JOISTS.
IF JOIST END SUPPORTS
ARE NOT ACCESSIBLE,
PROVIDE INDEPENDENT
SUPPORT AS SHOWN.

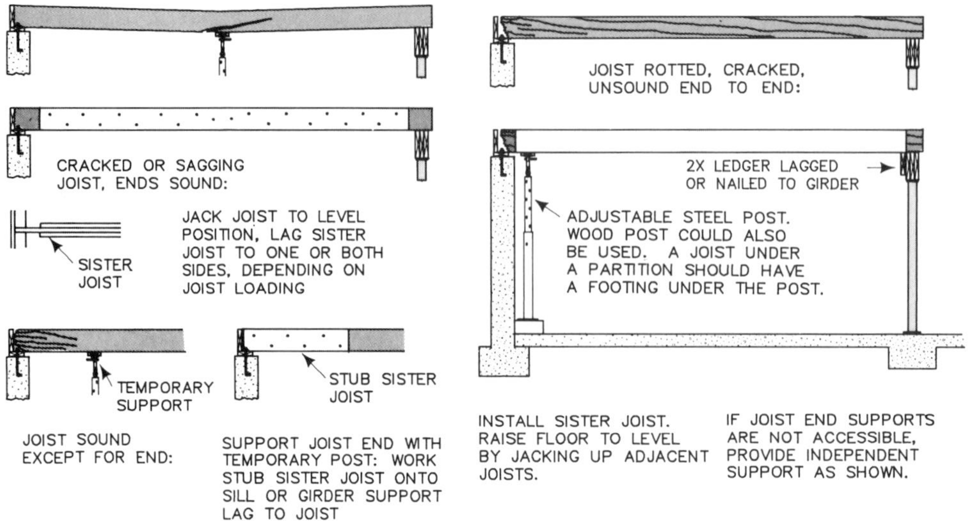

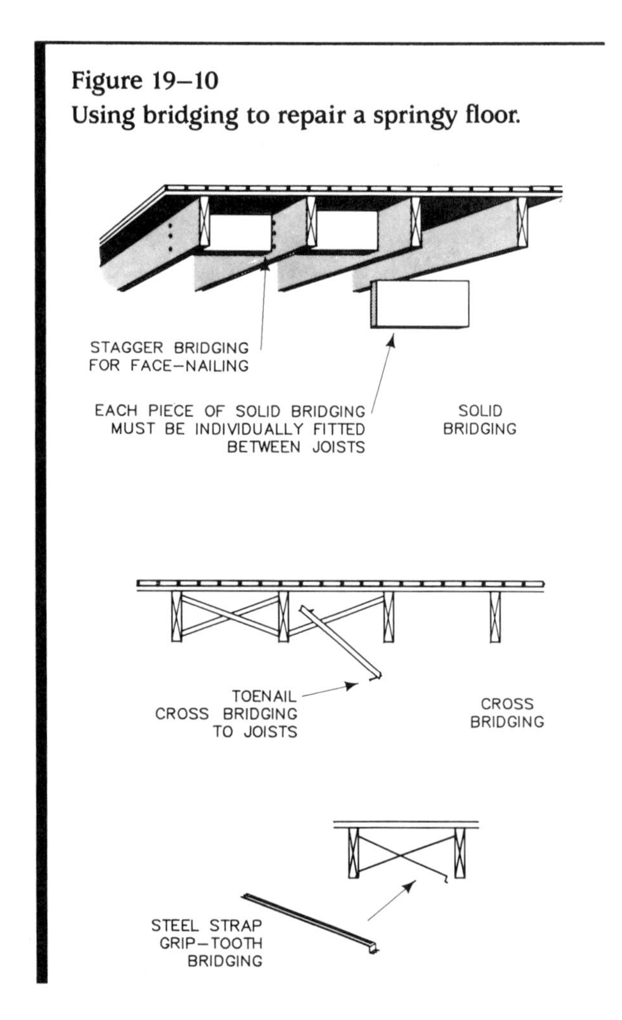

STAGGER BRIDGING
FOR FACE-NAILING
EACH PIECE OF SOLID BRIDGING
MUST BE INDIVIDUALLY FITTED
BETWEEN JOISTS
SOLID
BRIDGING
TOENAIL
CROSS BRIDGING
TO JOISTS
CROSS
BRIDGING
STEEL STRAP
GRIP-TOOTH
BRIDGING

Figure 19–10
Using bridging to repair a springy floor.

can be obtained with a short beam and adjustable posts, as shown in Figure 19–11.

If an existing wood girder has shrunk or twisted, nothing more may be needed than shims between the girder and the joists (see Figure 19–12). If you shim a post under a girder, be sure you maintain adequate attachment from the post through the shim into the girder. Don't depend on the girder to just rest there.

If you install a post for support on an upper floor, it cannot just rest on the flooring or on the floor joists below. The load must be transferred to the ground by a post on the floor below directly under it and another in the basement. The basement post is not needed if the load from above can be transferred to the main girder; however, it is often not possible to position the upper posts to accomplish this.

Other Causes of Sagging

A sag in the middle of the floor, away from posts or walls, is most likely a split joist or girder. The best cure, short of replacement, is to install a post under the split (the method is shown in Figure 19–13). After jacking up the damaged piece, attach sister beams on each side with lag screws.

Figure 19–11
Post-and-beam support under sagging joists. Tighten locking nuts and fasten top and bottom plates after correcting the sag.

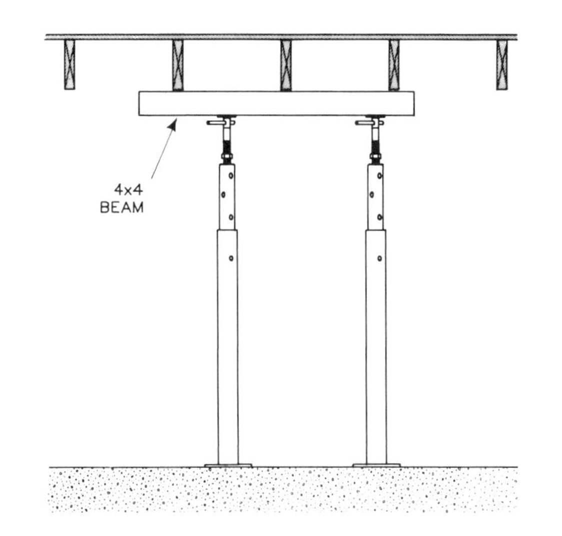

Figure 19–12
Shimming a warped girder with wedges on a post.

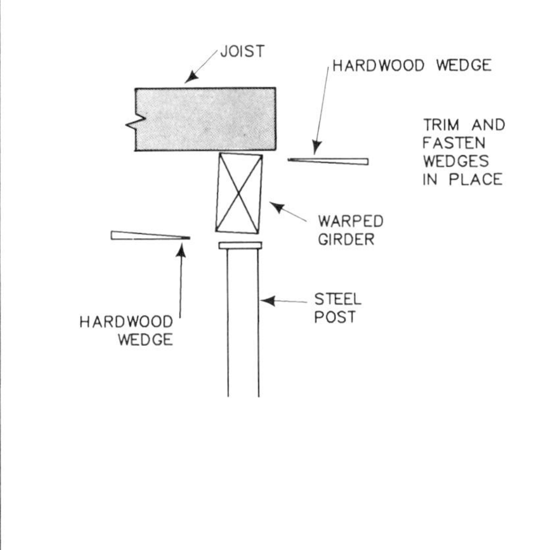

Figure 19–13
Installing a new post support for a girder. Place the jack posts equal distances on either side of a crack.

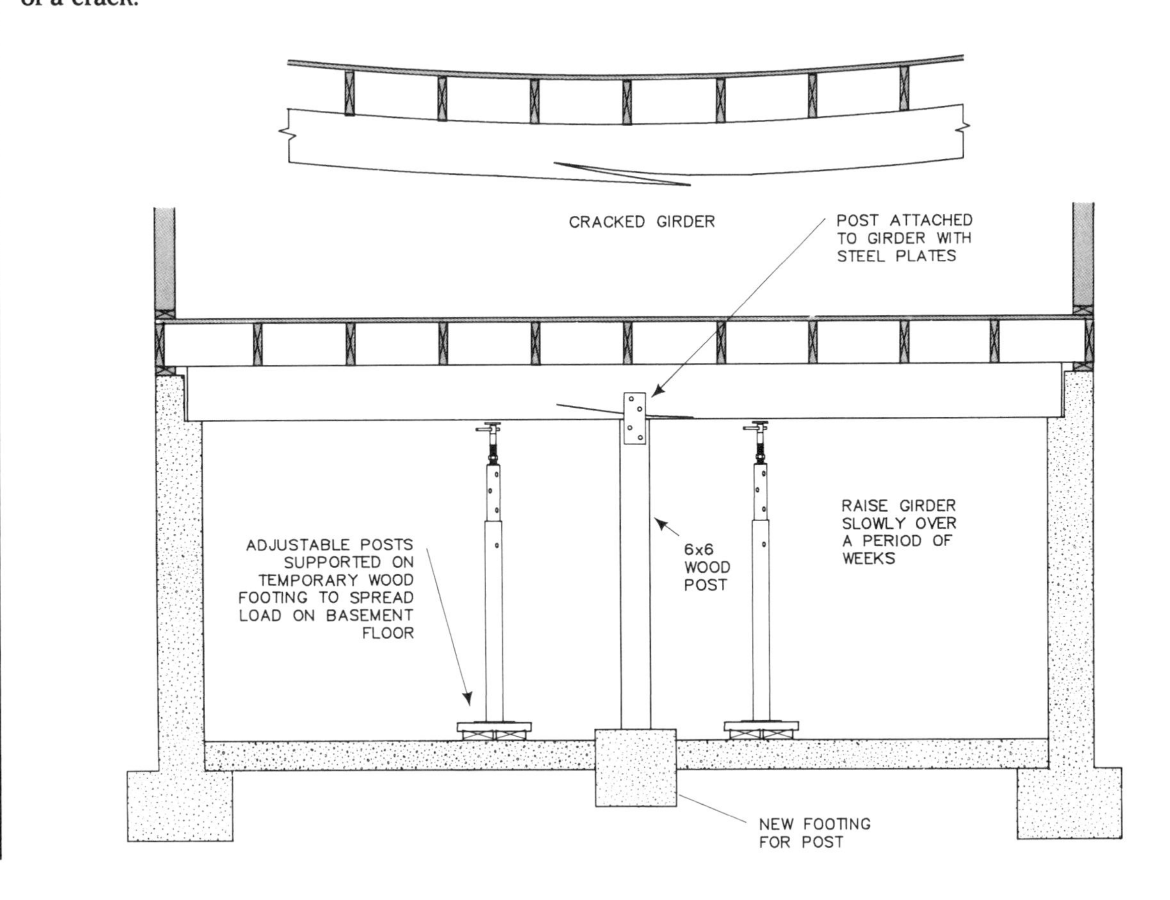

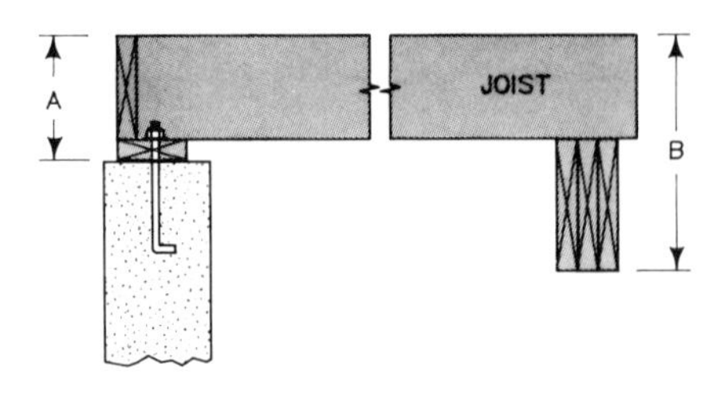

Figure 19–14
Shrinkage of support. Because there is more wood depth at B than at A, shrinkage will be greater and sagging more likely to occur.

When jacking up, do it slowly over days or weeks. Just take a bit of the load on the jack posts, then wait while the house parts readjust themselves in the next 24 hours. A half-turn a day on each post is a good rate to avoid cracking or splitting of other structural members on upper floors. Provide a footing under the new post and attach it to a girder as shown.

A sagging problem may be caused by differential shrinkage, because there is more wood in the structural stack than at another (see Figure 19–14). Another possible cause for sagging is rotten post or joist ends. In basement locations, check whether footing is crumbling under a post—or maybe there was no footing there in the first place. Also check upper stories, where rainwater may be getting in and doing damage.

20 Ceramic Tile

At one time ceramic tile meant little hexagonal tiles (white, perhaps with black) on bathroom or kitchen floors. Then bathroom walls were tiled: square tiles in a single, subdued color—a *plain* color. Today, ceramic tile can be used in virtually any room of the house and outdoors in an explosion of patterns, colors, and styles. Tile can be applied to almost any wall and floor construction, and over most finish wall and floor materials in remodeling. Today, tiles are used imaginatively, as seen in Figures 20–1 and 20–2.

KINDS OF CERAMIC TILE

Tiles come in thousands of shapes, sizes, and colors, and in matte, glossy, and textured surfaces. Tile is an excellent low-maintenance finish floor material for any room, especially entry areas, and of course it is an excellent wallcovering wherever food is prepared or water may splash.

Glazed tile. A coating applied to the tile before firing gives glazed tile its surface color and texture. Glazed tile comes in 1″ to 12″ squares in a wide variety of colors and patterns with either a gloss or matte finish. High-gloss tiles show scratches more than matte tiles and are slippery; thus they should be used primarily on walls. Textured glazed tiles are used on both walls and floors. Glazed tiles are made in many sizes and geometric shapes, as shown in Figure 20–3.

Mosaic tile. Ceramic mosaic tiles, shown in Figure 20–4, are small—1″×1″, 1″×2″, 2″×2″, and 2″ hexagonal. The color is dispersed through the tile. The tile is highly wear-resistant, and it is an excellent material for both interior and exterior walls and floors. Mosaic tile is sold in sheets of tiles mounted on mesh. There are solid colors and multicolored patterns, some combining tiles of different sizes. The whole sheet is set as a unit, greatly speeding tile setting (Figure 20–5).

Quarry tile. This unglazed natural clay or shale tile is available in warm earth shades ranging from beige to clay reds to dark brown. Quarry tile is thicker than other tiles (½″ to ¾″) and comes in 4″ to 9″ squares and geometric shapes (Figure 20–6). This tile is ideally suited for areas of heavy traffic as it is highly resistant to abrasion. The

Figure 20–1
Ceramic tile in the bathroom. The walls are faced with 6″ Suedetone Seacliff glazed tile; the floor is done with Suedetone Seacliff and Ivory Mist. (Photo courtesy of American Olean Tile Company)

Figure 20–2
Ceramic tile in kitchen. This kitchen floor is composed of three tile sizes—small square, large square, and rectangle—in three colors. (Photo courtesy American Olean Tile Company)

Figure 20–3
Glazed tile sizes and shapes. These are representative of a great number of variations.

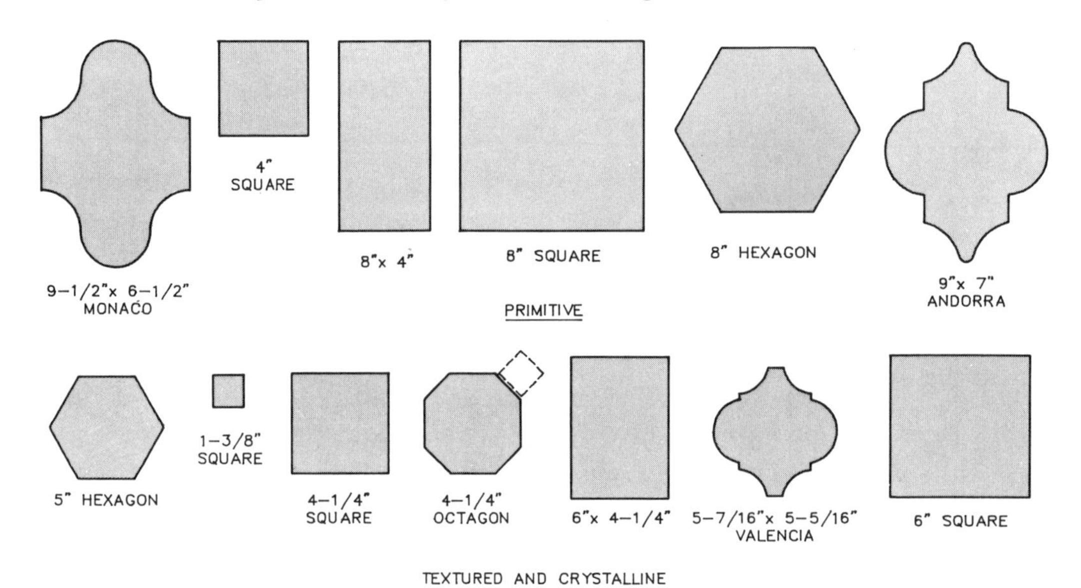

Figure 20–4
A mosaic tile kitchen floor. (Photo courtesy American Olean Tile Company)

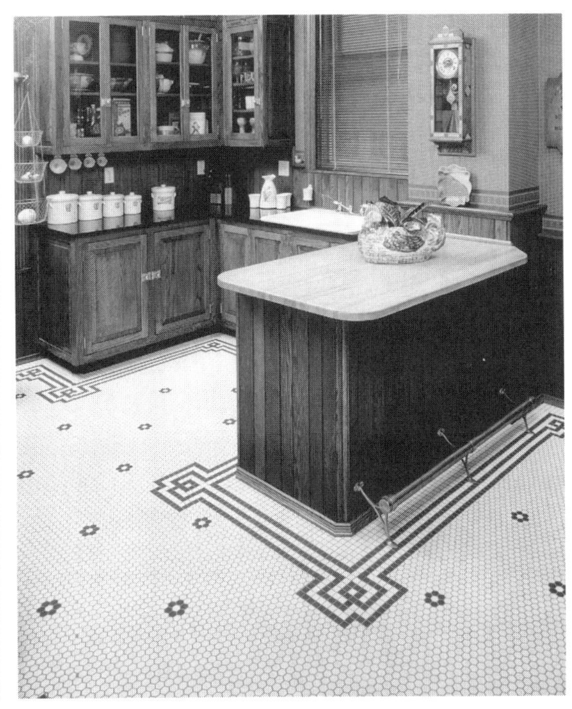

Figure 20–5
Mosaic tile sizes. Mosaic and glazed tiles are supplied mounted to a mesh to speed accurate setting.

1"
SQUARES
2"
1"X 2"
2"
HEXAGON

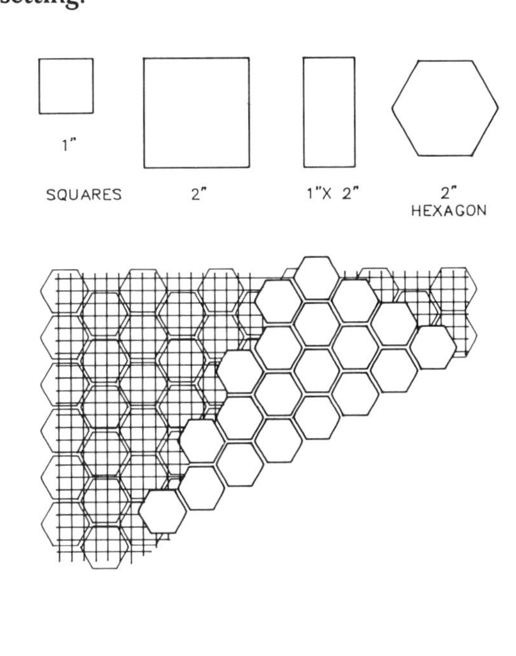

Figure 20–6
Quarry tile sizes and shapes.

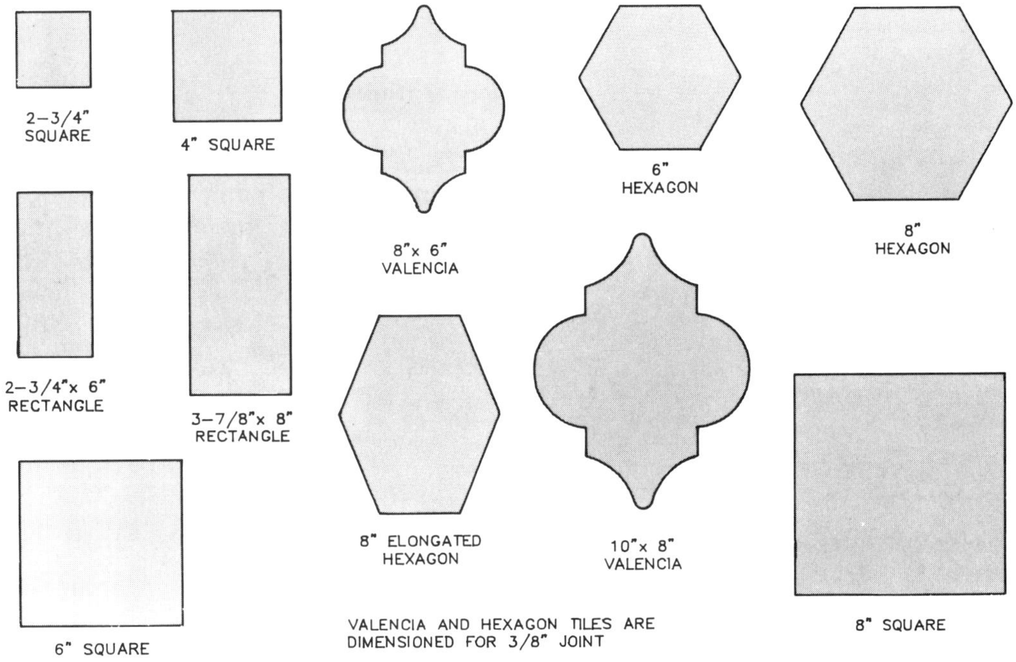

2-3/4"
SQUARE
4" SQUARE
8"x 6"
VALENCIA
6"
HEXAGON
8"
HEXAGON
2-3/4"x 6"
RECTANGLE
3-7/8"x 8"
RECTANGLE
8" ELONGATED
HEXAGON
10"x 8"
VALENCIA
6" SQUARE
VALENCIA AND HEXAGON TILES ARE
DIMENSIONED FOR 3/8" JOINT
8" SQUARE

surface of quarry tile is usually porous and normally needs a sealer for protection from staining.

Trim tiles. These are the special shapes used to finish off the edges of tiled areas, round corners, change surface planes neatly, form lips on counters, and otherwise make a tiling job look finished and professional. These shaped tiles have standard names, as shown in Figure 20–7. An important distinction in trim tiles is the difference in bullnose tile edges for thickset (mortar) and thinset (adhesive) application, illustrated in Figure 20–8.

When buying tiles for a project, be sure you can get matching trim tiles if you plan to use them. They are not available to match all tile lines, or may be available only in a limited number of shapes.

TILE-SETTING MATERIALS AND TOOLS

Laying ceramic tile is a two-step operation. First, a layer of mortar or adhesive is spread over the supporting subsurface; tiles are aligned and pressed into the wet surface. When the mortar or adhesive is dry, grout—a creamy mixture of filler material—is wiped on and forced into the spaces between the tiles, then cleaned from the surface of the tiles.

There are many kinds of mortars, adhesives, and grouts, each with advantages and disadvantages. The materials described below are all available premixed (some require water) and are suitable for countertops, walls, and floors.

Mortars and Adhesives

Modified epoxy emulsion mortar. This two-part mortar consists of emulsified epoxy resins, hardeners, portland cement, and silica sand. The mortar is used to set tile on walls and floors, indoors and outdoors. It has high bond strength, practically no shrinkage, and is easy to use. The mortar is used in a thin layer to set tile on plywood and concrete or as a bond coat. It is also used for grouting.

Organic adhesive. Packaged ready to use, organic adhesive is for interior use only. It is suitable for laying tile on walls, counters, and floors in both wet and dry areas.

Epoxy adhesive. This two-part resin and hardener adhesive is used for laying tile on floors, walls, and counters. It is designed primarily for high bond strength and easy application. The adhesive is applied by trowel in a thin layer, first covering with the flat edge, then going over with the notched edge for uniform thickness. Epoxy adhesive fumes are very irritating; this material is not like the epoxy glue that comes in two tubes.

GROUTING MATERIALS

The type of grout used with ceramic tile must match the tile and the application. Grouts are available in colors, and some neutral grouts can be tinted.

Dry-set grout. Also available as dry-set mortar, this packaged grout (to which you add water) is a mixture of portland cement, sand, and additives. It can be used for grouting all walls and floors.

Latex–portland cement. The latex emulsion additive helps curing, and the grout will absorb less water than regular cement grouts. The grout can be used with tile subject to ordinary use.

Epoxy grout. This two-part resin and hardener grout, with sand filler, is used for high bond strength and impact resistance. When used as both mortar and grout, the material gives added strength to the tile, especially on wood bases. Joint widths should be ¼″ or wider, with tiles no thicker than ½″ for full groove penetration. When still wet, the grout can be cleaned up with water; after drying, nothing will work.

Figure 20–7
Ceramic tile trim shapes and names.

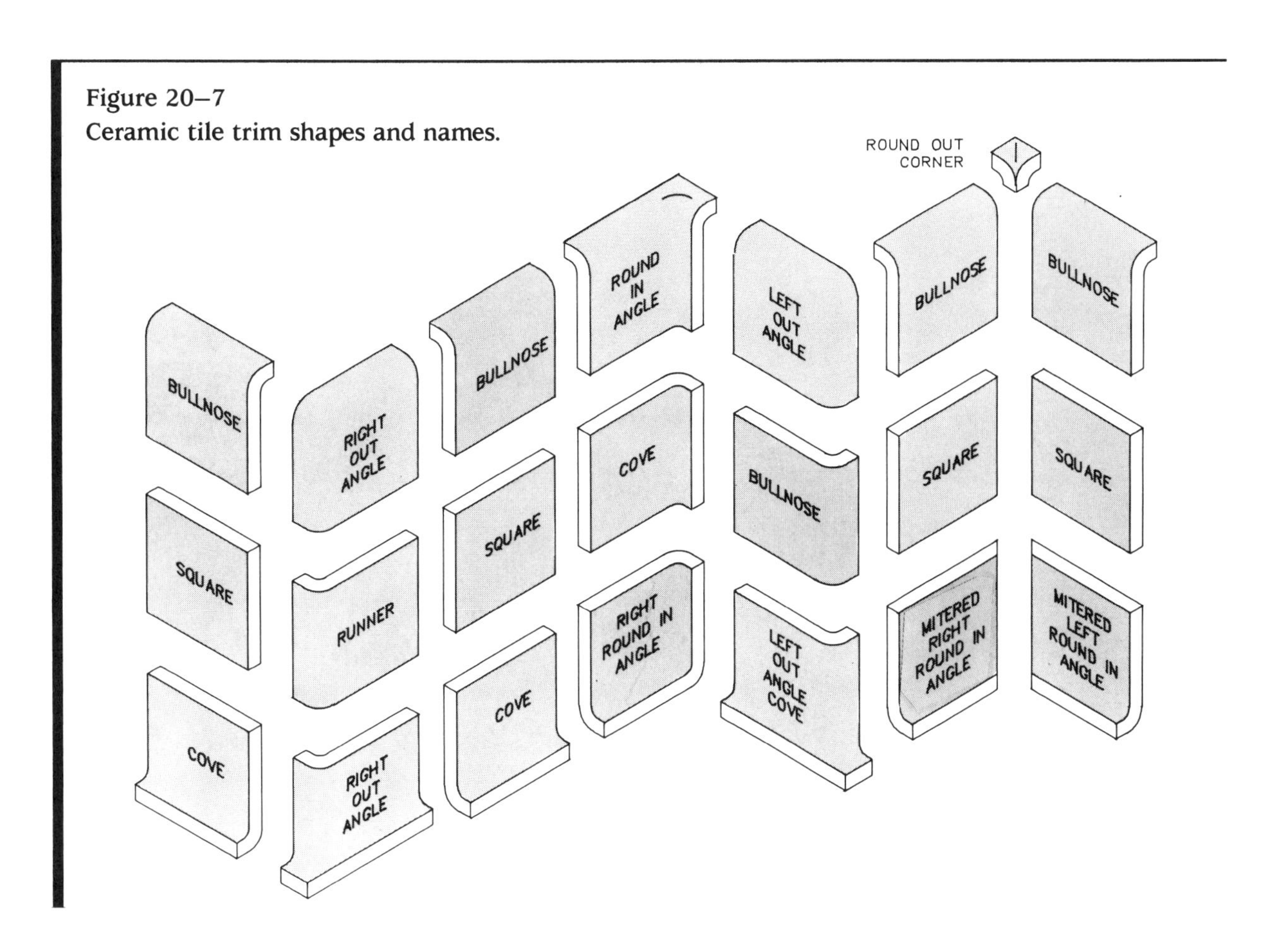

Figure 20–8
Thickset and thinset bullnose tiles.

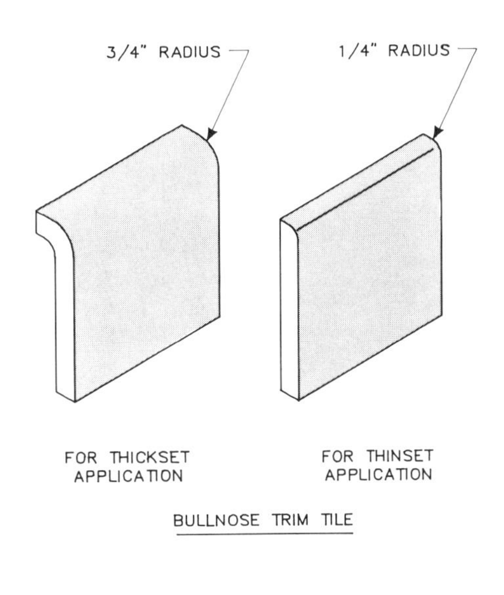

Tile-Cutting Tools

Tile cutter. This tool, shown in Figure 20–9, is used for straight cuts across the tile from edge to edge. Most tile retailers will rent or lend you a tile cutter. Cutters all work the same way: the tile is positioned on the bed of the cutter, and the handle is pulled along a guide bar so that a carbide wheel can score the surface of the tile. The handle is then depressed, causing "wings" at its base to press the halves of the tile, snapping it along the scored line.

Nippers. These are a form of sharp-jawed pliers used to cut into a tile in small bites, to fit around corners, pipes, electrical boxes, and the like. The secret of success is to take small bites, nibbling up to the final shape.

Carbide drills and saws. Small holes can be drilled in tile (before or after installation) with a

Figure 20–9
Tile cutter used for straight edge-to-edge cutting. First, score the tile by drawing the cutting wheel across the face of the tile from edge to edge. Then, press down on the handle to break the tile.

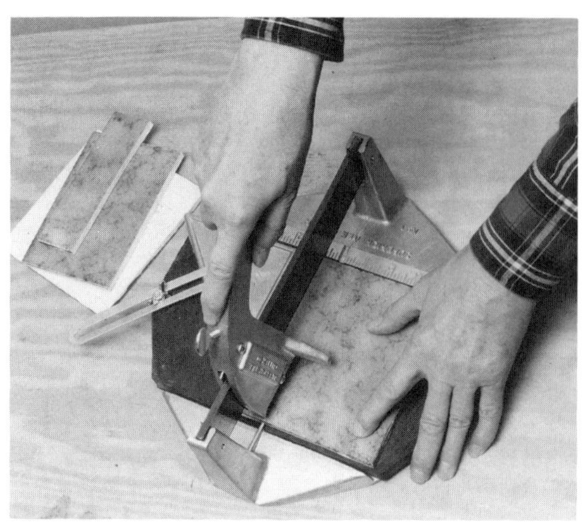

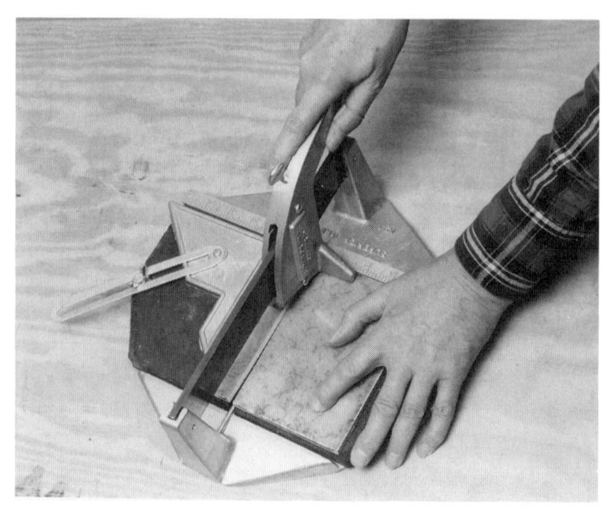

drill and a carbide masonry bit. Scratch a glazed tile at the center of the hole location to start the drill. Feed the bit slowly to avoid overheating the tile, which could cause it to crack. Neat large holes can be drilled with a carbide hole saw. As a great deal of heat stress will build up, this should be done with water cooling and a cordless drill *only*, to avoid electrical shock.

Other Tools

Beater board. A piece of ¾″ plywood somewhat larger than the tile you are laying is placed over positioned tile and hit with a rubber mallet to level the tile and force it into the adhesive. (Don't beat handmade tiles because they will break.)

Tile spacers and wedges. Some tiles are molded with spacing knobs on the edges that protrude and keep tiles a uniform distance apart when they are laid edge to edge. To get uniform spacing with other tiles, use spacers or wedges. Tile spacers, shown in Figure 20–10, are plastic pieces that can be inserted between tiles to maintain accurate tile placement and constant grout line width. Wedges are shims that hold wall tiles in position until the adhesive sets up. Both should be removed before grouting. They are made in a variety of sizes.

Figure 20–10
Spacers are available in various sizes. These separate tiles by ½″. Remove spacers before grouting.

GENERAL TILING PROCEDURE

The steps described here are typical for any countertop, dry floor, or dry wall in your home. Plan the job first on paper. Draw a plan and work out exactly how many full tiles, cut tiles, and trim tiles (if used) will be needed (Figure 20–11). Allow 10 percent more for breakage. Don't forget to include the width of the grout lines in the layout.

The room temperature should be between 65° and 80°F. If lower, setting will be too slow; if higher, too fast.

Preparing the Tiles

Adhesives set up quickly, so have all tools ready and tiles stacked at hand. If the job is going to have a row of cut tiles along an edge or adjoining wall (or walls), start by trimming all of them so you don't have to stop and cut each one as you go. Allow for grout line widths when you trim tiles. Use spacers to maintain an accurate grout line width.

Spreading Adhesive

Use a notched trowel for all setting materials; proper spreading technique is shown in Figures 20–12 and 20–13. First spread the adhesive with the flat edge. Hold the trowel at a 30° angle to force the adhesive into good contact with the bed. Spread only an area that you can cover with tiles in 15 minutes—the adhesive starts setting up in 30 minutes. Now comb the adhesive with the notched edge to form a pattern of ridges over the whole adhesive surface. V-notches should be used for setting small tiles and sheet-mounted tile. Square notches, which leave larger and higher ridges, are better for larger tiles.

When combing, hold the trowel at an angle of 45° or steeper, depending on how high you want the ridges. Keep the ridge height constant.

Testing the Setting Material

Inspect the adhesive. If the ridges have sagged, the mix is too wet. Press a tile in the adhesive, then remove it. The adhesive consistency and the combed thickness are right if the entire back of

Figure 20–11
Tile layout plan for a kitchen floor.

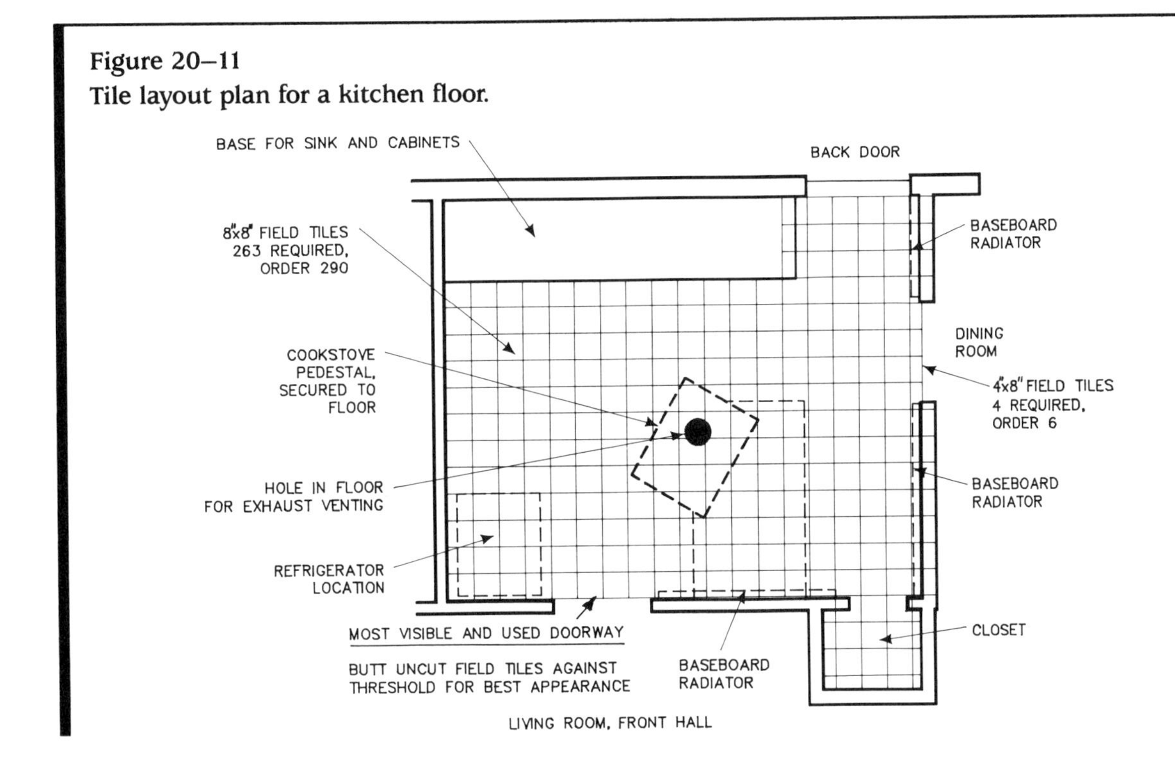

the tile is covered, with half the adhesive on the tile and half on the substrate. If the mix is too dry, adhesive will adhere to the tile only in spots, if at all. If only the tops of the adhesive ridges stick to the tile, the consistency is right, but it was combed too thin. Spread more adhesive and comb with the trowel at a steeper angle, or use a trowel with larger notches. Adhesive on the sides of the tile means the layer is too thick. Nose tiles and warped handmade tiles need extra adhesive buttered directly on the back of the tile.

As you lay the tiles, assure good adhesion by

Figure 20–12
Trowel notches and correct working angles for spreading and combing.

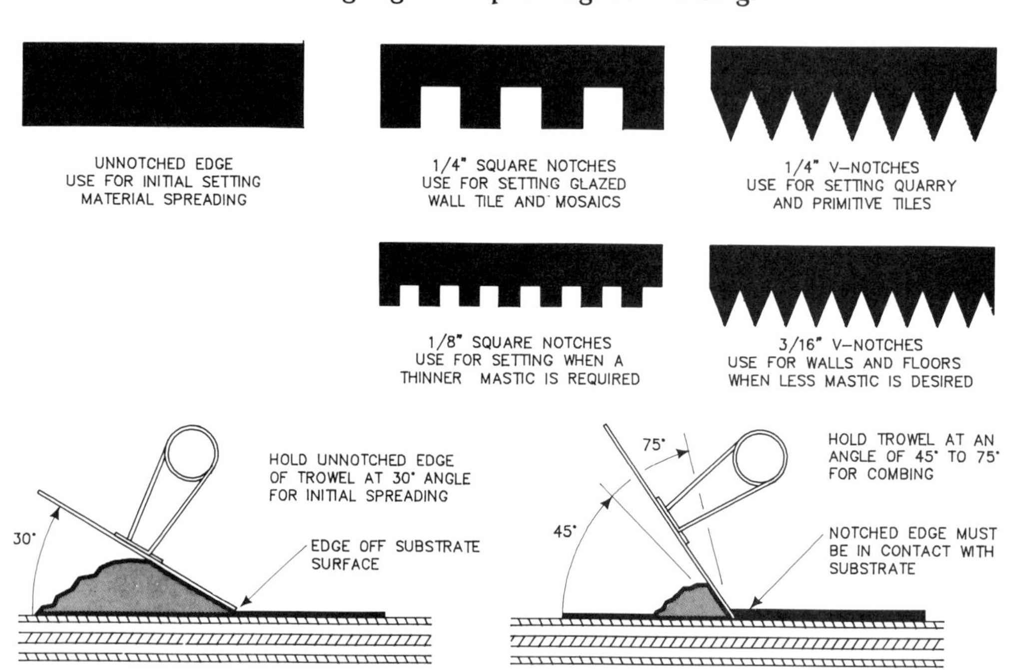
UNNOTCHED EDGE
USE FOR INITIAL SETTING MATERIAL SPREADING
1/4" SQUARE NOTCHES
USE FOR SETTING GLAZED WALL TILE AND MOSAICS
1/4" V–NOTCHES
USE FOR SETTING QUARRY AND PRIMITIVE TILES
1/8" SQUARE NOTCHES
USE FOR SETTING WHEN A THINNER MASTIC IS REQUIRED
3/16" V–NOTCHES
USE FOR WALLS AND FLOORS WHEN LESS MASTIC IS DESIRED
HOLD UNNOTCHED EDGE OF TROWEL AT 30° ANGLE FOR INITIAL SPREADING
30°
EDGE OFF SUBSTRATE SURFACE
75°
45°
HOLD TROWEL AT AN ANGLE OF 45° TO 75° FOR COMBING
NOTCHED EDGE MUST BE IN CONTACT WITH SUBSTRATE

Figure 20–13
Apply thinset mastic or adhesive in two steps. First, spread the mastic with an unnotched trowel edge. Then comb the mastic with a notched edge.

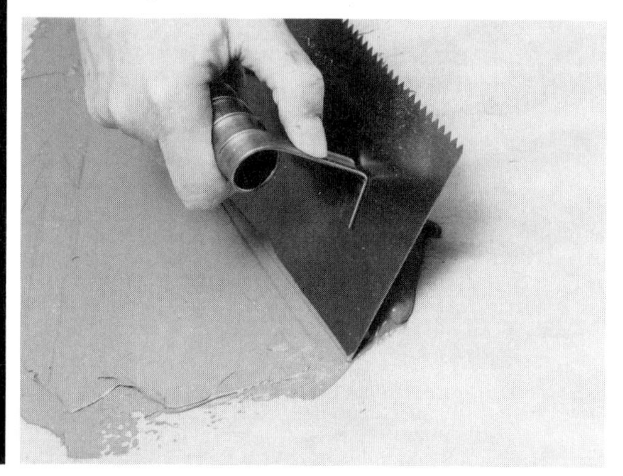

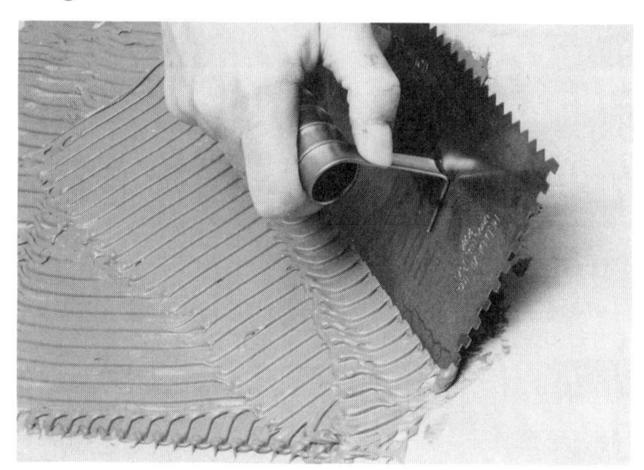

laying a beater board over them and tapping it with a rubber mallet. Clean off excess setting material while it's still plastic as you go, using a utility knife. Clean tile faces with steel wool, then vacuum.

Let the set tiles cure overnight before grouting. If you stop tiling before the job is done, scrape up and discard all adhesive not under tiles.

Grouting

The final step is filling the grout lines, or spaces between tiles, with grout. Before grouting, remove the plastic spacers if they have been used. Grout about 6 square feet at a time. Apply the grout with a rubber-faced grout trowel (Figure 20–14). Hold the float at a 30° angle and spread the grout diagonally across the face of the tiles. Work over the whole area three or four times, pressing the grout into the grooves. Pack it in firmly so there are no voids in the filling.

When the grout lines are packed full, scrape any excess grout from the surface of the tiles with the edge of the grout trowel—hold the trowel almost straight up. Then grout another area. Keep going, but check how the grout is drying and watch the clock.

Figure 20–14
Forcing grout into tile spaces with a rubber-faced trowel. Remove spacers or wedges from between the tiles before grouting.

The grout will start to set in 5 to 30 minutes. Grout on the tile surface will harden quickly and must be sponged off before it does. Use a damp sponge, as dry as you can wring it. When wiping the surface, be careful not to raise grout out of the grooves, where it will be soft longer (if it comes out, you are wiping the surface too soon). If you have to scrub really hard, you waited too long. If the grout has dried on the surface, use a Scotch-Brite pad, then sponge. Use an old bath towel or similar cloth to wipe off the hazy film left by the sponge, after the grout dries.

TILING OVER A WOOD SUBFLOOR

For a kitchen, remove any base cabinets and baseboard. In a bathroom, remove footed tubs, toilets, and basin pedestals. Take up the old finish flooring, then remove the plywood underlayment, baring the subfloor. Vacuum the floor thoroughly. Nail new ¼"-thick plywood underlayment with joints staggered from plywood subfloor joints. If the subfloor is boards instead of plywood, center underlayment joints on boards. See Chapters 2 and 18 for nailing information.

When done, run a straightedge all across the surface to discover any protruding nail heads; flatten them on the surface. Unless you want to tile under the base cabinets, reinstall them now.

The various layers of the floor are shown in Figure 20–15. Snap layout lines to locate the centers of the grout lines separating partial from full tiles along two adjoining walls. Be sure the lines are perpendicular. Apply tile as previously described.

TILING A KITCHEN COUNTERTOP

The instructions that follow apply to installing tile on a newly constructed counter or an existing counter (with the old top removed and new countertop underlayment installed). The specific example illustrated is a peninsula attached to one wall and accessible from three sides (Figure 20–16). An electric rangetop/grill was mounted in an angled pedestal supporting the end of the counter. The pedestal and countertop were constructed in the same manner as a base cabinet, except that the pedestal was finished on all four sides. The counter was to be covered and the pedestal trimmed with tiles as shown.

The pedestal was sized to match a particular Jenn-Air rangetop unit; the countertop was sized to use full tiles plus edge trim across its width. Because of a tight clearance to the oven door (beyond the end of the countertop), it was necessary to finish the countertop layout with a partial tile at the wall.

Figure 20–15
Tile applied over a wood subfloor with an organic adhesive.

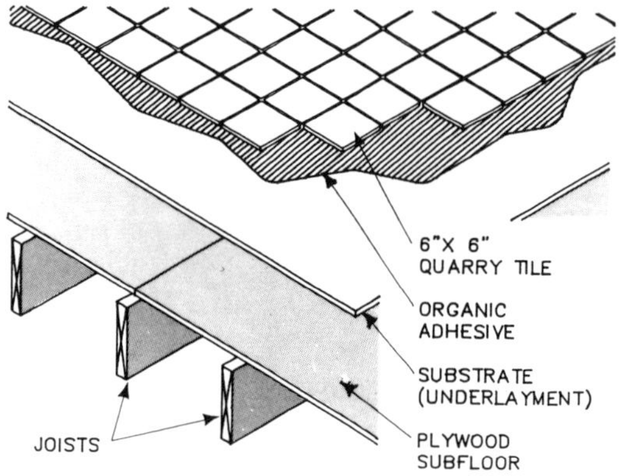

Installing Countertop Tile

If you are tiling a new counter, proceed to the layout step. If you are renovating an old counter, first remove the sink, outlet plates, and everything else in the way. Strip the wall to the gypsum board or plaster. Replace the old countertop underlayment with new and cut the sink opening.

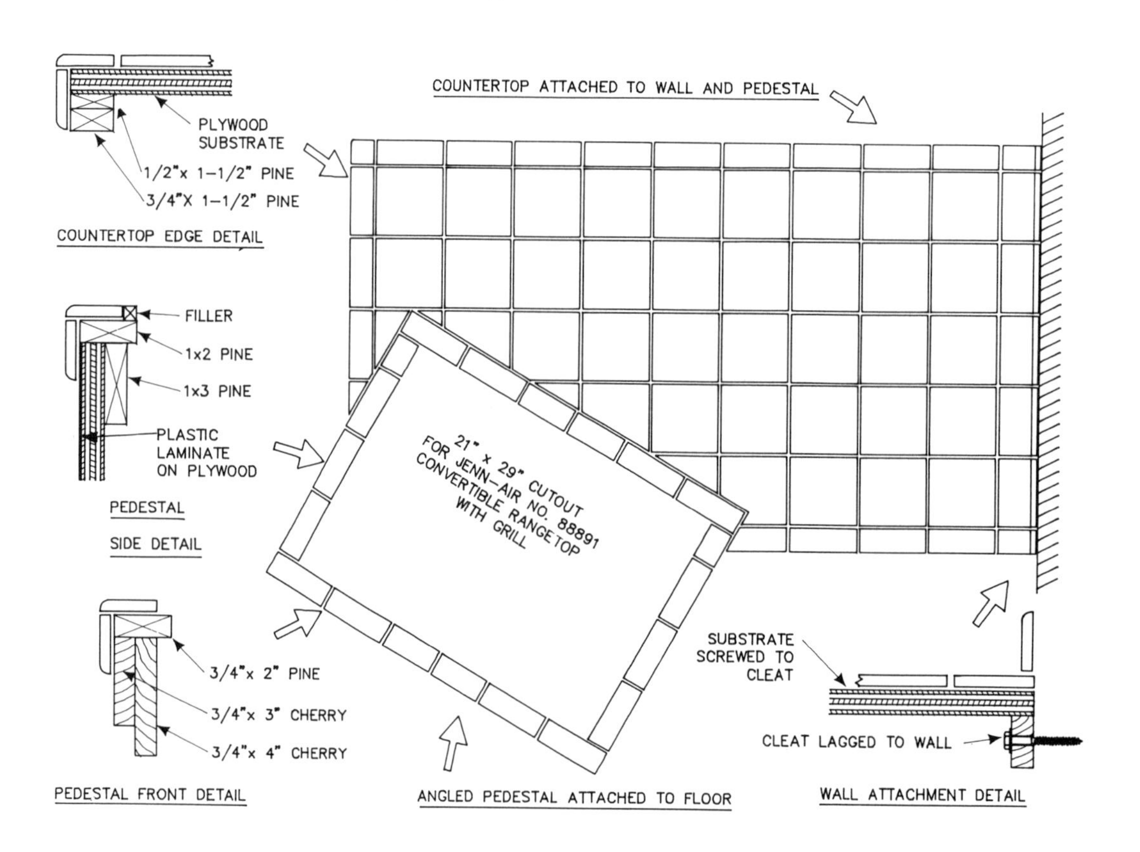

Figure 20–16
Countertop and angled pedestal: layout, edge, and attachment details.

Lay out the tiles dry (without adhesive) on the countertop and mark the center of the grout lines on the underlayment (Figure 20–17). Next, cut all partial-width tiles to the required size. If you are going to use wood bullnose edge trim, install it. If you plan to use bullnose tiles, position them and mark the edge locations. Apply adhesive and install the full field tiles first. Then butter and install the bullnose, trim, and cut tiles. Finish the wall edge with a single row of bullnose trim as shown in the figure, or lay tiles on the wall to form a backsplash of any desired height. The technique is the same as for installing wall tiles, as explained next.

TILING WALLS

To tile a wall completely or install a backsplash above a counter, first establish a tile layout that will avoid trimming at the sides or that will give an equal balance at each side. For a small area, lay out a row of tiles across the base. For a large area, draw a level line about one-third of the total distance up from the bottom. Mark out the tile widths on this line, adding in the width of each grout joint. Then draw a vertical line through the center of the area and work out the vertical spacing. If you are using a row of trim tile at the top, work downward so that all trimming is at the base. If the tile runs all the way to the ceiling, or to the bottom edge of wall cabinets in the case of a backsplash, work up from the bottom so the trimmed row is at the top. (If a trim row results in cutting more than two-thirds off a tile, respace to give equal trim rows at the top and bottom.)

Draw multiple guidelines horizontally and vertically. Then apply mastic adhesive or mortar, as shown in Figures 20–18 and 20–19. In wet areas (such as a kitchen sink, bathroom, or laundry room), a gypsum board wall should be made of water-resistant panels. Apply tile as for a counter-

Figure 20–17
Countertop tile layout and application.

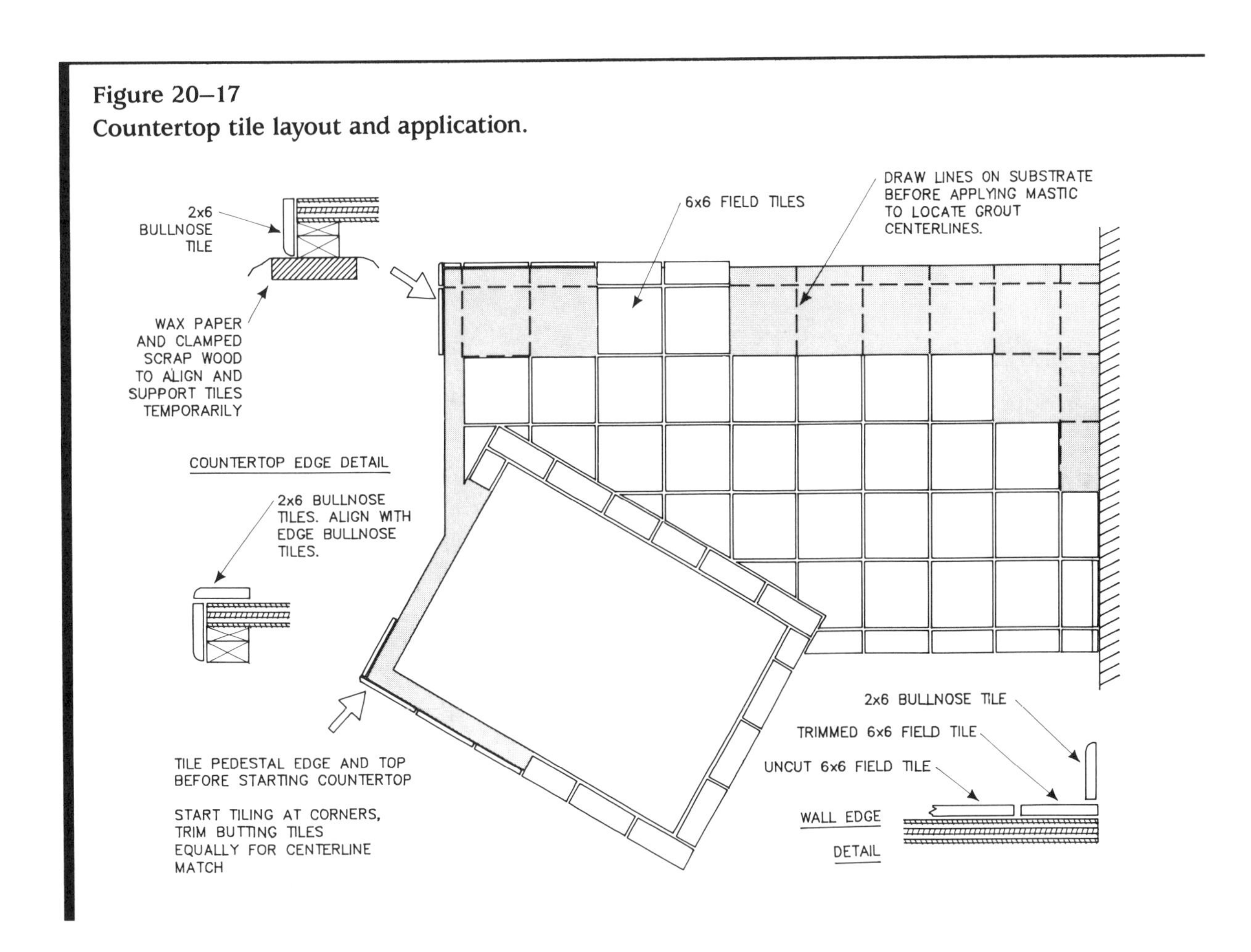

Figure 20–18
Tiling stud walls. Gypsum board must be at least ½″ thick to support tile. In damp locations, be sure the adhesive is suitable for that purpose.

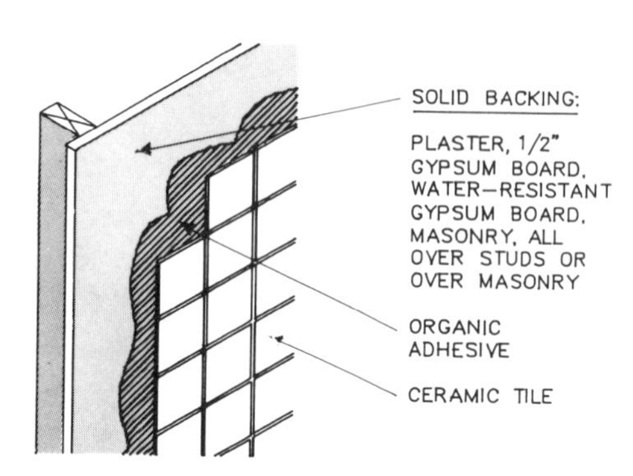

Figure 20–19
Use a latex–portland cement mortar to install tile directly on a dry masonry interior wall.

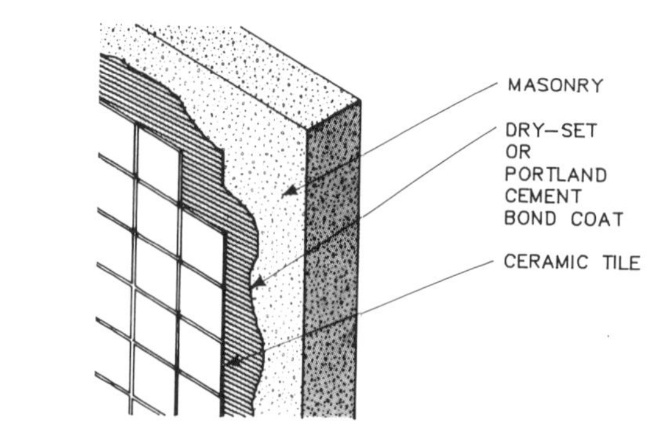

top. Use a level frequently to check horizontal and vertical alignment.

INSTALLING TILES OVER A SLAB FLOOR

Use epoxy or organic adhesive to set mosaic or glazed floor tile on a concrete slab indoors (Figure 20–20). The slab must be completely cured, with no cracks. The surface should have a steel trowel and fine broom finish. Use epoxy grout.

Figure 20–20
Ceramic tile applied to indoor slab.

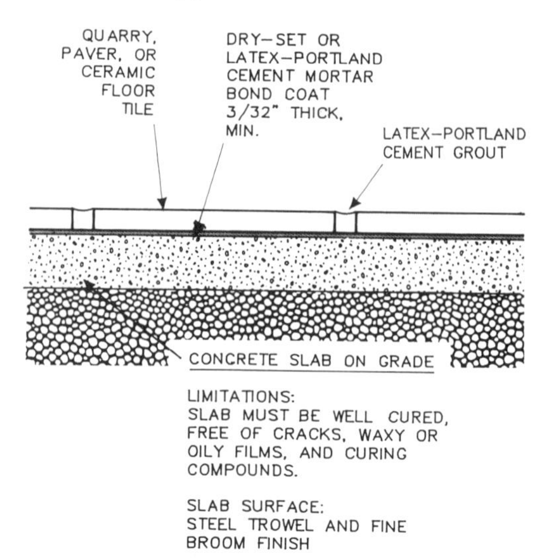

RENOVATING WITH CERAMIC TILE

Ceramic tile can be laid over many finish surfaces—paint, wood paneling, vinyl or asphalt tile, seamless resilient flooring, wood flooring, concrete, or old ceramic tile. It is always better to remove the old finish material and install a good underlayment for the tile; however, that is not always practical.

Tiling over a Smooth Tile, Terrazzo, Slate, or Stone Floor

Retiling a bathroom floor may involve laying tile on tile (Figure 20–21). The floor must be sound, well bonded, without structurally caused cracks. In a bathroom, remove a floor-mounted toilet or sink if possible. Tiling will be easier. A footed tub can be a problem, so it should be disconnected and removed. You will probably need a new threshold to match the new floor height.

Clean the floor with a detergent and roughen (scarify) the old floor with a Carborundum disk on an electric drill (wear eye protection and a dust respirator). Wash with clean water. Let the surface dry thoroughly before setting new tile. Use epoxy, latex–portland cement mortar, or organic adhesive.

Tiling over Wall Tile

Tile can be set over existing glazed tile or any other tile when a change in decoration is desired (Figures 20–22, 20–23, and 20–24). Tile can also be set over marble, slate, and stone walls. The wall must be sound, well bonded, and without large structural cracks. Epoxy, organic adhesive, or latex–portland cement mortar can be used. Prepare the surface as described above for a floor. Setting procedures will differ slightly, depending on the configuration of the old and the new tile. For example, it may be necessary to shim the wall above the old tilework with gypsum board to carry the new tile to a greater height.

Figure 20–21
Tiling over a ceramic tile bathroom or kitchen floor.

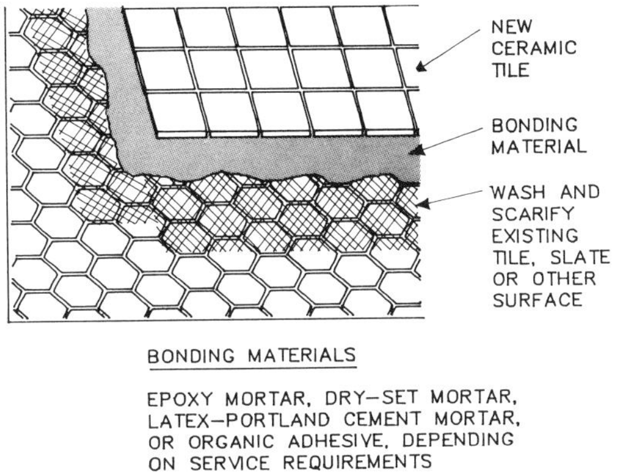

NEW CERAMIC TILE
BONDING MATERIAL
WASH AND SCARIFY EXISTING TILE, SLATE OR OTHER SURFACE

BONDING MATERIALS

EPOXY MORTAR, DRY–SET MORTAR, LATEX–PORTLAND CEMENT MORTAR, OR ORGANIC ADHESIVE, DEPENDING ON SERVICE REQUIREMENTS

Figure 20–22
Applying ceramic tile over existing wainscot wall tile, with the top edge finished with cut, thickset bullnose trim tile.

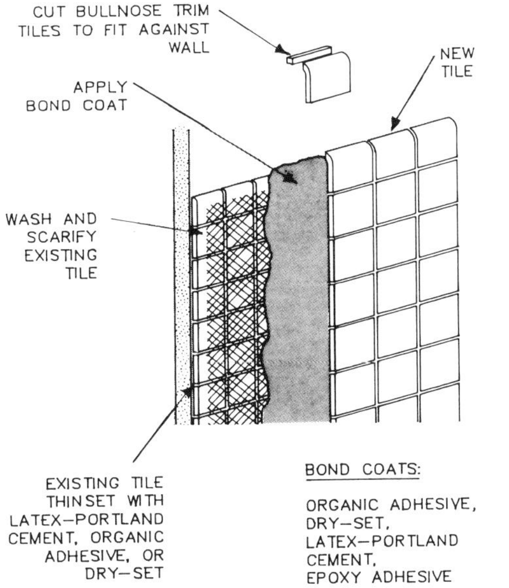

CUT BULLNOSE TRIM TILES TO FIT AGAINST WALL
NEW TILE
APPLY BOND COAT
WASH AND SCARIFY EXISTING TILE
EXISTING TILE THINSET WITH LATEX–PORTLAND CEMENT, ORGANIC ADHESIVE, OR DRY–SET

BOND COATS:

ORGANIC ADHESIVE,
DRY–SET,
LATEX–PORTLAND CEMENT,
EPOXY ADHESIVE

Figure 20–23
Applying ceramic tile over existing wainscot wall tile, with upper wall brought flush with old tile and wall retiled to ceiling.

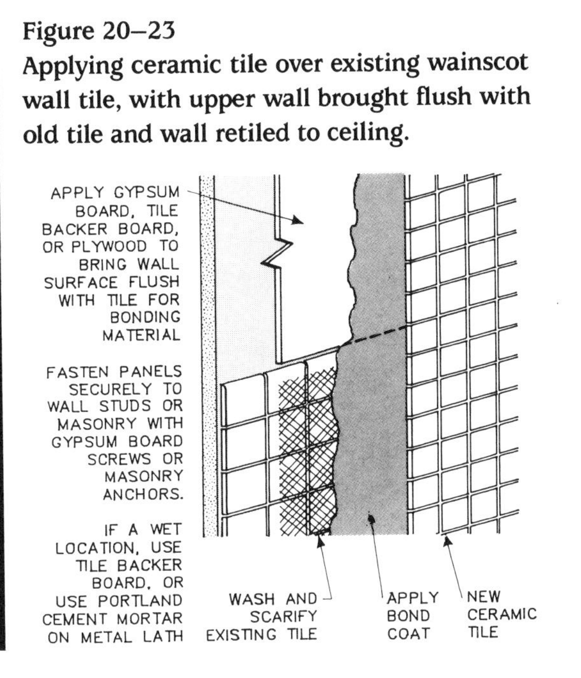
APPLY GYPSUM BOARD, TILE BACKER BOARD, OR PLYWOOD TO BRING WALL SURFACE FLUSH WITH TILE FOR BONDING MATERIAL
FASTEN PANELS SECURELY TO WALL STUDS OR MASONRY WITH GYPSUM BOARD SCREWS OR MASONRY ANCHORS.
IF A WET LOCATION, USE TILE BACKER BOARD, OR USE PORTLAND CEMENT MORTAR ON METAL LATH
WASH AND SCARIFY EXISTING TILE
APPLY BOND COAT
NEW CERAMIC TILE

Figure 20–24
Applying ceramic tile over existing thickset wainscot tile; top edge finished with thickset bullnose and ¼"-radius bullnose trim tile.

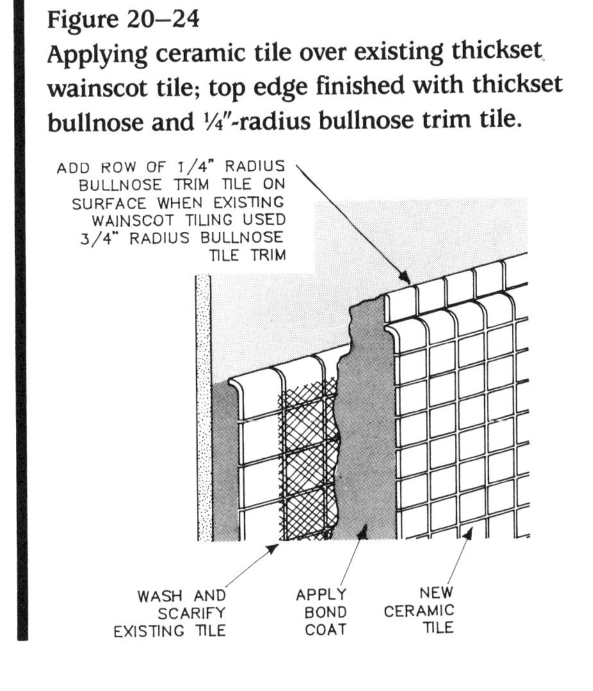
ADD ROW OF 1/4" RADIUS BULLNOSE TRIM TILE ON SURFACE WHEN EXISTING WAINSCOT TILING USED 3/4" RADIUS BULLNOSE TILE TRIM
WASH AND SCARIFY EXISTING TILE
APPLY BOND COAT
NEW CERAMIC TILE

21 | Shelves

Shelves get things up off the floor and out of boxes. You will probably want to install shelves in the laundry room, kitchen, garage, bathroom, closets, and wherever you might like to have books at hand. If shelves must fit a particular space, you almost always have to build them yourself.

A single shelf is usually supported by two steel or wood brackets. A shelf between parallel walls can be supported on cleats under the ends. A corner shelf can be supported by two cleats or a cleat and a bracket. Multiple rough shelves can be supported on Z-brackets as shown in Figure 21–1.

Figure 21–1
Rough shelves can be quickly hung using Z-brackets and precut plywood shelves.

SHELF MATERIALS

Pine boards, plywood, particleboard, and waferboard used for utility shelves should normally be ¾″ thick (Figure 21–2). For small, lightly loaded, or short shelves, ½″-thick material both looks better and is more economical. If good appearance with minimum effort is your goal, you can buy prefinished (white or wood-grain) shelves.

The grade of pine boards sold as shelving is Grade 2 or Grade 3 Common. Neither is very high

Figure 21–2
Shelf materials. From left: pine lumber, fir plywood, and particleboard.

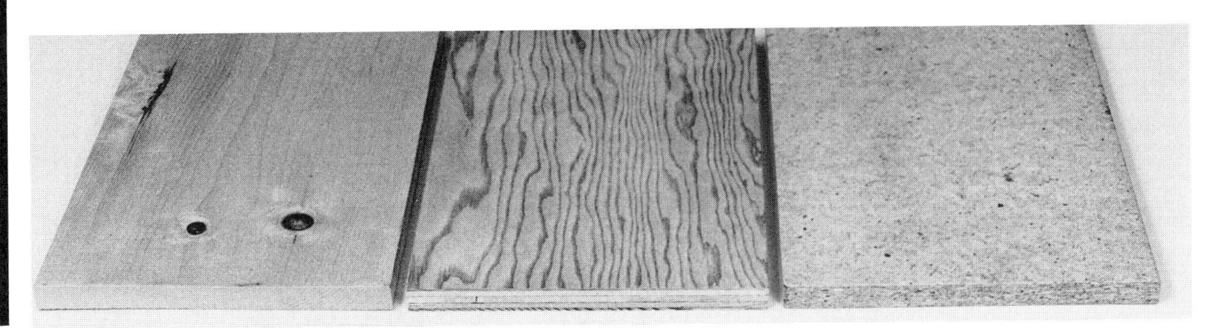

grade lumber. Grade 2 looks better, has fewer and smaller knots, will be less warped, and will make a stronger shelf. Knots can be a problem, as shown in Figure 21–3.

Fir plywood is a better shelf material than common pine, but less attractive unfinished than board shelving. Either A-D or A-C grade will do, unless the shelves will be exposed to moisture, when A-C Exterior plywood should be used. A problem with using plywood for shelving is that you must rip-saw the panel to usable widths. Lumberyards will rip the plywood for you, but they usually charge for the cutting.

Particleboard and waferboard are the lowest-cost shelving. Particleboard is heavy compared with lumber, fir plywood, or waferboard, and hard on your cutting tools, but it has a more finished-looking edge than the other materials. Particleboard and waferboard are available ripped to 12″ width in most home centers. The edges of both kinds of board should always be rounded over and sanded because the materials in the rough-cut stage are uncomfortably sharp.

Figure 21–3
Sound round knots (left) will not detract from a board's usefulness as shelving. Small knot-holes should be plugged to prevent things from falling through. Avoid shelving with spike knots across the face (right). Such a board will not support any weight and will break. The two spike knots shown together extend across the total width of the board.

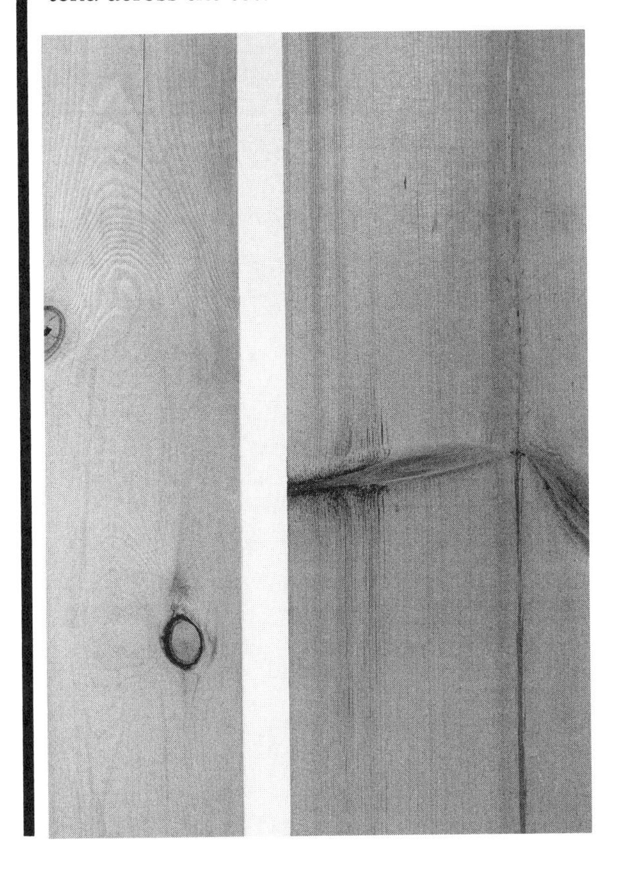

SHELF SUPPORTS

Brackets

You can buy a variety of metal brackets to support shelves; if you want wood brackets or cleats you must make them yourself. The most important decision is selecting the correct fastener to attach the shelf support to the wall or other vertical surface. There are three important considerations for any shelf: (1) the load you intend to put on it, (2) the supporting wall construction and whether it will bear the weight of the shelf and whatever is piled on it, and (3) how finished or crude you want the shelf to look.

The most common kinds of shelf brackets are shown in Figure 21–4. They take the horizontal weight of the shelf and its contents and transfer the load to the vertical surface of the wall. The bracket must do this without bending even slightly, because that would allow the shelf to tip forward and spill its contents. The best brackets for shelves are made of relatively thin stamped metal with a formed-in reinforcing rib. Large flat metal corner angles also are used for shelf brackets, but they tend to flex more as the corner is not reinforced and may cost more. You can make your own brackets out of scrap 1 × 2 lumber, and they will be as strong as the steel ones. You can also make more decorative light-duty brackets from ¾″ or 1″ lumber sawed in patterns such as those shown in Figure 21–5. Z-brackets

Figure 21–5
You can make shelf brackets from blocks of wood and mending plates. The brackets are appropriate for decorative or small shelves.

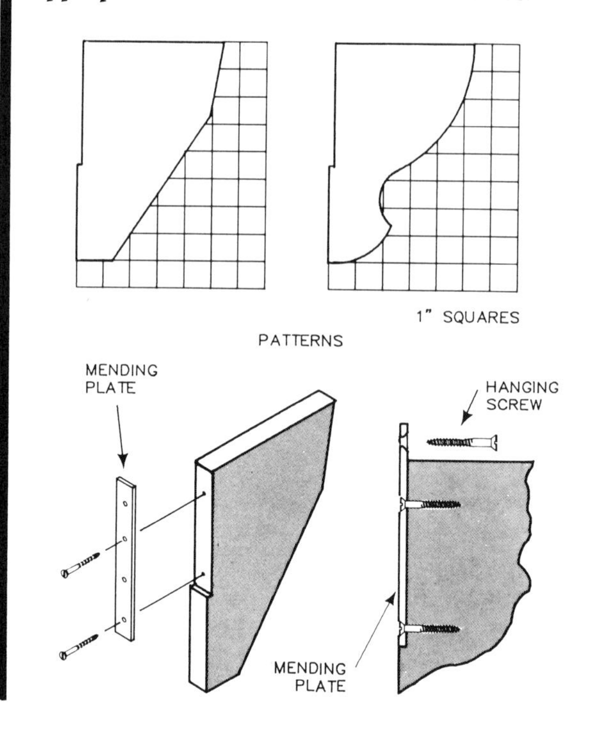

Figure 21–4
Utility shelf brackets. The stamped steel bracket (center) is stronger than the flat metal corner brace because it has a formed-in reinforcing rib.

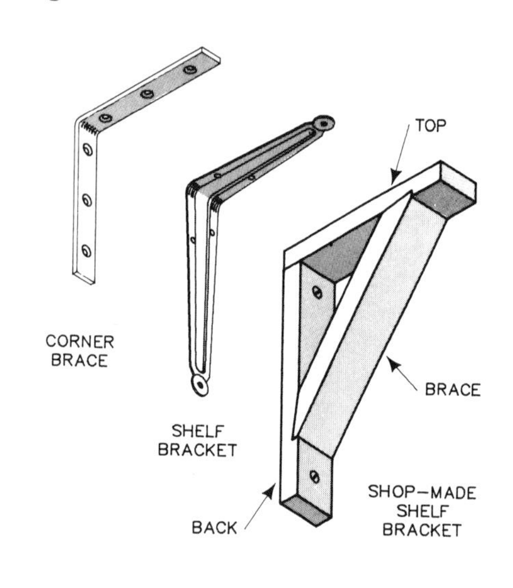

Figure 21–6
Stamped metal Z-brackets are used in pairs to support 12″ shelves. They are screwed to studs or to masonry walls using appropriate anchors. Each pair of brackets supports three shelves.

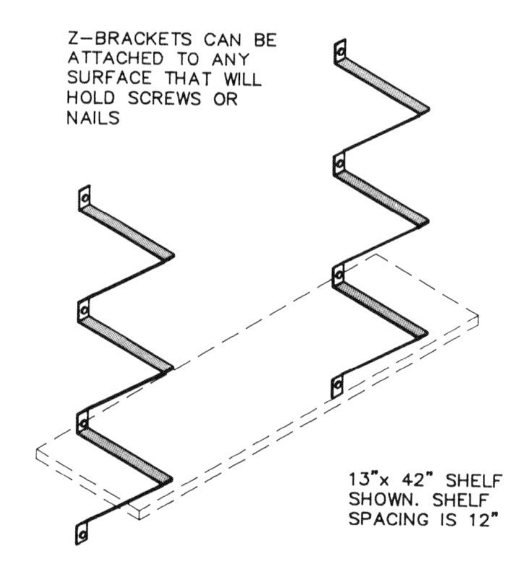

Figure 21–7
Bracket attachment and forces on screws.

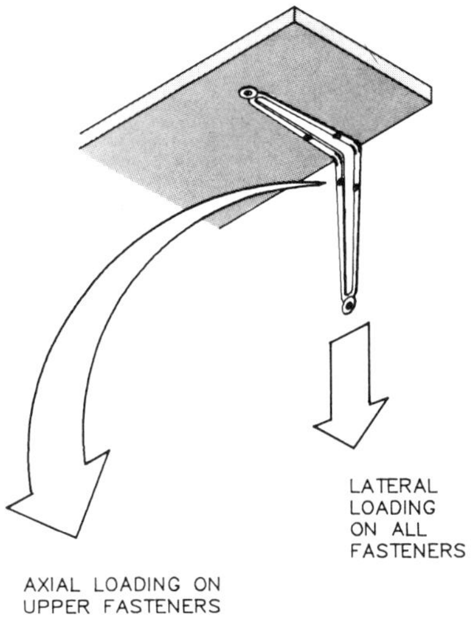

LATERAL LOADING ON ALL FASTENERS

AXIAL LOADING ON UPPER FASTENERS

(Figure 21–6) are a convenient way to hang several shelves one above another.

Single-shelf brackets are attached to the wall with one or two screws near the top end, plus one at the bottom. The load is not evenly shared by the screws. As shown in Figure 21–7, the top end screws have the heaviest load; they must resist being stripped out of the wall by axial load forces, in addition to being pulled down by the lateral load. The screw at the bottom need only resist being bent down; its main function is to keep the bracket vertical. The shelf should be attached to the bracket, so the bracket cannot rack sideways.

Figure 21–8
Match bracket size to shelf size.

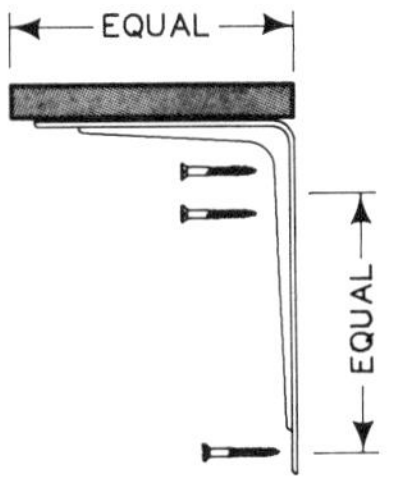

EQUAL

EQUAL

The required size of a bracket is determined by the shelf depth, the front-to-back dimension. The distance between the mounting screws at the top and the bottom of the bracket should be at least as great as the depth of the shelf, as shown in Figure 21–8.

Pilasters

If you are putting up shelves one above another, and they are not going to be too heavily loaded, one of the easiest ways to go about it is to use ready-made pilaster standards, or strips as they are sometimes called.

The simplest standards, shown in Figure 21–9, have horizontal slots spaced 1″ apart. Special clips can be snapped into the slots to support shelves at any height in 1″ increments. Four of these standards are usually required to support a stack of shelves, two at each end. There are different styles for surface and recess mounting, as shown.

Another type of standard mounts behind the

Figure 21–9
Pilaster (standard) shelf supports.

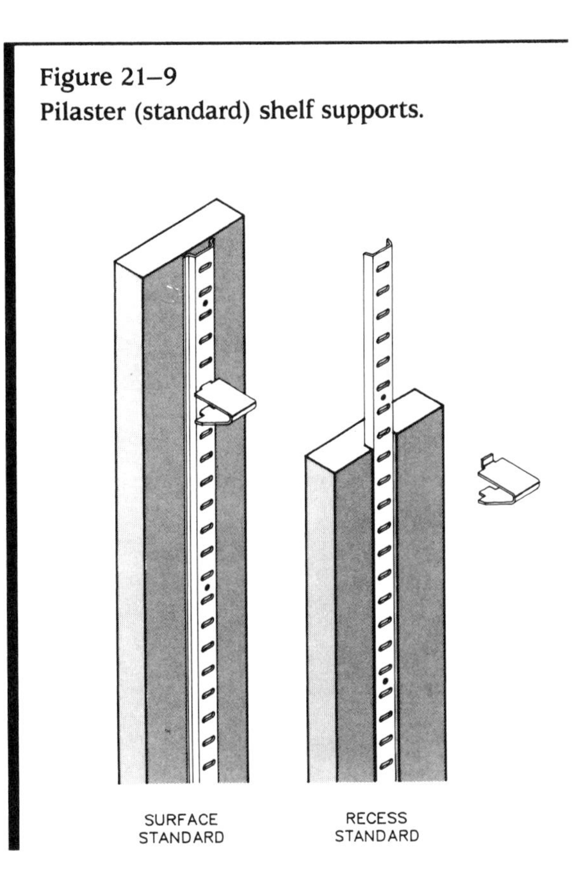

shelves and accepts special metal brackets that extend forward to support the shelves from below. Brackets are made for shelves 6″, 8″, 10″, and 12″ deep. Brackets and standards are usually not interchangeable between manufacturers. Both kinds of standard can be used on any wall, provided you use the proper fasteners.

Cleats

Cleats are strips of wood fastened to a wall or partition to support the ends or back of a shelf. They can be attached to any kind of wall with appropriate fasteners. In place of wood cleats, you can use short lengths of aluminum or steel angle strips. Proper cleat nailing is shown in Figure 21–10. If a shelf is to carry a heavy load, screw the end cleats to the wall and add a cleat at the rear, running the length of the shelf.

Figure 21–10
Nailing a cleat. Nails should be driven horizontally or at a downward angle to have any holding power. Use screws for a heavy-duty shelf.

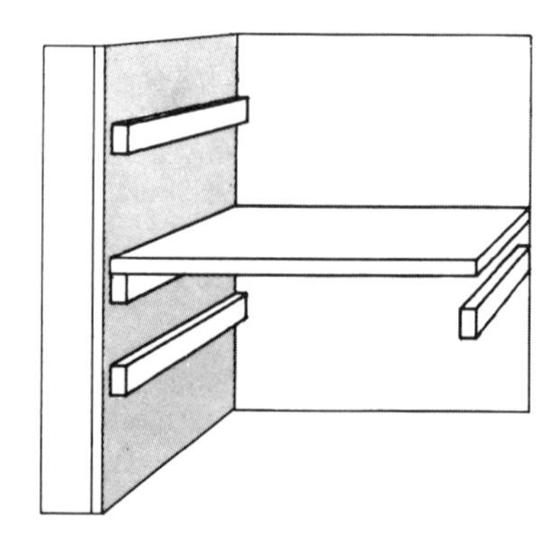

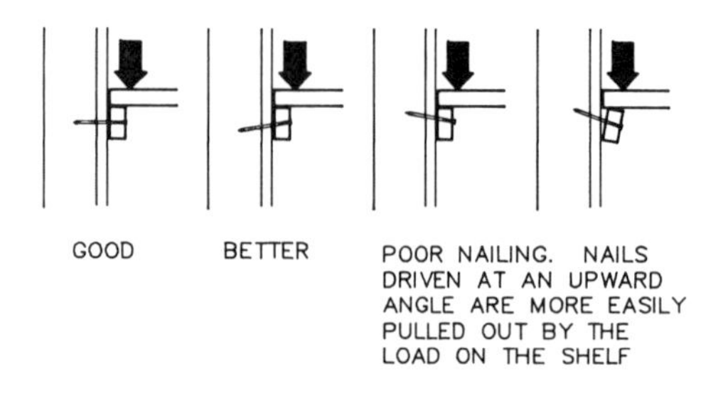
GOOD
BETTER
POOR NAILING. NAILS DRIVEN AT AN UPWARD ANGLE ARE MORE EASILY PULLED OUT BY THE LOAD ON THE SHELF

Spacing Shelf Supports

How far apart can shelf brackets be, or how long a shelf can you support between cleats at the ends? It is difficult to give a hard-and-fast rule. It depends on how much weight you are piling on the shelf, the shelf thickness and material, the kind of brackets or other support, how the support is attached to the wall, and the wall itself. For a heavily loaded shelf, 32″ would be a good maximum distance between brackets or cleats. Brackets should not be placed at the very ends of shelves, but 6″ to 8″ in from the ends (see Figure 21–11). This placement will reduce the tendency of the shelf to sag under the load. As a rule of thumb, if a ¾″-thick shelf sags more than ¼″ under its load when supported by brackets or cleats 32″ apart, the shelf is overloaded. To avoid sagging, you can use thicker shelf material, put an additional bracket in the middle, or take some of the load off the shelf. When supports are spaced more than 32″ apart, shelves, particularly particleboard shelves, will sag under their own weight.

Figure 21–11
Bracket spacing and shelf length.

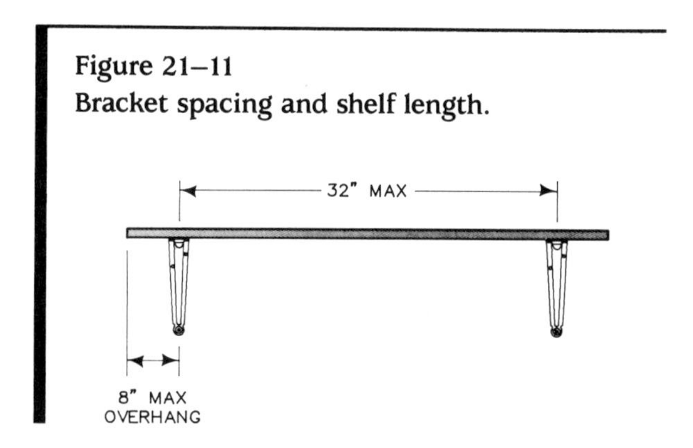
32" MAX
8" MAX
OVERHANG

Wall Construction and Fasteners

Fastener type and size are important considerations. The type of fastener—screw, nail, masonry anchor, expansion anchor (Molly bolt), toggle bolt—is selected on the basis of how the wall supporting the shelf is constructed. The size of the fastener used will depend on what is needed to carry the load of the shelf.

As just stated, in order to choose the right fastener to attach a shelf bracket or cleat to the wall, you must know for sure what the wall is made of. If you can't tell by looking at it, drill a small test hole in a spot that will be covered by the shelf. Use a carbide-tipped masonry drill rather than a twist drill for this job. Then refer to Table 21–1.

Note: Be careful when you drill into any hollow wall, because you could run your drill into electrical wiring or plumbing. You could also run into thermal insulation, which is good news overall, but if the insulation happens to be fiberglass, your drill may get entangled in it and that can mean trouble getting it out. If your drill has a reverse switch, use it when withdrawing the bit.

HANGING A SHELF ON A HOLLOW WALL

If possible, locate brackets over studs as shown in Figure 21–12. This will allow you to use ordinary inexpensive wood screws to attach the brackets to the wall. If you must locate a bracket between studs, you will have to use a hollow-wall anchor, either an expansion bolt or a toggle bolt. Stud support is stronger.

Figure 21–12
Hanging a shelf on a hollow wall. For the strongest support fasten brackets to the wall studs.

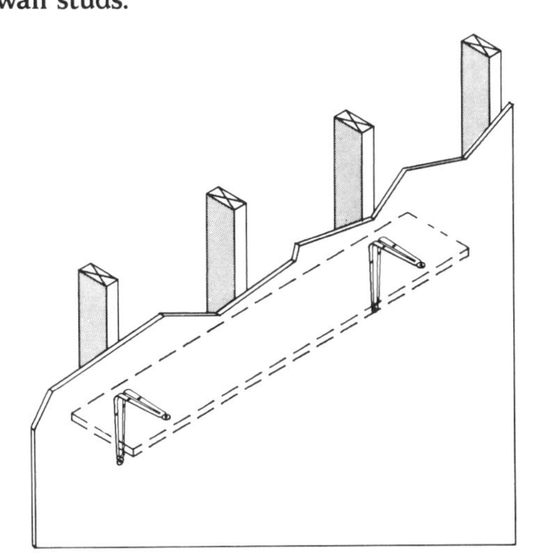

Locating Hidden Studs

Finding the studs in the wall is not too difficult. You can buy a magnetic stud finder that reacts to nails or screws in the studs, or an electronic stud finder that senses hollow space between studs. (It works well but is easily fooled by plumbing inside the wall.) You really need only a couple of finishing nails to find studs, as shown in Figure 21–13. Rap on the wall with your knuckles—studs are located where the wall sounds least hollow. At the height that will be hidden by the shelf, drive in one of the nails. If it goes through, drive another nail 1″ to the side. Once you have hit solid wood, you can easily find the centerline of the stud with one or two additional nails. Studs are 1½″ wide and normally are spaced 16″ on centers, but don't count on it: use 16″ or 32″ as a measurement to start looking for another stud,

Figure 21–13
Locating hidden studs with nails. It is important not only to locate the stud, but to find its center.

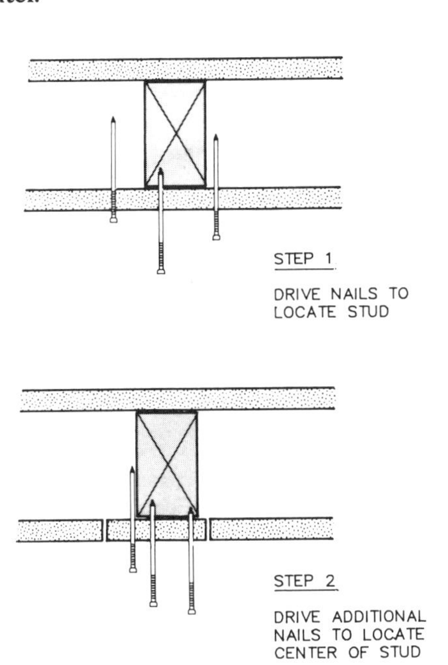
STEP 1
DRIVE NAILS TO LOCATE STUD
STEP 2
DRIVE ADDITIONAL NAILS TO LOCATE CENTER OF STUD

Table 21–1
WALL CONSTRUCTION AND FASTENER SELECTION

Wall Construction	Fastener to Use	Drilling Results	Notes
Gypsum board	Expansion bolt	White dust, quick breakthrough	—
Gypsum board on studs	Wood screw or lag screws	White dust, a bit of sawdust, little or no further penetration	You have found a stud. Drill pilot holes into stud for screws; they will hold in the stud, not in the gypsum board.
Plaster over wood lath (old house)	Expansion bolt or toggle bolt	White dust, gray dust, sawdust, breakthrough	Use toggle bolt if you can't get long-enough expansion bolt.
Plaster over expanded metal lath	Expansion bolt	White dust, gray dust, possible metal chips	—
Plaster over cinder block (if you break through, you have hit a cinder block core)	Masonry anchor or toggle bolt	White dust, dark gray dust, possible breakthrough	Breakthrough means you have reached block core; use toggle bolt. You may have to bed expansion bolt in epoxy cement.
Plaster over cement block	Masonry anchor	White dust, light gray dust, easy drilling	Do not use hardened masonry nails.
Plaster over poured concrete	Masonry anchor	White dust, light gray dust, hard drilling	Do not use hardened masonry nails.
Plaster over brick	Masonry anchor	White dust, red dust	Do not use hardened masonry nails.
Paneling over gypsum boards on studs (see second entry)	Expansion bolt	Sawdust, white dust, breakthrough, solid wall 3″ to 4″ behind	—
Paneling on studs (see second entry)	Expansion bolt	Sawdust, breakthrough, solid wall 3″ to 4″ behind	—
Paneling on furring	Hollow-door anchor	Sawdust, breakthrough, solid wall 1″ behind	—
Solid paneling (1″ thick) on studs, furring, or masonry	Wood screws or nails	Sawdust; little progress with masonry drill in solid wall behind	—
—	—	Sawdust, or white dust, little drill progress, possible metal chips	You have probably hit a nail, try somewhere else.

Figure 21–14
(Left) Twist drill bit for wood, metal. (Right) Carbide-tipped bit for masonry, plaster, gypsum board, brick. When going through a mixed-material wall, use the appropriate bit.

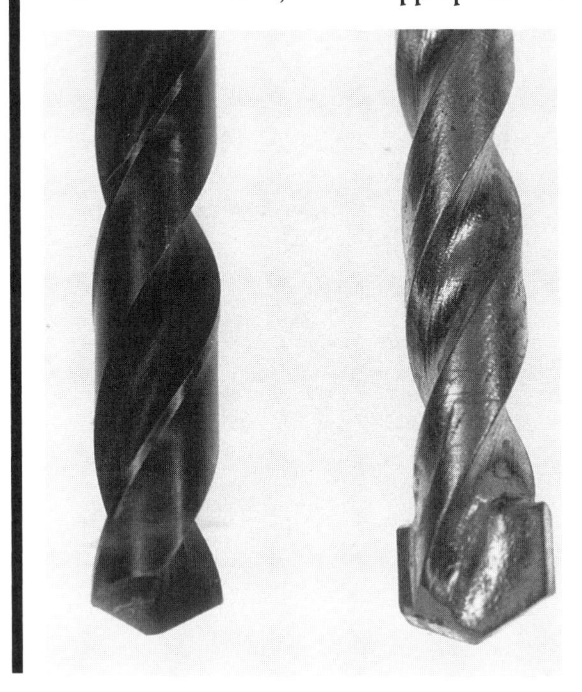

Figure 21–15
Hollow-wall anchors. From top: (1) Molly bolt blind expansion anchor, which can be used in any material but requires a pilot hole; (2) sharp-pointed expansion bolt, which can be driven into hollow gypsum board wall without drilling a pilot hole; (3) hollow door anchor, a special kind of blind fastener for thin (⅛″) materials; and (4) spring-wing toggle bolt.

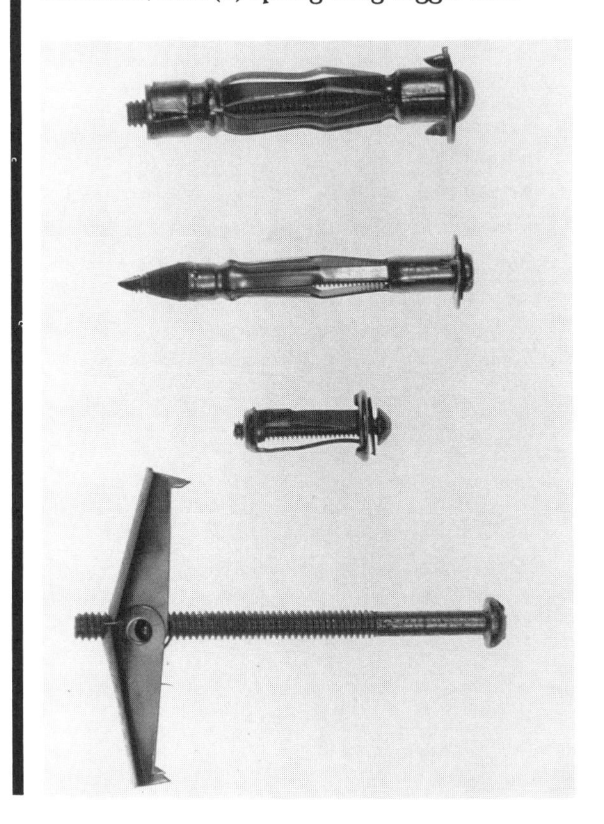

not as a sure spacing for a second shelf support.

You need two drill bits—a carbide-tipped masonry bit for a clearance hole in the gypsum board (don't be fooled by the seeming softness of the gypsum; it will destroy a twist drill bit fast) and a twist bit to drill a pilot hole for the screw in the stud. The shape of these bits is shown in Figure 21–14. If the finish wall is wood or hardboard (with no gypsum board underneath) you need just the twist bit to make a pilot hole.

HOLLOW-WALL ANCHORS

When it is not possible to fasten directly to studs, use hollow-wall anchors. There are two kinds: expansion bolts, which deform behind the wall, and toggle bolts, which open arms behind the wall (see Figure 21–15). The screw of an expansion bolt can be removed and reinserted in the fastener. But when you remove the screw of a toggle bolt, the wings drop inside the wall.

Installing Expansion Bolts

To install an expansion bolt, drill a hole the same size as the body and tap the bolt all the way into the hole. In gypsum board use a sharp-pointed expansion bolt that can be driven into the wall without a predrilled hole. Hold the body with pliers or a special bent-wire tool (available where you bought the bolts) to keep the bolt from turning while you insert the screw and start turning it. Turning will be hard at first, then easy. Keep turning the screw until you again meet resistance, which indicates that the legs have spread and are seated against the back of the wall surface. Installation of an expansion bolt is shown in Figure 21–16. Remove the screw to attach your shelf support.

An expansion bolt requires a flat, clear area behind the panel to seat properly; it cannot be installed close beside a stud or other obstruction. If an expansion bolt jams, it has to be removed. Remove the screw; grab the flange with pliers and twist it off. Then push through the hole with a screwdriver or other tool until the expanded housing drops into the wall cavity. A shelf attached to a gypsum board wall with expansion bolts should not be heavily loaded. It is not the bolts that are weak, but the gypsum board.

Figure 21–16
Installing an expansion bolt. First, the bolt is tapped through a predrilled hole (unless the pointed version is used). Turning the screw then deforms the expansion anchor. The anchor is secured when it presses firmly against the back of the panel. The bolt can now be unscrewed and inserted through the bracket or another piece to be mounted on the wall.

Installing Toggle Bolts

A toggle bolt has a nut with two spring-loaded wings. The nut is removed and the bolt inserted through the bracket; then the nut is replaced. The two wings are folded up toward the screw head and poked through a hole drilled in the hollow wall just large enough for them to pass. The spring opens the wings, which then bear against the back of the wall as you turn the bolt to draw the bracket or other shelf support securely against the wall surface.

The only times a toggle bolt should be used instead of an expansion bolt are: (1) When you can't get an expansion bolt long enough to match the wall thickness, or (2) when the wall is constructed of plaster on wood lath, as in an old house. The problem with toggle bolts is that they need an oversize hole to accommodate the folded wings going in. Toggle bolts cannot be reused or reinstalled.

HANGING A SHELF ON PLYWOOD OR HARDBOARD PANELS

When hanging a shelf on a paneled wall, it is especially important to determine how the wall is constructed. If the paneling is attached to studs, there is no problem, provided you attach the shelf supports to the studs. However, often paneling is not attached to studs but to furring strips, which may be 1 × 2s flat against the wall behind (Figure

21–17). To check, drill a small hole where the wall sounds hollow. If a small wire inserted in the hole goes in only 1″ or so before it hits a solid surface, the panels are on furring strips. You can't put too much trust in the load-bearing capacity of furring strips, because you do not have any way of knowing how well they are attached to the wall. Also, a wall paneled with ⅛″ or ¼″ grooved plywood or hardboard has even lower load-bearing capacity than gypsum board. One solution is to use a shelf unit with supports that rest on the floor or that can be attached securely to ceiling joists overhead.

Figure 21–17
Hanging a shelf on a paneled wall. Furring or framing strips behind the paneling may be only lightly attached to the underlying wall. For secure fastening, drill through and install anchors in the structural (masonry) wall behind.

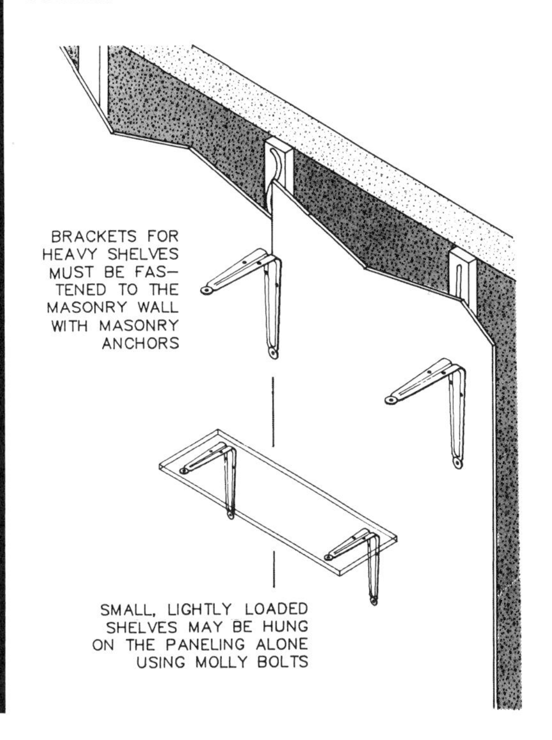
BRACKETS FOR HEAVY SHELVES MUST BE FAS–TENED TO THE MASONRY WALL WITH MASONRY ANCHORS
SMALL, LIGHTLY LOADED SHELVES MAY BE HUNG ON THE PANELING ALONE USING MOLLY BOLTS

HANGING A SHELF ON MASONRY

Masonry walls are constructed of cinder or cement block, brick, poured concrete, or stone laid up with mortar, like a brick wall. To attach anything to a masonry wall with screws, you must first install an appropriate anchor to give the screw something to thread into. Masonry anchors are made in many styles (see Figure 21–18).

Figure 21–18
Masonry anchors. From top: fiber Rawl plug, plastic anchor, lead sleeve, and expansion shield.

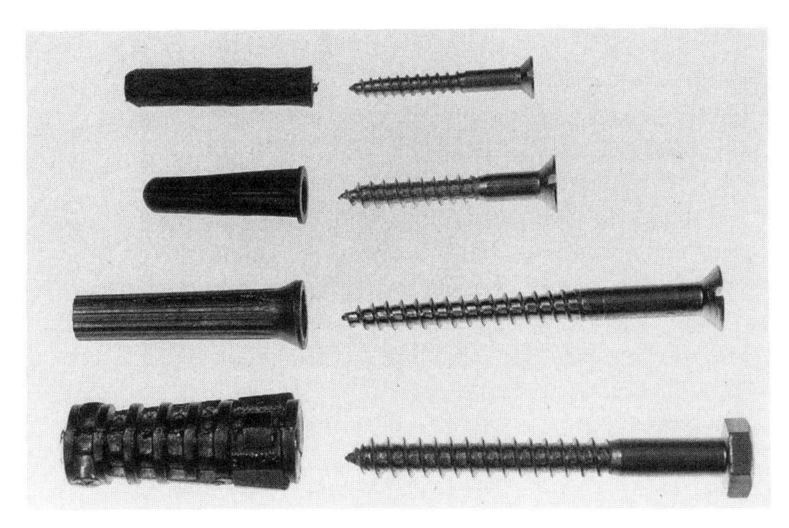

Table 21–2
MASONRY ANCHOR SELECTION

Anchor Type	Uses That Best For	Type and Size of Screw to Use	Hole Depth Required	Screw Length
Fiber sleeve (Rawl plug)	Light-duty: all masonry, tile, plaster, marble, slate, terra cotta, structural glass	Wood screws, #6 and up	Length of plug	Plug + fixture
Plastic anchor	Light- and medium-duty: all masonry	Self-tapping wood screws #6 to #16	Length of anchor + ¼″	Plug + fixture + ¼″
Lead sleeve	Heavy-duty: poured concrete, other masonry	Self-tapping wood screws, #6 and up	Length of plug	Anchor + fixture + ¼″
Expansion shield	Heavy-duty: all masonry	Lag screws	Length of anchor + ¼″	Anchor + fixture + ¼″

All those shown in Figure 21–18 and listed in Table 21–2 require an accurately sized hole in the masonry. The anchor is put in the hole and then made to deform in such a way that it expands in diameter to press tightly against the sides of the hole. This deformation is accomplished either with a special setting tool or by driving the mounting screw home.

It is important that you use the right screw with the masonry anchor. You cannot mix and match them.

If you are unable to get an anchor to seat properly, blow the dust out of the hole, coat the outside of the anchor with a filled (thickened) epoxy cement, and reseat it. Be careful to keep the epoxy away from the screw threads so you can remove the screw after setting.

You can also use special masonry nails to attach cleats to masonry walls (see Chapter 3). The nails are made of hardened steel. You cannot drive these nails with an ordinary claw hammer, because the hardened hammer head is liable to chip dangerously on impact and because you just can't hit the nail hard enough to make it move anyway. Instead, use a hand-drilling hammer available where the nails are sold. Be sure to wear gloves and face protection: hardened nails break more often than they bend, and tiny pieces fly like shrapnel.

Figure 21–19
How to install brackets that are vertical and horizontally level.

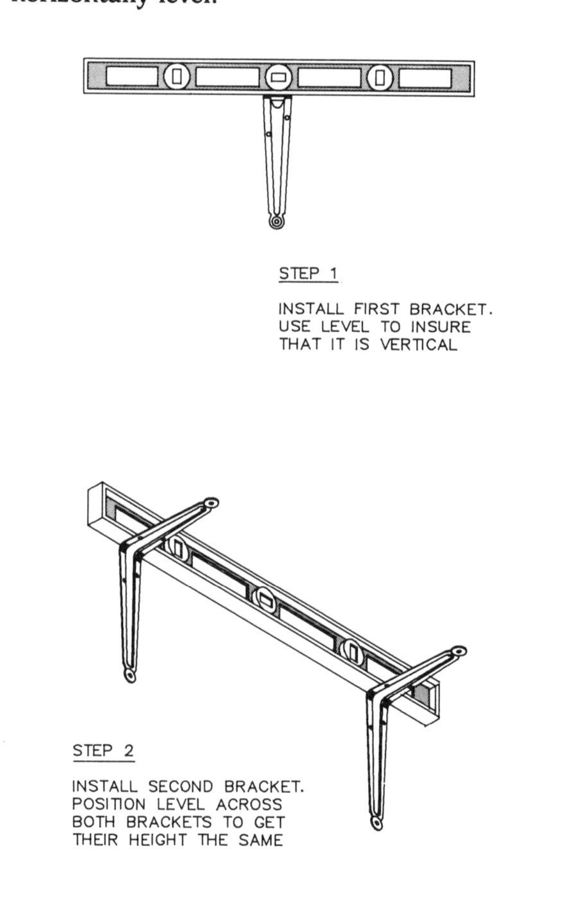

STEP 1
INSTALL FIRST BRACKET.
USE LEVEL TO INSURE
THAT IT IS VERTICAL
STEP 2
INSTALL SECOND BRACKET.
POSITION LEVEL ACROSS
BOTH BRACKETS TO GET
THEIR HEIGHT THE SAME

HOW TO INSTALL A LEVEL SHELF

For a shelf to be level, each bracket must be vertical, and their tops must be in a level line. To be sure that a bracket is vertical, use a level when marking locations for mounting hardware, as shown in Figure 21–19. Drill holes for the fasteners and mount the bracket. Use the level to position the second bracket at the same height as the first one. Getting the tops of all brackets at precisely the same height is important. If they are not at the same height, the shelf will not be level and whatever you put on it is likely to slide or rattle off.

22|Storage

A chest for storage, whether constructed to go flat against a wall or to fit across a corner, as in Figure 22–1, is basically a box. Similarly, when you attach sides to shelves so you can stack them one above the other in a single bookcase, you have made a box of sorts. Box building is basic to making many things you may want to add in your home—built-in bookcases and storage cabinets, window seats, under-bed storage, room dividers, shop storage, and kitchen cabinets.

There are right ways and wrong ways to build boxes. Right ways result in a box structure that is

Figure 22–1
Corner bed chest. Storage space is where you find it. A triangular lid provides access to the lower part.

strong enough to hold together rigidly while it is being used as intended. For example, a bookcase to stand on the floor should be constructed differently from one to hang on the wall.

There are times when building something yourself may not be the best choice. A hanging wall shelf is easy to build. The one shown in Figure 22–2 was not built from scratch but was purchased knocked down and assembled. It is a floor-standing bookcase, hung upside down to make a wall cabinet. It costs less completely finished than would the materials to duplicate it, without finish. While a ready-made unit is sometimes easier and cheaper than building your own, there are other considerations. A major consideration is that often a ready-made unit will not fill an available space exactly. Also, a ready-made that matches the style of other items in the room may not be available. Whatever the reasons for deciding to do it yourself, basic bookcases and storage cabinets are easy to construct.

BOX BASICS

The bookcase shown in Figure 22–3 is essentially a box open on one side with one or more shelves. The shelves can be either part of the box structure or loose and adjustable. A bookcase can rest on the floor (with feet, legs, or a base of some sort),

Figure 22–2
Hanging bookcase. This bookcase was purchased as a knocked-down floor-standing unit. After assembly it was turned over for mounting as a wall unit.

Figure 22–3
Basic bookcase construction. Joints are configured so that nails or other fasteners do not support loads axially, and for minimum visibility of the ends of boards or plywood.

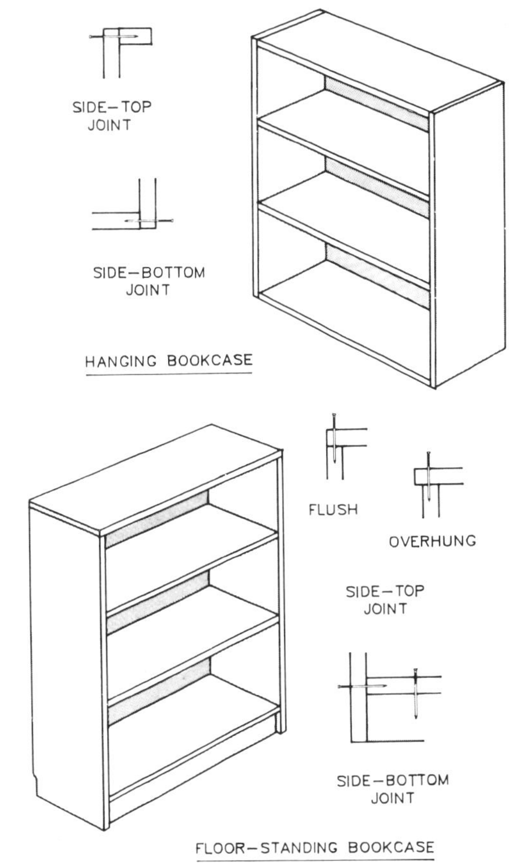

rest on another cabinet, or hang on the wall.

The top of a hanging bookcase is fastened between sides so that the ends of the top are not visible. From a structural standpoint, this construction is stronger, because the cabinet can be hung from a cleat under the top.

When a box or cabinet stands on the floor, the top is usually at or below eye level; therefore it should extend over the tops of the sides so they are not visible. This kind of construction is for appearance only; it does not add structural rigidity.

Butt joints must be glued but should also be nailed, screwed, or doweled. In the event of failure of the glue line (it happens), the joints will hang together on the nails or other fasteners, avoiding a catastrophe.

A back panel is a very important part of the structure of any box. A solid back is required to keep a boxlike bookcase rigid against racking caused by angular loads, often the result of moving the bookcase (Figure 22–4). The strength added by the back panel protects the glue joints from being overstressed, which will cause them to fail. The back panel need not cover the whole back to give the box the necessary strength. It is important that it fit into its opening snugly. The back panel should be nailed or screwed in place. Holes can be drilled in the panel for mounting screws.

The back panel of a floor-standing unit can be made of much lighter material than that of a hanging unit. Even ⅛″-thick hardboard can be used, provided it fits the opening well and the edge is fastened down all around. The easiest way to do this with a bookcase is to first nail ¾″ cleats to the panel all around, then nail or screw through the cleats to the bookcase sides, top, and bottom.

Figure 22–4
(Left) A box without a back panel can easily rack out of shape under angular forces. (Right) A tight-fitting back panel will keep a bookcase or cabinet rigid.

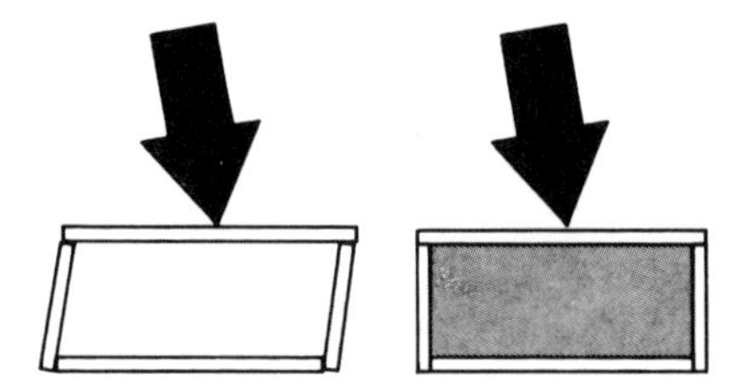

BUILDING A BOX

These step-by-step instructions are for the standing bookcase shown in Figure 22–3. They apply equally well to larger units with more shelves.

Planning

First, make an accurate drawing of what you want to build, with dimensions, as shown in Figure 22–5. Show three views: front, side, and top. In the drawing, work out and show how the pieces are to be joined—how the parts come together and what butts against what. This is important, because when figuring the dimensions of individual pieces you must make allowance for the thicknesses of other pieces.

Next, make up a materials list like that in Figure 22–5. List each part, its exact dimensions, and the number needed.

Decide what material you are going to use. The usual choice is between ¾″ fir plywood and ¾″ pine lumber. In the example bookcase, a depth of 9¼″ was chosen so that either plywood or 1″ × 10″ lumber (actually ¾″ × 9¼″) could be used. Figure the amount of material you will need by laying out the parts. For this cabinet you will need half a panel (4′ × 4′) of ¾″ A-C grade fir plywood as shown in the figure, or 12′ of 1″ × 10″ #2 Common pine board lumber. If you choose pine, allow some extra length because you must expect some waste working around knots.

Cutting

Mark off the dimensions of each piece carefully. Follow the professional carpenter's rule: measure twice, cut once. When you cut the pieces, make sure the cuts are square. Use a sander to remove

Figure 22–5
Plans and materials list for a simple bookcase.

PART	QTY.	DIMENSION
TOP	1	3/4 x 9–1/4 x 24
SIDE	2	3/4 x 9–1/4 x 21–1/4
SHELF	1	3/4 x 8–1/4x 22–1/2
BOTTOM	1	3/4 x 9–1/4 x 22–1/2
BACK	1	3/4 x 10–1/2 x 22–1/2

Figure 22–6
Assembling the bookcase.

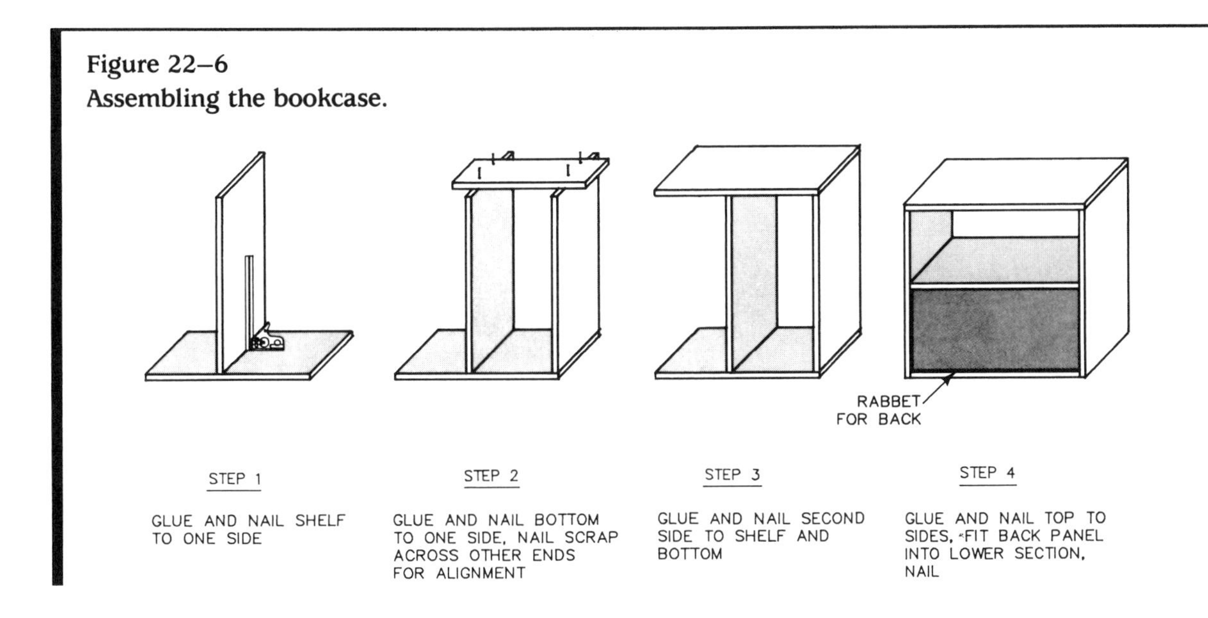
RABBET
FOR BACK
STEP 1
GLUE AND NAIL SHELF TO ONE SIDE
STEP 2
GLUE AND NAIL BOTTOM TO ONE SIDE, NAIL SCRAP ACROSS OTHER ENDS FOR ALIGNMENT
STEP 3
GLUE AND NAIL SECOND SIDE TO SHELF AND BOTTOM
STEP 4
GLUE AND NAIL TOP TO SIDES, FIT BACK PANEL INTO LOWER SECTION, NAIL

high spots, then do a trial assembly to be sure the parts go together correctly.

Assembly

Unless you know you can work fast and get all the six joints (except the back) nailed and glued in one shot, do the joints one or two at a time in the order shown in Figure 22–6.

Getting the first joint square is important. For this bookcase, the first joint should be between the shelf and one side. Align the back edge of the shelf with the back edge of the side. (If the back panel is exact size, it can be used to align the side and shelf, or it can be put on after the piece is assembled.)

Nail and glue the first joint, making sure it is square, then wait overnight while the glue dries. Next, glue the bottom to the same side, then nail and glue the second side. Tack-nail a piece across the free ends to keep the spacing the same as at the other ends. When the glue is dry, add the top.

Long clamps such as pipe or bar clamps are handy for assembling cabinets, but with glue and nail construction they are not absolutely necessary. The nails will hold the joint together while the glue dries. With clamps you can squeeze out more glue, and you will have a thinner glue line and faster drying time. With either white glue or aliphatic resin (yellow) glue, the joint will be equally strong either way.

Before fitting the back panel, be sure it and the bookcase are square. If the bookcase is not square, it can be forced to fit a squared back panel. Use nails or screws. Gluing a panel in an opening is a messy operation, and it will be difficult to clean up excess glue.

CHEST OF DRAWERS

A dresser or chest of drawers (Figure 22–7) is essentially a box of boxes. Construction of the basic chest is the same as for a basic bookcase, with the addition of movable drawers. In more advanced versions, however, solid-panel shelf-like drawer supports would be replaced by open center frames or by simple cleats at the sides.

Figure 22–7
Chest of drawers. Keep the drawers loose to avoid seasonal sticking. A better, but more involved drawer construction is shown in Chapter 23.

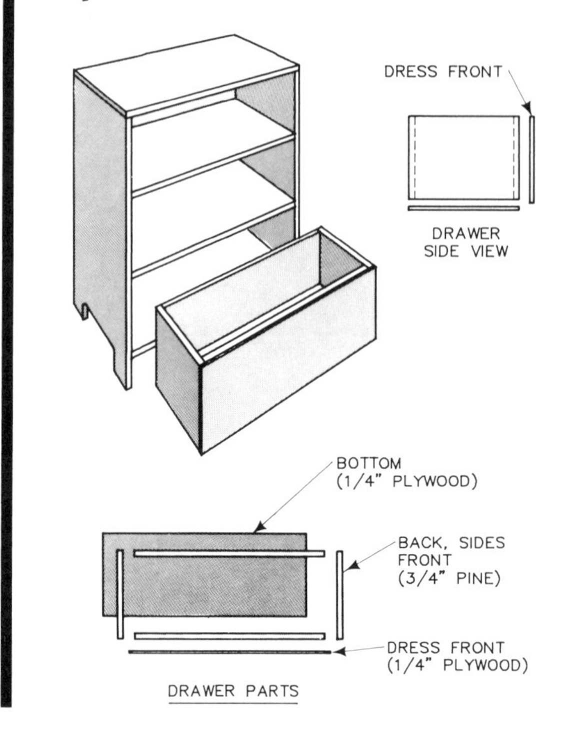

Planning

Draw a plan of the overall box that will be the outer case of the dresser, as explained above for a bookcase. The drawer supports are the same as full-depth shelves. A chest or bookcase that will rest on the floor should usually have feet, legs, or a base of some sort to raise it a few inches. The easiest way to do this is to make legs integral with the sides by extending the sides below the bottom, as shown in Figure 22–7.

Also draw a plan for a drawer. If you divide the interior space of the dresser equally in halves, thirds, or quarters, all the drawers will be the same size in all dimensions. If you want a shallower drawer at the top, draw two plans, one for the top drawer and one for the drawers of equal size that go below. Don't forget to subtract the thickness of the drawer supports when calculating interior space.

When planning drawers, do not size them to fit too snugly. If you build in winter or dry weather, the higher humidity in other seasons will swell the wood and the drawers will stick in the openings. The minimum gap, top and sides, for drawers more than 6″ high should be ⅛″; for smaller drawers, 1/16″. Allow an extra 1/16″ for drawers more than 24″ wide. Also make allowance in sizing the drawer sides, front, and back for the thickness of the bottom panel and an optional facing front panel, often called the dress or appearance front. The exposed edges of the sides can be covered by a dress front.

Make up a materials list from your drawings. Remember to list the proper number of parts for all the drawers, not just one.

Cutting and Assembly

Measure and cut each piece carefully. Because so many separate elements must fit together, absolutely square cuts are essential. When cutting out the drawer bottom plywood panel, have the surface grain run front to back rather than side to side. Drawer operation will be smoother.

Assemble the outer box as explained previously for a bookcase. Then build the drawers. Attach the drawer bottom to the sides first, then the front dress panel, and finally the back. The dress panel should overlap the front edge of the bottom.

The front and back of a drawer should be nailed and glued between the sides for strength. Figure 22–8 indicates why a drawer front should be secured between the sides. Every time a drawer is opened, it is equivalent to trying to pull the front off. If the front were fastened just to the front ends of the sides, it would be pulled off in no time. There is far less need for strength in attaching the back of the drawer.

Figure 22–9
Room divider. A single unit on a separate base like this can be built to any practical height.

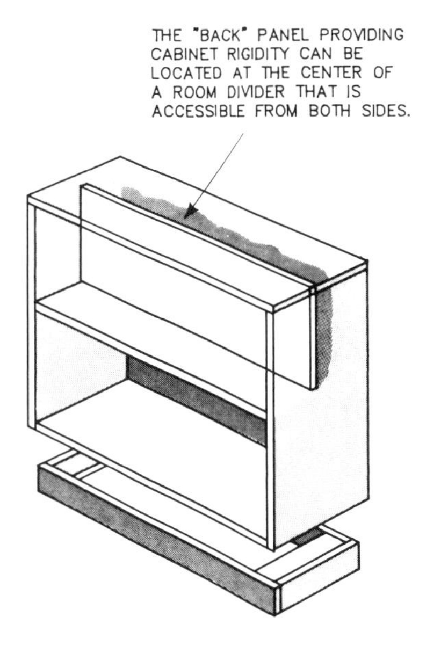

THE "BACK" PANEL PROVIDING CABINET RIGIDITY CAN BE LOCATED AT THE CENTER OF A ROOM DIVIDER THAT IS ACCESSIBLE FROM BOTH SIDES.

Figure 22–8
Stress on a drawer. Opening the drawer is equivalent to trying to separate the drawer front from the sides—the more load in the drawer, or sticking, the worse the problem.

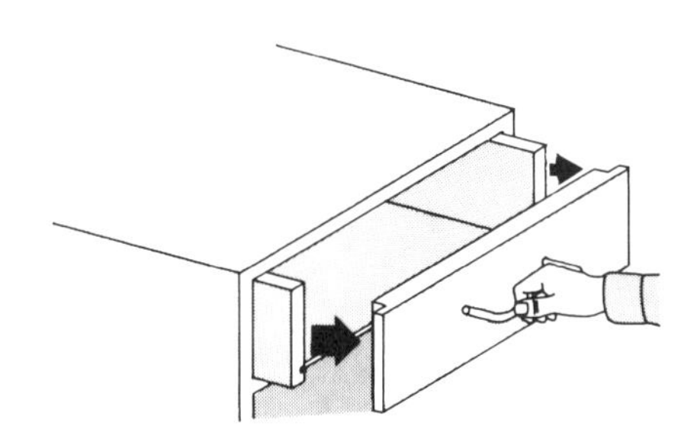

After each drawer is assembled, check its fit in the chest opening to be sure the bottom and dress panel allowances are sufficient. It is better that the drawer go in too far than stick out. You can glue wood blocks in the back of the chest to keep a drawer from going in too far; that is easier than removing wood from the drawer.

To mark the drawer fronts for the positions of handles or other pulls, put all the drawers in place in the chest. This will make it easier to align the mounting hole locations vertically.

ROOM DIVIDER

The room divider unit shown in Figure 22–9 is simply a bookcase or chest open at both front and back. The need for a solid "back" panel to provide the rigidity that will prevent racking is met by placing the panel at the center of the divider. This will provide access to the shelves from both sides.

Using the basic case dimensions, a variety of modular room dividers can be made (see Figure 22–10). Rather than building a single large divider, modular units can then be stacked two-high on a recessed common base to form a room-divider partition. A typical configuration is shown in Figure 22–11. The same units, constructed one-sided and only 12″ to 15″ deep, could form a modular storage wall.

BOX WITH LID

Another form of chest is a lidded box, such as a blanket chest, shown in Figure 22–12. In con-

Figure 22–10
Modular room dividers. Using the same basic case dimensions, a variety of modular room dividers can be made: (A) single-shelf open unit, (B) two-shelf unit with center dividing panel for small books, (C) single-shelf unit with doors on both sides, (D) desk unit with drop-down door in lower section (top section can be open or closed), and (E) three-drawer unit with drawers opening from both sides.

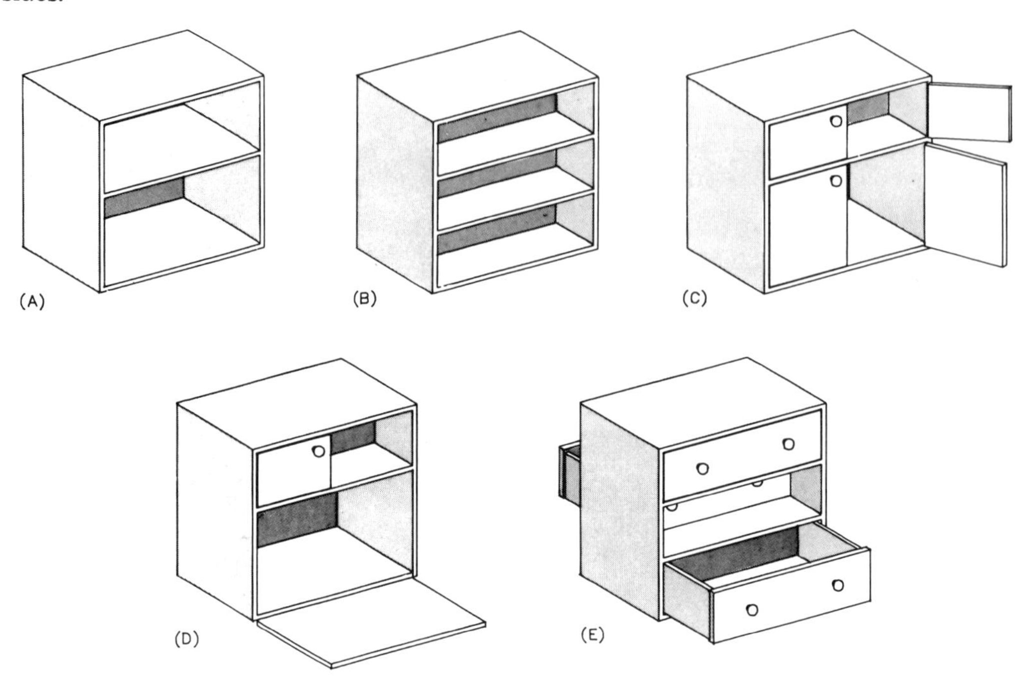

Figure 22–11
Room divider partition assembled from modular units.

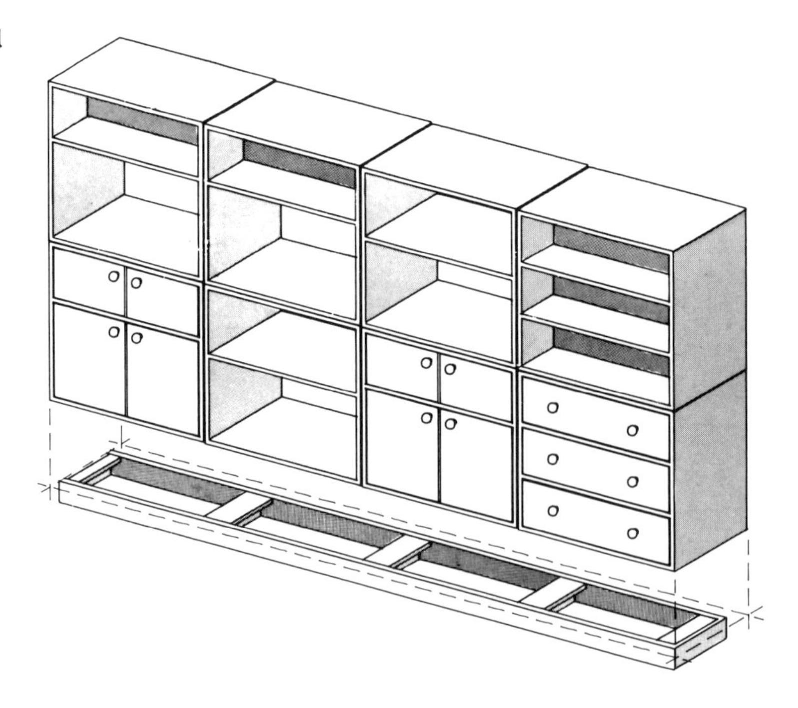

struction, butt joints at the corners are replaced with simple lap joints, which are much stronger but more work to make. These lap joints can be used on any boxlike cabinet or various containers such as a window box or window seat.

For a chest of any significant size, plywood should be used rather than pine, as warp-free pine boards are expensive. Use cleats to attach the bottom. Glue and nail (or screw) the cleats to all four sides. Hinge the top with a continuous (piano) hinge and use a self-balancing lid support for safety. Figure 22–13 shows a typical support.

Figure 22–12
Lidded box construction. Use plywood for all parts; glue and nail the corner joints. If the box will be used for toys or for linens, blankets, and the like, the lid should be equipped with a safety support.

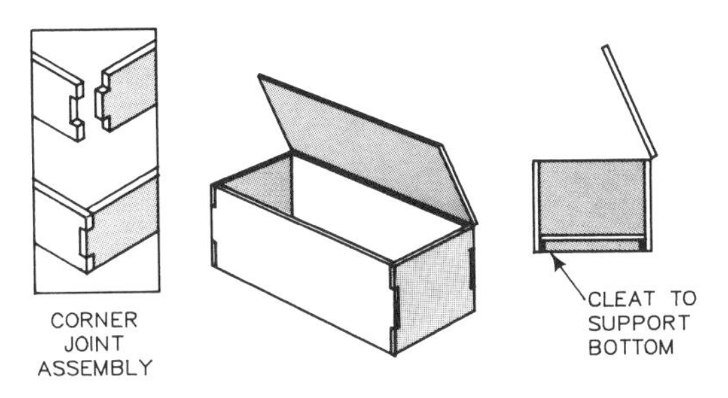

Figure 22–13
A safety support prevents a chest lid from suddenly dropping shut. The type shown works in conjunction with normal hinges; other types combine hinging and safety support.

SOME BUILT-IN STORAGE IDEAS

Armoire Closet

Customizing the interior of a closet is a good way to store clothes in a more orderly manner. Figure 22–14 shows an armoire false front that extends a

Figure 22–14
Armoire closet, installed in place of customary sliding closet doors.

closet into the room about 6″. The front was built as a unit and fitted into the closet door opening after the original sliding flush doors and track were removed. (Construction is shown in Figures 22–15 and 22–16.) The interior of such an extended closet can be finished with a wall-hung dresser (shoe and suitcase storage below) and a clothes pole front-to-back on the right-hand side.

Framed in Closet

A corner of a room, or anywhere in an attic or unfinished basement or garage, can be made into a closed storage area simply by framing it in, as shown in Figure 22–17. Framing with 2 × 3s and finishing inside and outside with gypsum board (Chapters 17 and 18) produces an inexpensive closet. The door can be hinged; or, if space in front of the closet is limited (as is likely in an attic), a lift-out door can be used. The closet can

Figure 22–15
Armoire closet false front assembly.

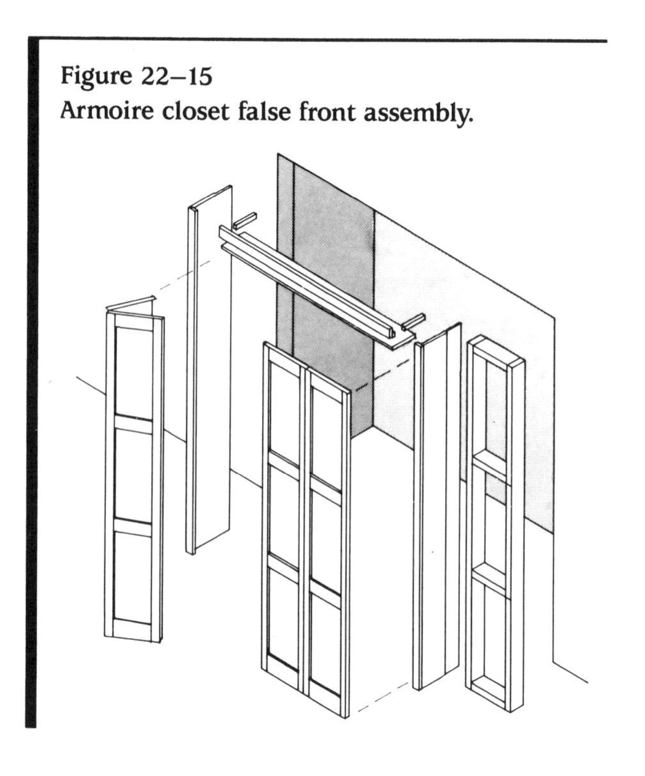

Figure 22–16
Armoire closet details.

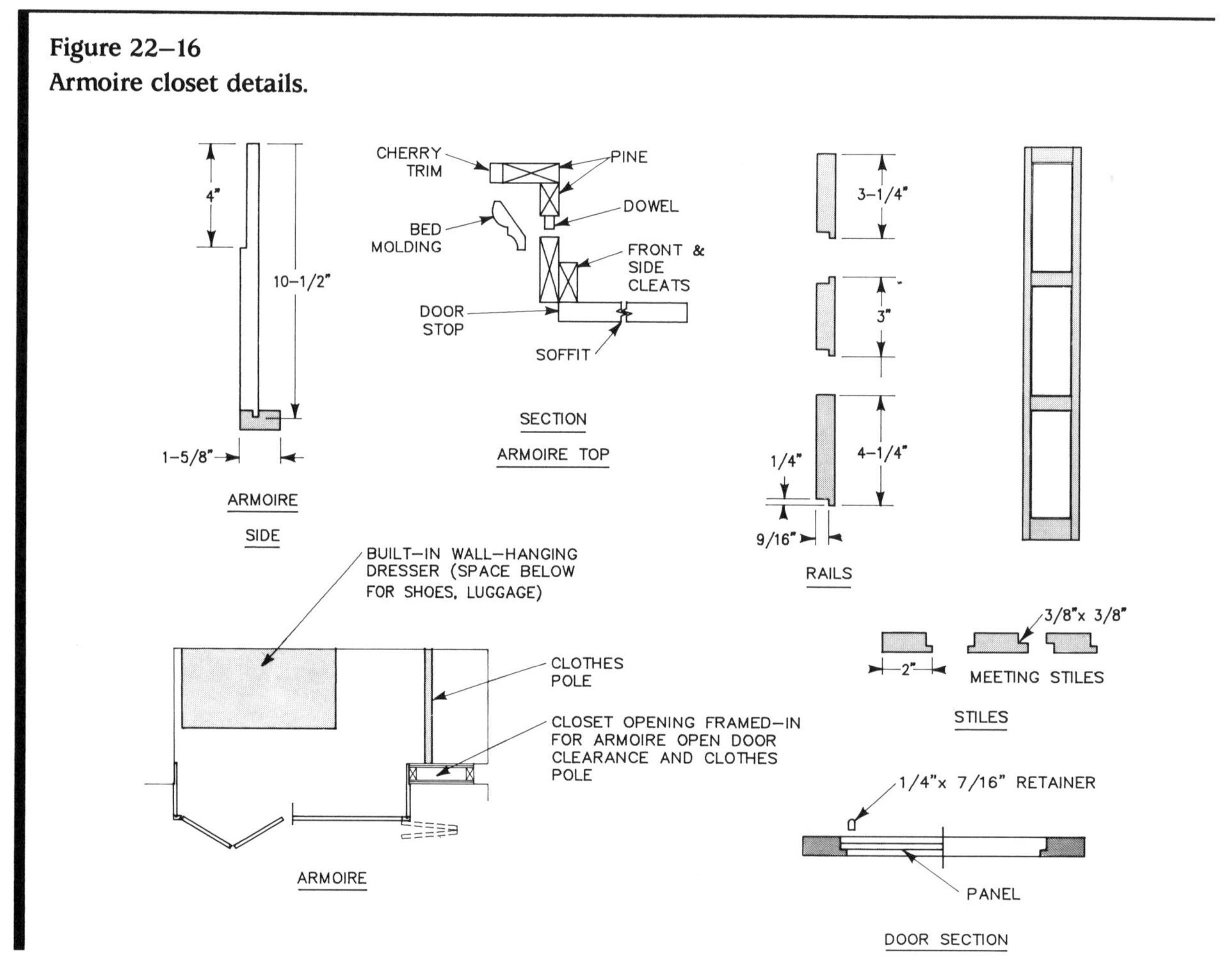

Figure 22–17
Framed-in closet. This kind of construction is suitable for an attic, a basement, or the corner of a finished room.

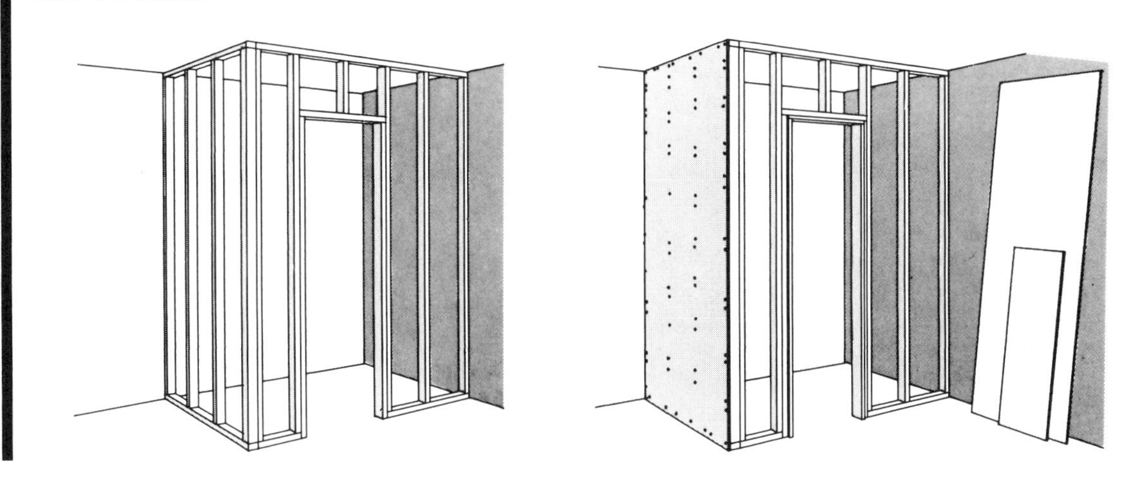

Figure 22–18
Attic storage shelves. The shelves can be sized and spaced for inexpensive standard storage boxes or objects of any size.

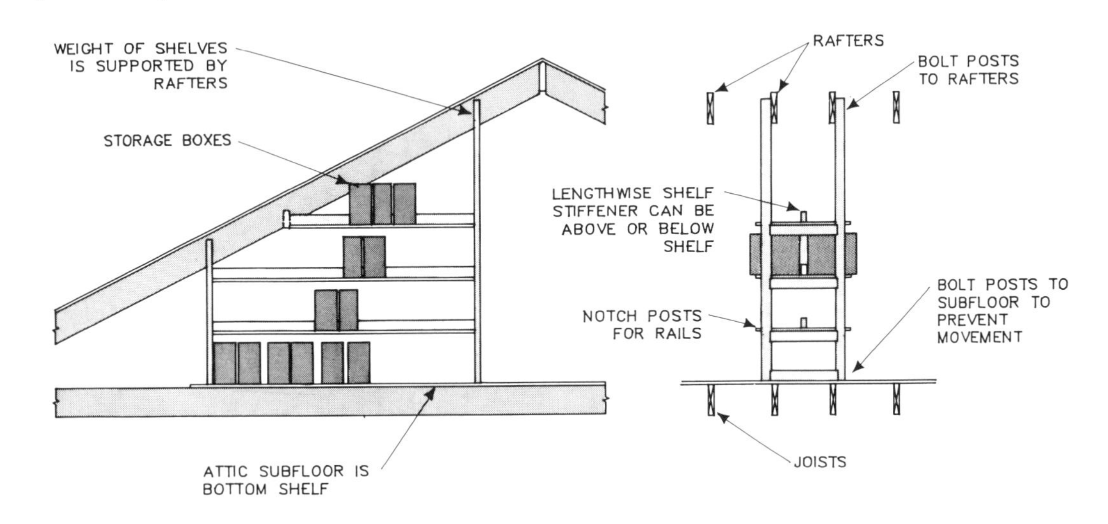

be lined with cedar, either tongue-and-groove boards or panels. If located in the attic, insulate the walls and door. If not opened too often, the closet will maintain a reasonably cool and stable temperature inside for clothes storage in summer.

Attic and Under Stair Shelves

The normally wasted space under the slope of a roof or stairway can easily be converted to useful storage, as shown in Figure 22–18. The shelves can be dimensioned and spaced to accommodate the specific objects to be stored.

Window Seat

A window seat is a useful built-in unit, especially in a dining room for the storage of linens and serving dishes that are used only on occasion.

Figure 22–19
Storage window seat.

Figure 22–20
Window seat construction.

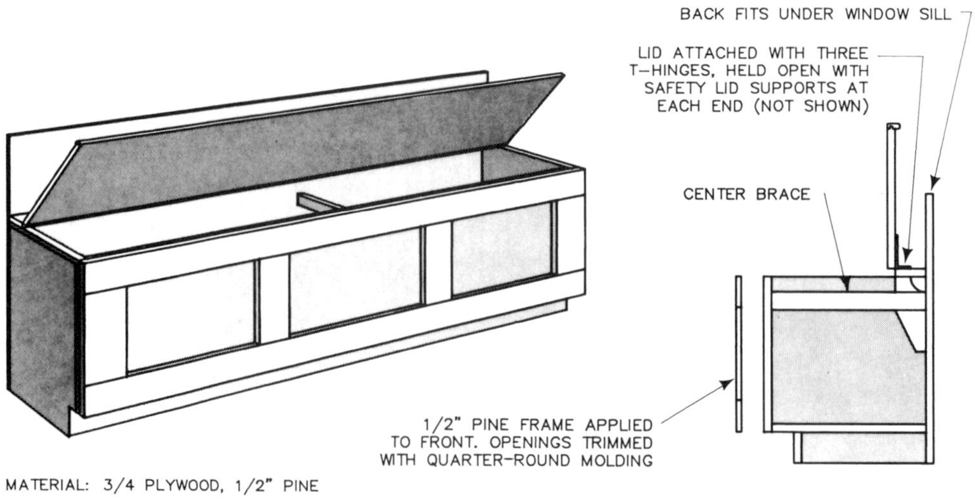

BACK FITS UNDER WINDOW SILL

LID ATTACHED WITH THREE T–HINGES, HELD OPEN WITH SAFETY LID SUPPORTS AT EACH END (NOT SHOWN)

CENTER BRACE

1/2" PINE FRAME APPLIED TO FRONT. OPENINGS TRIMMED WITH QUARTER-ROUND MOLDING

MATERIAL: 3/4 PLYWOOD, 1/2" PINE

Figure 22–19 shows a window seat ready for installation; Figure 22–20 shows its basic construction. The box should be equipped with safety hinges that will hold the lid open in any position, the kind now required on toy boxes. Do not put a window seat in a position that will block a radiator or a hot or cold air vent.

Corner Chest

Head-to-head twin beds in a corner are an attractive and useful arrangement, but what do you do with the corner? A table leaves the space underneath inaccessible. A square corner chest will provide both table space and storage. Figures 22–21 and 22–22 show construction of the corner chest seen in Figure 22–1. Once you get adjusted to the idea of building something with two fronts and two backs, and no sides, it's an easy project. The storage capacity is huge.

Figure 22–21
Corner chest. The upper and lower sections are built independently. The lower storage section has a triangular lid.

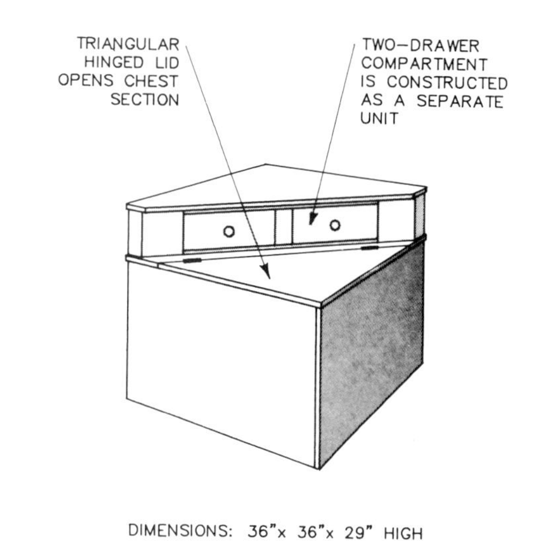

Figure 22–22
Corner chest parts and assembly. A 3′-square chest can be built from a single plywood panel if the chest top is constructed of three pieces, as shown. The long front piece laps the short front piece at the front corner.

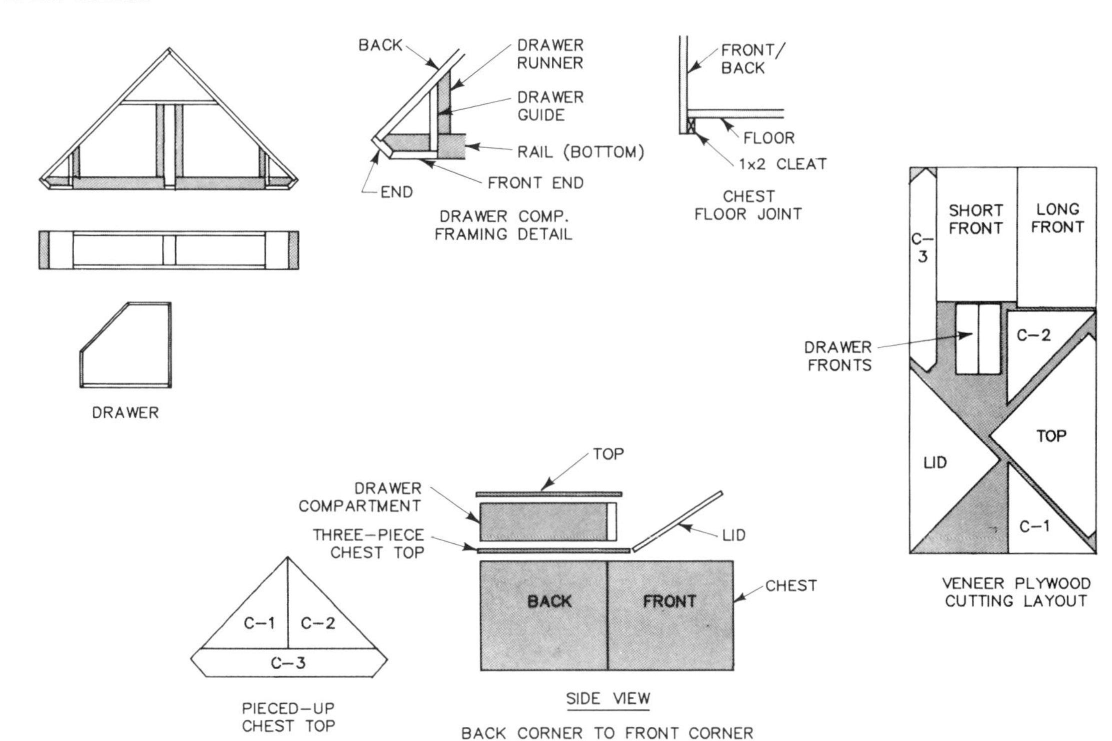

Figure 22–23
On-the-wall corner desk.

Corner Desk

Built-ins do not have to be plain, as shown in Figure 22–23. The desk and bookcase are built of cherry lumber and cherry veneer plywood. The one-piece L-shaped desk hangs on the wall, supported only by lag screws into studs. Figure 22–24 shows its construction.

Figure 22–24
On-the-wall corner desk construction. Desk is lag-screwed through its back into the wall studs.

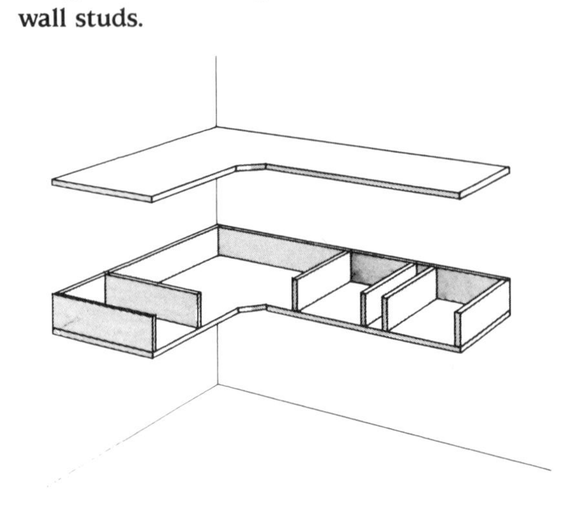

THE L–SHAPED DESK IS HUNG ON THE WALL WITHOUT THE TOP WITH LAG SCREWS INTO STUDS, THEN TAKEN DOWN FOR FINISHING AND ATTACHING THE TOP

23 Kitchen Cabinets

Should you build or buy? If you need a whole kitchen full of cabinets, it is probably more practical to buy them than build them—if standard cabinets will fit the space available and if they meet your style needs. If you are adding one or two cabinets that must match what's there, building may be the better choice.

It may not be necessary to buy new cabinets, even if you need several. Replacing only the doors and drawer fronts of existing cabinets can be a very economical and labor-saving alternative and will result in an all-new appearance. To do this, your old cabinets must be in reasonably good physical condition. The job is called refacing; you can have it done, buy the doors and do it yourself, build your own replacement doors, or restyle existing doors.

CABINET BASICS

Kitchen cabinets are essentially boxes with tops, sides (usually called ends), bottoms, and backs. The back and the reinforcing nailer rails hold the cabinet square. The cabinets may have fixed or movable shelves, partitions, or drawers. They can be built as separate cabinets or in cost-saving multiple units. The construction of kitchen base or counter cabinets and wall cabinets is essentially the same.

The best way to build kitchen cabinets is to use ¾″ plywood or particleboard for the ends, tops, bottoms, partitions, and shelves, and ¼″ plywood for the backs.

Cabinets are also made with lightweight ends (sides) and top panels—⅛″ to ¼″ plywood, and sometimes only plastic laminate ends—on hardwood frames. Both methods of construction are shown in Figure 23–1.

Styles

There are two styles of construction—American and European. In American-style or face-frame construction, the front of the cabinet is dressed with a nailed-on face frame of good wood that contributes to the rigidity of the cabinet (see Figure 23–2). This frame provides a place to attach the door hinges and covers the rough edges of the plywood used in the rest of the cabinet construction. The face frame is usually made a fraction of an inch wider than the cabinet

Figure 23–1
Box and frame kitchen cabinet construction compared. These are base cabinets with integral kickplates.

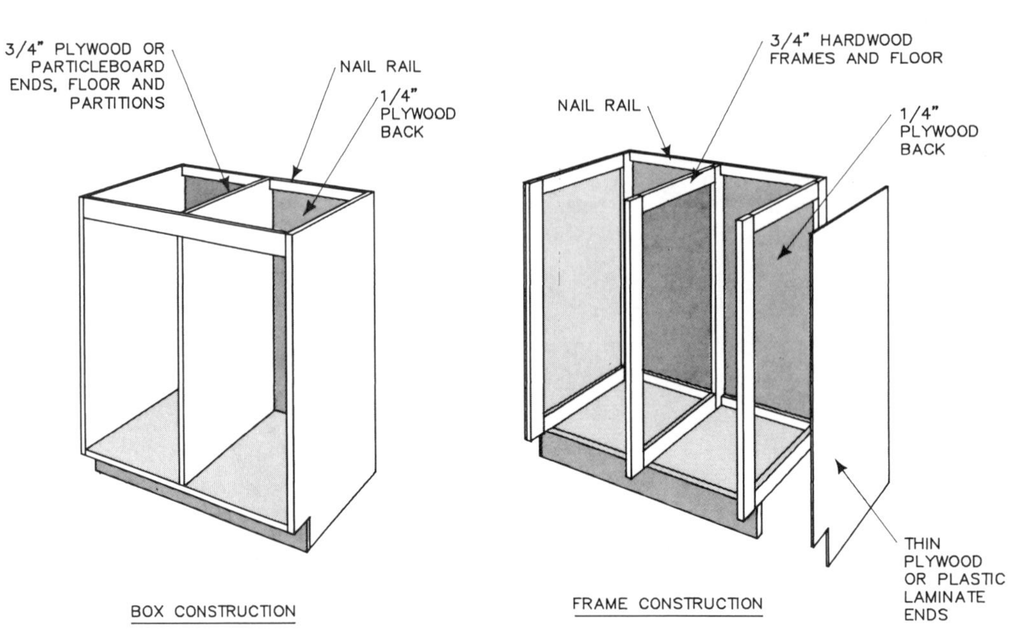

for a snug fit against a wall at the side. The major disadvantage of face-frame construction is a reduction in the size of the door opening.

In European-style construction, also called box or casework construction, face frames are not used, which allows cabinet openings of full width and height. The front edges of the top, bottom, and ends are finished and the doors are hinged directly to the end edges using European-style hinges.

Figure 23–2
Face-frame construction. Frame members are doweled together and attached to the cabinet with screws.

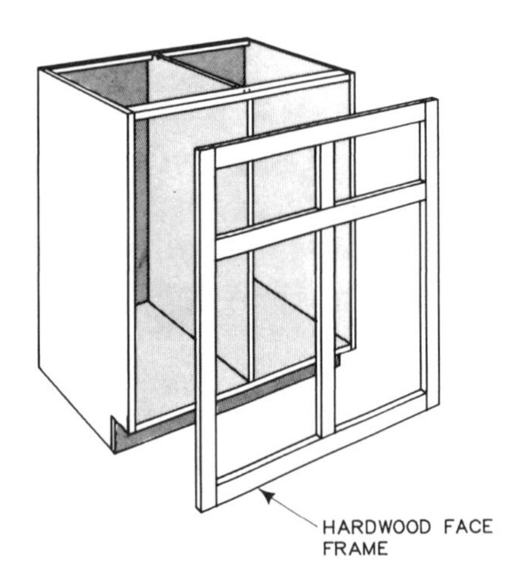

Countertops

The underlayment (core stock) of a countertop is attached across the base cabinets in one piece after they have been attached to each other and the kitchen wall. The countertop may be secured to the cabinet by screws down through the top or reaching up through the cabinet framing, as shown in Figure 23–3. Building and covering countertops with plastic laminate is covered in Chapter 24.

Cabinet Sizes

If you are adding a single cabinet, dimensions shouldn't be a problem—make the new cabinet match the rest. If you are starting a kitchen from scratch, stick close to the basic dimensions given in Figure 23–4. Kitchen cabinet dimensions are standardized. If you depart from them, appliances won't fit. If you change the counter height significantly, it just won't look right to someone buying the house. If you are making a cabinet to fit a new appliance, purchase the appliance first.

Figure 23–3
Countertop (core stock) attachment.

BUILDING CABINETS

Face-Frame (American-Style) Construction

Figure 23–5 shows a typical face-frame wall cabinet. Hidden ends and optional partitions should be cut ½″ narrower than visible ends. Nailers at the top and bottom are used to secure the cabinet to the wall, usually with screws (use lag screws where they won't show because they are easier to drive). The top nailer is notched into hidden ends and partitions, and rabbeted into finish ends. The top should be ¾″ plywood nailed and glued to the nailer, ends, and partitions for strength. The nailer at the bottom is nailed and glued to the bottom and back panel. Don't omit the ¼″ plywood back to save money—it gives the cabinet rigidity.

The face frame is built separately and attached to the cabinet as a unit with finishing nails. Use dowels (two per joint), or make mortise-and-tenon joints. One or two shelves should be provided.

Box (European-Style) Construction

Wall cabinet box construction is shown in Figure 23–6. It is similar to face-frame construction

Figure 23–4
Basic kitchen cabinet installation dimensions.

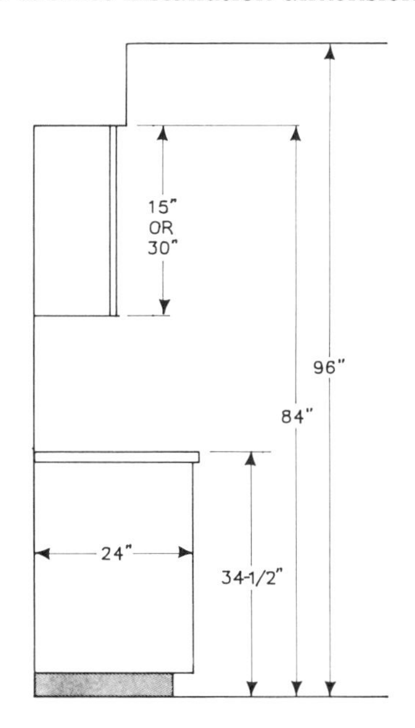

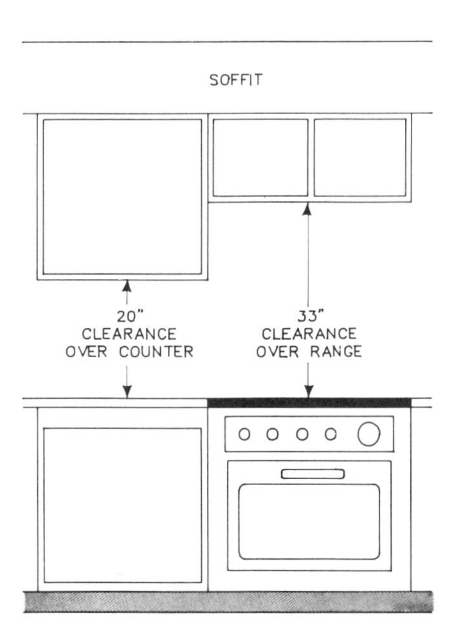

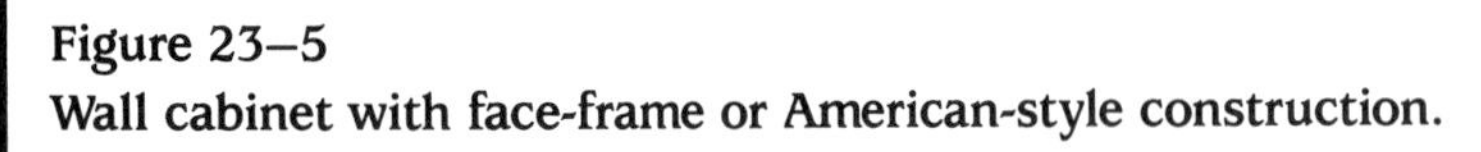

Figure 23–5
Wall cabinet with face-frame or American-style construction.

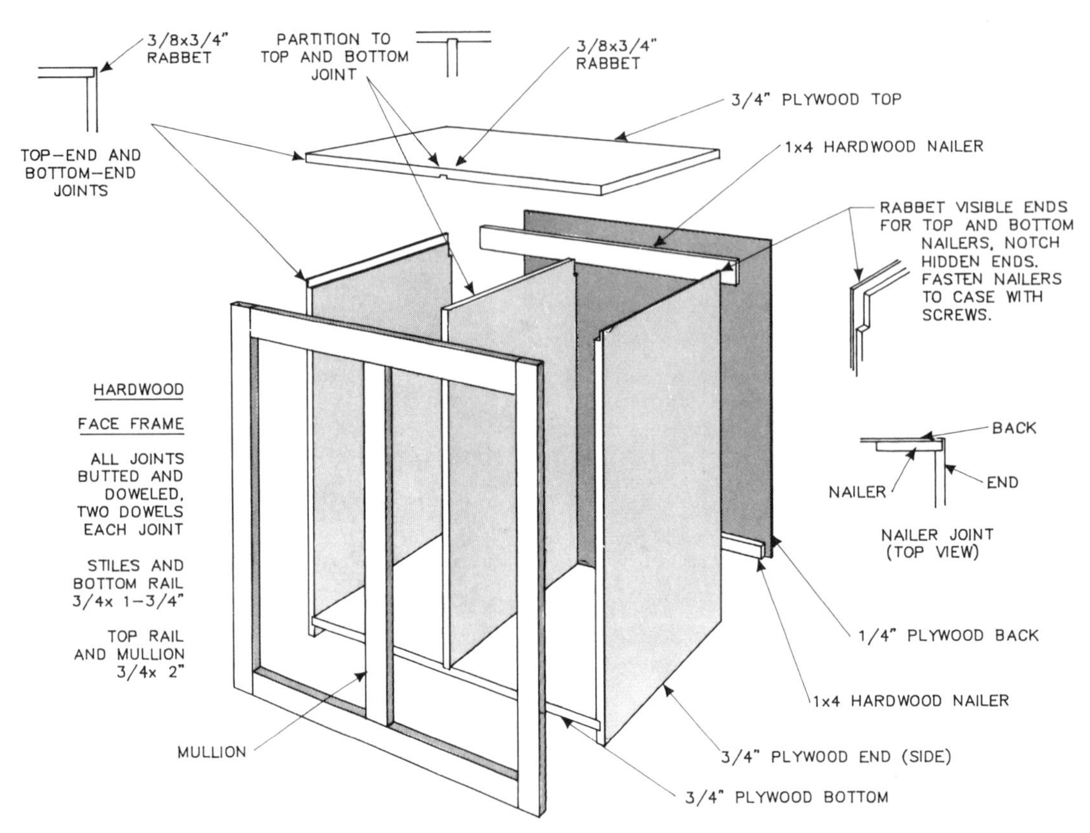

Figure 23–6
Wall cabinet box or European-style construction.

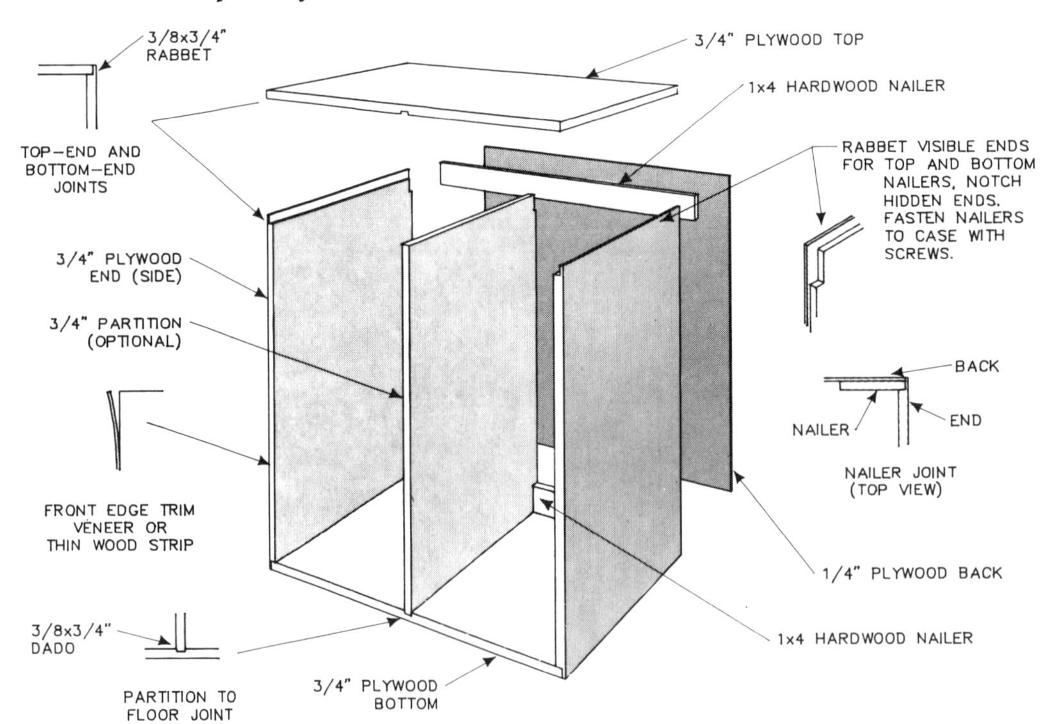

except that the bottom must be rabbeted flush rather than dadoed and the front edges of the box must be veneered.

Base Cabinet Construction

Base cabinets (Figure 23–7) can be built with integral kickplates, but installing the cabinet will be much easier if the cabinet and kickplate are built separately. The construction of a base cabinet is the same as for a wall unit, except that the top (which is also the countertop core stock underlayment) is not attached until after the cabinet or cabinets are installed. Either American- or European-style construction can be used, to match the wall cabinets.

Doors and Drawer Fronts

These are the most visible decorative features of kitchen cabinets. There are many different ways you can build and hang doors, both for new cabinets and to dress up or restyle existing cabinets. There are also four ways the doors can be hung on the cabinets, as shown in Figure 23–8. Lipped doors are widely used with face frames, as they can cover the opening with a minimum of installation difficulty, using special hinges designed to fit the rabbet. Reveal overlay doors are best hung with European-style hinges (see be-

Figure 23–8
Styles of cabinet door installation.

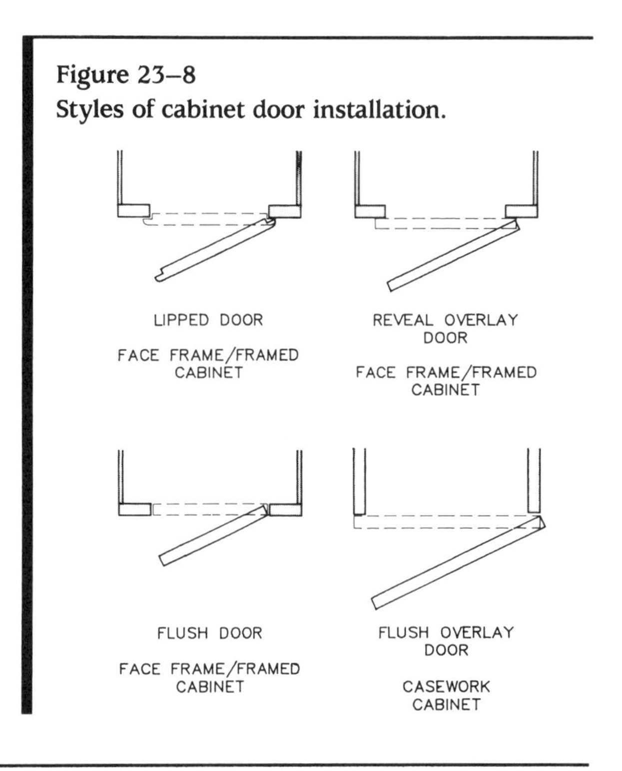
LIPPED DOOR
FACE FRAME/FRAMED CABINET
REVEAL OVERLAY DOOR
FACE FRAME/FRAMED CABINET
FLUSH DOOR
FACE FRAME/FRAMED CABINET
FLUSH OVERLAY DOOR
CASEWORK CABINET

Figure 23–7
Base cabinets can be built with face-frame construction, as shown, or box construction. Build kickplates as separate units to go under all the cabinets along a wall.

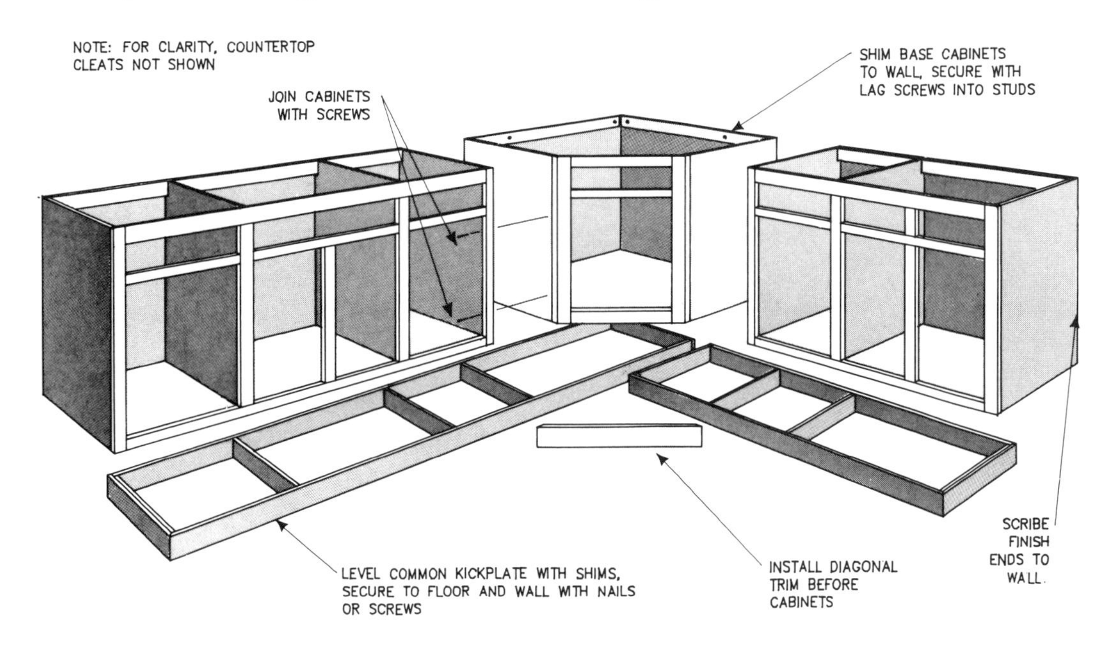
NOTE: FOR CLARITY, COUNTERTOP CLEATS NOT SHOWN
JOIN CABINETS WITH SCREWS
SHIM BASE CABINETS TO WALL, SECURE WITH LAG SCREWS INTO STUDS
SCRIBE FINISH ENDS TO WALL
LEVEL COMMON KICKPLATE WITH SHIMS, SECURE TO FLOOR AND WALL WITH NAILS OR SCREWS
INSTALL DIAGONAL TRIM BEFORE CABINETS

low). Flush doors set into an opening are difficult to hang neatly and are seldom used. Flush overlay doors provide a clean contemporary look. Drawer fronts should match the doors.

Today, the best hinges to use are European-style ones, shown in Figure 23–9. These hinges are widely used on kitchen and other cabinets despite their considerably higher cost. They can be used with or without face frames, provided the correct hinge model is selected. These hinges have three major advantages: (1) they are invisible when the door is closed, (2) the fit of the door on the cabinet can be adjusted after the door is hung with simple screwdriver adjustment, and (3) the door can be removed without disturbing fit adjustments by loosening one screw at each hinge.

Figure 23–9
European-style hinges. Top, for face-frame construction; bottom, for box construction. (Photo courtesy Julius Blum Inc.)

Drawer Construction

The drawer construction shown in Figure 23–10 is the method used by professionals; it requires the ability to make rabbets and dadoes. Drawers can also be built in the simplified method shown in Chapter 22. To make the drawer, rabbet the drawer front for the sides, and dado the sides for the back; also cut a dado in each piece for the bottom. The bottom can be slid under the drawer back, or the back can be dadoed and the bottom boxed in. Reinforce the glued front-to-side joints with nails or dowels. Allow 1⁄16″ all around when dimensioning the drawer—don't try for too snug a fit; it will give you nothing but trouble.

Drawer Installation

The drawer guides shown in Figure 23–11 work well with wide face framing. The guides should be positioned to hold the drawer free of the face framing on all sides. The kicker keeps the drawer horizontal when it is pulled partway out. Stop blocks (not shown) can be glued at the back of the drawer opening to keep the drawer from being pushed too far in. You can also buy drawer guide assemblies that attach to the drawer sides and the cabinet. The dimensions given are typical for lipped drawer fronts. Full-overlap drawer fronts are constructed in the same manner as cabinet doors.

Kickplate

The kickplate raises the cabinet 4″ off the floor so you can stand closer to the cabinets. It should be constructed of 1×4 lumber or 3⁄4″ plywood, as shown in Figure 23–12. Exposed corner joints should be mitered and blocked. Hidden sides and backs should be set in as shown to stay clear of wall irregularities and allow the ends of finish pieces to be trimmed for tight fits against the wall.

Figure 23–10
Drawer construction.

Figure 23–11
Drawer installation.

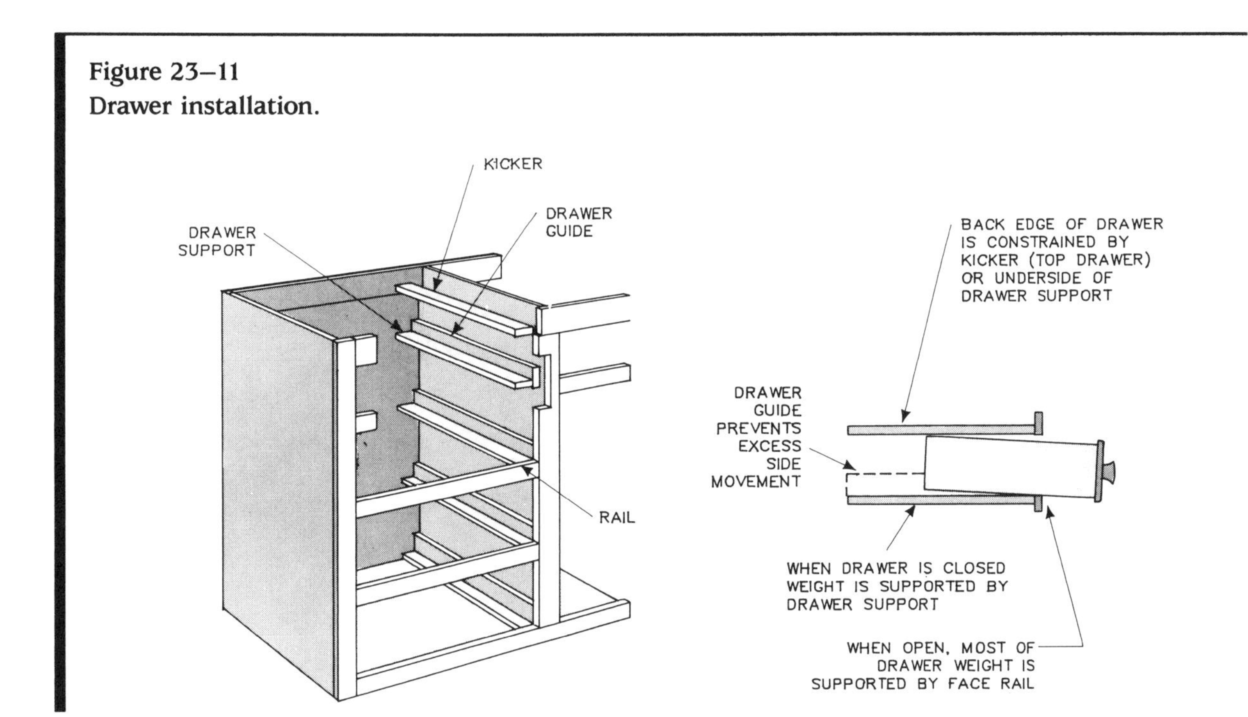

INSTALLING KITCHEN CABINETS

Preliminaries

If you are doing a major kitchen renovation, you should rip out everything to the bare walls instead of trying to work around the sink and appliances.

The new countertop and the bottoms of the wall cabinets must be level, whether your floor and ceiling are or not. The standard height for a countertop is 34″. Draw a level line on the wall to mark that. The line must be measured from the highest point on the floor, so your first step is to

Figure 23–12
Top views of various kickplate configurations. Broken lines indicate the cabinet overhang.

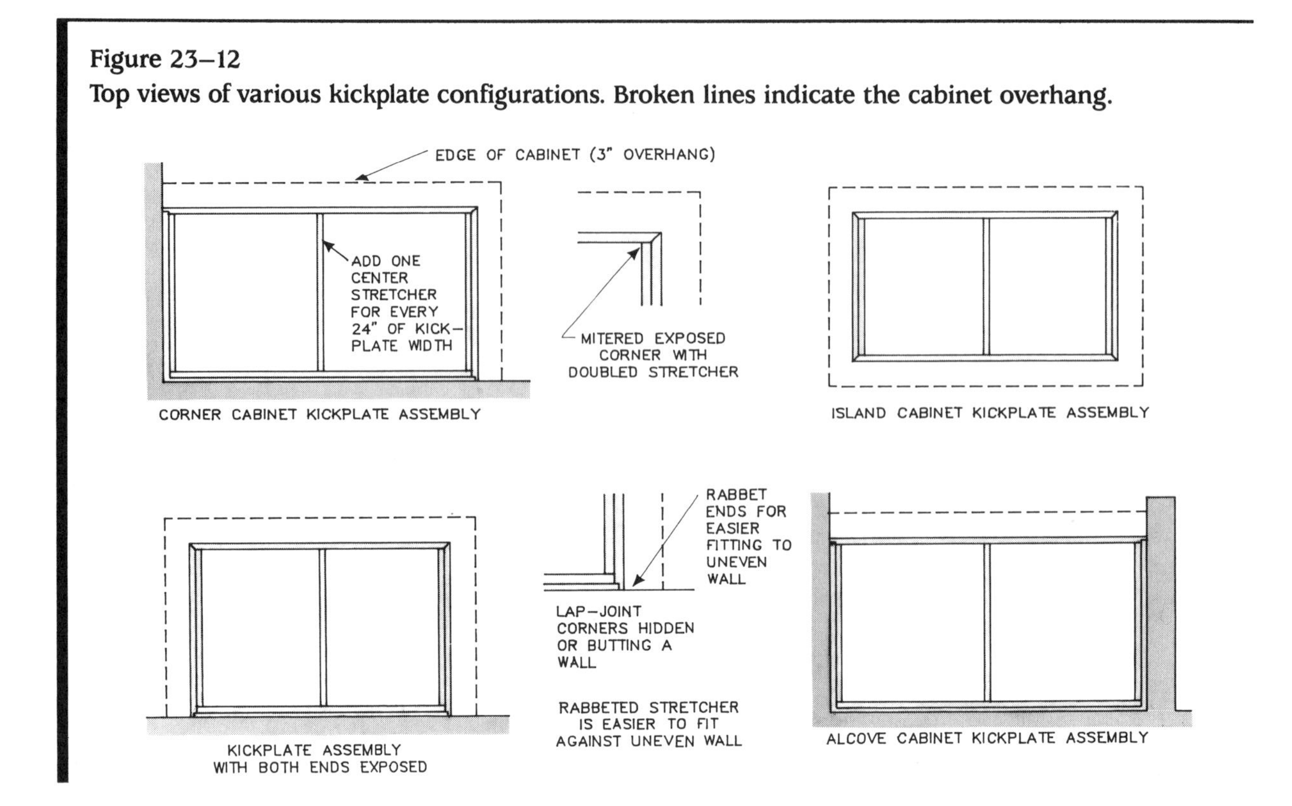

determine how level your floor is. Use a long straightedge and a level.

The underside of the wall cabinets should be 16″ to 18″ above the countertop surface. Next, check the walls to locate high spots that will interfere with cabinet installation. The cabinet could twist out of square as you screw it tight—resulting in door and drawer operating problems. Protruding areas wholly behind the cabinets will be less of a problem than high areas crossed by cabinet ends or butting sides.

Locate the studs and mark their locations, as the cabinets must be fastened to studs. Next draw the locations of the cabinets on the walls.

Should you install the base cabinets first or the wall cabinets? If you put the wall cabinets up first, you can get closer to them and there will be less chance of marring the base cabinets.

On the other hand, if you install the base cabinets first, it is easier to brace the wall cabinets in position against the wall. You won't be banging your head on the wall cabinets, and it may be easier to get the countertop installed. Overall, it is probably easier to do the base cabinets first.

Base Cabinet Installation

First install the kickplate. This usually involves little more than getting it level and scribing the exposed ends to the wall. If several cabinets are covered by a single countertop, or if they butt in a corner, the kickplates should be joined or made into a single unit before being leveled and installed.

Use wood shingles or similar wedges for leveling; tack them in place with construction adhesive. Toenail the kickplate to the floor, or use cleats and nails or gypsum board screws. Trim wedges flush with the outside of the kickplate; ends can be covered with molding later, or the kickplate can be faced with resilient flooring baseboard.

To install a base cabinet (Figure 23–13), place it on its kickplate and move the cabinet against the wall. If it doesn't fit flat to the wall, scribe a line on the cabinet matching the contour of the wall. (Scribing is explained in Chapter 14.) Plane the cabinet side to the line (use a file if the cabinet side is plastic laminate). Proceed cautiously, and

Figure 23–13
Base cabinet installation.

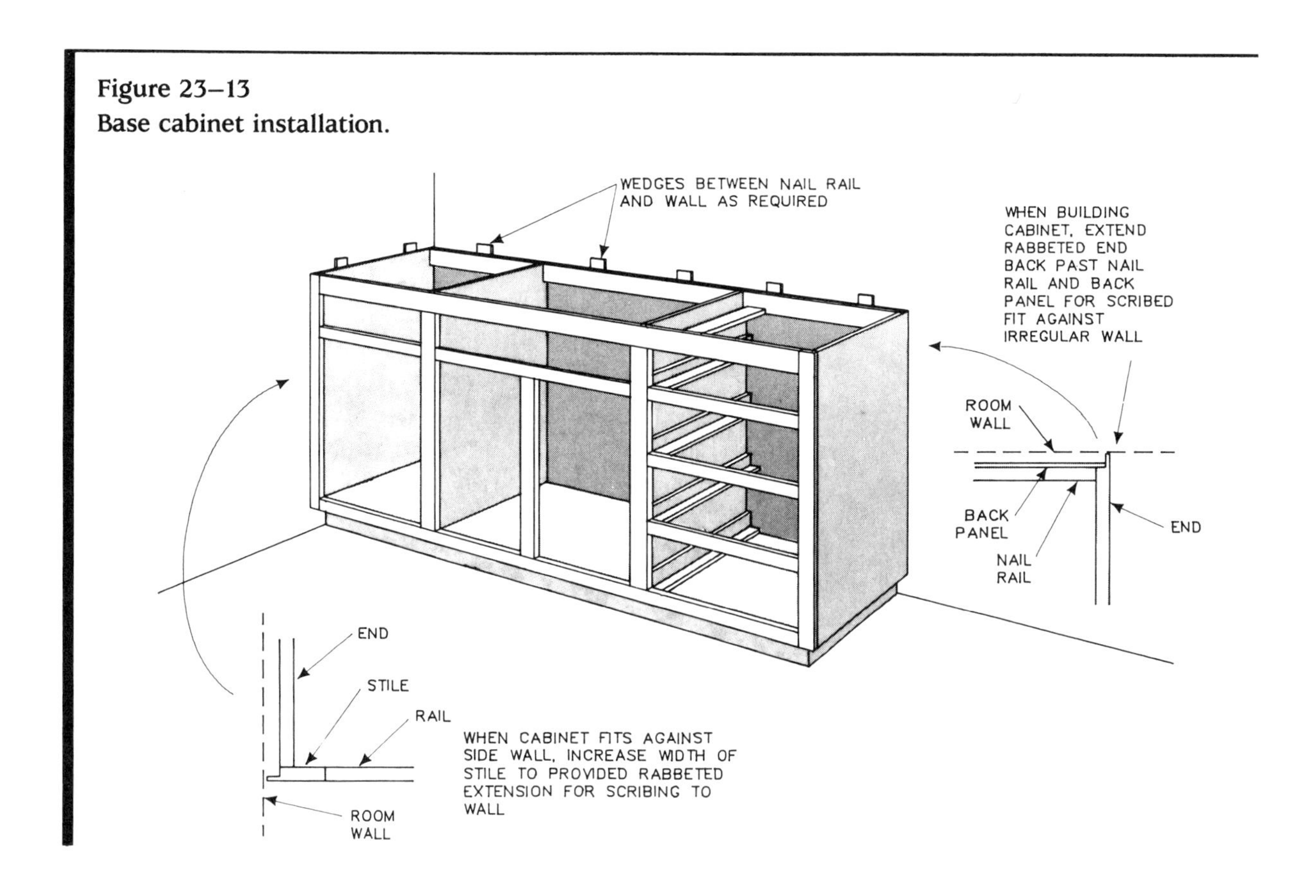
WEDGES BETWEEN NAIL RAIL AND WALL AS REQUIRED
WHEN BUILDING CABINET, EXTEND RABBETED END BACK PAST NAIL RAIL AND BACK PANEL FOR SCRIBED FIT AGAINST IRREGULAR WALL
ROOM WALL
BACK PANEL
NAIL RAIL
END
END
STILE
RAIL
WHEN CABINET FITS AGAINST SIDE WALL, INCREASE WIDTH OF STILE TO PROVIDED RABBETED EXTENSION FOR SCRIBING TO WALL
ROOM WALL

check often so you don't overshoot. Do the same on the cabinet frame if it butts a side wall.

Position the cabinet and drill screw clearance holes in the nail rails at stud locations previously marked on the wall. Drop in gap-filling wedges between the cabinet and wall, drill screw pilot holes through the gypsum or plaster, and start the screws or lag screws. Screws should be long enough to penetrate at least 1″ into the stud. If several cabinets are being installed together, they should be either joined with screws or bolts before any are attached to the wall or before driving the screws home. Open and close the doors and drawers to make sure they operate smoothly as you turn in the screws. Trim the wedges at the top before adding the countertop underlayment.

Position the countertop underlayment (see Chapter 24) and scribe it to the wall. A perfect fit is not necessary except at the ends. Before installing the countertop, make a final level check of the doors and drawers of the base cabinets. If they aren't level, go over your installation to find out what has been twisted out of square and adjust the shimming. Correction will be difficult later. The underlayment can be screwed to the cabinet tops now, unless you are installing a full laminate backsplash, which you should do first. Screw heads must be sunk below the countertop surface and filled over.

Large corner cabinets—the kind containing a lazy Susan—and undersink cabinets sometimes come without backs. Before installing them, put 1 × 2 cleats on the wall at the same height as the front of the cabinets to provide support for the back edge of the countertop underlayment. Use plywood for cleats, not 1 × 2 lumber, which can shrink or swell and throw the countertop off level.

Purchased base cabinets. These units have integral kickplates. They are installed with their tops aligned to a level line drawn on the wall. If you are installing several cabinets, and you can handle a group, C-clamp them together on the floor and join them permanently with screws first, so you can install them as a single aligned unit. Otherwise, you will have to tediously install and align them one by one.

Shim the base cabinets up from the floor so they align with the level line on the wall, and shim from the back wall until they are plumb, before securing them to the wall with screws. As an alternative to scribing cabinets to walls, filler strips cut and prefinished for the purpose can be used.

Wall Cabinet Installation

With the base cabinets in place, the wall units can be supported on boxes shimmed with scrap wood. Plan to shim; full-height supports may end up trapped. If the wall cabinets go to the ceiling, keep their bottom surfaces level and scribe to the ceiling, or plan to use a trim molding. Scribe to the wall too, and be sure to keep the front surfaces vertical.

Shim the wall cabinets and fasten them to the wall with screws or nails into studs the same as the base cabinets. Fasteners should be located for minimum visibility. Be sure to check the door operation before pulling cabinets up tight. If a soffit is desired, to fill the cabinet-to-ceiling space, construct it as shown in Figure 23–14.

Figure 23–14
Soffit construction.

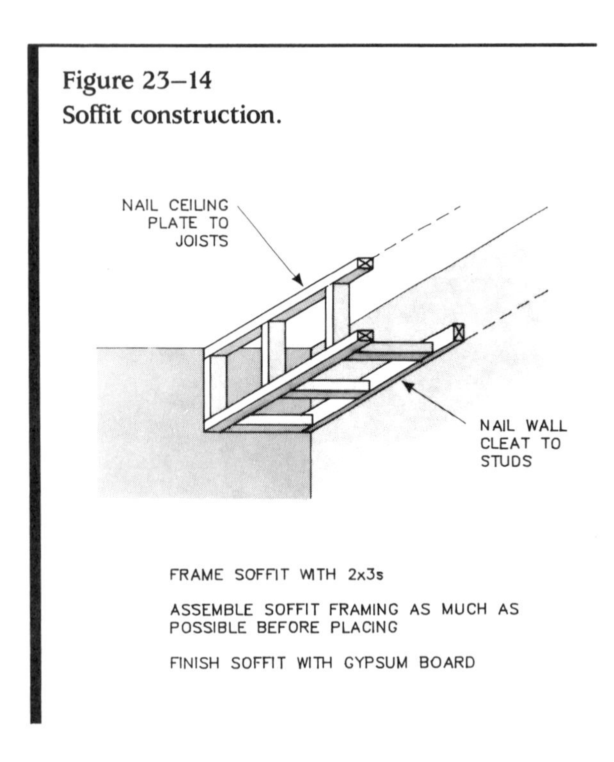
NAIL CEILING PLATE TO JOISTS
NAIL WALL CLEAT TO STUDS
FRAME SOFFIT WITH 2x3s
ASSEMBLE SOFFIT FRAMING AS MUCH AS POSSIBLE BEFORE PLACING
FINISH SOFFIT WITH GYPSUM BOARD

Countertop and Backsplash

The final step in kitchen cabinet installation is the countertop and backsplash, either plastic laminate or ceramic tile applied to core stock, or a ready-made laminate-covered countertop.

Note: If a full plastic-laminate backsplash, one that extends up to the wall cabinets, is desired it should be installed before hanging any wall cabinets. See Chapter 24 for details.

Countertop choices. The advantage of a ready-made laminate countertop is that the countertop and backsplash are combined without a seam; the disadvantages are cost, a long diagonal joint if the counter turns a corner, and limited options for backsplash height.

Plastic laminate is less expensive and less work to install than ceramic tile and produces an attractive and serviceable work surface; the best way to go, however, is ceramic tile. Plastic laminate installation is covered in Chapter 24 and working with ceramic tile is covered in Chapter 20.

REFACING AND RESTYLING KITCHEN CABINETS

Probably the biggest task in remodeling a kitchen is "doing something" with the cabinets. You can either rip out the old cabinets and buy new ones, or reface the existing cabinets by installing new cabinet doors and drawer fronts and facing the exposed cabinet surfaces with laminate or veneer to match the doors.

You should not be too quick to opt for discarding your old cabinets, unless the boxes (the parts behind the face frames) are irreparably shabby or falling apart, or the cabinets are the wrong size for your remodeled kitchen. Well-made old hardwood-frame cabinets may be of better quality than new replacement cabinets that will fit into your budget.

Figure 23–15
Before (below) and after (right) refacing. Doors and drawer fronts shown can be purchased custom-made to your dimensions. (Photos courtesy Quality Doors)

Refacing

The big reason for the popularity of refacing is simply that it costs less than ripping out the old cabinets and starting with new ones. It also can be done a lot faster. A new countertop and new additional cabinets can be included in the job. The procedure for refacing is simple and straightforward. Refacing is widely advertised by franchised contractors, but you can buy the parts and do it yourself at considerable savings. A refacing before and after is shown in Figure 23–15.

In refacing, the doors are replaced completely; the lips on old drawer fronts are usually trimmed off and new drawer fronts are attached to the drawers. Self-adhesive veneer is applied to the cabinet fronts.

There is a wide variety of style choices in refacing. You can rehang your cabinets with overlap (part and full) laminate-faced doors, molded polyurethane doors, or wood frame-and-panel doors. Visit kitchen cabinet showrooms and look at home magazines and ads to get an idea of what is new in cabinet styles.

Sources of Do-It-Yourself Doors

Quality Doors of Duncanville, Texas, will custom-manufacture doors and drawer fronts to your exact dimensions in red oak, ash, maple, or a paint-grade hardwood. Doors are available in twelve frame-and-panel styles, frame only (insert your own panel), frames with mullion inserts for glazing, and frame and leaded glass in several styles. They are normally unfinished, but optional finishing is available. You can order doors through home centers, using a "how-to" brochure giving complete measuring, preparation, and installation instructions, as well as pricing information. Delivery takes about three to four weeks.

Masonite Corporation offers solid oak cabinet doors and drawer fronts, and matching peel-and-stick oak veneer. Doors and drawers are not custom made—nine door and three drawer sizes and matching veneer are stocked at home centers. You may have to shim openings (Masonite's recommended practice) in order to get a proper fit.

Indoors by Interface Industries, Inc., sells Formica laminate doors and drawer fronts in twenty-four styles for do-it-yourself installation.

Veneer. What makes refacing or restyling so easy and economical today is the new adhesive-backed veneer that can be used to cover the face frames behind the doors. These veneers are much easier to work with than laminates and look far better than painting and "graining" the cabinet face frames to match the finish, grain, and figure of the doors.

If you have worked with standard veneer in the past, the new adhesive-backed veneers will be a surprise. The pieces are large—up to 24″×96″. And the veneer is sanded and thin—1/64″. Two types of adhesive are used: pressure-sensitive peel-and-stick and a heat-sensitive adhesive that allows you to install the veneer with a household iron set to the cotton temperature setting.

24 Plastic Laminate

Plastic laminate is widely used for kitchen and bathroom countertops. It is also used extensively for tabletops, workbenches, cabinets, and doors—anywhere an attractive, wear-resistant, easy-to-clean surface is needed. Common trade names for plastic laminate are Formica, Micarta, and Textolite, among others. Laminate comes in a wide range of surfaces, colors, and patterns. Smooth surfaces include gloss and matte; other surfaces are embossed to simulate materials such as slate, leather, stone, or basketweave. Patterns include solid colors, wood grain, fabric, and almost anything imaginable.

High-pressure plastic laminates consist of layers of kraft paper impregnated with phenolic resins covered by melamine resin and a pattern sheet and a final protective melamine resin layer. The sandwich is compressed between stainless-steel platens under extremely high heat and pressure.

General-purpose-grade plastic laminate is 1⁄16" thick. Standard widths are 24", 30", 36", 48", and 60". Standard lengths are 5′, 6′, 7′, 8′, 10′, and 12′. Sheets normally are supplied slightly oversize, so you do not have to make allowance for trimming when ordering. The cost of plastic laminate begins at about $1.50 per square foot for plain colors and surfaces, and goes up from there for complex patterns and textures. Plan new projects economically to utilize standard sheets, and by all means buy a sheet big enough to do a tabletop or counter in one piece—it is very difficult to get an invisible butt joint between large pieces.

A new type of plastic laminate that has integral solid color—the same color all the way through—is available at a premium price. The Formica product is called Colorcore (Figure 24–1). There are other brands. With this material, there is no dark core line to show at butted edges or corners, and any chips and scratches are not as visible as with ordinary laminate. However, with no visible dark core, glue lines will show: care must be exercised during installation to get the mounting adhesive applied thinly and evenly. Color and pattern ranges are limited. The solid-color laminate is also stiffer than ordinary laminate and must be handled carefully until cemented down.

Laminate is installed with contact cement. It can be applied to any flat surface, but the surface must be clean and free of old finish, such as paint or varnish—the solvents in the cement can loosen the finish, allowing the laminate to lift off.

Figure 24–1
Plastic laminate is ideal for use in bathrooms and kitchens because its smooth surface is easy to clean and is scratch-resistant. The laminate here, Colorcore, has color through its entire thickness. (Photo courtesy Formica Corporation)

You can apply new laminate over old, but the old surface, particularly a glossy surface, must be sanded with silicon carbide paper to break the gloss. Aluminum oxide, flint, and garnet papers are too soft to cut the hard laminate surface effectively. Dust thoroughly after sanding, before applying the contact adhesive.

COUNTERTOP MATERIALS

Core stock. The usual core stock for countertops is particleboard, flakeboard, or plywood, normally ¾″ thick. (In the countertop trade, core stock is called underlayment.) Interior grade A-D and exterior A-C fir plywood are not the best choices for core stock because the inside plies are not required by grading standards to be sound. You can also waste a lot of time filling the knotholes and cracks permitted in surfaces and edges to get the smooth solid surface required for laminating. When cutting core stock, the edges must be absolutely square (Figure 24–2). Otherwise it will be impossible to make precise edge and corner joints with the laminate.

Nosing and joints. If edges are to be thicker than ¾″, as is common for most countertops, blocking should be applied as shown in Figure 24–3. Joints in core stock should be avoided as they produce a potentially weak counter. If a joint is necessary, it must be backed along its entire length with a piece screwed on both sides of the joint, as shown.

Adhesives. Sheets of laminate are applied to the core stock with contact cement, which does not require clamping. There are two kinds of contact adhesive—flammable and nonflammable. The nonflammable kind is a lot safer to use, from

Figure 24–2
Core stock edges must be cut square with the top surface to achieve proper edge lamination.

Figure 24–3
Counter edge nosing and joint construction. Pine may be used for nosings, but plywood should be used for backing underlayment joints to minimize high-humidity joint expansion.

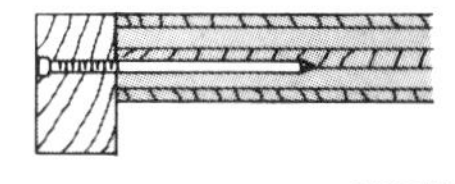
GOOD
UNDERLAYMENT
NOSING
NOSING (BANDING) LETS COUNTERTOP APPEAR THICKER THAN 3/4", AND OVERHANG BASE CABINET FRONT TO HIDE JOINT

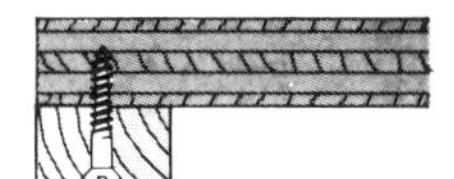
WEAK
UNDERLAYMENT
NOSING
APPLYING THE NOSING TO UNDERSIDE OF COUNTER–TOP IS NOT RECOMMENDED AS IT PRODUCES A WEAK BASE FOR THE LAMINATE EDGING

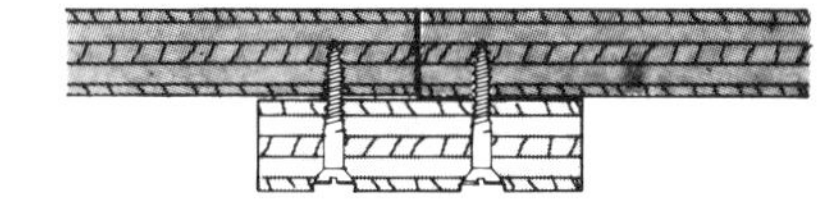
JOINT
CONSTRUCTION
UNDERLAYMENT JOINTS SHOULD BE AVOIDED, BUT IF NECESSARY SHOULD BE SOLIDLY BACKED

both fire and health hazard standpoints. If you use the flammable type (some installers believe it is superior to the nonflammable), be sure to have adequate cross-ventilation, and leave the room while waiting for the adhesive to dry.

BASIC LAMINATING TECHNIQUES

Installing plastic laminate involves three tasks: cutting the pieces, gluing (cementing) them in place, and trimming and finishing the edges.

Cutting

Pieces of laminate should be rough-cut ½" oversize, then trimmed to neat finished edges with a piloted carbide router bit after being cemented to the core stock. Laminate can be cut with a fine-toothed handsaw, hacksaw, circular saw with a plywood blade, or saber saw. Or you can score the face with a sharp knife and bend the laminate over a straight edge to break it along the scored line. (Scoring and bending is not the best method. It requires a carbide-tipped utility knife, practice, and luck.) No matter how you cut laminate, you must remember that it is a very brittle material that chips and shatters easily. It is also a very hard material and can dull saw teeth quickly.

If you cut laminate with a saber saw (Figure 24–4), use a Sears #28761 superfine finish blade. This blade cuts on both up and down strokes; the teeth slice the edges of the kerf rather than chisel it, as the teeth of a conventional blade do. Although the blade is initially dulled as any other noncarbide blade would be, it continues to cut cleanly with minimal chipping if you keep the feed slow. Put a strip of masking tape down each side of the saw base plate to avoid scratching the laminate surface. Another way you can cut laminate is with a router and a $\frac{3}{16}$" straight carbide router bit.

Figure 24–4
Cutting laminate with a saber saw.

CRAFTSMAN

When cutting laminate in any manner, be sure to adequately support the piece being cut off. If you use a saber saw, start the cut with a piece of scrap plywood under the cut.

Gluing

Follow the directions on the adhesive can when gluing. Both surfaces must be completely and evenly coated with contact cement. A porous surface may require a second coat. An otherwise useless but clean paintbrush is a good applicator, which can be tossed when you are finished. Or use a serrated steel trowel with 1⁄16" notches that can be cleaned (Figure 24–5). The glue must be allowed to dry before the surfaces will adhere. But then, once the surfaces are brought into contact, the cement bonds and they cannot be separated, so it is important to be sure of the alignment.

Trimming

Laminate edges can be trimmed with a router or a file. A router is best, but a file will be needed in hard-to-reach places. Final edge finishing can be done with silicon carbide paper. The techniques are described in the following sections.

Figure 24–5
Applying contact cement with a serrated steel trowel.

LAMINATING A COUNTERTOP OR TABLETOP

The proper sequence for covering a counter or similar surface with plastic laminate is as follows:

1. Glue laminate to the edges.
2. Trim the edge laminate.
3. Glue laminate to the surface.
4. Trim the surface-covering edges.
5. Finish the edges.

Laminating Edges

If all four edges will show, start with the back, then do the sides; finish with the front edge. This is to minimize the visibility of the edge of the laminate. (*Note:* If you have to laminate the back side or underside of the core material, do that first, before the edges.) Use this same procedure with integral-color laminate.

Cut strips of laminate ½" wider than the edge to be covered. Apply contact cement to the edges of the counter and to the backs of the laminate strips and let it dry. Press the strips into place with the excess width extending ¼" on each side of the counter edge. The contact cement grips instantly and permanently when the two coated surfaces are brought together, so proper alignment is very important from the outset. Use a slipsheet of waxed paper and move it along to allow the surfaces to come into contact as you work. Once a strip is in place, tap it with a hammer and protective wood block (Figure 24–6) to ensure adhesion along the entire edge.

Trimming Edges

Edges should be cut square when they are to be lapped over by another piece of laminate and either beveled 22½° or cut square when they will be exposed. Use ball-bearing piloted straight and bevel bits in a router, as shown in Figure 24–7. There is also a combination straight and

Figure 24–6
After applying each edge strip, tap it as shown to get complete adhesion.

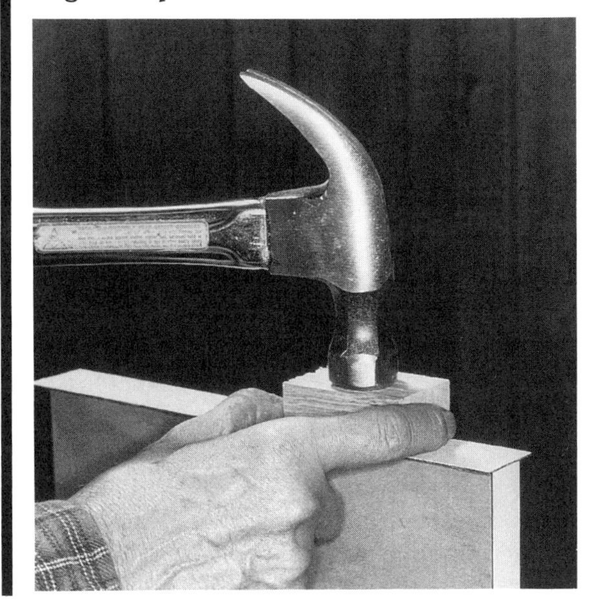

Figure 24–7
Trimming edges. If the laminate edge is to be overlapped by another piece of laminate, trim it square. A visible laminate edge is usually beveled, but may be trimmed square. Adjust the router carefully to avoid overcutting.

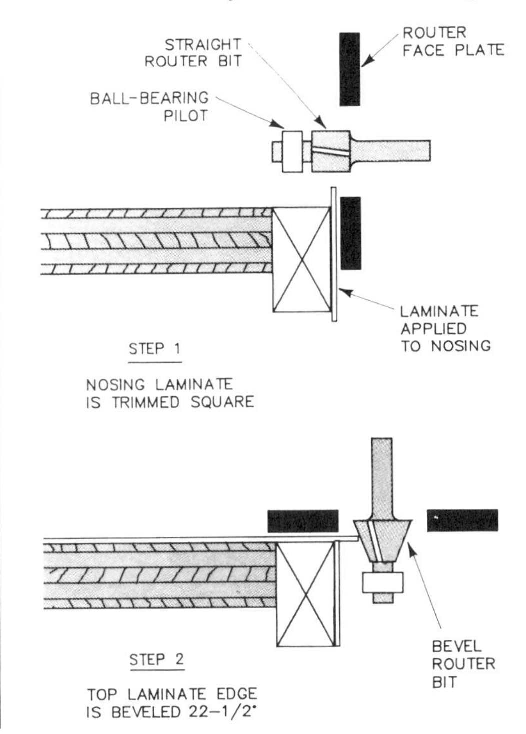

bevel bit that is used with a special Porter-Cable router base to trim edges.

Trimming a laminate edge with the base of the router riding flat on the laminate edge is a risky undertaking. The slightest tip of the router away from you will cut into the edge. Before routing, clamp a block of wood flush with the laminate surface (Figure 24–8) to provide a wider surface to support the router base plate.

When trimming the first piece, the pilot collar of either a straight or bevel trimming bit will be guided by the core stock surface. It is important that this surface be smooth, without depressions of any size along the guide path. Small dents and holes elsewhere in the core surface will not affect the laminating job.

Edges can also be cleaned up with a file, as shown in Figure 24–9. If you are laminating a countertop in place, you will have to do some trimming with a file, as a router can't trim much closer than 3″ from a wall because of the face plate. A file is also useful for trimming edge banding before you apply the top piece when you are laminating over a core that has irregularities along the track of the pilot collar that guides the

Figure 24–8
To prevent the router from accidentally tipping over a narrow edge, clamp straight and true scrap stock as shown to provide more support for the router base plate.

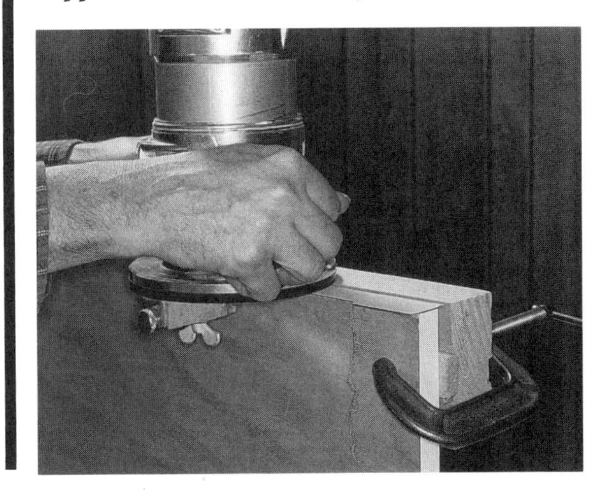

bit. Wrap a layer of masking tape around the end of the file to keep it from gouging the core surface.

Laminating Top Surfaces

The laminate for the top of the sheet should be rough-cut oversize ½″ at each edge and trimmed with the router after it is glued down. Coat the counter surface and the back of the laminate with a thin layer of contact cement and then let it dry.

There are several ways to get the laminate positioned accurately. Have help. Four hands and four eyes are better than two. Overlapping sheets of kraft paper or wax paper on the core stock will prevent sticking while you adjust the laminate. When you have the laminate in position, lift one edge, pull out a piece of the paper, and press the laminate onto the core stock. Then pull out the rest of the paper piece by piece.

Another way is to clamp scrap wood and the core stock to your bench, using spacers so the scrap pieces will position a corner of the oversize laminate. This method is shown in Figure 24–10. Another way is to put a row of dowels on the core stock and roll the laminate into position, then remove dowels one at a time as you press down the laminate.

You must now apply some quick pressure over the whole surface to ensure good contact and adhesion. Conventional clamping will certainly do it: clamp briefly, unclamp, move along to the next spot and clamp again. A small block of wood and a hammer are noisy but work quite well. You can also use a roller, but you must bear

Figure 24–9
Filing laminate edges. To keep an edge square, use the flat surface of core stock to guide the file.

Figure 24–10
Aligning laminate panel on core stock. Clamped scrap blocking locates one corner of the laminate.

down heavily to do any good. The shorter the roller, the better.

Trimming the Top

Use a straight bit in a router for a square edge cut or a bevel bit for an angled edge cut. Let the base plate rest on the top surface (protect the surface with smooth paper if the base plate has any nicks or gouges). The pilot collar of the bit rides against the laminate-covered edge of the countertop. If you choose to bevel the edge, be careful not to overcut into the edge laminate.

Figure 24–11 shows a way to trim laminate to fit against an irregular wall. Position the underlayment against the wall; clamp the oversize laminate in position close enough to the wall so that the router base plate will be guided by the wall as the trim cut is made.

Finishing Edges

Routed laminate edges are sharp. To finish the edges, break their sharpness with 320A wet or dry silicon carbide paper wrapped around a block of wood to round the very sharp edge, as shown in Figure 24–12.

Figure 24–11
Fitting a laminate panel to an irregular wall.

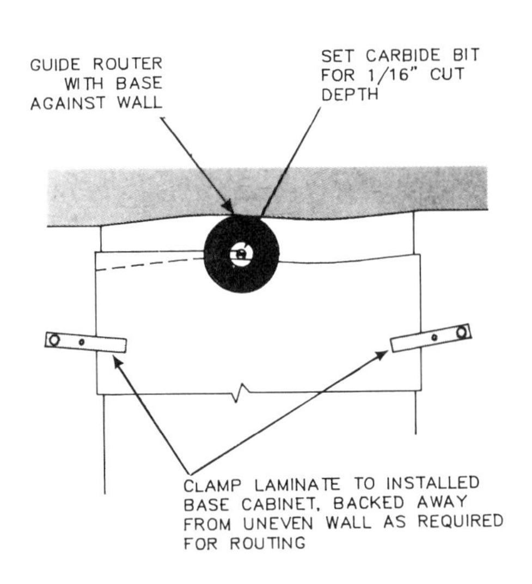
GUIDE ROUTER
WITH BASE
AGAINST WALL
SET CARBIDE BIT
FOR 1/16" CUT
DEPTH
CLAMP LAMINATE TO INSTALLED
BASE CABINET, BACKED AWAY
FROM UNEVEN WALL AS REQUIRED
FOR ROUTING

Figure 24–12
Finish laminate edges by sanding with silicon carbide paper to break the sharp corner along each edge.

MAKING A SEAM

Try to avoid having to use two pieces of laminate on a countertop. If that is not possible, place the seam inconspicuously—along the edge of a sink, for example. Cut the pieces so that factory edges will meet at the seam. After a trial fit, apply adhesive to the core stock and both sheets. Be sure there is no glue on the edges being joined. Adhere one of the sheets in place. Then position dowels as shown in Figure 24–13. Roll the second sheet into position on dowels and butt the edges at the seam. Press onto the core. Clamp the butted edges before removing dowels one by one, starting with the dowel closest to the seam (Figure 24–14).

INSTALLING A BACKSPLASH

A plastic laminate backsplash (Figure 24–15) is a durable and easily cleaned wall surface behind a counter. The best backsplash covers the wall sur-

Figure 24–13
Making a seam. After the first piece is cemented down, roll the second piece into position on dowels.

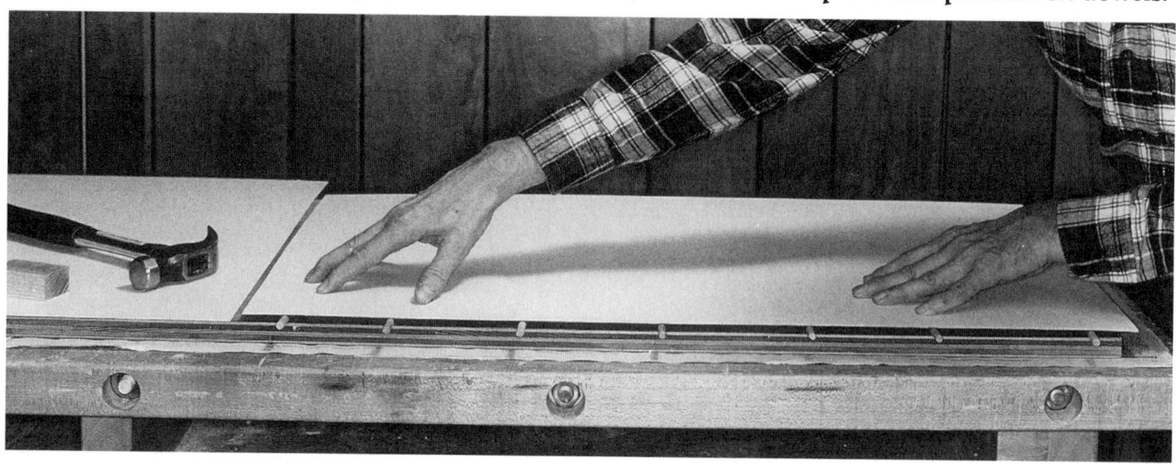

Figure 24–14
Clamp the seam with a batten before removing the dowels under the second piece. Work from the seam to the far end.

face between the counter and the bottom of any wall cabinets above. Partial backsplashes running 5″ to 12″ up the wall can be made several ways. The backsplash is joined to the countertop with an aluminum extruded flanged cove molding. Exposed backsplash edges are trimmed with an extruded aluminum cap molding. If the backsplash is installed before the wall cabinets, you can make it high enough so they overlap the upper edge, saving you a joint to be fitted.

You can also fit a laminate backsplash without the aluminum moldings, but it involves a lot of cut-and-try work.

Figure 24–15
Installing laminate backsplashes with extruded aluminum trim molding.

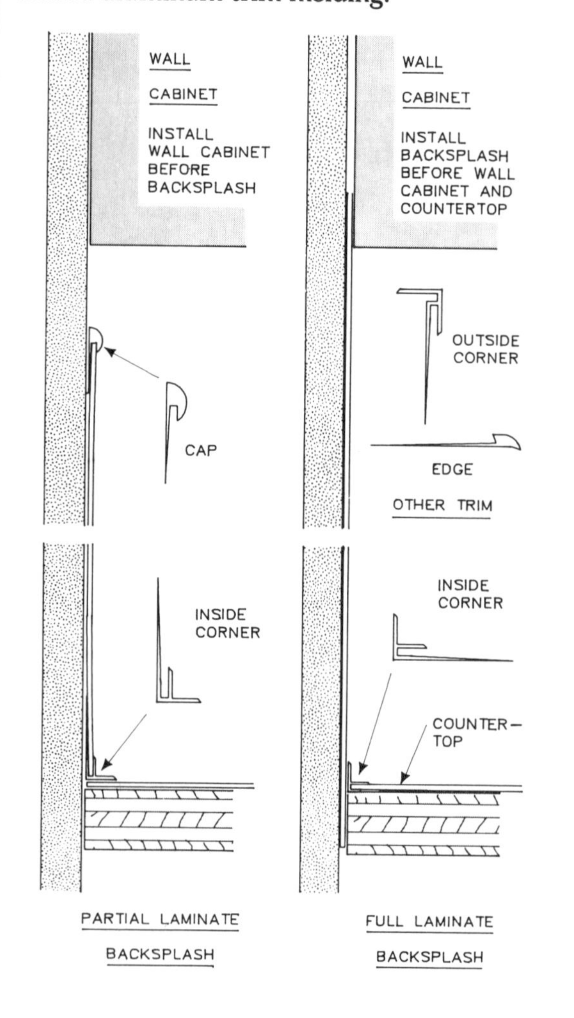

25 Built-in Beds

Bedrooms tend to be among the smaller rooms in the house, yet they contain some of the largest pieces of furniture. If a bedroom is also used for another purpose—study, playroom, den, home office—on a time-sharing basis, the bed is in the way more than it is used for sleeping.

Building a bed into a room, rather than having a separate piece of furniture, increases the usable space in the room. A built-in can also incorporate storage space more efficiently and neatly than shoving boxes under the bed. A built-in unit can put two beds in the space of one, as in the case of a double-deck bunk bed or a trundle bed, which rolls out of the way under another bed. Other built-in beds can be fitted into difficult-to-use space, such as under the eaves or in an alcove.

Perhaps the best-known kind of built-in bed is the Murphy bed, introduced in 1925, which tilts up from the foot to be hidden in a shallow cabinet or closet, or behind a partition. Murphy beds are available today in contemporary and traditional styles. You can purchase an entire cabinet unit, such as that shown in Figures 25–1 and 25–2, or you can buy just the folding, balanced bed frame and mechanism for installation in your own custom-built enclosure. You can even build a tilt-up Murphy-type bed frame of wood; details are given later in this chapter.

BUILT-IN BASICS

A built-in bed is basically a platform to support a mattress or a frame to support a spring and a mattress set. Although foam and other mattresses come in soft, medium, and firm, incorporating a set of springs makes a more comfortable bed. The platform or frame can be supported on legs, but it is more practical to support it on a box structure and incorporate storage drawers in the box.

In your design, you must accommodate the thickness of the mattress cover, pads, sheets, blankets, and pillows in addition to the basic spring and mattress dimensions. Allow necessary clearances. Don't put the side of a double bed against a wall—reaching across to tuck in sheets and blankets is very difficult; with a queen- or king-size bed, it's impossible. A 39″-wide standard twin bed is the largest size that should have one side placed against a wall.

Figure 25–1
This cleanly styled unit gives no hint that it conceals a full-size folding bed. (Photo courtesy Murphy Door and Bed Co., Inc.)

Figure 25–2
A Murphy bed can be folded away fully made up. The weight is counterbalanced by springs; the supporting legs at the foot unfold and retract as the frame is lowered and raised. (Photo courtesy Murphy Door and Bed Co., Inc.)

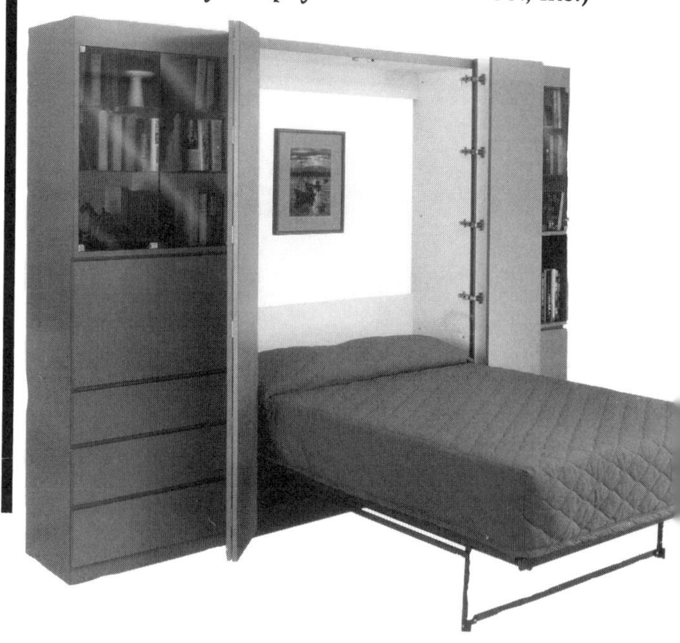

The first step in designing a built-in bed is to buy the mattress and the spring, if you are going to use one. Both come in standard sizes, as listed in Table 25–1.

SIMPLE BUILT-IN BEDS

The following three designs show you how to make the simplest platform frame (to support only a mattress) and box frames for platform and spring-and-mattress use. Each is intended to have one long side attached to a wall, but could also be used as a freestanding bed. The two box frames could even be mounted on casters so as to be movable.

Frame and Platform

A very basic built-in bed design is shown in Figure 25–3. It is a platform bed, with a frame of

Table 25–1
MATTRESS AND BEDSPRING DIMENSIONS

Size Name	Dimensions (approx.)
Bunk or Cot	30″ × 75″
Youth	33″ × 66″
Twin	39″ × 75″
Twin, Extra Long	39″ × 80″
Three-quarter	48″ × 75″
Full or Double	54″ × 75″
Queen	60″ × 80″
King	76″ × 80″
California King	72″ × 84″

Note: There are three types of bedsprings: (1) flat links attached to a steel angle frame with small coil springs, (2) large coil springs attached to the top of bottom frames made of an outer steel rod crisscrossed with flat link or heavy-wire lacing, and (3) type 2 attached to a bottom wooden frame and covered with cloth (this type is called a box frame).

Type 1 is the most suitable for foldaway beds because of its light weight and minimal thickness. It can be fastened to the bed frame with flange-and-bolt hardware or with lag screws inserted through holes in the steel angle frame. Types 2 and 3 are suitable for rollaway and other beds that are not tilted to a vertical position.

Figure 25–3
Simple frame and platform bed.

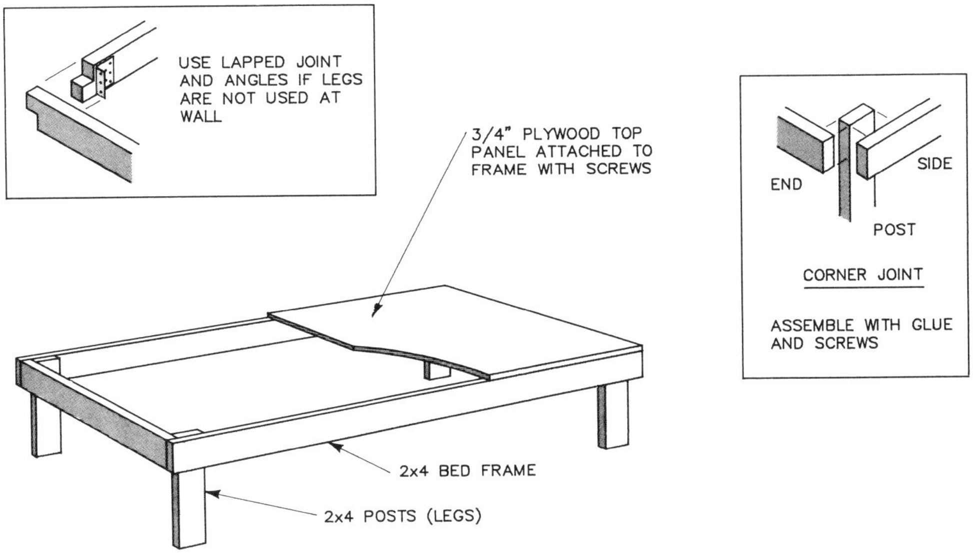

2 × 3s covered with a panel of ¾" plywood to support a mattress. The frame joints should be strongly made. In addition the bed should be attached to the wall to prevent racking the leg-to-frame joints. The frame could be enclosed with plywood if under-bed storage is not needed. The plywood could also extend up 1½" to provide a rim to keep the mattress from sliding off.

Box Frame

In the box design shown in Figure 25–4, cleats inside the top frame can support a ¾" plywood platform for mattress-only use or the frame of a spring, without a platform. As the corner joint detail shows, the legs act as posts to support the load on the sides and ends directly. This is much stronger and more rigid than the frame and platform construction, in which screws must carry the load at the corners.

Box Frame and Storage Drawers

In the improved design shown in Figure 25–5, the load is carried by divider frames in the center and post legs at the corners, constructed as in the previous design. A ¾" plywood platform is required for mattress-only support, but only ¼" plywood is needed if a bed spring is used. The plywood keeps dust out of the three drawers that fit in the spaces below. Drawer supports can be attached to the divider frames as shown, or commercial metal and nylon supports can be installed. (Drawer construction is covered in Chapters 22 and 23.) The end frames are covered with ¼" or ½" plywood; paneling or other rigid material could be used instead.

TRUNDLE BED

Two beds can be stored in the space occupied by one if one is a trundle bed that can be rolled under the other, as shown in Figure 25–6. The second bed can also be rolled under the first at a right angle so that part of its length protrudes.

The upper bed is constructed like the box frame shown in Figure 25–4, but without the long 2 × 3 bottom rail along one side. The lower trundle bed is built the same way, but without legs

Figure 25–4
Box frame bed. The post-type leg construction provides superior strength; the recessed support cleats ensure that a mattress or springs cannot slide off the bed.

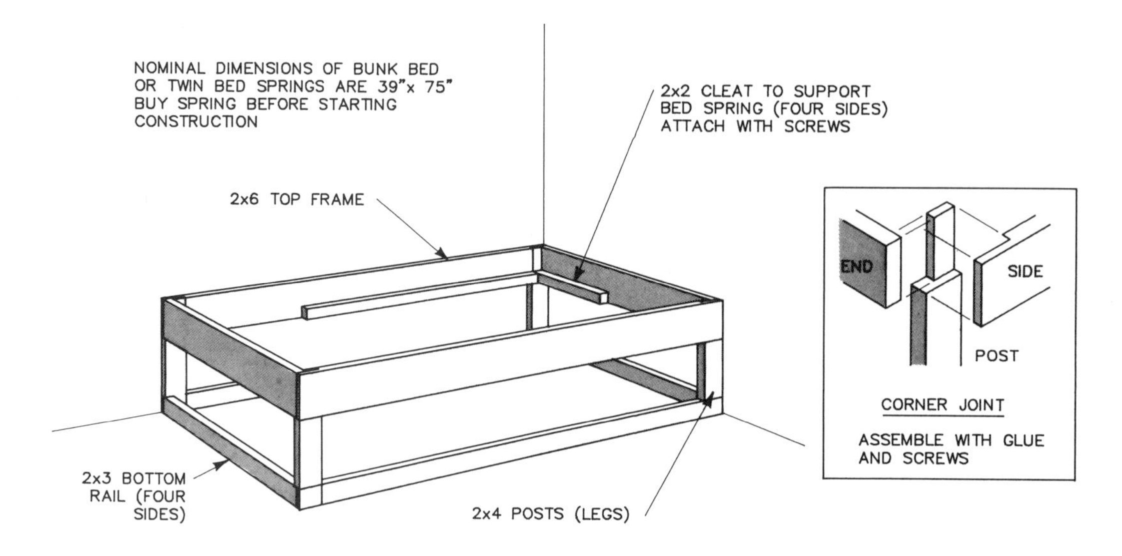

Figure 25–5
Box frame with storage drawers. The front frame and plywood end panels provide a finished appearance in a room.

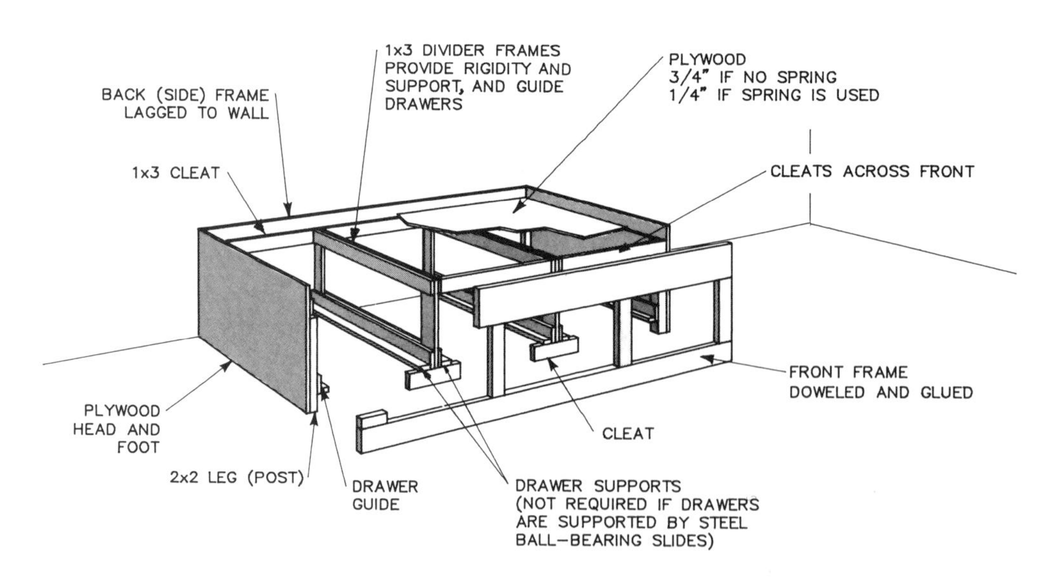

Figure 25–6
Trundle bed. The construction of each unit is a simple box frame with a full plywood panel to support the mattress.

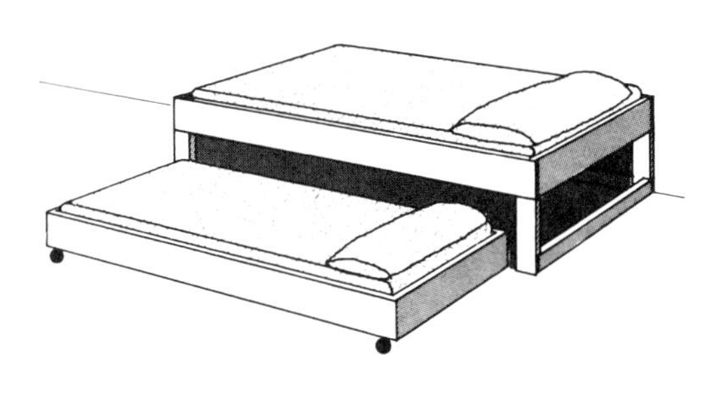

Figure 25–7
A twin-size, side-rolling trundle bed in an attic.

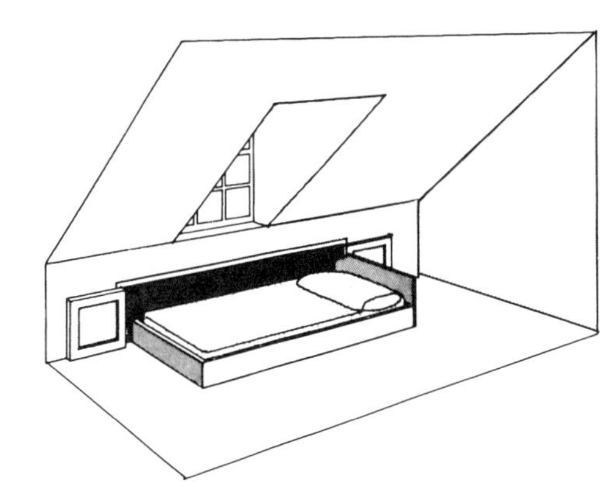

and with 2 × 3s instead of 2 × 2 cleats. A ¾″ plywood panel is screwed to the 2 × 3s to support a mattress. A platform caster, the kind with a flat top plate, is attached under each corner of the trundle unit.

BEDS IN SMALL SPACES

Attic Beds

An attic offers some interesting possibilities for built-in beds. Any bed can be located under a sloped ceiling, but the low side wall can also be opened and a low box-frame bed trundled into the wall, as shown in Figure 25–7. The wall enclosure should be insulated and the interior finished with gypsum board or paneling to keep the bed clean.

While a twin-size bed is relatively easy to tuck away in an attic wall space, a full-size double bed may present more of a problem. Figure 25–8, a section view of an attic room under a gable roof, shows a trundle bed that rolls out of the way in a deep under-roof compartment behind an attic wall. The double-bed inner spring and mattress are mounted on a wood frame equipped with nonswivel casters, so the bed can be effortlessly rolled in and out of the framed-in box. Tracks attached to the floor keep the bed movement in a straight line.

Alcove Beds

Many bedrooms have an alcove designed to hold a dresser. This can be an ideal spot for a child's bed. The bed can be a box frame or simply cleats on the side and end walls to support bed springs. A double-deck alcove bunk bed can be created by installing cleats to support a second set of springs at an upper level. The basic idea is illustrated in Figure 25–9.

An alcove bed is an ideal location for underbed storage drawers. They can be built as part of the bed structure or as units that roll independently on the floor (Figure 25–10). Rolling drawers are simply boxes sized to fit the space, with ½″ or ¾″ plywood bottoms and low casters at the corners. Some means to keep the drawers rolling straight under the bed is needed. If the floor under the bed is bare wood, nail wood strips in place to form guide tracks for the casters. If the floor is carpeted or has a covering you cannot or do not want to nail through, attach guide strips to the underside of the bed frame to guide the tops of the drawers.

Figure 25–8
Attic installation of a full-size, end-rolling trundle bed.

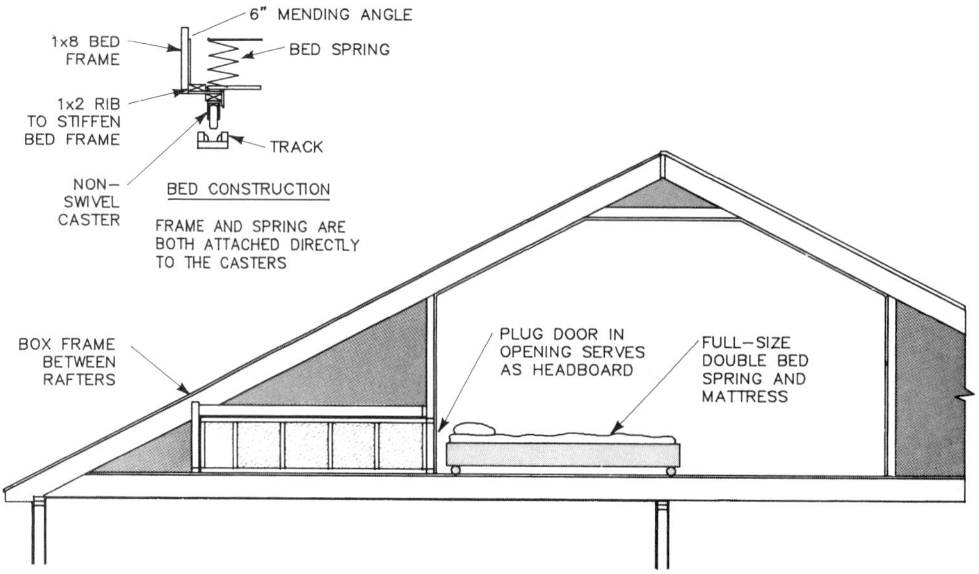

6" MENDING ANGLE
1x8 BED FRAME
BED SPRING
1x2 RIB TO STIFFEN BED FRAME
TRACK
NON-SWIVEL CASTER
BED CONSTRUCTION
FRAME AND SPRING ARE BOTH ATTACHED DIRECTLY TO THE CASTERS
BOX FRAME BETWEEN RAFTERS
PLUG DOOR IN OPENING SERVES AS HEADBOARD
FULL-SIZE DOUBLE BED SPRING AND MATTRESS

Figure 25–9
Child's alcove bed. A double-deck bunk bed should have a simple ladder for easy access.

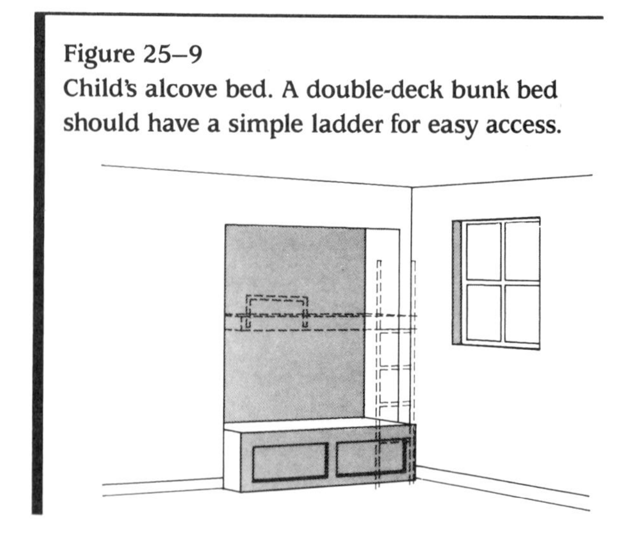

Figure 25–10
Rolling drawers are easy to build and provide very convenient under-bed storage.

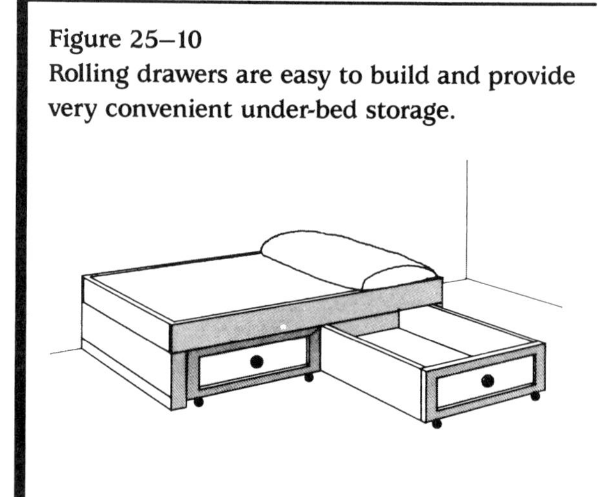

FOLDING BEDS

Although more than 65 years old, the Murphy bed concept is still one of the best ways to fold up a bed and get it out of sight. Most Murphy bed models fold into alcoves or closets at the head end; Figure 25–11 shows a side-mounted Murphy bed, another method of installation. Beds are attached to the floor only, with multiple screws. Cabinets do not provide any support for the bed—they merely enclose it. All commercial Murphy beds are supported in the raised position with balancing springs. Some models have integral bedsprings; others provide support for standard box spring and mattress combinations. You can store the bed fully made up with sheets and blankets. Figures 25–12 and 25–13 show typical cabinet and clearance dimensions for standard head-mounted and side-mounted Murphy beds.

Figure 25–11
A side-mounted twin-size Murphy bed is ideal for a child's room. (Photo courtesy Murphy Door and Bed Co., Inc.)

Figure 25–12
Typical installation dimensions for a head-mounted Murphy bed. Specific dimensions vary for units of various sizes and models.

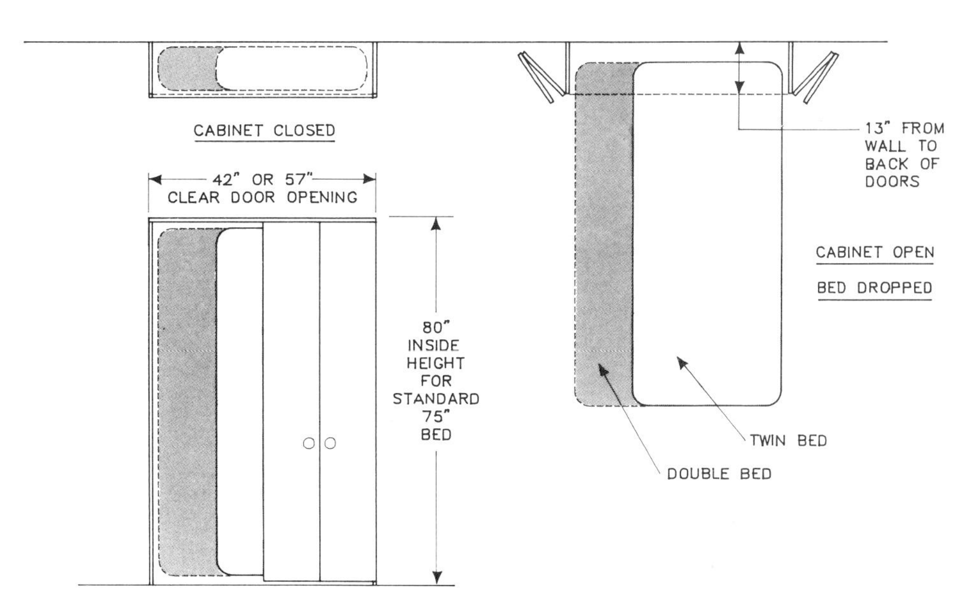

Figure 25–13
Typical installation dimensions for a side-mounted Murphy bed.

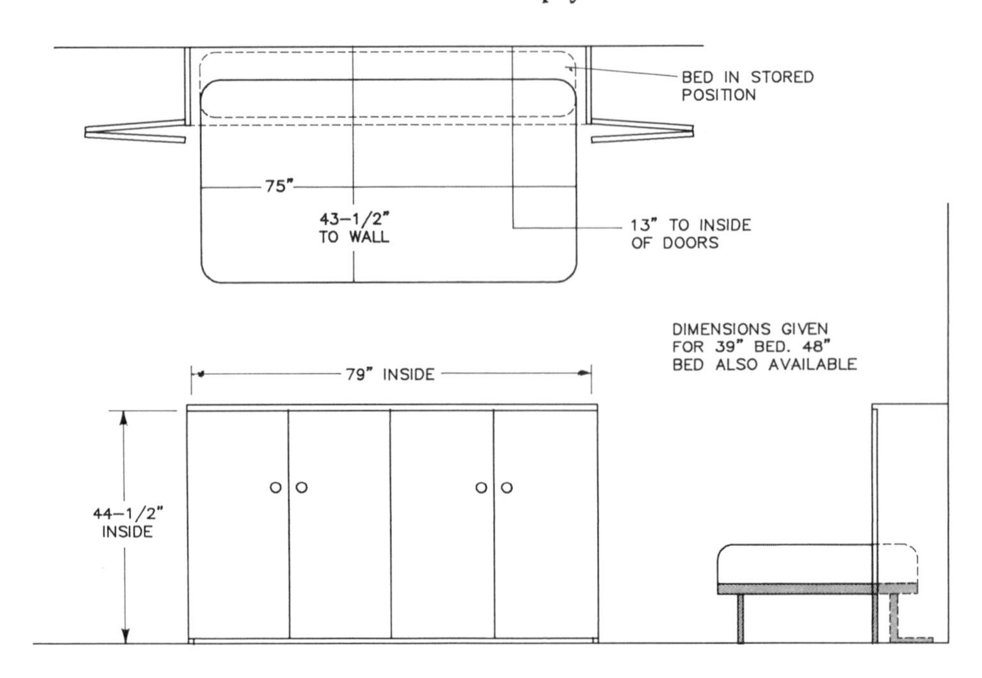

Figure 25–14
A home-built folding bed.

Figure 25–15
A panel decorated to match the walls covers the underside of the folding bed.

Do-It-Yourself Folding Bed

For any number of reasons, you may choose to build your own Murphy-style folding bed rather than buy a commercial unit. Figures 25–14 and 25–15 show a home-built unit made for a combination bedroom and study that measured only 12′ × 12′. To save space, cabinet doors were eliminated. A finished appearance was achieved by covering the bottom of the bed frame with a panel decorated to match the walls. The frame and springs of the bed are shown during construction in Figure 25–16. In any folding bed of this type, the springs must be screwed or bolted to the frame. The head was pivoted as close to the wall as possible; as a result, the closed depth, from the wall to the front of the decorated panel, is only 11″. The floor-to-ceiling side cabinets were built as separate units and were trimmed with crown molding at the top after they had been fastened in place.

Figure 25–16
Home-built folding-bed frame.

Index